High Performance Fiber Reinforced Cement Composites 6

RILEM BOOKSERIES

Volume 2

For other titles published in this series, go to
www.springer.com/series/8780

Gustavo J. Parra-Montesinos • Hans W. Reinhardt •
A.E. Naaman

Editors

High Performance Fiber Reinforced Cement Composites 6

HPFRCC 6

Springer

Editors

Gustavo J. Parra-Montesinos
University of Michigan
Department of Civil and
Environmental Engineering
2370 G.G. Brown Building,
2350 Hayward
48109-2125 Ann Arbor
USA
E-mail: gjpm@umich.edu

A.E. Naaman
The University of Michigan, Ann Arbor
Department of Civil &
Environmental Engineering
2340 G.G. Brown Bldg.
48109-2125 Ann Arbor Michigan
USA
E-mail: naaman@umich.edu

Prof. Hans W. Reinhardt
University of Stuttgart
Dept. of Construction Materials
Pfaffenwaldring 4
70569 Stuttgart
Germany
E-mail: reinhardt@iwb.uni-stuttgart.de

ISBN 978-94-007-2435-8 e-ISBN 978-94-007-2436-5
DOI 10.1007/978-94-007-2436-5
Springer Dordrecht Heidelberg London New York

Cover design: eStudio Calamar S.L.

Printed on acid-free paper

Springer is part of Springer Science+Business Media (www.springer.com)

Dedication

The Sixth International RILEM Conference on High Performance Fiber Reinforced Cement Composites (HPFRCC6) and its proceedings are dedicated to three outstanding members of our international community who have made significant and lasting contributions to broaden the safe application of fiber reinforced cement and concrete composites through fundamental understanding, testing, analysis, modeling and design.

Andrzej M. Brandt

Professor Andrzej Marek Brandt was born on November 15, 1930 in Bydgoszcz, Poland. He graduated from the Warsaw University of Technology, Faculty of Civil Engineering, in 1955, then joined the Institute of Fundamental Technological Research (IFTR) of the Polish Academy of Sciences in Warsaw, where he has been employed ever since. In 1958-1959, he completed advanced studies at the Centre de Hautes Etudes de la Construction in Paris, France. Professor Brandt holds two doctoral degrees, one in 1962, and one, a DSc, in 1967, both from IFTR. He was promoted Associate Professor in 1967 and Full Professor in 1979. His positions included Chairman of the Committee for Civil Engineering of the Polish Academy of Sciences, and Head of the Section on Strain Fields.

Professor Brandt's main interests have been focused on structural mechanics and bridge design, optimization of structures and materials, strain measurements and analysis, fracture mechanics and mechanics of brittle matrix composites. He is the author or co-author of more than 100 technical papers and 20 books among them a classic titled "Cement Based Composites - Materials, Mechanical Properties and Performance" published in 2009. He has been the main instigator and co-organizer of the continuous series of International Symposia on Brittle Matrix Composites (BMC) in Poland since 1985, a topic that encompasses fiber reinforced cement and concrete composites.

Professor Brandt received an Honorary Doctorate from the University of Paisley, Scotland, in 1997. He was a Visiting Professor at University of Poitiers in France, Delft University of Technology in The Netherlands, Koriyama and Tohoku Universities in Japan, and Drexel and North Carolina State Universities in the USA.

Professor Brandt has been active on numerous scientific committees and technical organizations in Poland and abroad, among them, the Polish Standardization Committee, the Euro-International Concrete Committee (CEB), the International Organization for Standardization (ISO), the American Concrete Institute (ACI) and RILEM. He also served as member of the editorial board of a number of national and international journals.

Professor Brandt is being honored for his contributions to the in-depth understanding of the mechanics of brittle matrix composites and optimization of their properties including high performance fiber reinforced cement composites.

Wei Sun

As a distinguished scholar, Professor Wei Sun has been engaged in teaching and scientific research at Southeast University, China, since her graduation from Nanjing Institute of Technology (previous name of Southeast University) in 1958. During her 53 years of teaching and research, she has supervised more than 60 master students and over 40 doctoral students. Professor Sun is currently director of the institute of fiber reinforced concrete and also director of the Jiangsu Key laboratory for Construction Materials. During her career she also was Head of the Department of Materials Science and Engineering, Southeast University, and deputy editor-in-chief of "Journal of the Chinese Ceramic Society". Due to her significant contributions to the theory and application of cementitious composites (including preparation, mechanical behavior, microstructure and modeling of ultra high performance fiber reinforced cementitious composites; service life prediction under the coupled action of load and environmental factors), she was selected to be a member of the Chinese Academy of Engineering in 2005. Together with over 20 of her team members, Professor Sun is still very active in research and teaching. With her team, she has successfully carried out over 50 important national or international projects and co-authored more than 400 technical publications.

Professor Sun is being honored for her long-term and numerous contributions to the investigation of HPFRCC materials, specifically their physical, mechanical, and impact properties, as well as their durability modeling.

Pietro G. Gambarova

Pietro G. Gambarova was born in Milan, on September 1st, 1941. He received his MS Degree in Aeronautical Engineering in March 1966 from the Politecnico di Milano. After two periods spent as an officer in the Corps of Engineers of the Italian Air Force (1966-67 and 1971), he joined the Department of Structural Engineering at the Politecnico di Milano, where he held the position of Assistant Professor from 1968 to 1974, Associate Professor from 1975 to 1980, and Full Professor since 1980. He teaches courses on Structural Analysis and Design, Reinforced Concrete Plates and Shells, and Fire Safety of Materials and Structures. He has been visiting professor (1976) at the National Somali University in Mogadisho, visiting scholar (1978 and 1982) at Northwestern University (Evanston, USA) and at EPFL-Ecole Polytechnique Fédérale de Lausanne (2006).

Pietro Gambarova has conducted research on non-destructive analysis of structures and materials, shear and punching shear in R/C, bond mechanics in R/C, high-performance and fiber-reinforced concrete, non-linear analysis of R/C structures, and lately high-temperature degradation of high-performance concrete and R/C structures. He is the author or coauthor of two books on structural analysis and R/C plates, and of more than 180 technical papers. In 2009, he coauthored the translation into Italian of the book "Structural Design for Fire Safety" by Andy Buchanan.

Pietro Gambarova is an active member of several Italian and international technical committees. He chaired the subcommittee for the preparation of the introductory chapter of fib Bulletin No.10 "Bond of Reinforcement in Concrete" (August 2000). He co-organized the workshop "Fire Design of Concrete Structures: What now? What next?" (Milan, December 2004) and the International Conference FraMCoS-6 (Catania, Italy, June 2007). He is one of the authors of fib Bulletin No.46 "Fire Design of Concrete Structures: Structural Behavior and Assessment" (2008).

Professor Gambarova is being honored for his outstanding contributions to the identification of the physical mechanisms that govern the behavior of high performance and fiber reinforced concrete subjected to high temperature, including fire.

Preface

HPFRCC6 will be the sixth workshop in a series dealing with High Performance Fiber Reinforced Cement Composites (HPFRCC). The five prior workshops have led to a definition of HPFRCC that mostly suggests a technical challenge. That is, composites that exhibit a strain hardening tensile stress-strain response accompanied by multiple cracking (and relatively large energy absorption capacity). Researchers have tried to reduce fiber content to a necessary minimum. By reducing fiber content, they are simplifying the production process, helping make standard mixing procedures acceptable, and opening the way to large-scale practical applications.

The first international workshop on High Performance Fiber Reinforced Cement Composites took placein June 1991 in Mainz, Germany, under the auspices of RILEM and ACI. It was funded in part by the US National Science Foundation (NSF) and the Deutsche Forschungsgemeinschaft (the German NSF). Other co-sponsors included the center for Advanced Cement Based Materials (ACBM), the University of Michigan, and the University of Stuttgart. The second workshop took place in Ann Arbor, Michigan, in June 1995, the third in Mainz Germany, in June 1999, the fourth in Ann Arbor, Michigan, in June 2003, and the fifth in Mainz, Germany, in July 2007, all supported bythe same sponsors. In each case hard-cover proceedings were published as a special RILEM publication. While the first workshop in 1991 included mostly US and German participants, subsequent workshops were opened to top researchers in the field from other countries. The last workshop in Mainz 2007 assembled researchers from 22 countries. The proceedings included 56 papers grouped in 6 different sections.

Since the first workshop in 1991, continuous developments have taken place in new materials, processing, standardization, and improved products for building and other structures. Also, enhanced theory and modeling techniques forHPFRCC now allow a better description oftheir behavior and reinforcing mechanisms. While in the first workshop HPFRCC implied relatively high fiber volume fractions (over 4%), today HPFRCC can be designedwith as little as 1% fiber volume content. While the root definition of HPFRCC is simplest (that is,fiber cement composites withstrain hardening and multiple cracking behavior in tension) to clearly differentiate them from other cement composites, this is not the only description of desirable performance. Durability, fire resistance, impact resistance, diffusion resistance, imperviousness, and constructability at reasonable cost are other important attributes that need to be further investigated.

In each workshop, a broad range of technical issues, ranging from microstructure characterization to design recommendations, are typically covered; however, some selected themes are emphasized. In this sixth workshop, the organizers identified the following themes for which research information is needed:

- Composite properties in the fresh and hardened states
- bond and pull-out mechanisms
- durability
- structural elements: design, detailing, shear, tension stiffening
- impact, cyclic and seismic loading
- ultra high performance fiber reinforced concrete
- textile reinforced concrete and hybrid composites.

Papers addressing these themes are grouped in seven separate sections of the proceedings.

The organizers hope that this new volume will help foster the continuous development and increasing utilization of HPFRCC in both stand-alone and structural applications.

G.J. Parra-Montesinos
H.W. Reinhardt
A.E. Naaman

HPFRCC6 - Workshop

Workshop Organization

Co-Chairman:	Gustavo J. Parra-Montesinos, Department of Civil and Environmental Engineering, University of Michigan, Ann Arbor, USA
Co-Chairman:	Hans W. Reinhardt, Department of Construction Materials, University of Stuttgart, Germany

Scientific Committee

Chair:	P. Balaguru, Rutgers University, USA
Deputy Chair:	B. Mobasher, Arizona State University, Tempe, USA

Members

- G. Balazs, Budapest University of Technology, Hungary
- J. Barros, University of Minho, Portugal
- M. Behloul, Lafarge, France
- A. Bentur, Technion - Israel Institute of Technology, Israel
- S. Billington, Stanford University, USA
- J. Bolander, University of California, Davis, USA
- G. Campione, Universita di Palermo, Italy
- G. Chanvillard, Lafarge, France
- S.H. Chao, University of Texas at Arlington, USA
- F. Dehn, University of Leipzig, Germany
- E. Denarié, EPFL, Lausanne, Switzerland
- A. Dubey, USG Corporation, USA
- H. Falkner, IBF Stuttgart, Germany
- E. Fehling, University of Kassel, Gemany
- L. Ferrara, Politecnico di Milano, Italy
- M.A. Glinicki, Polish Academy of Sciences, Poland
- P. Hamelin, University of Lyon, France
- W. Hansen, University of Michigan, USA
- A. Katz, Technion, Haifa, Israel
- D.J. Kim, Sejong University, South Korea

- K. Kosa, Kyushu Institute of Technology, Japan
- N. Krstulovic-Opara, ExxonMobil, USA
- A. Lambrechts, Bekaert, Belgium
- V.C. Li, University of Michigan, USA
- M. Lopez de Murphy, Pennsylvania State University, USA
- B. Massicotte, Ecole Polytechnique Montreal, Canada
- V. Mechtcherine, TU Dresden, Germany
- H. Mihashi, Tohoku University, Japan
- H.S. Müller, University of Karlsruhe, Germany
- C. Ostertag, University of California, Berkeley, USA
- A. Peled, Ben Gurion University, Israel
- G. Plizzari, University of Brescia, Italy
- P. Rossi, LCPC, France
- M. Schmidt, University of Kassel, Germany
- P. Serna Ros, Polytechnic University of Valencia, Spain
- Y. Shao, McGill University, Canada
- C. Sujivorakul, King Mongkut´s University of Technology Thonburi, Thailand
- L. Vandewalle, Katholieke Universiteit Leuven, Belgium
- K. Wille, University of Connecticut, USA
- F.H. Wittmann, Aedificat Institute, Freiburg, Germany

International Committee

Chair:	A.E. Naaman, University of Michigan, USA
Deputy Chair:	M. di Prisco, Politecnico di Milano, Italy

Members

- S.A. Altoubat, University of Sharjah, UAE
- N. Banthia, University of British Columbia, Canada
- W. Brameshuber, RWTH Aachen, Germany
- A.M. Brandt, Polish Academy of Sciences, Poland
- K.P. Chong, National Institute of Standards and Technology, USA
- M. Curbach, Technical University of Dresden, Germany
- G. Fischer, Technical University of Denmark, Lyngby
- H. Fukuyama, Building Research Institute, Tsukuba, Japan
- P. Gambarova, Politecnico di Milano, Italy
- R. Gettu, Indian Institute of Technology, Chennai, India
- M. Harajli, American University of Beirut, Lebanon
- B. Karihaloo, University of Cardiff, UK
- D. Lange, University of Illinois at Urbana-Champaign, USA
- C. Leung, Hong Kong University of Science and Technology, P.R. China
- C. Meyer, Columbia University, USA
- S. Mindess, University of British Columbia, Canada

- B.H. Oh, Seoul National University, Korea
- K.C.G. Ong, National University of Singapore
- K. Rokugo, Gifu University, Japan
- S.P. Shah, Northwestern University, USA
- H. Stang, Technical University of Denmark, Denmark
- L. Taerwe, Ghent University, Belgium
- K.H. Tan, National University of Singapore
- R.D. Toledo Filho, COPPE/UFRJ, Brasil
- T.C. Triantafillou, University of Patras, Greece
- J.C. Walraven, Delft University of Technology, The Netherlands
- J. Wastiels, Vrije Universiteit Brussel (VUB), Brussels, Belgium
- G. van Zijl, University of Stellenbosch, South Africa

Sponsorship

RILEM
American Concrete Institute
University of Michigan
University of Stuttgart
Deutsche Forschungsgemeinschaft, Germany

Local Organization

- Prof. G.J. Parra-Montesinos
- Prof. A.E. Naaman
- Prof. Sherif El-Tawil
- Prof. J.K. Wight
- Prof. W. Hansen
- Ms. S. Brueger

Production Assistant

- S. Stumpp, University of Stuttgart

Acknowledgements

This workshop was sponsored by RILEM (International Union of Laboratories and Experts in Construction Materials, Systems and Structures), and co-sponsored by the University of Michigan, the University of Stuttgart, the American Concrete Institute (ACI) and the Deutsche Forschungsgemeinschaft (DFG). The support of the above organizations is gratefully acknowledged.

The organizers would like to thank all members of the Scientific and International Committees for enthusiastically supporting the organization of this sixth workshop, and all the authors who have contributed with the valuable papers that make these proceedings. Special thanks are due to Ms. Simone Stumpp for preparing with great care and efficiency the camera-ready material for printing. Once more, H.W. Reinhardt and A.E. Naaman would like to express their deep gratitude to the Alexander von Humboldt Foundation for giving them the opportunity to initiate, in 1990, a long-term continually productive cooperation.

G.J. Parra-Montesinos
H.W. Reinhardt
A.E. Naaman

Contents

Part 1
Composite Properties in the Fresh and Hardened States

Strength Dependent Tensile Behavior of Strain Hardening Fiber Reinforced Concrete

D.J. Kim[1], K. Wille[2], A.E. Naaman[3], and S. El-Tawil[3]

[1] Civil and Environmental Engineering, Sejong University, Korea
[2] Civil and Environmental Engineering, University of Connecticut, USA
[3] Civil and Environmental Engineering, University of Michigan, USA

Abstract. The influence of matrix strength on the tensile behavior of Fiber Reinforced Cement Composites [FRCC] is investigated. The test parameters included four cementitious matrices with compressive strength of 28 MPa (4 ksi) [M1], 56 MPa (8 ksi) [M2], 84 MPa (12 ksi) [M3] and 190 MPa (28 ksi) [M4], respectively, two types of high strength deformed steel fibers, Hooked [H-] and Twisted [T-] fibers, and two volume fractions of fibers, 1% and 2%. It is observed that while the first cracking strength, post cracking strength and energy absorption capacity of FRCC are strongly influenced by the compressive strength of the matrix their strain capacity at peak stress and cracking behavior are not as much affected. While both H- and T- fibers led to improved performance when the matrix strength was increased, T- fibers take better advantage of higher strength matrices. A post-cracking tensile strength exceeding 15 MPa at a peak strain of 0.5% was achieved by using 2% T- fibers with an ultra-high strength matrix (M4).

1 Introduction

It is challenging to achieve strain-hardening behavior using relatively small amounts of fibers (1% to 2% by volume). In this study an attempt is made to address this issue through the use of deformed high strength steel fibers in cementitious mortar matrices with four different compressive strengths ranging from 28 MPa (4 ksi) to 190 MPa (28 ksi). Details about the ultra high performance concrete (UHPC) matrix with 190 MPa (28 ksi) compressive strength and the development of strain hardening ultra high performance fiber reinforced concrete (UHP-FRC) with low fiber content can be found in Wille et al. [1, 2]. Details about the three other mixtures and their use in developing high performance concretes are given in Kim et al. [3, 4]. This paper links both sets of studies and examines trends in the behavior of strain-hardening fiber reinforced cement composites over an extremely wide range of matrix strengths (28 MPa to 190 MPa). The study focuses in particular on the use of high strength deformed steel fibers used in quantities ranging from 1% to 2% by volume. The use of metallic steel fibers in such small

G.J. Parra-Montesinos, H.W. Reinhardt, and A.E. Naaman (Eds.): HPFRCC 6, pp. 3–10.

quantities is deemed to be economically viable and hence suitable for practical applications.

2 Research Background

There is a little experimental information about the influence of matrix characteristics on the overall pullout behavior of a single deformed steel fiber (Shannag et al. [5], Robins et al. [6] and Chan and Chu [7]); however, it is difficult to find its relationship to the overall tensile behavior of the composite. Kim et al. [8] performed a pullout test program designed to investigate the influence of matrix strength on the pullout behavior of deformed steel fibers, H- and T- fibers. They noted that the pull-out energy of T- fiber embedded in the highest strength matrix was more than twice that for the lowest strength matrix, while the pullout energy of H- fiber improved by only 44%. They attributed the superior pullout behavior of T- fibers in higher strength matrices to the un-twisting pullout mechanism of T-fiber, which leads to superior overall performance. The research, however, did not clarify whether the significant advantage observed in single fiber pull-out behavior translates into a similar advantage in the tensile response of the composite. An experimental program was therefore undertaken to investigate this issue.

3 Experiments

The test parameters included four cementitious matrices with compressive strength of 28 MPa (4 ksi) [M1], 56 MPa (8 ksi) [M2], 84 MPa (12 ksi) [M3] and 190 MPa (28 ksi) [M4], respectively, two types of high strength deformed steel fibers, Hooked [H-] and Twisted [T-] fibers, and two volume fractions of fibers, 1% and 2%. Table 1 provides the mortar mixture composition for the four matrices used and their average compressive strength. The key properties of the fibers used in this study are provided in Table 2. A Hobart type laboratory mixer was used to prepare the mix. For M1, M2, and M3, cement, fly-ash and sand were first dry mixed for about 2 minutes. Water mixed with superplasticizer and Viscosity Modifying Agent (VMA) was then added gradually and mixed for another 5 to 10 minutes. The detailed mixing procedure for M4 is described by Wille et al. [2].

Table 1. Composition of matrix mixtures by weight ratio and compressive strength

Matrix	M1	M2	M3	M4
Cement	0.70 (Type III)	1.00 (Type III)	0.80 (Type III)	1.00 (Type I)
Fly Ash°	0.30	0.15	0.20	-
Sand1	3.50	1.00	1.00	-
Sand2	-	-	-	0.28
Sand3	-	-	-	1.10
Silica Powder	-	-	-	0.25
Silica Fume	-	-	0.07	0.25
Superplasticizer A	0.009	0.009	0.04	-
Superplasticizer B	-	-	-	0.0054
VMA*	0.024	0.006	0.012	-
Water	0.65	0.35	0.26	0.22
f_c', ksi (MPa)	4 (28)	8 (55)	12 (84)	28 (190)

* Viscosity Modifying Agent; ° TYPE C;
1 #16 flint silica sand; 2 max. grain size 0.2 mm; 3 max. grain size 0.8 mm;
A Rhobuild 1000; B Polycarboxylate Ether;

Table 2. Properties of high strength Hooked and Twisted fibers

Fiber Type	Diameter in (mm)	Length in (mm)	Density g/cc	Tensile strength ksi (MPa)	Elastic Modulus ksi (GPa)
Hooked	0.015 (0.38)	1.18 (30)	7.9	420 (2900)	29000 (200)
Twisted	0.012 (0.3)*	1.18 (30)	7.9	304 (2100)**	29000 (200)

* Equivalent diameter ** Tensile strength of the fiber after twisting

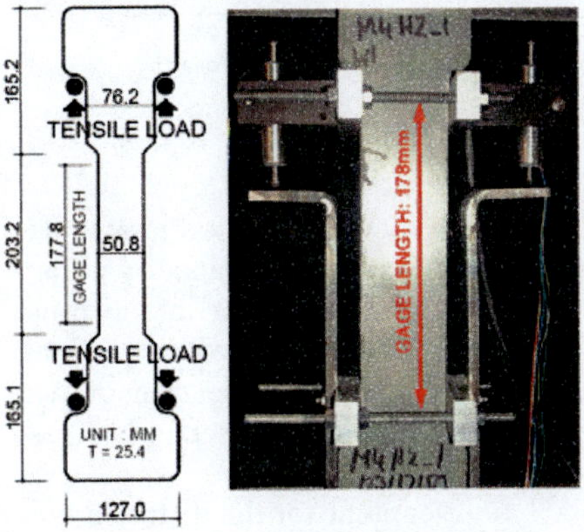

Fig. 1. Tensile test specimen and setup

The geometry of the specimen and test set up are shown in Fig. 1. Elongation of the tensile specimen under load was obtained from the average measurements of two LVDTs placed on each side of the specimen as shown in Fig. 1, and the load was obtained from the load cell pre-built into an MTS810 hydraulic machine.

Fig. 2 shows average tensile stress versus strain (up to peak stress) curves for each series. All the results show strain hardening behavior, which confirms that the use of slip hardening H- and T- fibers can produce strain hardening behavior at the low fiber contents considered. It is also observed that a higher tensile load resistance is generated as matrices with higher compressive strength are used regardless of the type of fiber employed. Moreover, even though a tensile strength increase is noticed with higher strength mortar, no reduction in strain capacity, i.e. strain value at maximum post cracking strength is noticed.

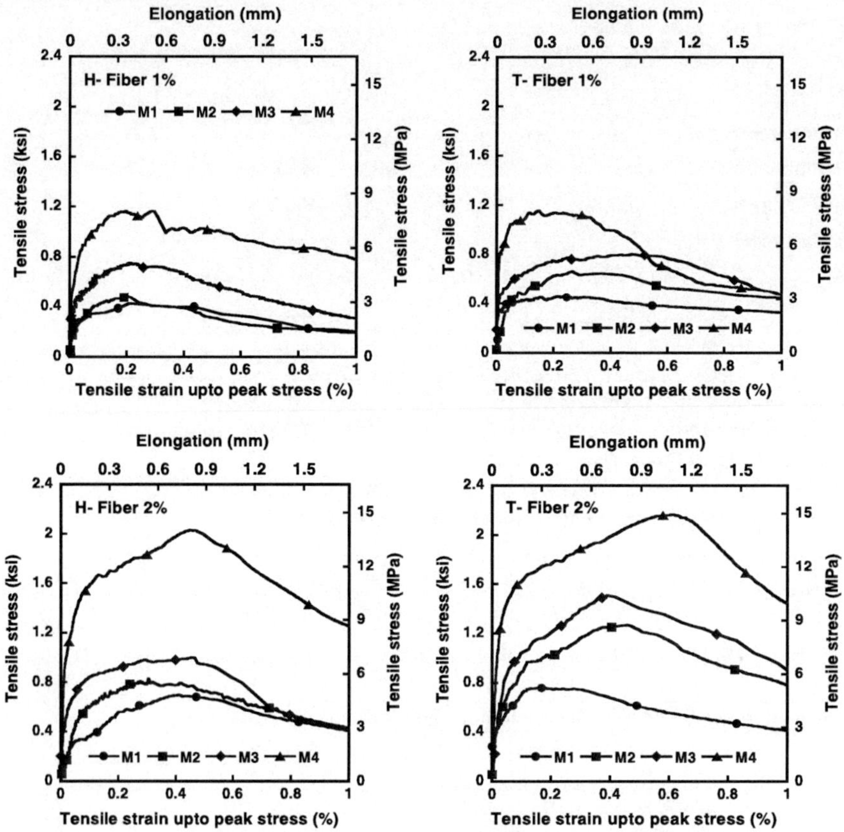

Fig. 2. Average tensile stress – strain curves

The tensile behavior of HPFRCC can be characterized by using several tensile parameters, e.g., first cracking strength, post cracking strength, strain capacity (strain value at post cracking strength), number of cracks within gage length and average crack width. To minimize the subjectivity of how first cracking is determined, the first cracking strength in this study is determined from the intersection of two lines, the first of which corresponds to the elastic region, while the second corresponds to the hardening region.

The cracking behavior of specimens with 2% fiber content for the different matrices is illustrated in Fig. 3. Average values of tensile parameters, each obtained from three to four specimens, and the influence of matrix strength on tensile parameters are graphically illustrated in Fig. 4. Since the process of estimating the number of cracks could be highly subjective, the number of cracks is counted from both front and back surface of specimens and the numbers are averaged.

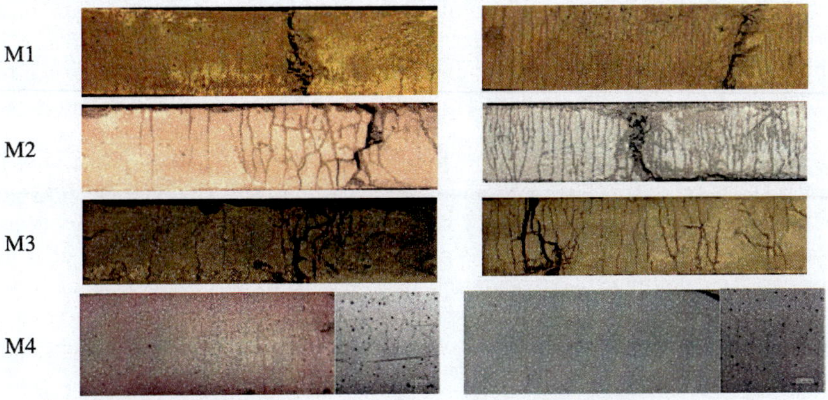

(a) Hooked Fiber 2% (b) Twisted Fiber 2%

Fig. 3. Tensile cracking behavior

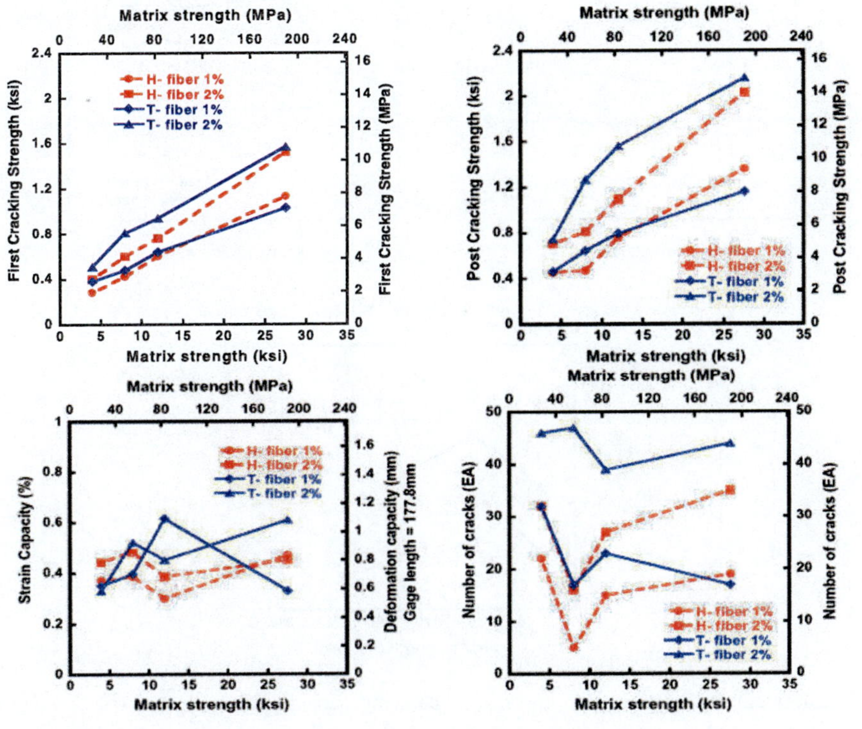

Fig. 4. Influence of matrix strength on tensile parameters: (a) first cracking strength; (b) post cracking strength; (c) strain capacity; and, (d) number of cracks

Fig. 4d shows the effect of matrix strength on the number of cracks within the gage length. As with strain at peak stress, there no clear dependency on matrix strength.

The first cracking strength is mainly influenced by the tensile strength of the matrix, which correlates strongly to its compressive strength as shown in Fig. 4a. Post cracking strength appears to be even more sensitive to the strength of the matrix than the first cracking strength as illustrated in Figs. 4a, 4b and Table 3. Also evident from the test data is that, in general, a higher post cracking strength is obtained by when T-fibers are used. The lower post cracking strength of M4 with 1% T-fibers (1.16 ksi, 8.0 MPa) in comparison to M4 with 1% H-fibers (1.36 ksi, 9.4 MPa) is attributed to fiber failure of some of the T-fibers during the test (which was audible); suggesting that higher strength fiber would have provided better performance. The strain capacities of the test specimens as a function of matrix strength are shown in Fig. 4c, from which it is clear that strain capacity at maximum stress that ranges from 0.35% to 0.6% with no clear dependency on matrix strength.

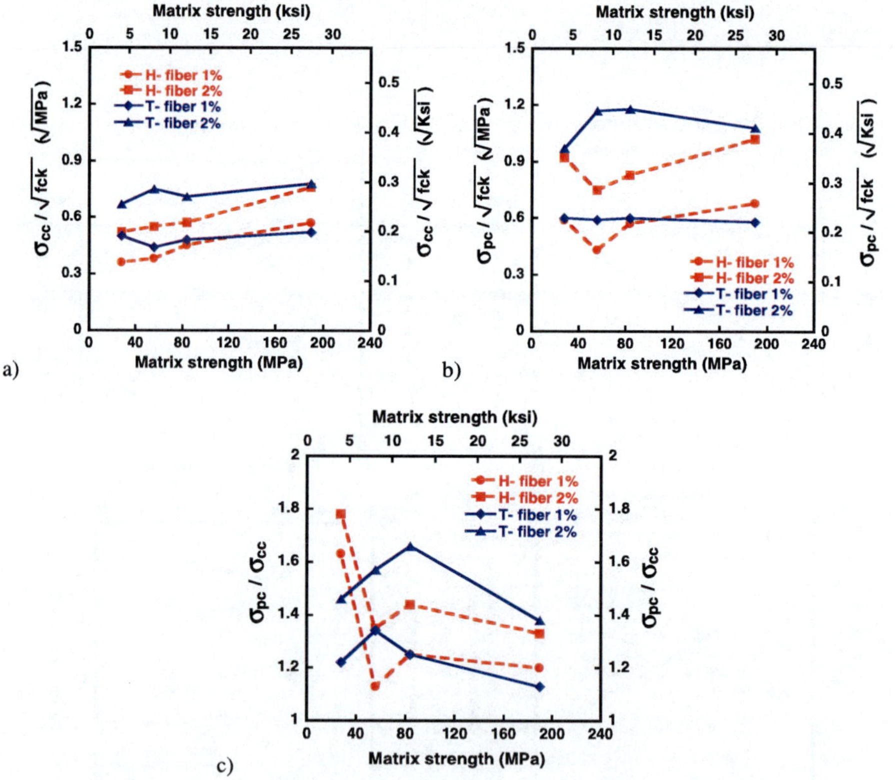

Fig. 5. Influence of matrix strength on load carrying capacity:(a) $\sigma_{cc} / \sqrt{f_{ck}}$, (b) $\sigma_{pc} / \sqrt{f_{ck}}$, (c) $\sigma_{pc} / \sigma_{cc}$

4 Evaluation of Experimental Results

Figs. 2, 4 and 5 clearly show the influence of matrix strength on the tensile behavior of HPFRCC. While an increase in matrix strength leads to increases in both first and maximum post cracking strength there is no clear influence on the strain capacity and cracking behavior. To quantify the effect of matrix strength, both first and post cracking strength are normalized by the square root of compressive strength ($\sqrt{f_{ck}}$) and their values are plotted versus the compressive strength in Fig. 5a and 5b, respectively. In addition, the ratio between the post cracking strength and the first cracking strength is illustrated in Fig. 5c.

As shown in Fig. 5a, the ratio $\sigma_{cc} / \sqrt{f_{ck}}$ slightly increases as the compressive strength of the matrix increases, e.g., composites with M1 generate a ratio $\sigma_{cc} / \sqrt{f_{ck}}$ ranging from 0.36 to 0.67, while composites with M4 produce higher ratios of $\sigma_{cc} / \sqrt{f_{ck}}$ ranging from 0.52 to 0.78. The influence of matrix strength on the ratio $\sigma_{pc} / \sqrt{f_{ck}}$ is graphically shown in Fig. 5b. There is no noticeable effect on the ratio $\sigma_{pc} / \sqrt{f_{ck}}$, whereas the ratio $\sigma_{pc} / \sigma_{cc}$ is dependent upon the type of fiber and volume contents of fiber as shown in Fig. 5c.

5 Conclusions

This study investigated the influence of matrix strength on the tensile behavior of HPFRCC using two deformed high strength steel fibers, namely Hooked (H-) fibers and Twisted (T-) fibers. The strengths of the four matrices employed are 28 MPa (4 ksi), 56 MPa (8 ksi), 84 MPa (12 ksi), and 190 MPa (28 ksi), respectively. One of the key results of the experimental program is that composites using high strength deformed steel fibers are sensitive to the strength of the matrix and that T-fibers take better advantage of higher strength matrices than H- fibers due to their unique un-twisting pullout mechanism. For example, a post-cracking tensile strength exceeding 15 MPa at a peak strain of 0.61% was achieved by using 2% T-fibers with the ultra-high strength matrix (M4), whereas only 14 MPa tensile capacity at 0.45% strain could be achieved with H-fibers. The results also showed that, as the compressive strength of the matrix increases, the first and post cracking strength of the composite correspondingly increase. However, the increase in matrix compressive strength does not lead to clear trends in either the maximum strain capacity or cracking behavior. In fact, the maximum strain capacity remains almost independent of matrix strength while energy absorption at peak load increases.

Acknowledgements. The research described herein was sponsored in part by the National Science Foundation under Grant No. CMMI 0754505; by the Basic Science Research Program through the National Research Foundation of Korea (NRF) funded by the Ministry of Education, Science and Technology (2010-0003161) and by the Human Resources Development of the Korea Institute of Energy Technology Evaluation and Planning (KETEP) grant funded by the Korea government Ministry of Knowledge Economy (No.

20104010100520). The second author, Kay Wille, was also supported by a fellowship from the Postdoc-Programme of the German Academic Exchange Service (DAAD). The opinions expressed in this paper are those of the authors and do not necessarily reflect the views of the sponsors.

References

[1] Wille, K., Kim, D., Naaman, A.E.: Strain-Hardening UHP-FRC with Low Fiber Contents. RILEM - Materials and Structures (2010); online first, doi:10.1614/S11527-101-9650-4

[2] Wille, K., Naaman, A.E., Parra-Montesinos, G.: Ultra High Performance Concrete with Compressive Strength Exceeding 150 MPa (22ksi): A Simpler Way. ACI Materials Journal 108(1), 46–54 (2011)

[3] Kim, D., El-Tawil, S., Naaman, A.E.: Correlation between single fiber pullout behaviour and tensile response of FRC composites with high strength steel fiber. In: Reinhardt, H.W., Naaman, A.E. (eds.) Proceeding of Rilem International Workshop on High Performance Fiber Reinforced Cement Composites – HPFRCC5, Rilem Proceedings, Pro. 53, S.A.R.L., Germany, Cachen, France, pp. 67–76 (2007)

[4] Kim, D., Naaman, A.E., El-Tawil, S.: High Tensile Strength Strain-Hardening FRC Composites with Less Than 2% Fiber Content. In: Fehling, E., Schmidt, M., Sturwald, S. (eds.) Proceedings of the Second International Symposium on Ultra High Performance Concrete, March 10, pp. 169–176. Kassel University Press GmbH, Germany, Heft 10, No. 10 (2008)

[5] Shannag, M.J., Brinker, R., Hansen, W.: Interfacial (Fiber-Matrix) Properties of High-Strength Mortar (150MPa) from Fiber Pullout. ACI Materials Journal 93(5), 1–7 (1996)

[6] Robins, P., Austin, S., Jones, P.: Pullout Behavior of Hooked Steel Fibers. Materials and Structures 35, 434–442

[7] Chan, Y.-W., Chu, S.-H.: Effect of Silica Fume on Steel Fiber Bond Characteristics in Reactive Powder Concrete. Cement and Concrete Research 34, 1167–1172 (2004)

[8] Kim, D.J., El-Tawil, S., Naaman, A.E.: Effect of matrix strength on pull-out behaviour of high strength deformed steel fibers. ACI Special Publication 272(7), 135–150 (2010)

Tailoring SHCC Made of Steel Cords and Plastic Fibers

A.P. Fantilli[1], H. Mihashi[2], T. Naganuma[3], and T. Nishiwaki[3]

[1] Politecnico di Torino, Torino, Italy
[2] Tohoku Institute of Technology, Sendai, Japan
[3] Tohoku University, Sendai, Japan

Abstract. Some Fiber-Reinforced Concrete (FRC), commonly called Strain-Hardening Cement-based Composite (SHCC), can show a very ductile behavior under tensile actions. Specifically, in the post cracking stage, several cracks develop before complete failure, which occurs when tensile strains localize in one of the formed cracks. To better understand the high mechanical performances of SHCC, an analytical model was previously proposed. The model is here used to analyze the strain-hardening behavior of a more cost-effective FRC, made with steel cords and plastic fibers. By combining direct uniaxial tensile tests, performed on dumbbell-shaped specimens, and the theoretical results of the model, the critical value of the fiber volume fraction can be evaluated. It should be considered as the minimum amount of steel cords which can lead to the formation of multiple cracking and strain hardening under tensile actions. The aim of the present paper is to reduce such volume as much as possible, in order to improve the workability and reduce the final cost of SHCC.

1 Introduction

According to Banthia and Sappakittipakorn [1], in hybrid Fiber Reinforced Concrete (FRC) <<*...there is positive interaction between the fibers and the resulting hybrid performance exceeds the sum of individual fiber performances. This phenomenon is often termed "Synergy".*>>

Combining fibers of different geometry is a possible manner to create this synergy. Indeed, short fibers, generally called microfibers, enhance the fracture toughness of cementitious matrix in tension. Thus, after the first elastic behavior, the mechanical response of FRC can show a delayed microcracking stage, because of the bridging action performed by the short thin fibers [2]. Conversely, the beneficial effects of long fibers are particularly evident in the third stage, and consist of arresting and delaying the growth of macrocracks.

G.J. Parra-Montesinos, H.W. Reinhardt, and A.E. Naaman (Eds.): HPFRCC 6, pp. 11–18.

If the amount of long and short fibers is appropriately evaluated, it is possible to obtain a strain hardening cementitious composite (SHCC), capable of developing more than one crack prior to the localization of tensile strains [3]. To achieve such ductile behavior, hybrid fiber reinforced composites should be tailored on the basis of experimental analyses [1], and/or through theoretical approaches, which often enable investigations on the parameters not covered by tests [4].

In what follows, a particular class of hybrid SHCC, made of short plastic fibers and long steel cords, is taken into consideration. In such tailor-made concretes, crack pattern is experimentally investigated by means of uniaxial tensile tests on "dumbbell type" specimens. The observed crack spacing is also predicted by a cohesive interface model, already introduced to calculate the critical fiber volume fraction [5]. It should be considered as the minimum amount of fibers (in a mono-fiber composite) which can lead to the formation of multiple cracking and strain hardening under tensile actions. The aim of the present paper is to extend the application of this model to hybrid SHCC, in order to reduce the volume of long fibers (i.e. steel cords) and improve the workability.

2 Modeling the Strain Hardening and Multiple Cracking of SHCC

Fiber can be considered as the reinforcing bar (A_f = area of the fiber) of an element in tension (Fig. 1a) having a single crack in the concrete area $A_c = A_f / V_f$ (in this way the fiber volume fraction V_f is equal to the geometrical reinforcement percentage $\rho = A_f / A_c$).

By increasing the crack width w, the mechanical response of this structure, in terms of N-w (Fig. 1c), depends on slips between fiber and matrix, whose distribution $s(z)$ is qualitatively drawn in Fig. 1b. In the case of fibers symmetrically situated with respect to the crack, the maximum slip is located in the cracked cross-section (where $s = w/2$), from which it progressively vanishes with the increase of the distance z from the crack. In particular, $s(z) = 0$ if $z \geq l_{tr}$ (where l_{tr} = transmission length). As z increases, the tensile stresses of fiber, $\sigma_s(z)$, continuously transfer to the cementitious matrix, because of bond stresses $\tau(z)$ acting on the interface between materials. As a consequence, within the domain $0 \leq z \leq l_{tr}$, tensile stresses in the matrix, $\sigma_c(z)$, are higher far from the crack surfaces (Fig. 1b). Beyond the transmission length, neither slips nor bond stresses exist, and, therefore, the condition of perfect bond is verified when $z > l_{tr}$ (i.e., σ_c = const and σ_s = const). In this zone, the stress σ_c reaches the maximum value $\sigma_{c,max}$.

At onset of cracking, when $w \rightarrow 0$, $\sigma_{c,max}$ is generally lower than the strength f_{ct} and the N-w diagram shows a softening branch. If these conditions persist for higher crack width, the failure of the structural element depicted in Fig. 1a occurs in the presence of a single tensile crack.

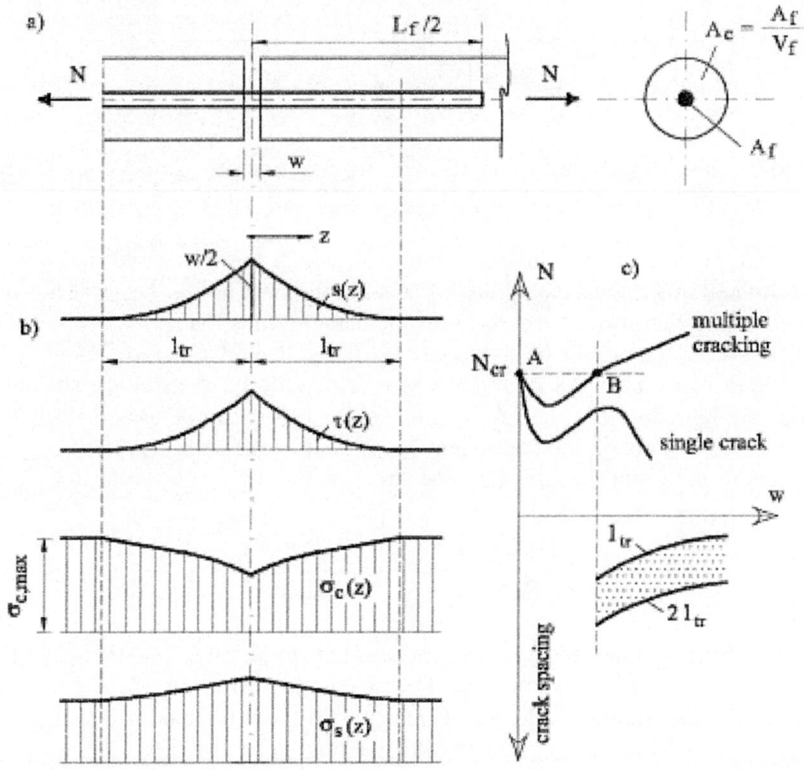

Fig. 1. The model for fiber-matrix interaction [5]: a) an element with a single crack; b) slip $s(z)$, bond stress $\tau(z)$, concrete stress $\sigma_c(z)$, and fiber stress $\sigma_s(z)$ distributions around the crack; c) mechanical response in terms of normal force *vs.* crack spacing *vs.* crack width

Conversely, when $\sigma_{c,max} = f_{ct}$, which corresponds to the point B of Fig. 1c, new cracks appear and, with the increase of w, strain hardening characterizes the N-w diagram. According to Fantilli et al. [5], during this stage, the average crack spacing ranges between l_{tr} and $2\ l_{tr}$ (Fig. 1c).

Under the condition of symmetry depicted in Figs. 1a-b, the multiple cracking regime is possible if the semi-length of the fiber is higher than the maximum crack spacing (i.e., $2\ l_{tr} < L_f/2$). Thus the definition of l_{tr} at cracking load (i.e., at point B of in Fig. 1c) is of primary importance in evaluating the presence of multiple cracking and strain hardening response in FRC composites.

The following value of the transmission length is obtained by adapting the tension-stiffening equations of RC structures to the fiber-matrix tie illustrated in Fig. 1a (see Fantilli et al. [5]):

$$l_{tr} = -\frac{\ln\left[\left(p_f\ k_B\ V_f - 2A_f\ k_C\sqrt{\alpha}\right)/\left(p_f\ k_B\ V_f + 2A_f\ k_C\sqrt{\alpha}\right)\right]}{\sqrt{\alpha}} \quad (1)$$

where,

$$\alpha = \frac{p_f\, k_B}{A_f}\left(\frac{1}{E_f} + \frac{V_f}{E_c}\right) \tag{2}$$

In both the equations p_f = perimeter of fiber cross-section; E_c , E_f = Young's moduli of cement-based matrix and fiber, respectively; k_C = cohesive parameter; and k_B = bond parameter.

Since Eq. (1) is obtained in the situation in which the first crack is growing (i.e., $w \rightarrow 0$) and other cracks are going to develop (point B in Fig. 1c), k_C is used to approximate the fictitious crack model of the cementitious matrix. As depicted in Fig. 2a, this coefficient can be associated to the so-called Fracture Energy G_f , which differs from the work of fracture G_F [6]. Similarly, the tension stiffening equations are here applied when $s \rightarrow 0$, thus bond stresses can be assumed to be in direct proportion (through the coefficient k_B - Fig. 2b) with slips [5].

The condition of multiple cracking and strain hardening do not occur in FRC if:

$$V_f \le V_{fcr} = \frac{A_f\, 2\,(k_C)^2}{E_c\, p_f\, k_B}\left(1 + \sqrt{1 + \frac{(E_c)^2\, p_f\, k_B}{A_f\, E_s\,(k_C)^2}}\right) \tag{3}$$

where V_{fcr} = fiber volume fraction that produces the condition $l_{tr} \rightarrow \infty$ (see Eq. (1)).

In order to obtain a length from Eq. (1) and a number from Eq. (3), the parameter α should be the square of a length. This is possible when the bond parameter k_B (and k_C as well) is assumed to be a stress over a length (see Fig. 2).

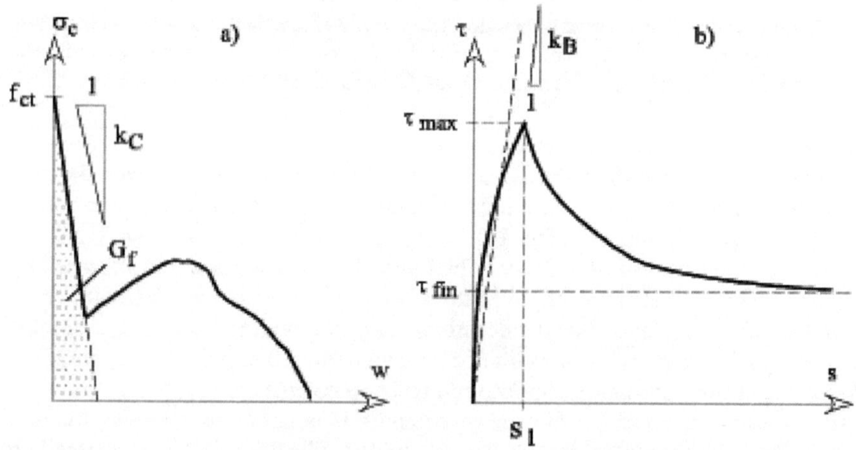

Fig. 2. The cohesive and bond parameters: a) fictitious crack model σ_c-w of cementitious matrix; b) bond-slip τ-s relationship of fibers in cementitious matrix

3 Experimental Study

In order to verify the applicability of the Eqs.(1)-(3) to hybrid SHCC, an experimental campaign has been performed on two series of composites. The PE series includes specimens reinforced with short Polyethylene (PE) fibers (0.75% in volume) and different amounts of long steel cords (SC). In the same way, the specimens of the second series contain 0.75% of short Poly Vinyl Alcohol (PVA) fibers, and different percentage of long SC. The main geometrical and mechanical properties of all the fibers are summarized in Table 1.

In each specimen, the fiber volume fraction V_f is only referred to the amount of SC, whereas plastic fibers are assumed to be a part of the cement-based matrix, whose constituents are reported in Table 2. Young's modulus E_c, cohesive parameter k_C, and tensile strength f_{ct} define the uncracked and cracked stages of the matrixes, both of PE series ($E_c = 16.4$ GPa, $k_C = 5.72$ MPa/mm, $f_{ct} = 2.82$ MPa), and PVA series ($E_c = 18.6$ GPa, $k_C = 22.7$ MPa/mm, $f_{ct} = 2.73$ MPa).

Table 1. Geometrical and mechanical properties of the fibers

Fiber type	Symbol	Diameter (mm)	Length (mm)	Density (g/cm^3)	Tensile strength (MPa)	E_f (GPa)
Steel cord	SC	0.4	32	7.84	2850	200
Polyethylene	PE	0.12	6	0.97	2580	73
Poly Vinyl Alcohol	PVA	0.04	6	1.3	1600	40

Table 2. Composition of the cementitious matrixes

Material	Symbol	Density (g/cm^3)	Tensile strength
Cement	C	3.14	Early strength Portland Cement
Silica fume	SF	2.2	-
Sand	S	2.61	mean diameter 180 μm
Super plasticizer	SP	1.05	Polycarboxylic acid system

The geometrical dimensions of the "dumbbell type" specimens tested in the present work, and depicted in Fig. 3a, are in accordance with the Recommendations of the Japan Society of Civil Engineers for HPFRCC composites [7].

Loads were vertically applied with a 30 kN capacity Instron testing machine, using "fix-fix" support conditions (Fig. 3b). Each test was controlled by vertical displacement at a velocity of 0.5 mm/min. Average extension was measured over the central gauge length (80 mm) by means of two LVDTs, placed on the opposite side of the member and attached to mounting frames, firmly clamped on to the specimens. In this zone, before strain localization, average crack spacing was also measured and compared with the theoretical prediction of the cohesive interface model [5].

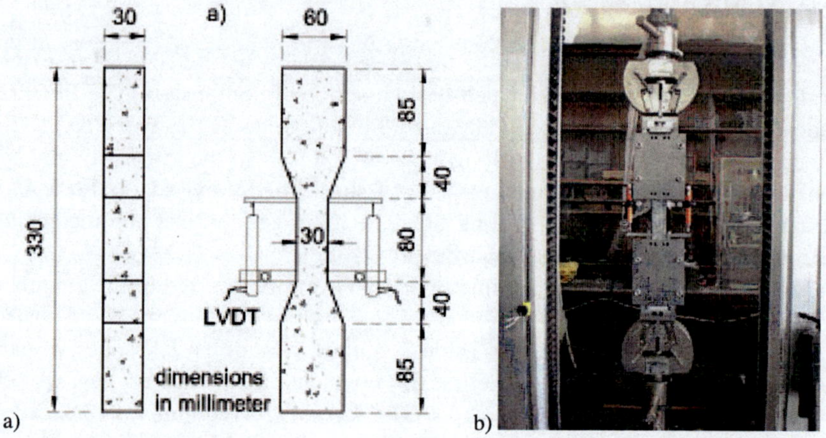

Fig. 3. Uniaxial tests on hybrid SHCC: a) geometrical properties of dumbbell specimens [7];b) the Instron testing machine

Table 3. Average crack spacing measured in the central part of dumbbell specimens

	Name	V_f steel cord (%)	Crack distance (mm)
PE series	PE_075_1	0.75	4.5
	PE_100_1	1.00	4.6
	PE_100_2		4.1
	PE_100_3		5.0
	PE_100_4		4.8
	PE_100_5		5.9
	PE_125_1	1.25	5.6
	PE_125_2		6.2
	PE_125_3		6.3
PVA series	PVA_100_1	1.00	11.0
	PVA_100_2		9.6
	PVA_100_3		8.6
	PVA_100_4		13.9
	PVA_100_5		11.9
	PVA_100_6		9.1
	PVA_150_1	1.50	8.9
	PVA_150_2		10.7
	PVA_175_1	1.75	10.0
	PVA_175_2		8.6
	PVA_175_3		8.3
	PVA_175_4		5.7
	PVA_175_5		7.6
	PVA_175_6		8.6

4 Experimental Results and Theoretical Prediction

For the sake of brevity, only the average distances between the cracks are reported in Table 3. For the specimens of each series, the higher the amount of steel cords, the higher the number of cracks which affect the central part (80 mm in length) of the dumbbell specimen depicted in Fig. 3a.

As shown in Fig. 4, the measured crack spacing is compared with the values of transmission length (Eq. (1)). However, to make this comparison possible for all the mixtures, the parameter k_B (related to bond and snubbing phenomena of steel cords) has to be defined from the best-fitting of the experimental results.

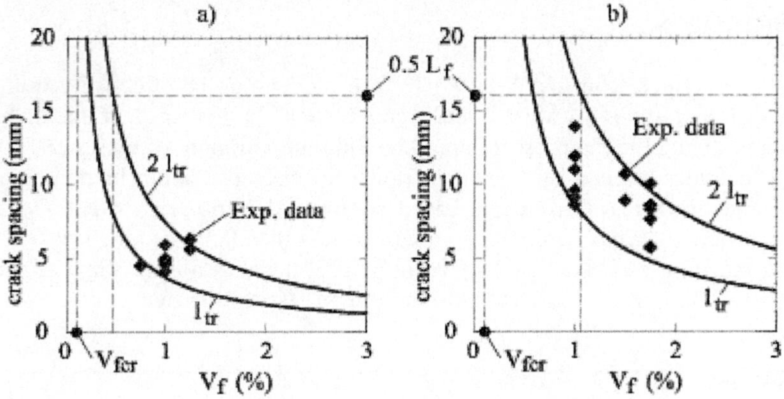

Fig. 4. Comparison between the average crack spacing predicted with the cohesive interface model [5] and those experimentally measured: a) specimens of the PE series; b) specimens of the PVA series

4.1 PE Series

If $k_B = 62$ MPa/mm is assumed for the 9 specimens of the PE series (Fig. 4a), all the measured crack spacings fall within the range bordered by the curves l_{tr}-V_f and 2 l_{tr}-V_f (Eq. (1)). Moreover, Eq. (3) gives $V_{fcr} = 0.1\%$ for the specimens of this series. Nevertheless, when $V_f = V_{fcr}$, only steel cords longer than 32 mm can generate multiple cracking and an average crack spacing ranged by the minimum and the maximum theoretical distance (l_{tr} and 2 l_{tr}, respectively).

Conversely, Fig. 4a seems to suggest $V_f = 0.5\%$ as the minimum volume fraction of steel cords having a length of 32 mm. As in the present experimental campaign the lowest volume is 0.75% (specimen PE_075_1 in Table 3), the minimum amount of SC can be further reduced of about 30%.

4.2 PVA Series

If $k_B = 110$ MPa/mm is assumed for the 14 specimens of the PVA series (Fig. 4b), most of the crack spacings experimentally measured are inside the range bordered by the curves l_{tr}-V_f and 2 l_{tr}-V_f (Eq. (1)). Moreover, for the composites of this series, Eq. (3) gives $V_{fcr} = 0.3\%$, even if, at $V_f = V_{fcr}$, only steel cords longer than 32 mm can generate an average crack spacing ranged by l_{tr} and 2 l_{tr}.

Conversely, Fig. 4b seems to suggest $V_f = 1\%$ as the minimum volume fraction of the adopted steel cords (Table 1). This value coincides with the lowest volume of SC added to the specimens reinforced with short PVA fibers (Table 3). Thus, for PVA series, the volume of steel cords cannot be reduced.

5 Conclusions

Eqs.(1)-(3), obtained from a cohesive interface model, can be effectively used to analyze different types of very ductile composites. In the case of the hybrid SHCCs investigated in the present paper, the minimum amount of long steel cords can be effectively defined by these equations. Specifically, strain hardening and multiple cracking occurs in cement-based composites reinforced with Polyethylene short fibers (0.75% in volume) and with no less than 0.5% of long steel cords. A similar behavior can also be observed in SHCC made with Poly Vinyl Alcohol short fibers (0.75% in volume) and with more than 1% of long steel cords.

References

[1] Banthia, N., Sappakittipakorn, M.: Toughness enhancement in steel fiber reinforced concrete through fiber hybridization. Cement and Concrete Research 37, 1366–1372 (2007)

[2] Stahli, P., van Mier, G.M.: Manufacturing, fibre anisotropy and fracture of hybrid fibre concrete. Engineering Fracture Mechanics 74, 223–242 (2007)

[3] Kawamata, A., Mihashi, H., Fukuyama, H.: Properties of Hybrid Fiber Reinforced Cement-based Composites. Journal of Advanced Concrete Technology 1, 283–290 (2003)

[4] Bolander, J.E., Duddukuri, S.R., Li, Z.: Hybrid-fiber reinforced composites: modeling issues. In: High Performance Fiber Reinforced Cement Composites (HPFRCC 5), Mainz, Germany, July 10-13 (2007)

[5] Fantilli, A.P., Mihashi, H., Vallini, P.: Multiple cracking and strain hardening in fiber-reinforced concrete under uniaxial tension. Cement and Concrete Research 39, 1217–1229 (2009)

[6] Bazant, Z.P., Becq-Giraudon, E.: Statistical prediction of fracture parameters of concrete and implications for choice of testing standard. Cement and Concrete Research 32, 529–556 (2002)

[7] Japan Society of Civil Engineers, Recommendations for Design and Construction of HPFRCC with Multiple Fine Cracks (2007) (in Japanese)

Model of Hooked Steel Fibers Reinforced Concrete under Tension

C. Sujivorakul

Department Civil Technology Education, King Mongkut's University of Technology Thonburi, Thailand

Abstract. The main objective of this research was to develop a model for predicting stress-strain relationship of hooked steel fibers reinforced concrete (HF-FRC) under tension. The research was divided into 2 parts. The first part was extensive tensile tests of HS-FRC. The hooked steel fibers having two plastic hinges at both ends were used. The size of tensile specimens was 5 x 5 cm in cross section and 50 cm in length. Parameters used in the study were: (a) L_f/D and L_f of the fibers (65/60, 65/35 and 80/60); (b) volume fractions of the fibers (0, 0.25, 0.5, 1.0 and 1.5%); and (c) compressive strength of concrete (30, 40 and 50 MPa). The main results obtained in test were stress-strain relationship of HS-FRC, first-peak strength, and post-cracking strength. The second part focused on statistical prediction. The basic model of stress-strain relationship was proposed in this research, i.e. linear relationship up to first-peak strength and assumed constant post-cracking strength. In statistical prediction, the dependent variables were first-peak strength and post-cracking strength of HF-FRC, while the independent variable were L_f/D, L_f, volume fraction of fibers, and compressive strength of concrete. A good prediction was obtained from the model, and the equations for using in the design were recommended. It was seen in the model that an increase in L_f/D ratio, L_f, volume fraction of the fibers, and compressive strength of concrete significantly enhanced the tensile performance of hooked steel fiber reinforced concrete, and could lead to high performance fiber reinforced concrete.

1 Introduction

It is generally known that the addition of discontinuous fibers to matrices (mortar or concrete), resulting in fiber reinforced concrete (FRC), enhances the ductility and the tensile strength (post-cracking strength) of the composites due to the fibers bridging the cracks through bond between fibers and matrix. The ductility and the post-cracking strength of these composites is dependent on volume fraction of fiber, fiber orientation, length of fiber, tensile strength of fiber, compressive strength of matrix, fiber-matrix bond [1-4]. However, fiber-matrix bond is recognized as a key element affecting the post-cracking strength of FRC

G.J. Parra-Montesinos, H.W. Reinhardt, and A.E. Naaman (Eds.): HPFRCC 6, pp. 19–26.

composites, whereby different types of fibers present different bond mechanism between fiber and matrix. Hooked steel fiber is a type of discontinuous fibers which is widely used in reinforced concrete structures, such as industrial floor, slab, beam, and column. The bond mechanism of hooked steel fibers consists of the contribution of two plastic hinges and the friction along the length of fibers [5]. However, the contribution of bond from the friction is very small comparing to the plastic hinges. There are different length and size of hooked steel fibers available in market. Table 1 is the code of hooked steel fibers used by a supplier, and those fibers were used in this research. Several researchers [6, 7] had study the mechanical properties in bending and tension of cement composites reinforced with hooked steel fibers. However, no research tried to predict stress-strain relationship of hooked steel fibers reinforced concrete under tension.

2 Objective and Research Significance

The main objective of this research is to develop a model for predicting stress-strain relationship of hooked steel fibers reinforced concrete (HS-FRC) under tension. Parameters included in the research are volume fraction of the fibers, aspect ratio of the fiber, length of fibers, and compressive strength of concrete. The research significance is to apply stress-strain model of HS-FRC recommended in this research in the design of reinforced concrete structures when hooked steel fibers (V_f = 0% to 1.5%) are added in concrete (f'_c = 30 to 50 MPa). This model would help structural engineers to predict the nominal resistance as well as their behavior for flexural RC members when the compatibility procedure is used.

Table 1. Properties of hooked steel fibers used in this research

Fiber ID	Length, L_f [mm]	L_f/D	Diameter, D [mm]	Tensile Strength (MPa)
HF-65/60	60	65	0.92	
HF-65/35	35	65	0.58	1,100
HF-80/60	60	80	0.75	

Table 2. Mix Proportion by weight and compressive strength of concrete

ID	Portland Cement Type I [kg/m^3]	Fine Aggregate [kg/m^3]	Coarse Aggregate [kg/m^3]	Water [kg/m^3]	Target f'_c [MPa]	Actual f'_c [MPa]
C300	409	953	751	166	30	33
C400	523	851	751	171	40	41
C500	593	790	751	174	50	51

3 Experimental Program and Test Setup

The experimental program in tensile tests of steel fiber reinforced concrete consisted of 39 series of tests. Three specimens were tested in each series of tests. Parameters used in this research were: (a) L_f/D and L_f of the fibers (HF-65/60, HF-65/35 and HF-80/60 as shown in table 1); (b) volume fractions of the fibers (V_f = 0%, 0.25%, 0.5%, 1.0% and 1.5%); and (c) compressive strength of concrete (f'_c = 30, 40 and 50 MPa as explained in table 2).

The size of tensile specimens was 5x5 cm in cross section and 50 cm in length. Fig. 1 shows tensile test setup. The top and bottom ends of the specimens are held by specially designed grips attached to the testing machine. Load versus elongation relationship were obtained from the tests. The elongation of the tensile specimens was measured using the average recording of 2 symmetrically placed LVDTs over a gauge length of 20 cm. The tensile stress σ_t and tensile strain ε_t were calculated from the following equations:

$$\sigma_t = \frac{T}{A} \tag{1}$$

$$\varepsilon_t = \frac{\Delta L_g}{L_g} \tag{2}$$

where T is the tensile load obtained from the load cell, A is the cross sectional area of the specimens, ΔL_g is elongation of the tensile specimen, and L_g is the gauge length of the tensile specimens.

Fig. 1. Tensile test setup

4 Experimental Results and Proposed Model of Stress-Strain Relationship

Fig. 2 shows an example of tensile stress and strain relationship of hooked steel fiber reinforced concrete. It is seen that the addition of hooked steel fibers to concrete significantly enhances the ductility and the tensile strength (i.e. first-peak strength and post-cracking strength) of the composites. Fig. 3 shows the proposed typical tensile stress and strain model of HF-FRC, which comprises of linear relationship with elastic modulus of composites (E_{com}) up to first-peak strength and assumed constant post cracking strength. The first-peak strength, post-cracking strength, and elastic modulus of composites are defined as follows

$$\sigma_F = k_1\sqrt{f_c^{'}} \tag{3}$$

$$\sigma_P = k_2\sqrt{f_c^{'}} \tag{4}$$

$$E_{com} = E_C(1-V_f) + E_f V_f \tag{5}$$

where σ_F, σ_P, and E_{com} are first-peak strength, post-cracking strength, and elastic modulus of composites, respectively; k_1, and k_2 are first-peak coefficient and post-cracking coefficient, respectively; E_c is elastic modulus of concrete; E_f is elastic modulus of steel fibers; and V_f is volume fraction of hooked steel fibers. It is noted that in case of $\sigma_F < \sigma_P$, thus the value of $k_1 = k_2$.

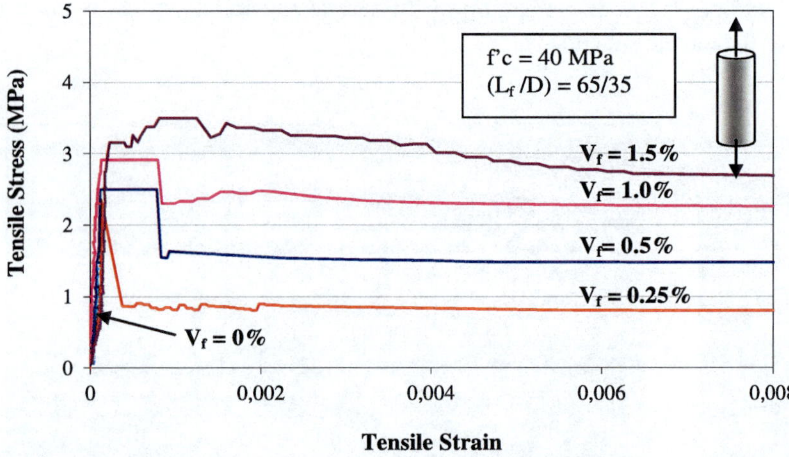

Fig. 2. An example of tensile stress and strain relationship of HF-FRC

Table 3 is a summary of first-peak strength and post-cracking strength from tensile tests of HF-FRC. It is seen that an increase in volume fraction of fibers and compressive strength increases first-peak strength and post-cracking strength. It is also observed that when the strength of concrete and volume fraction of fibers increases, post-cracking strength of composites would be higher than first-peak strength of composites, thus it leads to high performance fiber reinforced concrete. Moreover, it is seen that different fiber aspect ratio, L_f/D, and fiber length, L_f, also affect post-cracking strength of composites. An increase in L_f/D and L_f tends to increase post-cracking strength of composites, which is similar to the results reported by Sujivorakul and Naaman [8].

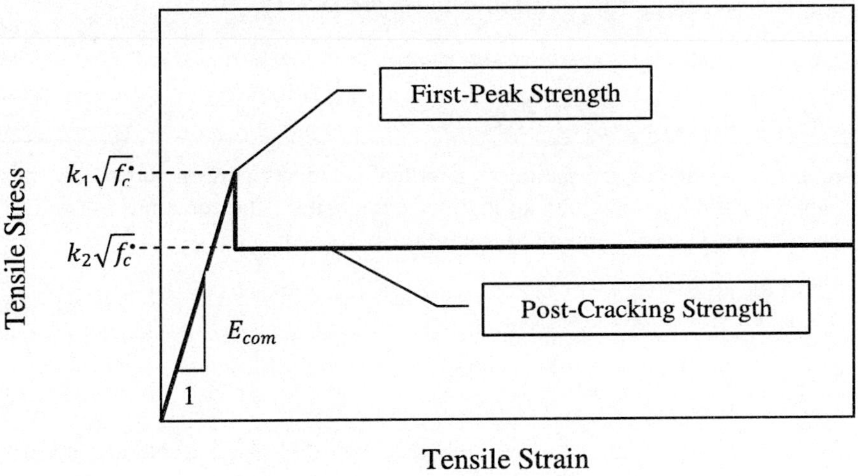

Fig. 3. Proposed model of tensile stress and strain relationship of HF-FRC

Table 3. First-peak strength and post-cracking strength from tensile tests of HF-FRC

f'_c [MPa]	V_f [%]	HF-65/60 First-Peak Strength [MPa]	HF-65/60 Post-Cracking Strength [MPa]	HF-65/35 First-Peak Strength [MPa]	HF-65/35 Post-Cracking Strength [MPa]	HF-80/60 First-Peak Strength [MPa]	HF-80/60 Post-Cracking Strength [MPa]
30	0	1.77	0	1.77	0	1.77	0
	0.25	2.29	0.87	2.16	0.52	2.32	1.20
	0.50	2.36	1.51	2.30	1.21	2.81	2.17
	1.00	2.75	2.21	2.64	1.97	3.03	2.56
	1.50	2.95	2.96	2.85	2.66	3.15	3.52
40	0	1.88	0	1.88	0	1.88	0
	0.25	2.55	1.05	2.35	0.90	2.76	1.59
	0.50	2.63	1.74	2.50	1.62	3.04	2.64
	1.00	2.96	2.81	2.87	2.47	3.33	3.61
	1.50	3.27	3.20	3.16	3.49	3.51	4.02
50	0	2.33	0	2.33	0	2.33	0
	0.25	2.75	1.44	2.45	1.28	2.97	1.82
	0.50	3.01	2.45	2.89	2.03	3.38	2.90
	1.00	3.37	3.71	3.21	2.75	3.58	3.95
	1.50	3.74	4.33	3.61	3.89	3.92	4.73

5 Analytical Prediction for First-Peak Load and Post-Cracking Strength

As explained in the previous section, first-cracking strength and post-cracking strength of HF-FRC are dependent on volume fractions of the fibers, compressive strength of concrete, fiber aspect ratio (L_f/D) and fiber length (L_f). These relationships can be shown in the following expressions:

$$\sigma_F = k_1\sqrt{f_c^{'}} \quad \propto \quad (V_f)^{\beta_1} \tag{4}$$

$$\sigma_F = k_2\sqrt{f_c^{'}} \quad \propto \quad (V_f)^{\lambda_1}, (L_f / D)^{\lambda_2}, (L_f)^{\lambda_3} \tag{5}$$

where β_1, λ_1, λ_2, and λ_3 are parameters indicating to affecting level to the first-peak strength and post-cracking strength of composites. The equations (4) and (5) can be rewritten in term of k_1, and k_2 as follows:

$$k_1 = \frac{\sigma_F}{\sqrt{f_c^{'}}} \quad = f\left((V_f)^{\beta_1}\right) \tag{6}$$

$$k_2 = \frac{\sigma_P}{\sqrt{f_c^{'}}} \quad = f\left((V_f)^{\lambda_1}, (L_f / D)^{\lambda_2}, (L_f)^{\lambda_3}\right) \tag{7}$$

It is observed in equations (6) and (7) that k_1 is the function of V_f, while k_2 is the function of V_f, L_f/D, and L_f. In order to obtain the predicted values of k_1 and k_2, the relationship between k_1, k_2 versus V_f, L_f/D, and L_f, are plotted in Figs. 4 and 5, respectively. The following equations express the statistical results of predicted k_1 and k_2 which are separated into 2 parts: mean values obtained from regression analysis; and lower-bound value recommended by author to use in the design of structural members.

For mean values:

$$k_1 = \frac{\sigma_F}{\sqrt{f_c^{'}}} \quad = 0.3481 + 0.1329V_f \tag{8}$$

$$k_2 = \frac{\sigma_P}{\sqrt{f_c^{'}}} \quad = (-0.0014V_f^{\ 2} + 0.0046V_f) \times (L_f / D) \ \times (L_f)^{0.2} \tag{9}$$

For lower-bound values (recommended by author):

$$k_1 = \frac{\sigma_F}{\sqrt{f_c^{'}}} \quad = 0.25 + 0.16V_f \tag{10}$$

$$k_2 = \frac{\sigma_P}{\sqrt{f_c^{'}}} \quad = (-0.001V_f^{\ 2} + 0.0038V_f) \times (L_f / D) \ \times (L_f)^{0.2} \tag{11}$$

It is observed that a good prediction is obtained, because the values of R^2 in equations (8) and (9) are 0.8389 and 0.963, respectively, which approach to 1.0.

However, there is some error in the prediction of equation (8), because the effect of L_f/D ratio does not include into this equation.

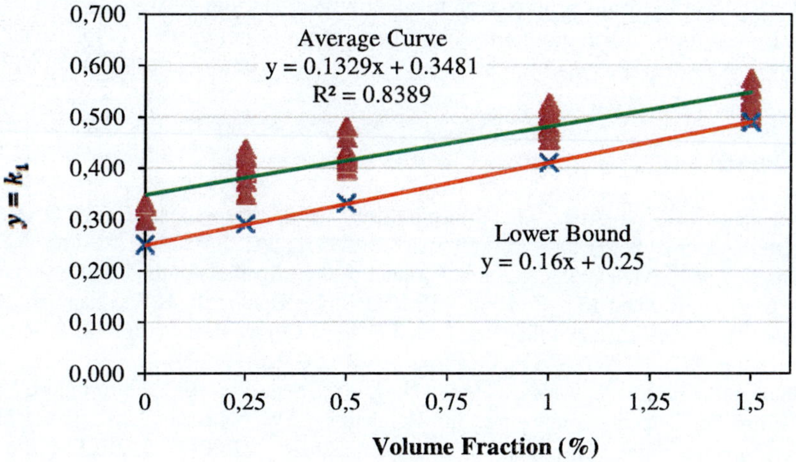

Fig. 4. Relationship between volume fraction of fibers versus k_1 coefficient

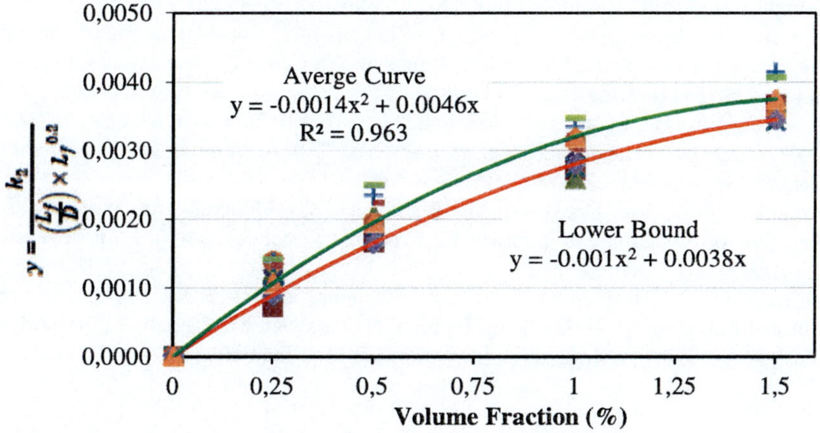

Fig. 5. Relationship between volume fraction of fibers versus $\frac{k_2}{\left(\frac{L_f}{D}\right)\times {L_f}^{0.2}}$

6 Conclusions

A model for predicting stress-strain relationship of hooked steel fibers reinforced concrete under tension was developed in this research with the inclusion of affected parameters, i.e. volume fraction of fibers, aspect ratio of fibers, length of fibers, and compressive strength of concrete. This model was assumed linear relationship up to first-peak strength and constant post-cracking strength. Equations for predicting stress-strain relationship were purposed here, and they would be a useful tool to help structural engineers in the design of reinforced concrete structures when hooked steel fibers with V_f between 0% to 1.5% are added into the normal to medium strength concrete (f'_c between 30 to 50 MPa).

References

[1] Naaman, A.E., Homrich, J.R.: Tensile Stress-Strain Properties of SIFCON. ACI Material Journal 86(3), 244–251 (1989)

[2] Oluokun, A.F., Malak, S.A.J.: Some Parametric Investigations of the Tensile Behavior of Slurry Infiltrated Mat Concrete (SIMCON). In: Reinhardt, H.W., Naaman, A.E. (eds.) High Performance Fiber Reinforced Cement Composites - HPFRCC 3, RILEM Pro 6, pp. 271–297. RILEM Publications S.A.R.L, Cachan (1999)

[3] Shah, S.P., Peled, A., Aldea, C.M., Akkaya, Y.: Scope of High Performance Fiber Reinforced Cement Composites. In: Reinhardt, H.W., Naaman, A.E. (eds.) High Performance Fiber Reinforced Cement Composites - HPFRCC 3, RILEM Pro 6 , pp. 113–129. RILEM Publications S.A.R.L, Cachan (1999)

[4] Sujivorakul, C., Naaman, A.E.: Modeling Mechanical Bond in FRC composites. In: Reinhardt, H.W., Naaman, A.E. (eds.) High Performance Fiber Reinforced Composites (HPFRCC 4), E & FN SPON, London, pp. 35–48 (2003)

[5] Alwan, J.M., Naaman, A.E., Guerrero, P.: Effect of Mechanical Clamping on the Pull-Out Response of Hooked Steel Fibers Embedded in Cementitious Matrices. Concrete Science and Engineering 1, 15–25 (1999)

[6] Li, V.C., Wu, H., Maalej, M., Mishra, D.K.: Tensile Behavior of Cement Based Composites with Random Discontinuous Steel Fibers. Journal of the American Ceramic Society 79(1), 74–78 (1996)

[7] Kooiman, A.G., Van Der Veen, C., Walraven, J.C.: Modelling the Post-Cracking Behavior of Steel Fibre Reinforced Concrete for Structural Design Purpose. Heron 45(4), 275–307 (2000)

[8] Sujivorakul, C., Naaman, A.E.: Tensile Response of HPFRC Composites Using Twisted Polygonal Steel Fibers. In: Banthia, N., Criswell, M., Tatnall, P., Folliard, K. (eds.) ACI-Special Publication on Innovations in FRC for Value, SP216-11 (2003)

Use of Double Punch Test to Evaluate the Mechanical Performance of Fiber Reinforced Concrete

S.-H. Chao, N.B. Karki, J.-S. Cho, and R.N. Waweru

Department of Civil Engineering, University of Texas at Arlington, USA

Abstract. Mechanical properties of fiber reinforced concrete (FRC) determined by material test methods can be used to ensure that the FRC mixture is batched properly and to give indications of performance if used in structural members. An ideal material test method for FRC should give low variability in the measurement of properties such as peak and residual strengths. ACI 318-08 uses results from the third-point bending test [1] as the performance criteria for FRC. Experimental evidence, however, has indicated that this bending type test method has potential problems in the reliability of determining the peak and residual strengths. The coefficient of variation of residual strength is typically very high and generally greater than 20%. The considerable scatter results make it difficult for quality control, particularly when such properties are intended to be used to estimate the strength of structural members. In addition, the complex test setup and the requirement of using a closed-loop servo-controlled machine often prevent its use in small laboratories. Other test methods such as the direct tensile test experience similar problems. As a consequence, these test methods are generally time consuming and expensive as they require more specimens to obtain reliable test data.

The double punch test (DPT) was originally developed, based on the plasticity theory of concrete, to determine the tensile strength of plain concrete but has rarely been used since the 1980s. This test method can be carried out by using conventional compression test machines with small capacity and simple test setup. This study investigates the feasibility of using DPT for the evaluation of mechanical properties and quality control of FRC. Experimental results showed that DPT generally gives much lower variability compared to other test methods. DPT also has the ability to identify important FRC characteristics such as strain-softening, strain-hardening, stiffness, tensile strength, residual strength, and toughness.

1 Introduction

Mechanical properties of FRC, such as peak and residual tensile/flexural strengths, are essential if they are to be used for evaluating the use of FRC in structures. An

G.J. Parra-Montesinos, H.W. Reinhardt, and A.E. Naaman (Eds.): HPFRCC 6, pp. 27–34.

ideal material test should be able to exhibit low variability in the measured properties [2]. ACI 318-08 [3] requires the third-point bending test [2] as performance criteria for FRC. Experimental evidence, however, shown by Bernard [4] has indicated that the residual strength of a third-point loaded beam is the least attractive parameter because it displays very poor reliability, in that the coefficient of variation is generally greater than 20%. This large scatter in the residual strength can be a major problem if characteristic values have to be determined. An alternative test method for FRC is a direct tensile test with a dog-bone shaped specimen. However, the direct tensile test method also normally shows high variability in the measured peak and residual strengths, which can make it difficult for quality control, especially when it is intended to be used in structural members. In addition, those test methods require complex test setup with a closed-loop servo-controlled machine which is often not available in most small laboratories.

A potential material test method for FRC, the double punch test (DPT), was originally developed as an indirect tensile test method for plain concrete, which was introduced by Chen [5, 6]. It stood on the base of the theory of plasticity. Chen and Yuan [7] conducted extensive research, such that they obtained idealized dimensions of specimens and steel punches as well as test procedures to optimize reliable results. Based on Chen and Yuan [7], a cylinder specimen with dimensions of 152.4×152.4 mm (6x6 in.) was chosen, which is created by cutting half of a 152.4×304.8 mm (6x12 in.) cylinder. Compression is applied through two steel punches, having 25.4 mm (1 in.) height and 38.1 mm (1.5 in.) diameter, placed at the top and bottom surfaces of the cylinder along its central axis (Fig. 1).

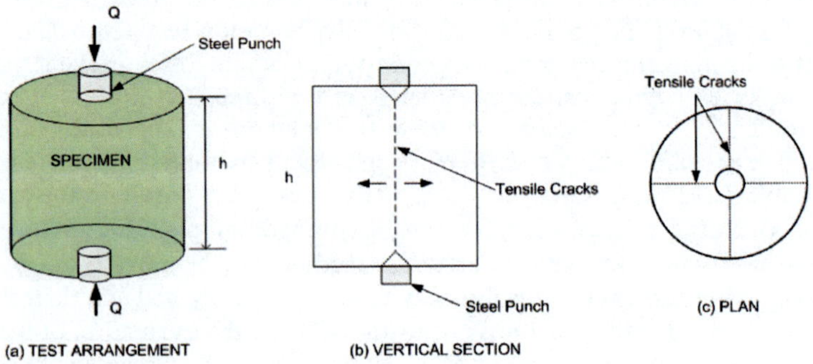

Fig. 1. Schematic view of DPT

Deformation of the specimen can be measured by LVDTs (linear variable differential transformers) placed next to the specimen. By applying compressive load through the punches, uniform tensile stresses are generated over diametric planes, and tensile cracks occur along these diametric planes. The equivalent tensile stress, derived from the assumptions of plastic material and multiple tension cracking failure mechanisms, is calculated by Chen and Yuan [6]:

$$f_t = \frac{Q}{\pi\left(1.20bh - a^2\right)} \tag{1}$$

where f_t is equivalent tensile stress, Q is applied load, b is radius of cylinder, h is height of cylinder, and a is radius of punches.

Chen and Yuan [7] applied the DPT to steel as well as polymer fiber reinforced concrete, and compared those with the split-cylinder test. They concluded that the DPT was a better test method since a DPT specimen failed at the weakest sections while a split-cylinder test specimen would fail at a predetermined failure plane. Molins et al. [8] compared experimental results between the DPT and beam test with various types and content of fibers, and concluded that the DPT provided more consistent results because the coefficient of variation (C.O.V.) for the peak and residual strengths were smaller than those from the beam test. In this study, three different material evaluation methods (ASTM C1609 bending test, direct tensile test, and DPT) were compared by using three different FRC mixtures. The main focus is on the consistency of the test results.

2 Experimental Program and Results

2.1 Materials

Properties of the steel hooked and twisted fibers used in this study are listed in Table 1. Three FRC mixtures were used in this study. The first two mixtures (ML-075 and ML-150) had steel hooked fibers of 0.75% and 1.5% by volume, respectively. The third mixture (Hybrid) is a hybrid mixture which included 0.75% steel twisted fibers and 0.75% steel hooked fibers, with a total volume fraction of 1.5%. Detailed information regarding the specimens is given in Table 2. The compressive strengths for the three mixtures were 64 MPa, 43 MPa, and 67MPa, respectively.

Table 1. Properties of steel fibers

Fiber type	Length (L, mm) [1]	(Equivalent) Diameter (D, mm) [1]	Aspect ratio (L/D)	Tensile strength[2], MPa
Twisted	25.4	0.40	63.5	2000
Hooked	47.5	0.76	62.5	1100

Note: [1] measured; [2] provided by manufacturer

Table 2. Specimen information

Specimen	Steel fiber type	Volume fraction of fiber	Number of specimens
ML-075	Hooked	0.75%	6
ML-150		1.50%	
Hybrid	Hooked + Twisted	0.75% + 0.75%	

2.2 Specimen Details and Test Setup

The detailed dimensions of specimens and test setups for the direct tensile and ASTM C1609 bending test can be found elsewhere [9]. For the double punch test, a 267-kN (60-kip) Baldwin hydraulic testing machine was used without the need of a closed-loop servo-controlled machine. A pair of LVDTs was used to measure the vertical deformation of the specimens. The loading rate was 445 N (100 lb)/min prior to the first crack, and increased to 1335 N (300 lb)/min in the post-cracking stage. The test setup for each of the tests is shown in Fig. 2.

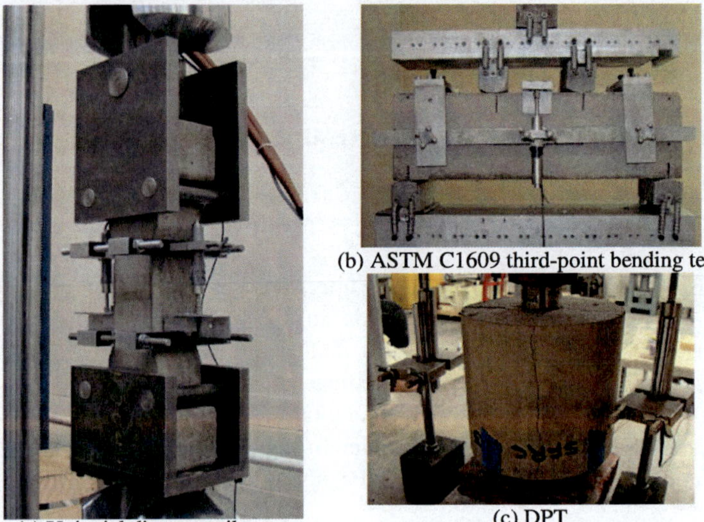

(a) Uniaxial direct tensile test

(b) ASTM C1609 third-point bending test

(c) DPT

Fig. 2. Test setup for each test method

2.3 Experimental Results

The load-versus-deformation curves for direct tensile test, ASTM C1609 bending test, and DPT are presented in Figs. 3, 4, and 5 ((a), (b), and (c) only), respectively. The equivalent tensile stresses for DPT specimens as shown in Fig. 5 were estimated according to Eq. (1). It should be mentioned that only the results from the bottom halves of the cylinders for DPT are shown (will be discussed later). Also, not all test data for the direct tensile test were valid due to the formation of primary cracks outside the gauge length in some of the specimens, even though both ends of the specimens were reinforced with steel meshes. It is clearly seen that variability of the residual strength is relatively higher for both direct tensile test and ASTM C1609 bending test, as compared to that of DPT. Tables 4 and 5 show the comparison of C.O.V between the three test methods at peak strength as well as residual strength at a specified deformation. It is also interesting to note that the load-versus-deformation responses of all DPT specimens are quite consistent throughout the entire curves. One the other hand, the C.O.V of the strengths varies considerably for the responses of both the direct tensile test and ASTM C1609 bending test, especially at the post-cracking stage. As a consequence, individual curves can significantly deviate from the average one and the use of averaged parameter values could lead to conclusions that significantly differ from true behavior.

Fig. 5 is also used to illustrate the ability of DPT in distinguishing between specimens with different ductility, residual strength, toughness, as well as strain-hardening or softening behavior. A fourth FRC mixture, T-350, with 3.5% twisted fiber by volume and compressive strength of 55 MPa was used for this investigation, and Fig. 5(d) shows the load (stress)-versus-deformation curve as well as photo of one of the specimens. All tests were stopped when the residual strength was approximately 20-25% of the peak strength. While ML-075, ML-150, and Hybrid mixtures exhibited strain-softening behavior, T-350 showed strain-hardening behavior with multiple cracking. ML-075 had the least ductility (or deformation capacity) compared to the other three mixtures. The performance of the four mixtures can also be distinguished by a residual strength at specified deformation value. For example, the residual strength-to-peak strength ratios of ML-075, ML-150, Hybrid, and T-350 mixtures at 3 mm deformation are 27%, 51%, 73%, and 99%, respectively. Finally, the toughness which represents the energy absorption capability can be evaluated by calculating the area of the load-deformation curves up to specified deformation values. It should be mentioned that, while the other test methods can also distinguish between those properties, DPT has the simplest test procedure and setup. Specimens for DPT are the same as those for routine compression tests thus only standard cylinder molds are needed.

Table 3. Summary of peak loads and C.O.V for each test method

Mix	Direct Tensile Test		ASTM C1609 Bending Test		DPT	
	Load (kN)	C.O.V	Load (kN)	C.O.V	Load (kN)	C.O.V
ML-075	38.8	14%	49.0	23%	178.3	7%
ML-150	34.3	28%	69.9	12%	184.6	5%
Hybrid	42.3	22%	76.0	10%	209.2	12%

Table 4. Summary of residual loads and C.O.V for each test method

Mix	Direct Tensile Test[3]		ASTM C1609 Bending Test [4]		DPT[5]	
	Load (kN)	C.O.V	Load (kN)	C.O.V	Load (kN)	C.O.V
ML-075	16.8	40%	15.8	36%	56.1	17%
ML-150	22.4	18%	16.7	24%	106.3	3%
Hybrid	23.4	39%	13.3	48%	153.1	4%

Note: [3] measured load at deformation of 1.0 mm (0.04 in.); [4] measured load at deformation of 3.0 mm (0.12 in.); [5] measured load at deformation of 2.5 mm (0.1 in.)

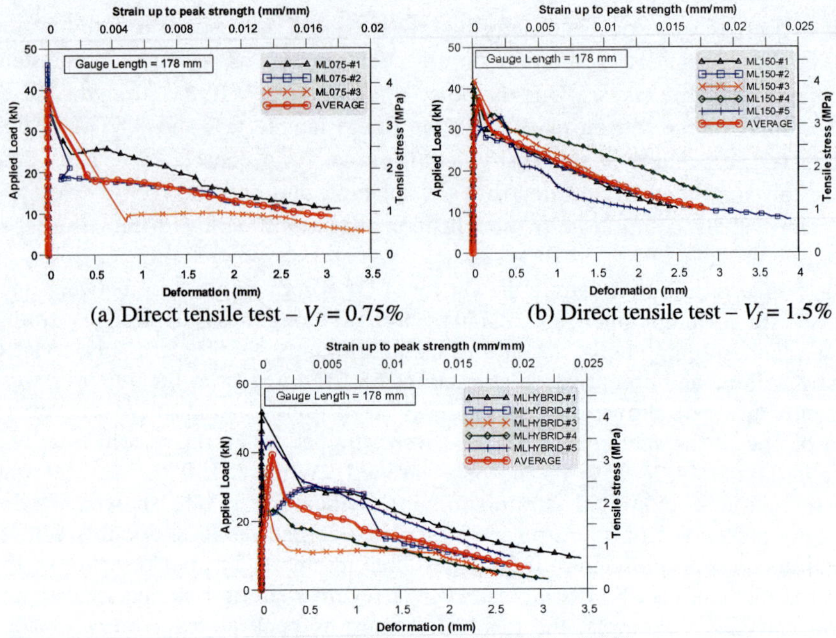

(a) Direct tensile test – $V_f = 0.75\%$ (b) Direct tensile test – $V_f = 1.5\%$

(c) Direct tensile test – Hybrid ($V_f = 1.5\%$)

Fig. 3. Load versus deformation curves for direct tensile test

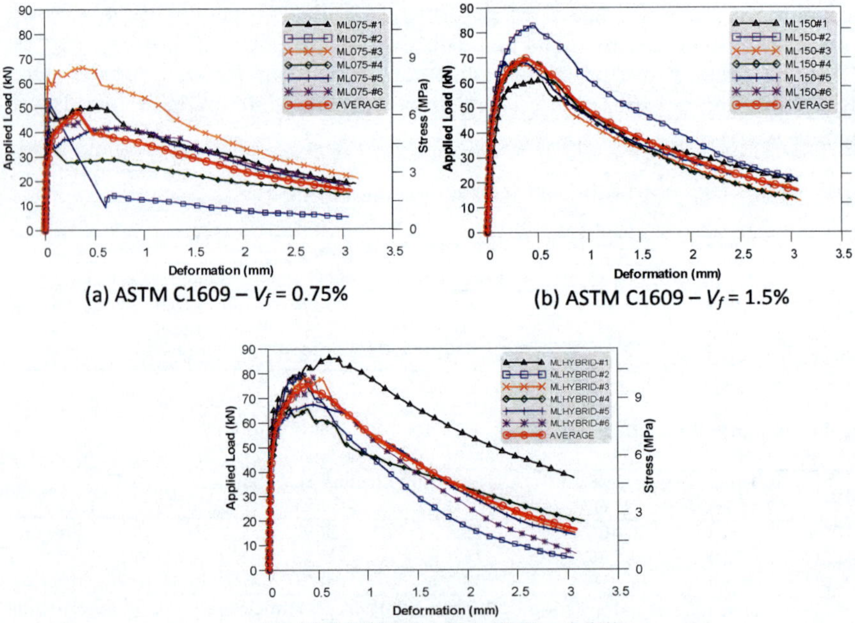

(a) ASTM C1609 – $V_f = 0.75\%$ (b) ASTM C1609 – $V_f = 1.5\%$

(c) ASTM C1609 – Hybrid ($V_f = 1.5\%$)

Fig. 4. Load versus deformation curves for ASTM C1609 test

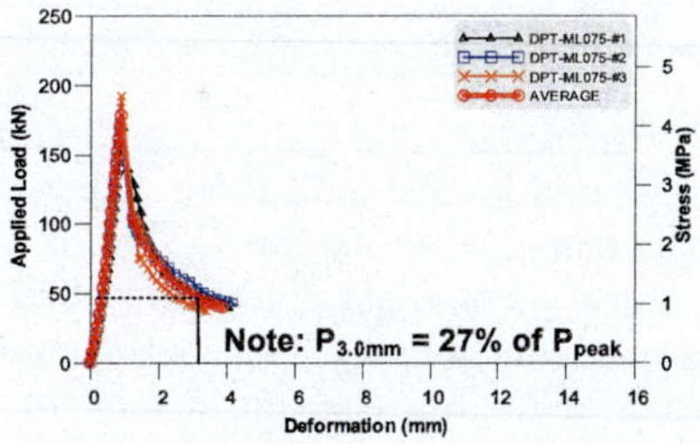

(a) DPT – $V_f = 0.75\%$

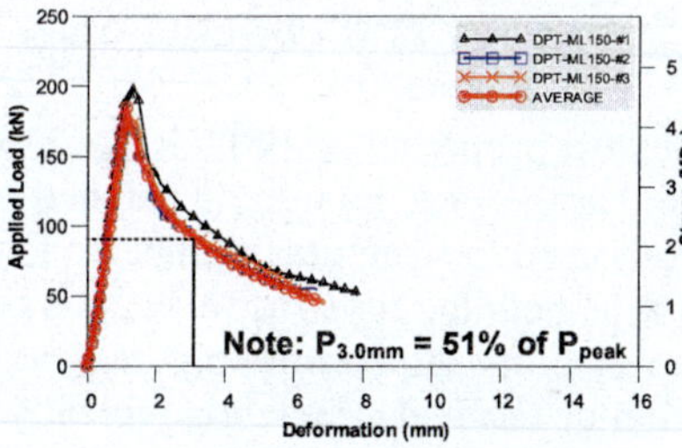

(b) DPT – $V_f = 1.5\%$

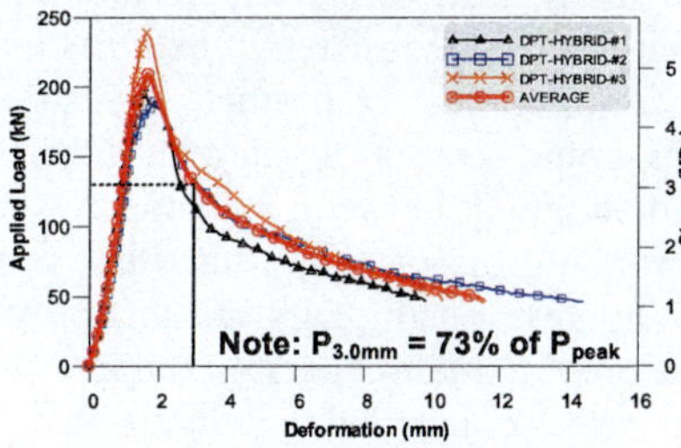

(c) DPT – Hybrid ($V_f = 1.5\%$)

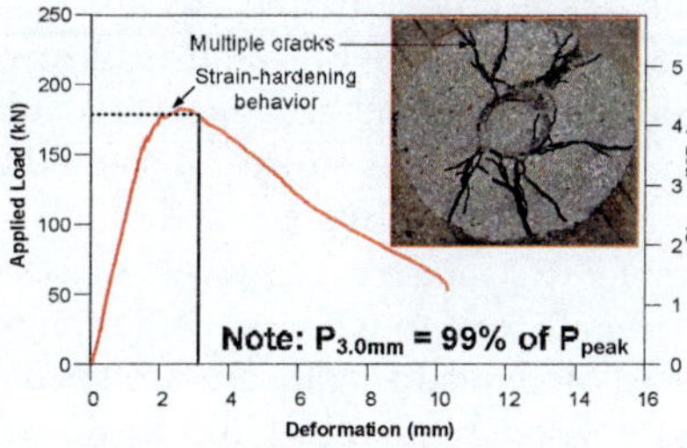

(d) DPT – Twisted fibers only ($V_f = 3.5\%$)

Fig. 5. Load versus deformation curves for DPT (bottom halves of the cylinders)

As stated earlier, only the bottom halves of the DPT specimens were used for evaluation purposes. This is due to the fact that it is consistently observed that the bottom halves of the DPT specimens gave less variability and slightly higher peak strength than that of the top halves of the cylinders. For example, Fig. 6 shows the DPT results for specimens with the ML-150 mixture. The C.O.V for the residual strengths at deformation of 2.5 mm (0.1 in.) was 21% and 5% for top and bottom halves, respectively. It is believed that the higher C.O.V for the top halves of the cylinders resulted from the water bleeding near the top of the cylinder surface during the compactness and vibration of making the cylinders. The quality of the concrete near the top surface can affect the performance because the loading is applied through the surface (Fig. 1). As a consequence, it is suggested the bottom halves of the DPT cylinders be used for material evaluation.

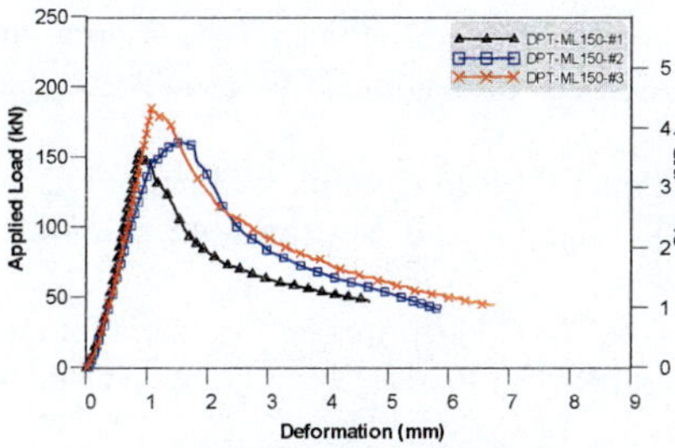

(a) DPT –1.5%; Top halves

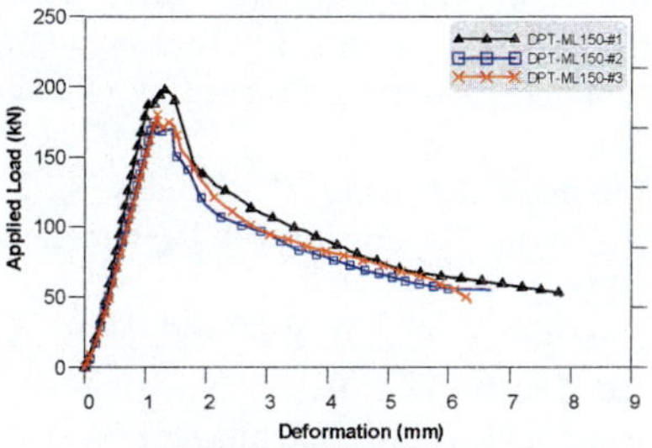

(b) DPT –1.5%; Bottom halves

Fig. 6. Comparison of top and bottom halves of the specimens

3 Summary and conclusions

Mechanical properties of FRC determined by material test methods can be used to ensure that the FRC mixture is batched properly and to give indications of performance if used in structural members. The most commonly used test method is the beam-type bending test. However, this type of test method has potential problems in the reliability of determining the peak and residual strengths. The coefficient of variation of residual strength is typically greater than 20%. The considerable scatter results make it difficult for quality control, particularly when such properties are intended to be used to estimate the strength of structural members.

This study compares the ASTM C1609 bending test, direct tensile test, and double punch test (DPT), and it is observed that the DPT consistently exhibits low variability along the entire load-versus-deformation curves. Preliminary investigation carried out in this study indicated that DPT could serve as a potential material test method. The results show that DPT can distinguish between specimens with different FRC mixtures, in terms of strain-hardening or softening, ductility, residual strength, and toughness. Other major advantages of using DPT are that only a few ASTM standard cylinders are needed and only a small-capacity compression testing machine (no need for a closed-loop servo-controlled machine) that is commonly available in laboratories is required.

References

[1] ASTM C1609/C1609M-10, Standard Test method for Flexural Performance of Fiber-Reinforced Concrete (Using Beam with Third-Point Loading). ASTM International, West Conshohocken, PA, p. 9 (2010)

[2] Mindess, S., Young, J.F., Darwin, D.: Concrete, 2nd edn. Prentice Hall, Upper Saddle River (2003)

[3] ACI Committee 318, Building Code Requirements for Reinforced Concrete and Commentary (ACI318-08/ACI318R-08). American Concrete Institute, Detroit (2008)

[4] Bernard, E.S.: Correlations in the Behaviour of Fibre Reinforced Shotcrete Beam and Panel Specimens. Materials and Structures 35, 156–164 (2002)

[5] Chen, W.F.: Double Punch Test for Tensile Strength of Concrete. ACI Journal, 993–995 (December 1970)

[6] Chen, W.F.: Limit Analysis and Soil Plasticity, ch. 11, pp. 501–541. Elsevier Scientific Publishing Company, Amsterdam (1975)

[7] Chen, W.F., Yuan, R.L.: Tensile strength of concrete: Double-Punch Test. Journal of Structural Division, Proceeding of American Society of Engineers 106(ST8) (August 1980)

[8] Molins, C., Aguado, A., Saludes, S.: Double Punch Test to Control the Energy Dissipation in Tension of FRC (Barcelona Test). Material and Structures 42, 415–425 (2009)

[9] Chao, S.-H., Cho, J.-S., Karki, N. B., Yazdani, N.: FRC Performance Comparison: Direct Tensile Test, Beam-Type Bending Test, and Round Panel Test. ACI Special Publication (2011, to be published)

Determining Specimen Size Influences on FRC Response Using the Digital Image Correlation Technique

L.N. Talboys, A.S. Lubell, and V.S. Bindiganavile

Department of Civil and Environmental Engineering, University of Alberta, Canada

Abstract. The global response of FRC members is known to be influenced by the force transfer mechanisms occurring at cracks, including the contributions from bridging fibers. These transfer mechanisms depend in part on the width of the cracks, which can vary with overall specimen size. This paper reports the response in flexure for notched specimens fabricated using normal weight concrete containing 1% volume fraction of hooked end steel fibers. Nominal specimen dimensions that varied by a factor of up to 2 were used to allow for the quantification of size dependent influences on the response. Instrumentation based on the digital image correlation technique allowed for a detailed study of the variation of crack width over the member depth, and its relation to the specimen size and loading. Companion tests were conducted on cylinders in compression using a size factor of 1.5. The results showed that the flexural strength decreased in relation to the overall specimen depth. From the crack width analysis, it was possible to develop relationships between the crack width and mid-span deflection.

1 Introduction

The size effect phenomenon in shear has been well documented in the quasi-brittle behavior of reinforced concrete members [1]. At this scale, as the member depth increases, the crack sizes increase, aggregate interlock action decreases and subsequently, the shear stress at failure decreases [2, 3]. Adding steel fibers in a random three dimensional (3D) orientation to a concrete mix has been shown to improve the crack distribution and the post peak tensile strength [4, 5] at the material scale. The fibers provide an alternate force transfer path across the cracks, thereby reducing crack widths and improving the aggregate interlock. However, size effects on shear strength are interestingly still prominent in beams constructed using fiber reinforced concrete (FRC) [1] and demonstrate the need to quantify this effect at the material scale.

Data acquisition methods during typical beam testing involve cumbersome instrumentation and physical measurements. Recently, the digital image correlation

G.J. Parra-Montesinos, H.W. Reinhardt, and A.E. Naaman (Eds.): HPFRCC 6, pp. 35–42.

(DIC) technique has been examined as a method to accurately measure 3D surface strains during testing without the need for additional instrumentation [6, 7].

The objective of this study was to examine the size effect on the flexural response of FRC prisms under ASTM C1609-10 through the use of the DIC technique. Notched specimens of three different sizes were tested according to the ASTM flexural test method. Deflection and crack growth were measured through the DIC system, and were evaluated against each member size.

2 Experimental Investigation

2.1 *Mix Details*

Laboratory trial batches were used to develop a FRC mix that could be easily replicated with consistent results in a ready mix plant. The trials aimed for a nominal compressive strength close to 40 MPa, and a slump near 100 mm with 1% volume fraction of steel fibers. InterCem™, a pre-blended mix of Portland cement and Class F fly ash, was chosen in order to develop a product suitable for applications where sulphate resistance is a priority. Hooked end steel fibers with a length of 30 mm, aspect ratio of 55 and yield strength of 1100 MPa were added to the mix at the batching plant. The laboratory produced mix contained no admixtures, while the ready mix product used a water reducing agent and a super plasticizer. The mix proportions are shown in Table 1.

The concrete for the specimens reported in this paper was supplied by a local ready mix concrete supplier within a 5.5 m^3 batch as part of a larger study. Superplasticizer was added at the batching plant and also on site to aid in mixing and to reduce fiber balling. Immediately prior to casting, the slump was 130 mm and the air content was 1.7% according to the ASTM C231 pressure method.

Six cylinders were cast for the ASTM C39-09a compressive strength tests; three 100 mm diameter x 200 mm long and three 150 mm diameter x 300 mm long. Nine prisms were cast for the ASTM C1609-10 flexural tests; three 100 x 100 x 300 mm span, three 150 x 150 x 450 mm span, and three 200 x 200 x 600 mm span. All cylinders and prisms were compacted in two layers with a vibrating table, save for the largest prisms which were compacted with a vibrating needle.

All cylinders and prisms were de-moulded after 48 hours and cured under wet burlap and a plastic covering for 28 days before testing. The 150 mm diameter cylinders were tested at 28 days, while the smaller ones were tested at 29 days. The prisms were tested between 31 and 37 days.

Table 1. Concrete mix proportions

Ingredients	Quantity
InterCem™ Cement (kg/m^3)	391
Fine Aggregate (kg/m^3)	796
14 mm Coarse Aggregate (kg/m^3)	990
Water (L/m^3)	157
Steel Fibers (kg/m^3)	78.6
Water Reducer (L/m^3)	0.25
Superplasticizer (L/m^3)	1.8

2.2 Test Set-Up

2.2.1 Compression

A 2600 kN capacity MTS Universal Loading Frame was used to perform the compression tests according to ASTM C39-09a. The cylinders were sulphur capped prior to testing and outfitted with a yoke containing three LVDTs at 120° separation to measure the vertical deformation at mid-height. The LVDTs were connected to a data acquisition system along with the MTS output. The test setup is shown in Fig. 1. A displacement controlled rate of 0.2 mm/min was used for both cylinder sizes, allowing capture of the full load-displacement relationship including the post-peak response.

2.2.2 Flexure

A 1000 kN capacity MTS Universal Loading Frame was used to perform the flexural tests according to ASTM C1609-10. A notch equal to 25% of the prism height was saw cut into the tension face of each prism at mid-span. The purpose of the notch was to pre-determine the crack location and prevent multiple small cracks from forming. A displacement controlled rate of 0.075 mm/min was used for the prisms with heights of 100 mm and 150 mm, while a rate of 0.1 mm/min was used on the 200 mm high prisms.

A DIC measurement system was used to track the crack growth and specimen deformation throughout the load history. One face of each prism was painted with flat white interior latex paint, and subsequently speckled with black acrylic paint to create a random pattern of dots of varying size up to approximately 3 mm. Two 5.0 megapixel digital cameras with 35 mm focal length lenses were set up at a distance of approximately 2.8 m from the prisms and at an angle of approximately 4.5° in order to view the entire prism face. Vic-Snap 2007 software [8] recorded pairs of images every 1 second during the test on the 100 mm and 150 mm prism sizes. To keep the amount of recorded data manageable with the 200 mm prisms, images were recorded every 3 seconds for the initial linear portion of the load-stroke diagram, then every 2 seconds at an average of 75% of the peak load, and finally back to 3 seconds post peak. The resulting 3D images obtained from Vic-3D 2007 [8] provided information on the specimen strains, deformations and rotations throughout the test by comparing the grey value pattern in each image with the initial reference image. Rigid body motion occurring from support settlement was analytically removed by the software before the calculations were performed.

A single LVDT was placed below each prism to one side of the notch by an epoxied aluminum plate on the underside of the prism and a magnetic base attached to the underside of the support plate. Due to the setup restrictions, only a single LVDT with ± 5 mm of travel could be used. Fig. 1 illustrates the test setup and LVDT location.

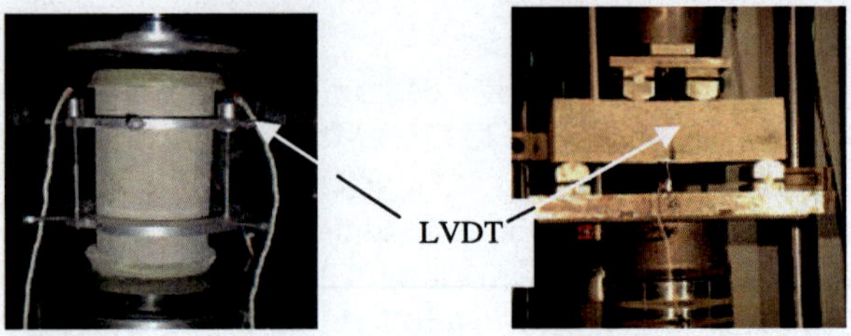

Fig. 1. (a) Compression test setup and (b) Flexural test setup

The MTS load and stroke data and the LVDT output were captured by the DIC control computer to allow time synchronization. The tests were stopped after the MTS stroke reached a deflection limit of L/100 where L equalled the prism span, in order to ensure the recorded DIC and LVDT deflections exceeded the L/150 limit in ASTM C1609-10.

3 Results and Discussion

3.1 Compression

The resulting best fit cylinder stress-strain curves are shown in Fig. 2. The compressive strengths showed a decrease with increase in specimen size. The 100 mm diameter cylinders exhibited an average compressive strength of 36.6 MPa, while the 150 mm diameter cylinders exhibited a 12% lower average compressive strength of 32.3 MPa. Results from one of the 150 mm cylinders (C150-4) were excluded due to a faulty sulphur cap. As shown in the figure, the strain at peak stress for the 100 mm diameter cylinders was between 0.0015 and 0.0018, and between 0.0025 and 0.0032 for the larger cylinders.

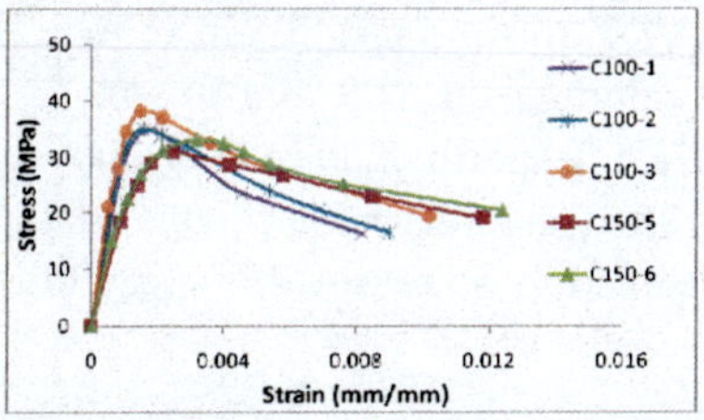

Fig. 2. Compressive stress-strain curves

3.2 *Flexure*

The DIC system was used to evaluate the relative movement of the prisms at the six locations shown in Fig. 3. These six locations correspond to a point immediately above the LVDT location, on each side at the base and top of the notch, and at mid-height of mid-span.

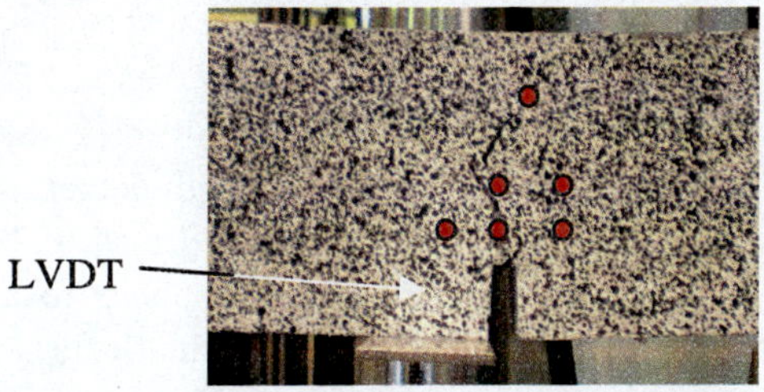

Fig. 3. DIC data points

The mid-span vertical deflection was calculated at the top and bottom of the notch by using the average from the data points located on both sides of the notch. These two values were then compared with the mid-height vertical deflection. In all three specimen sizes the deflections at all three locations were comparable, varying by less than 10%. The LVDT consistently measured a greater deflection than the DIC system in all specimens, owing to the fact that the DIC data had been corrected for rigid body motion while the LVDT had not. All subsequent references to the member mid-span deflection refer to the DIC deflection calculated at the bottom of the notch.

A comparison between the crack mouth opening displacement (defined as the crack width at the base of the notch and subsequently referred to as the CMOD) and vertical deflection in each size range showed a higher rate of increase in the crack opening than in the deflection. The change in CMOD is plotted against the change in deflection in Fig. 4. The result from all three specimen sizes based on the DIC data produced a ratio of approximately 2.5 and confirmed that the relationship is size-independent. Comparing the change in CMOD to the change in deflection based on the LVDT showed a ratio closer to 1.28, but produced scatter in the 200 mm prisms. Excluding the 200 mm prisms from the results increases the ratio to approximately 1.5. Experimental results analysed by Armelin and Banthia [5] on six 75 x 75 x 250 mm steel FRC prisms produced a ratio of approximately 1.5, which they deemed to be reasonably close to their analytical assumption of 1.33.

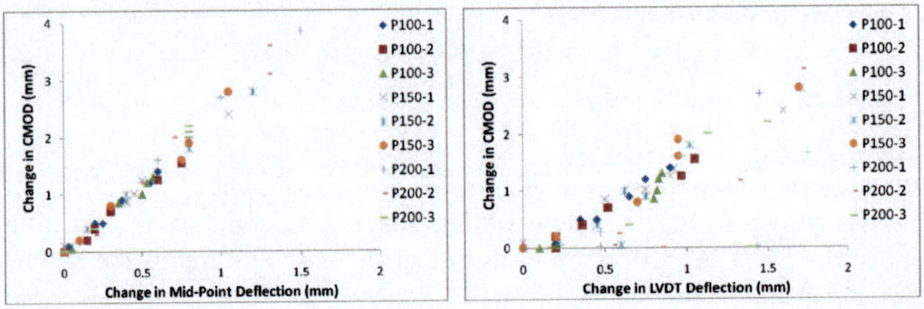

Fig. 4. Change in CMOD versus change in deflection based on (a) mid-span DIC data and (b) LVDT data

Fig. 5 shows the corresponding average load curves for all three specimen sizes. As expected, as the prism size increased, the tensile strength decreased. The 100 x 100 mm prisms experienced an average maximum applied load of 11.7 kN at an average vertical displacement of 0.38 mm, which corresponds to a modulus of rupture (MOR) of 6.25 MPa. This corresponds well with the flexural response obtained for members of similar compressive strength and fiber fraction by Shoaib et al. [9]. The 150 x 150 mm prisms demonstrated an average maximum load of 19.0 kN at a vertical displacement of 0.27 mm and MOR of 4.50 MPa. The 200 x 200 mm prisms experienced 30.6 kN at 0.9 mm of deflection and an MOR of 4.15 MPa.

The decrease in strength with increase in specimen size is clearly shown in Fig. 5(b). The CSA A23.3 and ACI 318-08 recommended values for MOR based on the compressive strength in plain concrete are also shown for comparison. The compressive strength used in this case was taken as the average of the 150 mm diameter cylinders.

Fig. 5(c) plots the average peak tensile strength against the corresponding average CMOD for each specimen size. The 150 x 150 mm prisms experienced a lower CMOD at peak load than the other two specimen sizes, and no overall trend was observed. Testing of an additional series of identical prisms is planned to help clarify whether a trend exists between peak tensile strength and corresponding CMOD.

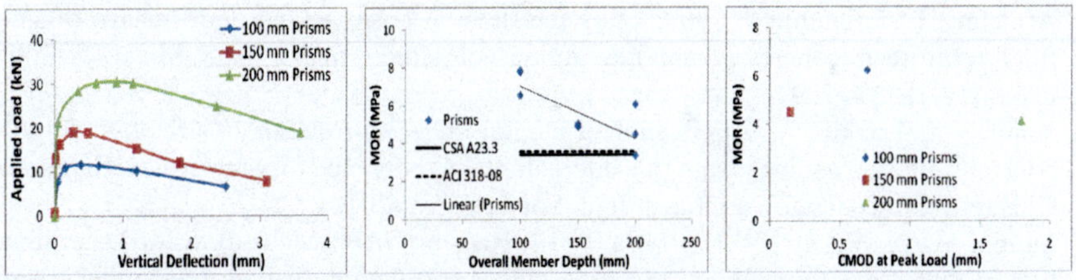

Fig. 5. (a) Average load-deflection curves based on DIC deflection at the bottom of the notch (b) MOR-size relationship and (c) MOR and corresponding CMOD relationship.

4 Conclusions

- Even with the addition of 1% volume fraction of hooked end steel fibers, a size effect is clearly shown in the compressive tests where the strength decreased as the size increased.
- Use of the DIC system allowed for comparisons between the vertical member deflections measured at mid-height, top of the notch, and base of the notch. The variation in the deflections was found to be minimal.
- The crack width was found to increase at a greater rate compared to the vertical deflection. The ratio between the change in CMOD and change in vertical deflection was approximately 2.5 based on the DIC data, and approximately 1.28 based on the LVDT data for all prism sizes.
- A size effect was also observed for the tensile strength of the notched flexural prisms. As the size of the prism increased, the MOR decreased. However, there was no apparent trend between the CMOD at the peak flexural load and the specimen size.

Acknowledgements. The authors gratefully acknowledge the financial support and material donations provided by the Natural Sciences and Engineering Research Council (NSERC) of Canada, AECOM Canada, Lehigh Hanson Materials Limited and Bekaert.

References

[1] Shoaib, A., Lubell, A.S., Bindiganavile, V.S.: Size effect in the shear response of steel fiber reinforced concrete members without stirrups. In: Proceedings from 8th International Conference on Short and Medium Span Bridges, Niagara Falls, ON, Canada (2010)

[2] Collins, M.P., Mitchell, D., Adebar, P., Vecchio, F.J.: A general shear design method. ACI Structural Journal 93(1), 36–45 (1996)

[3] Vecchio, F.J., Collins, M.P.: The Modified Compression-Field theory for reinforced concrete elements subjected to shear. ACI Journal 83(2), 219–231 (1986)

[4] Nguyen-Minh, L., Rovnak, M.: New formula for the estimation of shear resistance of fibre reinforced beams. Canadian Journal of Civil Engineering 38(1), 23–35 (2010)

[5] Armelin, H.S., Banthia, N.: Predicting the flexural postcracking performance of steel fiber reinforced concrete from the pullout of single fibers. ACI Materials Journal 94(1), 18–31 (1997)
[6] Alam, S.Y., Loukili, A.: Application of Digital Image Correlation to size effect tests of concrete. In: Proceedings from 7th International Conference on Fracture Mechanics of Concrete and Concrete Structures, Jeju, South Korea (2010)
[7] Sutton, M.A., Wolters, W.J., Peters, W.H., Ranson, W.F., McNeill, S.R.: Determination of displacement using an improved digital correlation method. Image and Vision Computing 1(3), 133–139 (1983)
[8] Correlated Solutions Inc. Vic-Snap 2007 and Vic-3D 2007 [Software] (2007), `http://www.correlatedsolutions.com/index.php/home`
[9] Shoaib, A., Lubell, A.S., Bindiganavile, V.S.: Effect of steel fibers on the shear response of structural lightweight concrete. In: Proceedings from CSCE 2009 Annual General Conference, St. John's, NL, Canada (2009)

Connecting Non-destructive Fiber Dispersion Measurements with Tensile HPFRCC Behavior

L. Ferrara[1], M. Faifer[2], M. Muhaxheri[1], S. Toscani[2] and R. Ottoboni[2]

[1] Department of Structural Engineering, Politecnico di Milano, Italy
[2] Department of Electrical Engineering, Politecnico di Milano, Italy

Abstract. Successful application with advanced fiber reinforced cement based materials relies on reliable design prescriptions and design-oriented test methodologies for the identification of relevant material properties, together with consistent and effective quality control procedures. The "randomly uniform" dispersion of fibers within a structural element is a crucial issue to be tackled, which, if not achieved, may lead to the activation of unanticipated failure mechanisms, thus lowering the load bearing capacity and jeopardizing the structural performance. It is henceforth evident that non-destructive techniques for fiber dispersion monitoring need to be developed and calibrated, as it has been going on for at least the last lustrum, and their outcomes to be effectively correlated to the mechanical performance of the material. In this paper a non-destructive method based on the magnetic properties of the composite has been applied to monitor the fiber dispersion and orientation in HPFRCC slabs. The results have been correlated to the tensile behavior, measured according to the newly conceived "Double Edge Wedge Splitting" technique. The work herein presented stands as a step towards the assessment of non-destructive methods for fiber dispersion monitoring and their inclusion into a sound quality control procedure in a design oriented perspective.

1 Introduction

Fifty years of continuing research have led to consolidated knowledge about the multifold aspects of the physical and mechanical behavior of fiber reinforced cementitious composites (FRCCs) and to continuous development in advanced materials and improved products, as well as in terms of standardization. The recently issued fib 2010 Model Code has made available to the world engineering community a set of internationally recognized design guidelines to confidently employ these materials to build structures and infrastructures able to meet the higher and higher demands of our society.

The dispersion of fibers inside a structural element has been long time recognized as a crucial issue for a safe and reliable design. A spotty dispersion of fibers

G.J. Parra-Montesinos, H.W. Reinhardt, and A.E. Naaman (Eds.): HPFRCC 6, pp. 43–50.

can in fact seriously affect the element load bearing capacity as well as trigger unanticipated failure mechanisms. In very recent years it has been shown that the synergy between the FRC and the Self-Compacting Concrete (SCC) technology, due to the rheological stability of the SCC matrix, may be effective at guaranteeing a randomly uniform dispersion of the fibers within a structural element [1-2]. Furthermore, thanks to both the suitably balanced fresh state properties of the mixture and a carefully designed casting process, it is possible to orient the fibers along the direction of the casting flow [3-4], thus achieving, along the same direction, superior mechanical behavior of the material and structural performance.

The assessment of fiber dispersion and orientation related issues through a non-destructive method is hence of the utmost importance in the framework of the aforementioned integrated material and structure design approach. The topic has received lots of attention in the very last lustrum and dramatic progresses have been made since, e.g., the early applications of X-rays [5], also thanks to the use of sophisticated image analysis techniques.

Computer Axial Tomography scanning [3] is able to provide nice 3D visualization of the fiber arrangement within a specimen, but the need of a dedicated equipment (like in the case of X-rays) and software for the quantitative processing of the collected data still stands as a major drawback to a wider use of the method.

Electrical methods have been developed, based on the effects of the conductive fiber reinforcement on either the resistive [6] or the capacitive [7] properties of the composite, as well as on the whole impedance [8]. The sensitivity of each method to either the preferential orientation of the fibers [6] or their local concentration [7], or both [8], has been shown through comparison with destructive monitoring methods [8] as well as with the mechanical performance of the same non-destructively monitored material samples [9]. Even if some industrial scale applications have been attempted [10], clearly and widely assessed quantitative correlations between non-destructive measurements and actual local fiber concentration and orientation are still lacking. The major drawbacks highlighted so far stand in the influence of the electrical coupling between the electrodes of the test set-up and the specimens, as well as in the sensitivity of the matrix resistance to moisture content and gradients and to the presence of electrolytes in the pores. Measurements can be highly affected by the aforementioned artefacts and the effect of fiber dispersion related parameters jeopardized.

In this paper a new method is employed [11], which monitors fiber dispersion and orientation through a probe sensitive to the magnetic properties of the composite. The method is based on the effects of the fibers, as a function of their concentration and relative orientation, on the magnetic circuit of the probe, when leant-on the element/structure surface, and hence on the variation of its inductance. Garnered measurements, processed for both the fiber concentration and orientation, will be correlated to the tensile behavior of the HPFRCC under investigation, measured by means of a novel testing technique [12-13].

2 Experimental Programme

In order to validate the proposed method for the non-destructive monitoring of fiber dispersion and further establish quantitative correlation with the tensile fracture toughness properties of the material a HPFRCC, already thoroughly investigated in a previous research project [5], has been assumed as a reference. Its mix design is shown in Table 1. Two further mixes, with either 50 kg/m^3 fibers and no fibers at all (plain matrix) have been designed to the purpose of this study, the compositions of which are also detailed in Table 1. The fresh state performance, as measured by means of the slump-flow and V-funnel tests, irrespective of the fiber content, also detailed in Table 1, highlights a self-consolidating ability.

Table 1. Mix design and fresh state performance of the employed HPFRCCs

Mix design (dosages in kg/m^3)			
Constituent	HPFRCC 100	HPFRCC 50	HPFRCC 0
Cement type I 52.5	600	600	600
Slag	500	500	500
Water	200	200	200
Superplasticizer	33 (l/m^3)	33 (l/m^3)	33 (l/m^3)
Sand 0-2 mm	983	1000	1017
Straight steel fibers (l_f = 13 mm;d_f = 0.16 mm)	100	50	0
Fresh state performance			
Slump flow diameter (mm)	750	760	775
T_{50} (s)	6	6	5
V funnel flow time T_V (s)	30	29	25

A thin slab 1 m long x 0.5 m wide x 0.03 m thick has been cast with each of the three aforementioned mixes: fresh concrete was poured at the center of one short edge and, after spreading to the whole width of the slab, completely filled the formwork by flowing parallel to the long side. A preferential orientation of the fibers along the aforementioned direction was with high probability obtained.

Fiber dispersion was monitored by means of the inductance variation based method detailed in the following chapter and according to the schematic shown in Fig. 1. For each of the eighteen cells the slab was "divided" into, measurements were garnered parallel and orthogonal to the flow direction and at ± 45° to it.

The slabs were then cut into square tiles according to the same schematic as the one employed for the inductance measurements; the tile specimens were employed to assess the tensile fracture toughness of the composite according to a novel testing methodology called the Double Edge Wedge Splitting (DEWS) test [12-13], as sketched in Fig. 2. The groove and notch cutting was done in such a way that the preordained fracture plane resulted either parallel or perpendicular to the preferential flow-induced fiber alignment (Fig. 1). In this way a thorough characterization of the fiber orientation and local concentration dependant tensile fracture toughness was obtained.

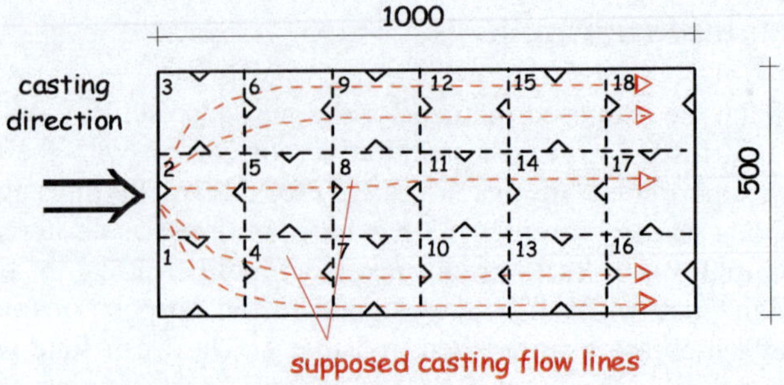

Fig. 1. Slab casting schematic and grid for magnetic measurements and DEWS specimen cutting

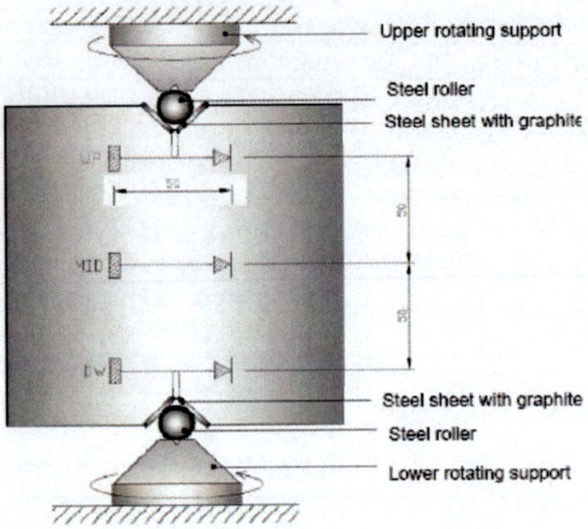

Fig. 2. Scheme of the DEWS specimen geometry and test set-up

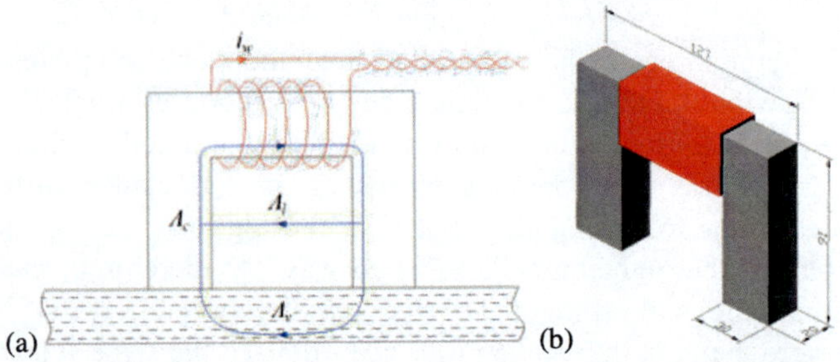

Fig. 3. Scheme of the measurement set-up (a) and of the magnetic probe (b)

3 Magnetic Inductance Method: Data Acquisition and Processing

Steel fibers employed in most of FRCC applications exhibit a ferromagnetic behavior. A SFRCC may hence be regarded as consisting of two phases with strongly different magnetic permeabilities: the cement based matrix and the fibers.

Neglecting interactions between them, because of the generally quite low volume fraction, the effective permeability of the composite is likely to depend on the permeability of the matrix and of the fibers, on the fiber volume fraction and on their aspect ratio. The measurement of a parameter affected by the effective composite permeability can be hence used to assess both the local average concentration and preferential orientation of fibers in a FRCC specimen.

This concept has been implemented by means of the experimental set-up shown in Fig. 3: it consists of a C-shaped N87 ferrite core with a 78-turn coil wounded around it, through which a 100 mA current flows, providing the magnetic flux. When the magnetic core is leant against a SFRCC specimen, the magnetic flux flows also through the specimen and the impedance measured at the terminals of the equivalent electrical circuit can be expressed as:

$$Z = R_c + j\omega\,[L_l + L_v] \tag{1}$$

where R_c denotes the resistance of the coil and L_l and L_v respectively denote the magnetic inductance associate to a magnetic flux outside and through the SFRCC specimen. The inductance L_v can be split into a matrix contribution L_{v0} and an incremental contribution $\Delta L_{v,fibers}$, which, for a given type of fibers, solely depends on their volume fraction and can thus be employed to evaluate their local concentration. Information about the local orientation of the fibers can be obtained by rotating the magnetic probe around its axis and assessing the angular direction where the measured incremental contribution is maximum. This incremental contribution will be further referred to as compensated inductance and was shown to be sensitive to both the nominal and local average fiber concentration and to the preferential orientation of the fibers. Whereas inductance L_v depended on the measurement frequency, compensated inductances resulted independent of it.

Quantitative assessment of local concentration of fibers, for each of the measured slab cells, was performed through the calibration of the relationship between the "nominal average compensated inductance" and the nominal fiber content in the mix. The former is defined as the average, computed over the whole slab, of the averages of the compensated inductance as measured along the four directions for each cell. By means of such a calibrated law, the average values of compensated inductance for each cell could be processed to assess whether and to what extent the local concentration of fibers differed from the assumed nominal value. Results shown in Fig. 4 highlight a reasonably homogeneous fiber dispersion inside the elements and coherent with the hypothesized casting flow kinematics.

For a quantitative assessment of fiber orientation "fractional" compensated inductances along the directions both parallel and orthogonal to the mean flow direction [15] have been calculated (Fig. 5):

$$f_{//} = \frac{\Delta L_{//}}{0.5\left(\Delta L_{//} + \Delta L_{\perp}\right)} \text{ and } f_{\perp} = \frac{\Delta L_{\perp}}{0.5\left(\Delta L_{//} + \Delta L_{\perp}\right)} \tag{2}$$

0.5 is the expected value in the case of perfectly in-plane isotropic dispersion.

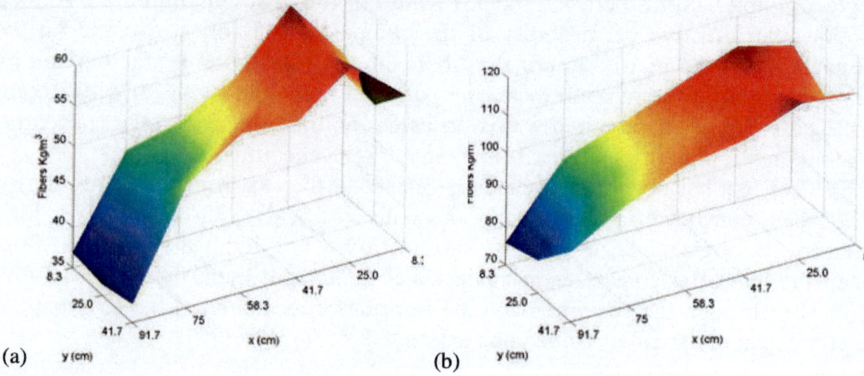

Fig. 4. Estimated fiber density in HPFRCC50 (a) and HPFRCC100 (b) slabs

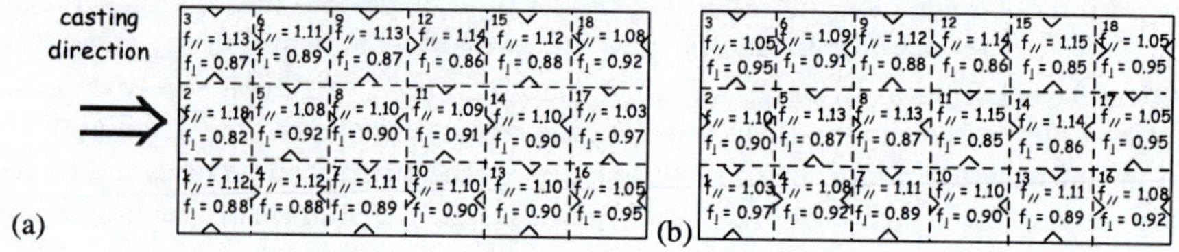

Fig. 5. Fractional compensated inductances: HPFRCC-50 (a) and -100 (b) slabs

4 Tensile Behavior and Correlation to ND Measurements

Tensile behavior of the two HPFRCCs at issue, differing by the nominal fiber content, has been investigated by means of the recently proposed Double Edge Wedge Splitting test, which is deemed to directly provide the stress-strain and/or stress vs. crack opening tensile response of HPFRCCs. The test geometry features a transverse tensile stress induced along the ligament by a longitudinally applied compressive load; thanks to the wedge grooves, the principal compressive stresses are always kept aside from the highly stretched region across the ligament.

Figs. 6a-b show the stress vs. crack opening curves. The plotted crack-opening value is the average of the measurements taken at different locations along the ligament (Fig. 2) and on both faces of each specimen. As for the transverse tensile stress σ_{ct} induced along the ligament by the applied longitudinal compressive load *P*, the following relationship has been employed [13]:

$$\sigma_{ct} = \frac{P}{t\, h_{ligament}} \frac{\cos\vartheta - \mu \sin\vartheta}{\sin\vartheta + \mu \cos\vartheta} \tag{3}$$

where t and $h_{ligament}$ are the specimen thickness and ligament height respectively, θ is the inclination angle of the groove edges (45°) and μ ($\cong 0.06$ as from [18]) is the friction coefficient between the steel sheets glued to the groove edges, and the steel roller through which the machine load is transferred to the specimen (Fig. 2).

Besides the obvious influence of the fiber content, it clearly appears that, due to the flow induced orientation of the fibers, the same material inside the same structural element, can not only provide different residual post-cracking bearing capacity (Fig. 6a) but, for higher fiber dosages (Fig. 6b), can even exhibit either a post-cracking hardening or softening (conventional FRC) behavior, whether it is stressed respectively parallel or orthogonal to the aforementioned fiber alignment.

Just for example, the residual stresses at a COD value equal to either 0.25 mm (f_{R1}) or 1.25 mm (f_{R3}) has been taken as representative of material behavior, e.g. at SLS and ULS. For specimens exhibiting a pre- peak multiple cracking, the aforementioned COD limits were measured with reference to the COD value at peak stress [11]. The aforementioned residual stresses have been plotted (Fig. 7) vs. a dispersion/orientation factor, calculated as the product between the fractional compensated inductance in the direction orthogonal to the ligament and the ratio between the estimated local and nominal fiber concentration in the specimen. The reliability of the proposed ND method clearly appears.

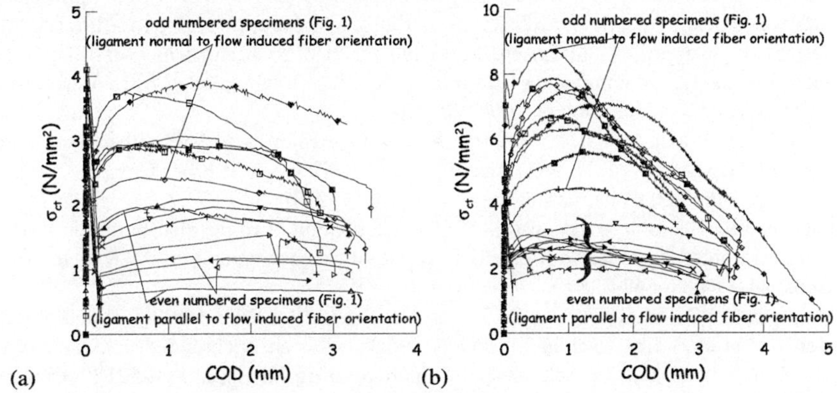

Fig. 6. σ-COD curves: DEWS tests from HPFRCC-50 (a) and -100 (b) slabs

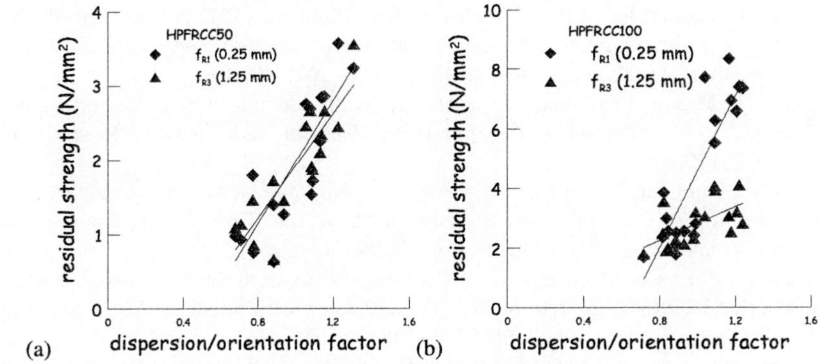

Fig. 7. Residual stresses vs. dispersion factor: HPFRCC-50(a) and -100 (b) slabs

5 Concluding Remarks

A non-destructive method based on the magnetic properties of SFRCCs has been proposed and applied to monitor fiber dispersion and orientation in HPFRCC slabs. The results have been correlated to mechanical tensile behavior, measured on specimens extracted from the same slabs according to the Double Edge Wedge Splitting Test technique. Correlation highlights the method reliability and its potential for application in quality control of HPFRCC structural elements.

References

[1] Ferrara, L., Meda, A.: Relationships between fibre distribution, workability and the mechanical properties of SFRC applied to precast roof elements. Materials and Structures 39(4), 411–420 (2006)

[2] Ferrara, L., Park, Y.D., Shah, S.P.: Correlation among fresh state behaviour, fiber dispersion and toughness properties of SFRCs. ASCE Journal of Materials in Civil Engineering 20(7), 493–501 (2008)

[3] Stahli, P., Custer, R., van Mier, J.G.M.: On flow properties, fibre distribution, fibre entation and flexural behaviour of FRC. Materials and Structures 41(1), 189–196 (2008)

[4] Ferrara, L., Ozyurt, N., di Prisco, M.: High mechanical performance of fiber reinforced cementitious composites: the role of "casting-flow" induced fiber orientation. Materials and Structures 44(1), 109–128 (2011)

[5] Stroeven, P., Shah, S.P.: Use of radiography-image analysis for steel fiber reinforced concrete. In: Swamy, R.N. (ed.) Testing and Test Methods of Fiber Cement Composites, pp. 345–353. Construction Press, Lancaster (1978)

[6] Lataste, J.F., Behloul, M., Breysse, D.: Characterisation of fibres distribution in a steel fibre reinforced concrete with electrical resistivity measurements. NDT & E International 41(8), 638–647 (2008)

[7] Van Damme, S., Franchois, A., De Zutter, D., Taerwe, L.: "Nondestructive determination of the steel fiber content in concrete slabs with an open-ended coaxial probe. IEEE Transactions on Geoscience and Remote Sensing 42(11), 2511–2521 (2009)

[8] Ozyurt, N., Woo, L.Y., Mason, T.O., Shah, S.P.: Monitoring fiber dispersion in fiber reinforced cementitious materials: comparison of AC Impedance Spectroscopy and Image Analysis. ACI Materials Journal 103(5), 340–347 (2006)

[9] Barnett, S., Lataste, J.F., Parry, T., Millard, S.G., Soutsos, M.N.: Assessment of fibre orientation in ultra high performance fiber reinforced concteye and its effect on flexural strength. Materials and Structures 43(7), 1009–1023 (2010)

[10] Ozyurt, N., Mason, T.O., Shah, S.P.: Non destructive monitoring of fiber orientation using ACIS: an industrial scale application. Cement and Concrete Research 36, 1653–1660 (2006)

[11] Faifer, M., Ottoboni, R., Toscani, S., Ferrara, L.: Non-destructive Monitoring of Steel Fiber Reinforced Concrete using a Magnetic Approach. IEEE Transactions on accepted for publication on Instrumentation and Measurement

[12] Ferrara, L., di Prisco, M., Lamperti, M.G.L.: Identification of the stress-crack opening behavior of HPFRCC: the role of flow-induced fiber orientation. In: Oh, B.H., et al. (eds.) Proceedings FraM-CoS 7, pp. 1541–1550 (2010)

[13] di Prisco, M., Lamperti, M.G.L., Lapolla, S.: On Double Edge Wedge Splitting test: preliminary results. In: B.H. (ed.) Proceedings FraMCoS 7, pp. 1533–1540 (2010)

Improved Tensile Performance with Fiber Reinforced Self-compacting Concrete

S. Grünewald[1], F. Laranjeira[2], J. Walraven[1], A. Aguado[2], and C. Molins[2]

[1] Section of Concrete Structures, Delft University of Technology, Delft/The Netherlands
[2] Department of Construction Engineering, Universitat Politècnica de Catalunya, Barcelona/Spain

Abstract. The use of self-compacting concrete (SCC) eliminates the need for compaction, which has benefits related to economic production, the durability, the structural performance and working circumstances. SCC is able to transport fibers which can replace in some structures conventional reinforcement. By taking into account tailor-made concrete characteristics, new fields of structural application can be explored. This paper discusses the potential for an improved performance of fibers in self-compacting concrete. In flexural tests significant differences were observed between conventional and self-compacting concrete at a given fiber type and dosage concerning the variation of results and the flexural performance. Mechanical testing and image studies on concrete cross-sections indicate how the flow influences the performance, the orientation and the distribution of the orientation of fibers. Differences between traditionally compacted and flowable concrete are pointed out.

1 Introduction

The structural performance of fiber reinforced concrete depends on material properties, production effects and structural boundary conditions. When fiber orientation is understood and can be controlled an improved tensile performance of structural elements can be accounted for. The effect of fibers on rheological characteristics has to be taken into account for the mix design. The maximum fiber content is the critical fiber dosage at which the workability drastically decreases. The size, the shape and the content of the coarse aggregates as well as the geometry and the volume fraction of steel fibers affect the workability of fiber reinforced concrete [1]. Mix design methods have been proposed for steel fiber reinforced concrete, SFRC [2] and flowable fiber concrete [3, 4, 5]. In the hardened state, fibers contribute to the mechanical performance of concrete; their effect depends on parameters like the dosage, the fiber type, the characteristics of the matrix and fiber orientation and distribution. Hooked-ends improve the pull-out behavior of

G.J. Parra-Montesinos, H.W. Reinhardt, and A.E. Naaman (Eds.): HPFRCC 6, pp. 51–58.

steel fibers; the pull-out behavior was modeled by Laranjeira [6, 7] and Van Gysel [8]. The production process also affects the structural performance: compaction of the concrete [9] and the flow of concrete [3] can cause differences in the distribution and the orientation of fibers. With the same material tensile-hardening or tensile-softening behavior might be obtained. The orientation of fibers depends on factors like the mixture composition (aggregates hinder the rotation of fibers) and the rheological characteristics which determine the mobility of a fiber in a matrix. A preferred fiber orientation in the flow direction was observed by Groth & Nemegeer [10] applying the X-ray technique on hardened concrete slices. Martinie & Roussel [5] simulated the flow and the orientation of fibers. With a model fluid containing stiff fibers they obtained a preferred orientation in the direction of the flow caused by the walls; the degree to which fibers orient depended on the rheological characteristics.

2 Flowable Fiber Concrete

The mixture composition and the production method affect the structural performance. Fig. 1 compares results of three-point flexural tests of a conventional, steel fiber reinforced concrete (SFRC) [11] and self-compacting fiber reinforced concrete (SCFRC) [3].

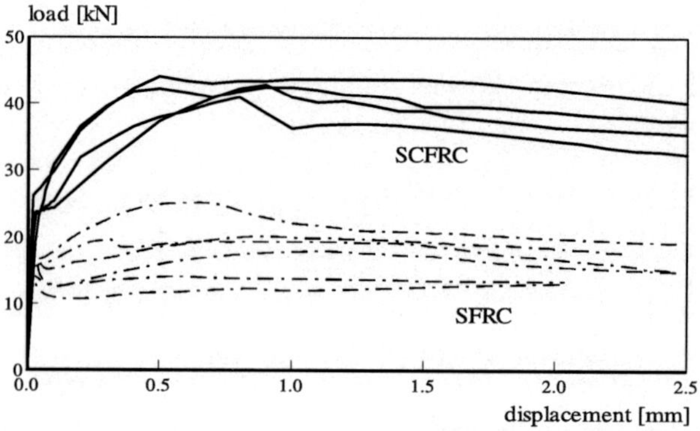

Fig. 1. Test results of three-point flexural tests: SFRC versus SCFRC; strength class: C55/67, Dramix 80/60 BP, V_f=60 kg/m^3.

The same fiber type and content (Dramix 80/60 BP; 60 kg/m^3) were applied. The compressive and splitting tensile strengths of the SFRC were 81.1 MPa and 4.6 MPa, respectively, whereas those of the SCFRC were 75.1 MPa and 8.6 MPa, respectively. The rate of deformation in the study of Kooiman was 1 μm/s up to a displacement of 5 mm and 50 μm/s beyond 5 mm (SCFRC: constant 50 μm/s). Referring to a study of Gopalaratnam & Shah [12], a minor influence of this difference on the maximum flexural load can be expected. The fibers increased the splitting tensile strength of SCFRC compared to the plain SCC reference mixture ($f_{ctm,spl}$ = 5.3 MPa). This is also reflected by an improved flexural performance of SCFRC. An increase of the load after first cracking of SFRC was observed for one specimen only; the scatter in the results was much higher. Table 1 compares the fracture energy $G_{F,2\ mm}$ (up to 2 mm) of both mixtures and its variation. Kooiman [9] showed that by increasing the specimen size the variation of the fracture energy significantly decreased (Table 1). However, the average fracture energy of the SFRC compared to SCFRC was lower (44 % at a specimen width of 150 mm and 69 % at a specimen width of 450 mm, respectively). In order to identify what caused the differences in flexural performance, additional tests were carried out.

Table 1. Flexural tests: Fracture energy and its variation

Specimen width	SFRC [9]		SCFRC [3]	
	$G_{F,2\ mm}$ [kN/mm]	Variation [%]	$G_{F,2\ mm}$ [kN/mm]	Variation [%]
b = 150 mm	1.83	21.3%	4.16	6.9%
b = 300 mm	2.79	9.0%	-	-
b = 450 mm	2.87	4.2%	-	-

Mixture 'SCFRC' (Fig. 1) was designed to be self-compacting and contained a steel fiber dosage (60 kg/m^3) close to the maximum dosage. Tensile hardening can be obtained with this fiber dosage as the experimental result of an uni-axial tensile test of Fig. 2 shows.

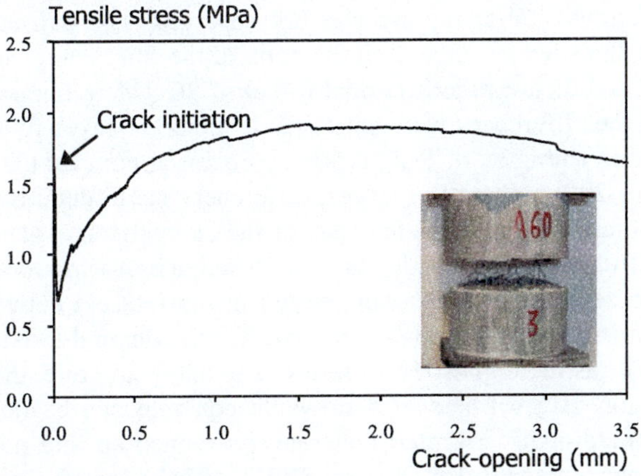

Fig. 2. Uni-axial tensile test [6]: Tensile stress versus crack-opening diagram (strength class: C40/50, Vf=60 kg/m3, Drampix 80/60 BP)

The height and the diameter of the cylinder were 150 mm; the circumferential notch had a depth of 15 mm. The rate of testing was increased during testing from 5 µm/min up to 1 mm/min. In order to obtain hardening behavior of a structure, fiber reinforced concrete not necessarily has to be tensile hardening; the type of structure and loading conditions are affecting parameters. In order to compose a concrete with self-compacting characteristics, the fiber content has to remain below the maximum fiber content. Due to this fact each fiber is fully embedded in the matrix and the interaction with coarse aggregates is minimized. Photographs of cross-sections of beams of SCFRC showed that only a few fibers were connected. The maximum fiber dosage can be determined with the MFC (Maximum Fiber Content)-volume, a factor that depends on the content and the distribution of the aggregates [3]. Below the predicted fiber dosage, fiber reinforced concrete can be self-compacting. To compose SCC at a higher fiber factor, the paste content has to be increased and/or the content of larger aggregates (relative to the size of the fibers) decreased.

3 Single Fiber Performance

The pull-out behavior of single hooked-end steel fibers from conventional (CC) and self-compacting concrete was studied by Markovic et al. [13] in order to quantify causes for the differences in flexural performance. The parameters of this study were the concrete type (SCC/CC), compressive strength (C35/45, C55/67 and C90/105), the fiber type (Dramix 80/60 BP/Dramix 80/30 BP) and the embedded length (Dramix 80/60 BP: 10 and 30 mm). The slump flow of SCC was in the range of 650-710 mm, whereas the slump of CC was in the range of 170-210 mm. Cylinders (diameter: 65 mm, height: 50 mm) were drilled from larger specimens and were glued on a steel plate (Fig. 3a). The pull-out force was recorded, while the test was controlled on the slip of the fiber relative to the matrix (rate of displacement: 5 μm/s). Fiber rupture was not observed. The pull-out forces (the maximum force, the average force until the hook of the fiber completely enters the straight channel and the average frictional force) of SCC were higher in most cases (Fig. 3b), but the differences were not large. It should be taken into account that the splitting tensile strengths of SCC (without fibers) at comparable compressive strengths were 17-29% higher. The largest differences were found for the strength class C55/67, which was the strength class for the comparison of the flexural performance of SCFRC and SFRC (Fig. 1). The difference in variation of the maximum pull-out forces (3 specimens were tested in most cases) between SCC and CC are smaller (variation: 2.0-12.3 %) compared to the flexural tests; the variation of CC (C90/105) was the highest of all series. The microstructure, the distribution and the orientation of the fibers in a flowable concrete can be rather different compared to a traditionally vibrated concrete. Entrapped air and neighboring fibers affect the performance of a fiber in SFRC probably more than in SCFRC. Therefore, the single fiber pull-out test might give a better indication of the actual performance of a fiber in SCC than in conventional concrete.

The angle of a fiber between an applied force and the cracked surface affects the pull-out performance. Fig. 4 shows that the embedded length and the fiber orientation angle significantly influence the fracture energy (at δ=1.20 mm). The compressive strength of the concrete at 28 days was 72 MPa.

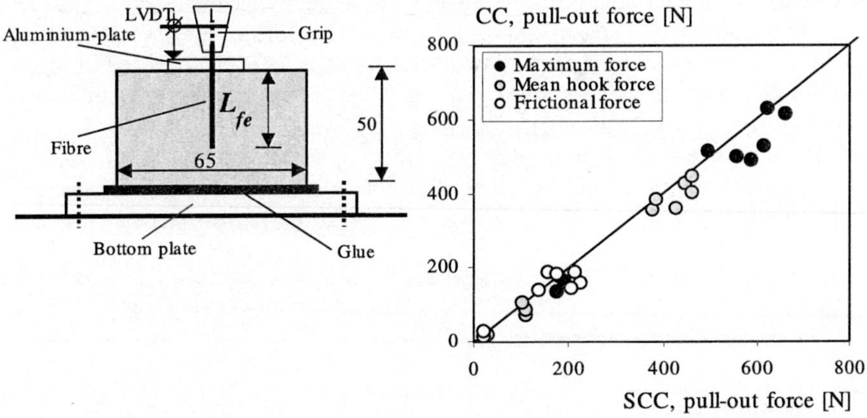

Fig. 3. Single fiber pull-out testing: a) Test set-up, b) Comparison of characteristic pull-out forces of steel fibers from SCC/CC matrices

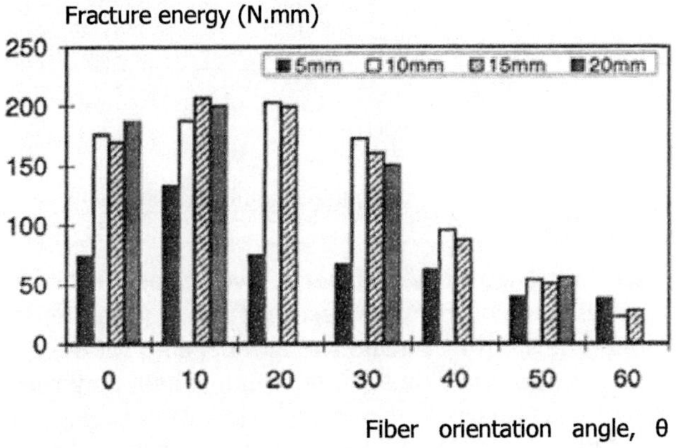

Fig. 4. Fracture energy of single-fiber pull-out tests (δ = 1.20 mm) depending on the fiber orientation angle and the embedded fiber length; hooked-end steel fibers with 30 mm length and 0.5 mm diameter [14]

The maximum fracture energy was obtained at a fiber inclination of 10-20°. Fig. 4 indicates that fibers that are almost aligned in the direction of the force are most effective, as long as fiber rupture is counteracted.

4 Fiber Orientation

Two tunnel segments (Element 1: 60 kg/m^3- Dramix 80/60 BN, L_f=60 m; Element 2: 60 kg/m^3- Dramix 45/30 BN; L_f=30 mm) were cast with SCFRC to determine how the flow affects the orientation of the fibers and if the fiber orientation relates to mechanical characteristics of the concrete [3]. Fifteen cylinders from one half of each tunnel segment were drilled to determine the splitting tensile strength (at different positions and in different directions). Fifteen additional cylinders were drilled point-symmetric to the point of casting to take X-ray photographs of slices. Deformation-controlled splitting tensile tests (deformation rate: 1.0 μm/s) were carried out and slices (d=18 mm) were sawn from cylinders in order to take X-ray photographs. The orientation of the fibers was determined according to the routine described by De Keukelaere [15] and was related to the performance of the cylinders (Figs. 5a/b).

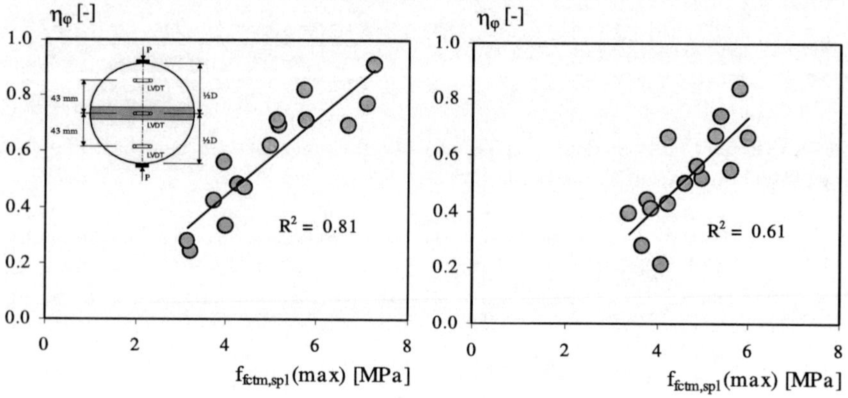

Fig. 5. Splitting tensile strength versus orientation number a) Segment A: Dramix 80/60 BN, b) Segment B: Dramix 45/30 BN

The 'splitting tensile strength' is the higher of two defined loads, the load when first cracking occurred (a load drop was recorded) or the maximum load in the post-cracking region. The splitting tensile strength depends on the position of the cylinders in the tunnel element and on the direction in which they were tested. The post-cracking strength was higher at increasing orientation number. The range of splitting tensile strengths was wider (higher strengths were measured) and the correlation between the maximum splitting tensile strengths and orientation numbers was better for 60 mm fibers - they are more susceptible to the orientation during the flow. The fibers were homogenously distributed which was determined by fiber counting. The analysis of the results indicated that the fibers oriented during the flow through the pipe of the truck mixer, whereas in the absence of walls (in the tunnel segment), they remained aligned and were mainly oriented perpendicular to the flow direction. The flow of concrete in the tunnel segment more or less was comparable with a plug flow in a pipe, which was the result of the rheological characteristics of the applied SCFRC. The variation of structural performance has different reasons, as the previous discussion indicated. Laranjeira [6] studied the performance and the variation of tensile strength of fiber reinforced concrete; image studies as well as mechanical tests were executed. By analyzing images of cross-sections of specimens subjected to bending and results from literature, Laranjeira found that the distribution of the fiber orientation can be approximated with a normal distribution [16]. The variation of fiber orientation decreased at increasing average orientation number (Fig. 6). Due to the alignment a fiber has less possibilities to orient, which decreases the variation. A low variation of flexural performance of homogenous flowable fiber concrete is both an intrinsic characteristic (due to orientation) and the result of consequently following production procedures (effects caused by production and the flow).

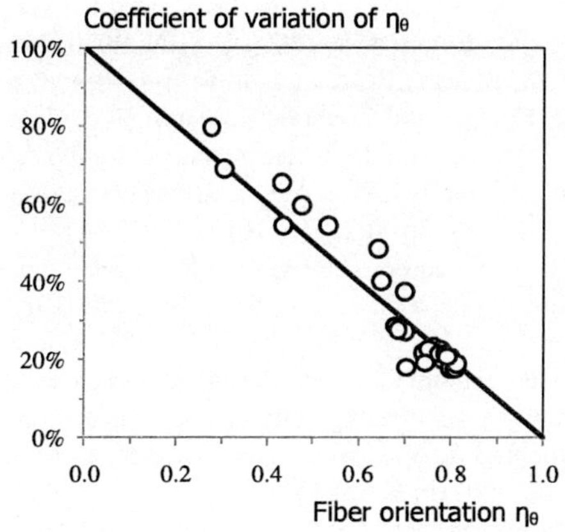

Fig. 6. Relation between the average fiber orientation factor η_θ and the coefficient of variation of the fiber orientation

5 Conclusions

Several parameters are discussed in this paper that affect the structural performance of flowable fiber reinforced concrete. Based on the results and the discussion the following conclusions can be drawn:

- The mixture composition and related rheological characteristics determine the flow behavior and possible fiber orientation and distribution. In a highly flowable concrete, the interaction of fibers is minimized.
- Similar pull-out responses of aligned fibers (single fiber testing) concerning the maximum pull-out force and its variation were obtained with self-compacting and conventional concrete considering a slightly higher tensile strength of SCC.
- The highest single fiber pull-out forces were obtained with inclined fibers (10-20° inclination); the variation of orientation is lower at a high average orientation number, which further increases the efficiency of the fibers.

References

[1] Swamy, R.N.: Fibre reinforcement of cement and concrete. Materials and Structures 8(45), 235–254 (1975)

[2] Rossi, P., Harrouche, N.: Mix Design and Mechanical Behaviour of some Steel-Fibre-Reinforced Concretes used in Reinforced Concrete Structures. Materials and Structures 23, 256–266 (1990)

[3] Grünewald, S.: Performance-based design of self-compacting fibre reinforced concrete. PhD-thesis, Delft University of Technology, Delft University Press, Department of Structural and Building Engineering (2004); ISBN: 9040724873

[4] Ferrara, L., Park, Y.D., Shah, S.P.: A method for mix-design of fiber reinforced self compacting concrete. Cement and Concrete Research 37, 957–971 (2007)

[5] Martinie, L., Roussel, N.: Fiber-Reinforced Cementitious Materials: From Intrinsic Properties to Fiber Alignment. In: Design, Production and Placement of Self-Consolidating Concrete. Rilem Bookseries, vol. 1, pp. 407–415 (2010); ISBN: 978-90-481-9663-0

[6] Laranjeira, F.: Design-oriented constitutive model for steel fiber reinforced concrete. PhD-thesis. Universitat Politècnica de Catalunya, Spain (2010)

[7] Laranjeira, F., Molins, C., Aguado, A.: Predicting the pullout response of inclined hooked steel fibers. Cement and Concrete Research 40(10), 1471–1487 (2010a)

[8] Van Gysel, A.: Studie van het uittrekgedrag van staalvezels ingebed in een cementgebonden matrix met toepassing op staalvezelbeton onderworpen aan buiging, PhD Thesis, Ghent University (2000) (in Flemish)

[9] Kooiman, A.G.: Modelling Steel Fibre Reinforced Concrete for Structural Design, PhD-thesis, Department of Structural and Building Engineering. Delft University of Technology (2000)

[10] Groth, P., Nemegeer, D.: The use of steel fibres in self-compacting concrete. In: Skarendahl, Petersson (eds.) First Int. Symposium on SCC, Stockholm, pp. 497–508. RILEM publications PRO 7, Cachan (1999)

[11] Kooiman, A.G.: Schaaleffecten in het nascheurgedrag van staalvezelbeton, Stevin-report 25.5-98-9, Department of Structural and Building Engineering. Delft University of Technology (1998) (in Dutch)

[12] Gopalaratnam, V.S., Shah, S.P.: Strength, deformation and fracture toughness of fiber cement composites at different rates of flexural loading. In: Shah, Skarendahl (eds.) Steel Fiber Concrete, US-Sweden Joint Seminar, Stockholm, pp. 299–332. Elsevier Applied Science Publishers, New York (1986)

[13] Markovic, I., Grünewald, S., Walraven, J.C., Van Mier, J.G.M.: Characterization of bond between steel fibres and concrete - conventional fibre reinforced versus self-compacting fibre reinforced concrete. In: Third Int. Symposium Bond in Concrete - from research to standards, pp. 520–528. Publishing Company of Budapest University of Technology, Budapest (2002)

[14] Robins, P., Austin, S., Jones, P.: Pull-out behavior of hooked steel fibres. Materials and Structures 35, 434–442 (2002)

[15] De Keukelaere, G.: Studie van de vezelverdeling en de invloed ervan op de mechanische eigenschappen van staalvezelbeton, Master thesis, Laboratory Magnel of Reinforced Concrete. University of Ghent (1993) (in Flemish)

[16] Laranjeira, F., Grünewald, S., Walraven, J., Blom, C., Molins, C., Aguado, A.: Characterization of the orientation profile of steel fiber reinforced concrete. Journal of Materials and Structures (2010b), doi:10.1617/s11527-010-9686-s

The Impact of Rheology on the Mechanical Performance of Steel Fiber-Reinforced Concrete

G.P.A.G. van Zijl and S. Zeranka

Department of Civil Engineering, University of Stellenbosch, South Africa

Abstract. Fibers are added to concrete to bridge cracks, in order to improve mechanical behavior from brittle to pseudo-plastic. Uniform dispersion of fibers is known to be critical, to avoid weak planes and the influence of self-compacting ability of fiber-reinforced concrete (FRC) on fiber dispersion has been studied and reported in the literature to some extent. A self-compacting (SCFRC) and a normal workability FRC (NFRC) was designed and tested systematically. Indirect assessment of fiber dispersion and orientation was performed by mechanical testing on small specimens, to study the influence on mechanical performance. The results show superiority of SCFRC in flexure.

1 Introduction

Fibers are added to concrete to arrest cracks, control brittle fracture, and provide reliable post-cracking strength. Important considerations for the application of FRC are the fiber type and geometry and the distribution, orientation and concentration of fibers within the concrete matrix [1].

Steel FRC (SFRC) provides a greater energy absorption capacity in concrete and hence is also being used more frequently in conjunction with normal reinforced concrete in structural members [2]. High-Performance-Self-Compacting-Fiber-Reinforced-Concrete (HPSCFRC) seems to have the greatest potential for the construction industry. The stress distribution and ductility of high performance concrete can be improved with the addition of fibers. The addition of fibers could also partly replace the steel reinforcement, limit crack widths and improve reinforcement spacing [3].The greatest benefit of fiber addition is the improvement of tensile stress and strain capacities and crack control [1].

The uniform dispersion of fibers is critical for the wider structural use of fiber-reinforced concrete. It can be achieved by taking advantage of the superior workability of self-compacting concrete (SCC) [4]. The matrix of the SCC provides a greater compactness through the increased amount of fines present. This may improve the properties of the interface zone and thus also the fiber-matrix bond, which may in turn enhance the post-cracking toughness and capacity to absorb energy [4]. Recent studies have investigated the distribution of steel fibers in

G.J. Parra-Montesinos, H.W. Reinhardt, and A.E. Naaman (Eds.): HPFRCC 6, pp. 59–66.

fiber reinforced self-compacting concrete (FR-SCC), showing that fiber orientation is related to the flow properties of the FR-SCC. The fibers are prone to be aligned in the flow direction, and this effect increases with the flowability of the FR-SCC. The flexural strength is influenced by the distribution and orientation of the fibers and there is a significant improvement when the fibers are oriented in the direction of the tensile stresses. This preferential orientation can only really be achieved through good workability. The opposite is true for concrete with a poor workability, where the fibers do not have an adequate orientation. As a result, the fibers contribute poorly to the flexural behavior [5].

Fiber addition is known to have a negative impact on concrete workability [6], [1]. The extent to which the workability will be negatively impacted depends on the type and amount of fibers used, the matrix and the properties of the components of the matrix on their own. The passing ability is also affected and consequently a larger bar spacing compared with normal SCC is needed to avoid the blocking of fiber-reinforced mixtures [6]. In this paper the effective utilization of steel fibers in concrete will be analyzed by comparing the benefits of steel fiber addition to normal workability – and self-compacting concretes.

2 Experimental Program

2.1 Concrete Mix Design

Two compressive strength classes were chosen: a normal strength (~ 30 MPa) and high performance (HP) (~90MPa) mix was desired in order to test the influence of self-compactability over a wide range in FRC strength. For each strength class, a normal (N) and self-compacting concrete (SC) was designed. Four fiber contents were chosen, namely 0%, 0.5%, 1.0% and 1.5%. The water/cement ratio was kept constant within each strength class. The same base mixture was used for varying fiber content and is included in Table 1. The coarse and fine aggregate were exchanged against the same volume of fibers. The aggregate content (sand & stone) was adjusted proportionately to account for the change in fiber content. A 6.7 mm stone and a 13.2 mm stone were used in a 60:40 ratio respectively, for all the concrete mixes and provided a reasonable grading combination of all the aggregates. Superplasticizer was then added to achieve the desired workability. A 30 mm Dramix® steel fiber was used. The ends of the fiber are crimped to increase fiber pull-out resistance or anchorage.

Table 1. Base concrete mix designs (0% fiber). All units kg per m^3 of concrete

Mix constituents	HPNFRC	HPSCFRC	NFRC	SCFRC
Cement:				
OPC (CEM I 42.5) / SB (CEM II 32.5)	450(OPC)	500(OPC)	292(SB)	323(SB)
Fly Ash	64	210	89	237
13.2 mm Greywacke stone (40%)	387	275	406	259
6.7 mm Greywacke stone (60%)	581	413	610	388
Malmesbury sand	778	809	816	953
Water	180	200	190	210
Superplasticizer	2.25	6.5	0	2.262

2.2 *Workability Tests*

Standard slump flow tests, T_{500} tests and L-box tests were performed on the self-compacting mixes in order to measure the flowability and passing ability of the material. Standard slump tests were performed on the normal workability mixes. Additionally the air content was also measured. The air content would indicate to which degree air entrainment is being caused by the addition of fibers.

2.3 *Mechanical Tests*

The standard mechanical tests performed and the specimen size in brackets, included: 7-day compressive cube strength (100mm cubes), 28-day compressive cube strength (100mm cubes), 28-day tensile splitting strength (100mm cubes) and 28-day beam flexure strength (500x100x100mm beams).

2.3.1 Specimen Preparation

Three specimens were cast and cured for each test. The moulds for the cube specimens were filled with concrete and in the case of the normal concrete mixes; the moulds were placed on a vibrating table to compact and de-air. The self-compacting concrete was placed in the moulds without any further external compaction. The specimens were left to set for 24 hours, after which they were removed from their moulds and cured, with a water temperature ranging between 22°C and 25°C.

In order to achieve a preferential orientation of fibers and utilise the enhanced workability of self-compacting concrete; beams were cast from one end. Provided that the mix has the correct workability, the fibers will tend to orientate in the direction of flow.

2.3.2 Instrumentation

Instrumentation was applied to the standard compressive strength test, in order to record and plot the compressive stress-strain relationships for the different fiber contents. A frame was fixed to each specimen at a specified gauge length (Lg = 60mm). Two LVDT's were positioned on either side of the specimen. Only the normal strength, normal workability mix was tested in such a manner due to the limited capacity of the load cells available for the specimen size used. A 500 kN load cell was used.

A cubical specimen is subjected to compressive forces applied along two diametrically opposed lines. The principal tensile stress in the plane joining the loaded lines, causes splitting in this plane. The splitting tensile strength of the concrete is calculated according to elastic theory:

$$f_s = 2P/\pi a^2 \tag{1}$$

where P = compression load at failure; a = cube dimension.

As illustrated in Fig. 1(b), the materials testing machine, Zwick Z250 was used to perform the test. A 200 kN capacity HBM Load cell and two 10mm LVDT's were used to record the applied load and total deformation of the specimen during the test. The LVDT's were attached to an aluminium frame which was firmly secured to the specimen.

A two-point loading arrangement was used to determine the flexural strength of concrete. For expression of flexural resistance, the notion of modulus of rupture is used, which assumes that stress increases linearly from the neutral axis. Therefore for two-point loading (at third points) the flexural stress in the mid-zone is:

$$f_f = PL/bh^2 \tag{2}$$

where P = ultimate load; L = distance between axes of supporting rollers [300 mm]; b = width of specimen [100 mm]; h = depth of specimen [100 mm].

The material testing machine, the Zwick Z250 was used to perform all flexure tests, at a controlled midspan deflection rate of 0.5mm/minute. The Zwick was set to continue applying a load to the specimen until a load 90% below the maximum applied load was reached, in order to safely measure the post-peak response as well. In order to measure the centre point deflection of the beam, two 10mm LVDT's were used to record the deflection on either side of the beam. An aluminium frame, securely tightened to the specimen was used to position the LVDT's as can be seen in Fig. 1(a). The LVDT's were positioned to touch the bottom of cast aluminium angles glued to the top centre point of the beam, before the commencement of the test.

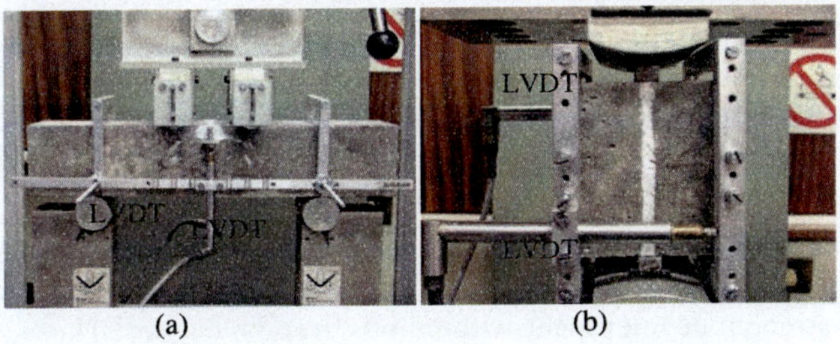

Fig. 1 (a) Flexural test setup; (b) Splitting test setup

3 Experimental Results

3.1 Workability Results

A reduction in workability occurred with increased fiber content for both the normal and self-compacting concretes, consequently greater quantities of superplasticizer had to be added with increased fiber content to achieve the same workability. The specified slump flow of 600 mm could not be achieved for all the self-compacting mixes and these mixes had already reached the limit of superplasticizer addition (see Table 2). A clustering of fibers at the centre of the slump flow and an unequal spread was observed for higher fiber contents. The rate of flow or viscosity was however within the specified range. None of the self-compacting mixes could achieve the passing ability required. The grid opening size to fiber length ratio may be a limiting factor, and kept in mind for application of SCFRC. The passing ability of the base mixtures i.e. without fibers is inadequate. In future more time must be invested to develop better mix designs. The air content remained relatively consistent around a 2% value for all the concrete mixes. The air void content of concrete can increase with increased fiber content. There is however no significant increase in air content with increased fiber content for these mixes. One possible way to improve resistance to segregation, flowability and passing ability may be to increase the air content to about 5%.

Table 2. Workability results for self-compacting mixes (HPSCFRC I SCFRC)

HPSCFRC I SCFRC % Fibers	Slump flow [mm]	T500 [s]	Passing-ability (3· bar)	Air content
0%	680 I 550	2 I 1	68.2% I 33.3%	2.00% I 2.20%
0.5%	500 I 545	4 I 2	28.7% I 39.3%	2.30% I 2.20%
1.0%	600 I 525	4 I 3	0.00% I 0.00%	1.60% I 2.40%
1.5%	600 I 560	3 I 4	0.00% I 0.00%	2.80% I 2.70%

3.2 Compressive Strength Results

The average 7 and 28-day compressive cube strengths are tabulated in Table 3. The two high performance mixes, HPSCFRC and HPNFRC did not achieve the desired 90 MPa strength, but it is clear that the strength development of the normal concrete mix is poor. The two normal strength mixes, SCFRC and NFRC in contrast exceeded the desired 30 MPa strength. The self-compacting concretes have greater strength development within both strength classes. These results were accepted due to the limited timeframe. In future contributions, mix designs should be developed that produce concretes with similar strengths.

In general there is a significant increase in compressive strength with increased fiber content. In the current tests the control of workability and air content explains the increased compressive strength with increased fiber content. Fibers are properly dispersed and their larger resistance than that of the concrete matrix is not offset by an increase in voids in the matrix.

Table 3. Average compressive cube strengths (MPa)

% Fiber	HPNFRC		HPSCFRC		NFRC		SCFRC	
	7-day	28-day	7-day	28-day	7-day	28-day	7-day	28-day
0%	57.2	60.3	62.8	81.9	21.8	31.8	32.3	46.4
0.5%	63.4	63.3	63.1	85.5	24.9	34.9	29.5	57.0
1.0%	62.3	71.0	56.2	85.9	36.3	42.6	34.2	56.7
1.5%	65.5	75.9	64.0	91.8	32.2	42.2	36.1	58.4

3.3 Splitting Tensile Strength Results

Table 4 summarizes the average tensile splitting strengths (f_s). Both mix types show an increased splitting strength with increased fiber content. On average, the normal mix performs slightly better than the self-compacting mix. The benefits of fiber alignment are not realized in the confined dimensions of the cube specimens. Fig. 2 illustrates typical splitting tensile stress–deformation curves. An increase in the maximum splitting stress and improved ductile behavior and energy absorption are observed for increased fiber contents.

3.4 Flexural Strength Results

All mix types have significant increases in flexural strength with increased fiber content. The results for the two high strength mixes are relatively similar, whereas the self-compacting mix exhibits greater flexural strength and improved post peak behavior for the normal strength concretes, as can be observed in Table 4.

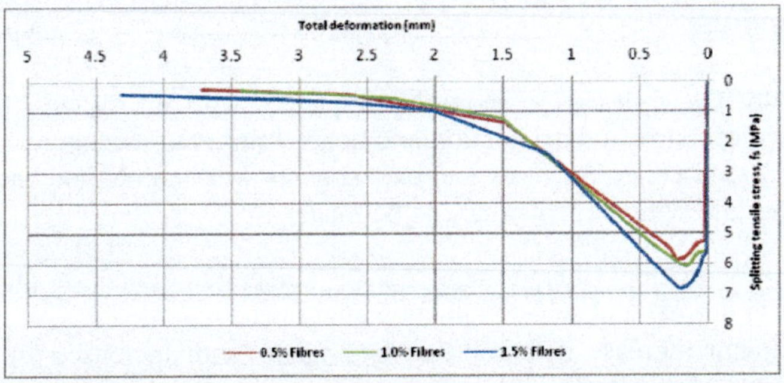

Fig. 2. Typical HPNFRC splitting tensile stress [MPa] – total deformation [mm]

Table 4. Average flexural strength (f_f), splitting tensile strength (f_s) and f_f/f_s ratio

	HPNFRC				HPSCFRC			
% Fiber:	f_f (MPa)	f_s (MPa)	f_f/f_s	f_s/f_c	f_f (MPa)	f_s (MPa)	f_f/f_s	f_s/f_c
0%	7.03	5.220	1.35	0.09	7.92	4.384	1.81	0.05
0.50%	7.37	5.682	1.30	0.09	7.78	5.480	1.42	0.06
1.00%	10.50	7.346	1.43	0.10	9.79	6.628	1.48	0.08
1.50%	13.38	6.794	1.97	0.09	13.11	6.642	1.97	0.07
	NFRC				SCFRC			
% Fiber:	f_f (MPa)	f_s (MPa)	f_f/f_s	f_s/f_c	f_f (MPa)	f_s (MPa)	f_f/f_s	f_s/f_c
0%	4.76	3.799	1.25	0.12	5.64	3.484	1.62	0.08
0.50%	4.86	3.767	1.29	0.11	6.82	4.434	1.54	0.08
1.00%	7.50	4.102	1.83	0.10	11.07	4.683	2.36	0.08
1.50%	7.87	5.108	1.54	0.12	10.87	4.475	2.43	0.08

Fig. 3 illustrates the flexural stress-deflection curves for the two normal strength concretes for 1.5% fibers. The other fiber contents for the normal strength class exhibit similar behavior whereas for the high strength class, the curves are relatively coincident. The higher f_f/f_s ratios achieved by the self-compacting mixes, as opposed to the normal mixes may be confirmation of better fiber alignment or utilization in self-compacting mixes, a property which is not observed for the cubes in splitting.

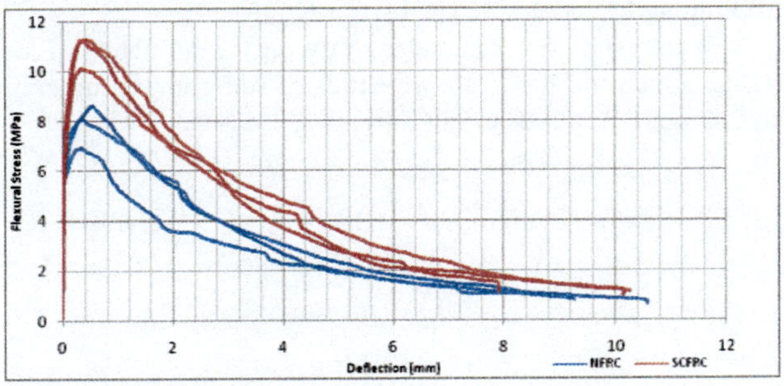

Fig. 3. NFRC & SCFRC flexural stress – deflection for 1.5% fiber

4 Conclusions

The aim of this research was to show that improved workability will result in enhanced fiber dispersion and orientation and hence improved mechanical properties of fiber-reinforced concrete. Tests and experiments were conducted and based on the results the following conclusions can be made:

- The addition of steel fibers to concrete has a negative impact on the workability of both normal and self-compacting concrete.
- A significant increase in compressive strength with increased fiber content was observed for all mixes.
- Both mix types showed a similar increase in splitting strength with increased fiber content, proportional to the increase in compressive strength. The benefits of fiber alignment are not being realized in the confined dimensions of the cube specimens.
- All mix types have significant increases in flexural strength and improved post-peak behavior with increased fiber content. The normal strength mixes provide the most convincing results with regard to the superiority of the self-compacting concrete over the normal workability concrete.

References

[1] Zollo, R.F.: Fibre-reinforced Concrete: an Overview after 30 Years of Development. Cement and Concrete Composites 19, 107–122 (1997)
[2] Özcan, D.M., Bayraktar, A., Sahin, A., Haktanir, T., Türker, T.: Experimental and finite element analysis on the steel fiber-reinforced concrete (SFRC) beams ultimate behaviour. Construction and Building Materials 23, 1064–1077 (2009)
[3] Ding, Y., Zhang, Y., Thomas, A.: The investigation on strength and flexural toughness of fibre cocktail reinforced self-compacting high performance concrete. Construction and Building Materials 23, 448–452 (2009)
[4] Ferrara, L., Park, Y., Shah, S.P.: A method for mix-design of fibre-reinforced self-compacting concrete. Cement and Concrete Research 37, 957–971 (2007)
[5] Boulekbache, B., Hamrat, M., Chemrouk, M., Amziane, S.: Flowability of fibre-reinforced concrete and its effect on the mechanical properties of the material. Construction and Building Materials 24, 1664–1671 (2010)
[6] Grünewald, S., Walraven, J.C.: Parameter-study on the influence of steel fibres and coarse aggregate content on the fresh properties of self-compacting concrete. Cement and Concrete Research 31, 1793–1798 (2001)

Quantification of Fresh and Mechanical Properties of HFRCC by Excess Paste Thickness

H. Mihashi[1] and N. Ishikawa[2]

[1] Tohoku Institute of Technology, Japan
[2] Tohoku University, Japan

Abstract. Influence of material components and mix proportions on the fresh properties and mechanical ones of Hybrid Fiber Reinforced Cementitous Composites (HFRCC) was experimentally studied. It was shown that the relative yield stress and the relative plastic viscosity of HFRCC decreased as the excess paste thickness increased. However, changing tendency of the ductility index was not simply described by the excess paste thickness. Then, the influence of volume content of each fiber and the geometry of the steel fiber was introduced to modify the excess paste thickness and all three properties were well quantified.

1 Introduction

In order to improve the poor mechanical properties of concrete, various types of fiber have been applied to produce fiber reinforced cementitious composites (FRCC). Among them, Hybrid Fiber Reinforced Cementitious composites (HFRCC) was developed as one of the high performance FRCC, in which rather longer steel cords (SC) and the rather shorter polyethylene (PE) fibers were mixed together. In this way, cracks on meso and macro levels are bridged by the longer fibers while micro and meso cracks are bridged by the shorter fibers. Generally it is more difficult to find a suitable mix proportion for HFRCC than FRCC containing a single kind of fibers.

Since ductility of FRCC is enhanced by fiber bridging, higher volume content of fibers usually achieves ductile properties. However, the workability becomes worth and the cost increases as the volume contents of fibers increases. Hence the workability problem is one of the difficult subjects for finding the suitable mix proportion of FRCC with the required mechanical performance. The mixture composition of FRCC is a compromise between acceptable workability in the fresh state and improved efficiency such as strength and ductility in the hardened state.

Workability of fresh concrete has been extensively studied for plain concrete, though no test methods to measure the fresh properties of FRCC have been standardized yet.

G.J. Parra-Montesinos, H.W. Reinhardt, and A.E. Naaman (Eds.): HPFRCC 6, pp. 67–74.

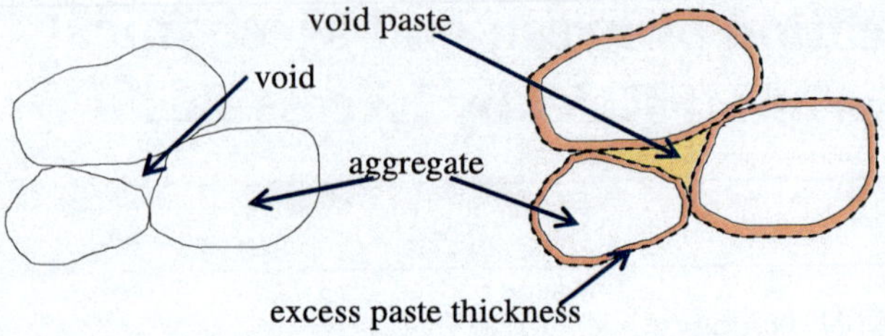

Fig. 1. Schematic description of excess paste thickness: t_e

Kennedy [1] proposed a "workability factor" for concrete mix design with respect to workability. The logical basis of this concept is schematically described in Fig. 1. When cement and water are added, the voids surrounded by aggregates are filled with cement paste and furthermore the surface of aggregates is covered by a film of the excess part of cement paste. The former was called 'void paste' and the latter was called 'excess paste'. The thicker the cement paste film, the lower the internal friction among aggregates. As a result, workability increases. Obviously the effect is dependent on the consistency of the paste itself. Thus fresh properties of concrete are determined by the thickness of the excess paste: t_e.

The workability of fresh concrete, however, is remarkably changed by adding either steel fiber or polymeric fiber. In general, workability of FRCC is determined by water-binder ratio (*W/B*), geometry and aspect ratio of fiber, fiber contents, superplasticizer, additives such as fly ash, and aggregate properties.

There are numerous test methods available for measuring the fresh properties of concrete such as rheometer test, slump flow test, flow-cone test, L-box test, and V-funnel test. Some of them were applied to FRCC in previous studies (e.g. Ding et al. [2]). Furthermore the rheological behavior of concrete is often characterized by using Bingham model and the model was applied also to FRCC in some of previous works (e.g. Bui et al. [3] and Yang et al. [4]). In order to describe the rheological property by the Bingham model, two parameters namely yield stress and plastic viscosity of the test sample need to be determined by measuring the rheological data of fresh state of the test sample. According to the Bingham model, the fresh FRCC can start to flow when the yield stress is overcome. Once the FRCC starts to flow, the shear strain rate increases linearly as the shear stress increases. The slope of the linear relation is defined as plastic viscosity. In other words, the plastic viscosity means the resistance of fresh FRCC to flow.

In the present paper, applicability of excess paste thickness for relating the mix proportion of HFRCC both to rheological properties of the fresh state and to mechanical properties of the hardened state is studied. Once the relation is established, it will be useful for optimizing the mix proportion of HFRCC which can satisfy the required performance both of workability and mechanical one. This new approach may reduce the huge amount of tests for determining empirical formulae for each material parameter.

2 Experimental Program

2.1 *Materials*

Material components of matrix used in this experimental study is rapid hardening Portland cement (C), fine silica sand (S), fly ash (FA) and superplasticizer (SP). For fiber reinforcement, four different types of hooked steel fiber (D), steel cord (SC) and polyethylene fiber (PE). Material properties of used fibers are shown in Table 1.

Table 1. Properties of fiber

Name	Code	Density	Length	diameter	Aspect ratio	Tensile strength	Young´s modulus
		g/cm^3	mm	µm	-	MPa	GPa
Steel fiber	D(30N)	7.84	30	620	48.4	1192	206
	D(35N)	7.84	35	550	63.6	1192	206
	D30P)	7.84	30	380	78.9	2830	206
	D(40P)	7.84	40	500	80.0	2830	206
	SC	7.84	32	400	80.0	2850	-
Synthesis fiber	PE	0.97	6	12	500.0	2850	73

2.2 *Mix Proportions and Mixing Procedure*

Table 2 shows mix proportions. As shown in the table, two levels of S/B and four levels of coupling of steel fiber and polymeric one were varied.

For mixing the material components, an omni mixer of five liters of the capacity was used. At first, cement, fly ash and fine silica sand were mixed for 1 minute. Then water and superplasticizer were mixed for 4 minutes. After mixing PE for 4 minutes, D or C was mixed for 4 minutes.

Table 2. Mix proportions of test series

Test series	W/B	SP/B	S/B	FA/B	Steel fiber	Synthesis fiber
	wt.%	wt.%	wt.%	wt.%	vol.%	vol.%
D(30N)-1	30	0.9	45,65	30	0.75	0.75
D(30N)-2a					1.00	0.75
D(30N)-3					1.00	1.00
D(35N)-1					0.75	0.75
D(35N)-2a					1.00	0.75
D(35N)-2b					0.75	1.00
D(35N)-3					1.00	1.00
D(30P)-1					0.75	0.75
D(40P)-1					0.75	0.75
SC-1					0.75	0.75

2.3 Calculation of Excess Paste Thickness and Rheological Testing

Excess paste thickness was calculated by Eq. (1).

$$t_e = V_e/S \tag{1}$$

where t_e (mm) is the excess paste thickness, V_e (mm^3) is excess paste volume and S (mm^2) is surface area of solids of fibers and aggregates. V_e was calculated by Eq. (2).

$$V_e = V_{total} - 100\, V_{solid}/G \tag{2}$$

where V_{total} (mm^3) is the total volume, V_{solid} (mm^3) is the solid volume and G is the absolute volume percentage of combined solids including fibers and aggregates. G was manually measured in one-liter cylindrical containers in accordance with a JIS (Japan Industrial Standard) test method after mixing the fibers and aggregates in the omni mixer for 30 seconds.

For measuring the rheological properties of HFRCC, J-shaped tube apparatus which is schematically shown in Fig. 2 was used. J-shaped tube apparatus was developed by Yamamoto et al. [5] and it has been used to measure the rheological properties of self-compacting concrete. Fresh HFRCC filled in the tube flows out by the self-weight. Based on the scheme as shown in Fig. 3, the relation between shear stress and the strain rate was determined according to Eqs. (3) and (4).

$$P_i = \{\rho\cdot(h_{i-1} + h_i)/2\}\,(R/2\,)/\,\{L + (h_{i-1} + h_i)/2\} \tag{3}$$

$$V_i = 4(h_{i-1} - h_i)\,/\,(t_iR) \tag{4}$$

where P_i (Pa) is shear stress in the interval i, R (cm) is the radius of the tube, L (cm) is the length of the curved part of the tube, $\rho\cdot$(g/cm^3) is the density of the HFRCC, h_i (cm) is the height of the sample in the interval i, V_i (1/sec) is the strain rate in the interval i, t_i (sec) is the flowing time in the interval i.

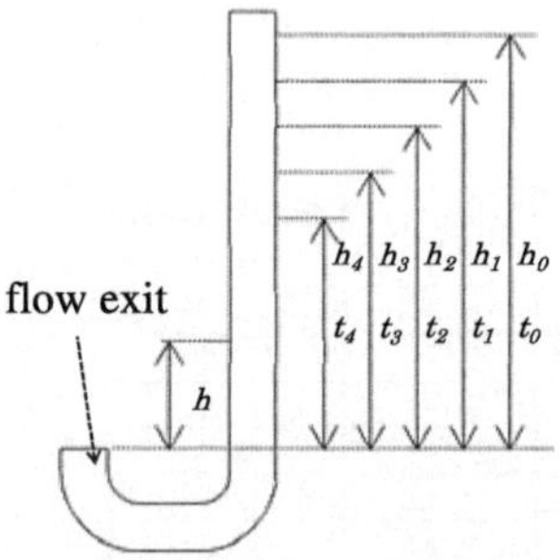

Fig. 2. Schematic description of J-shaped

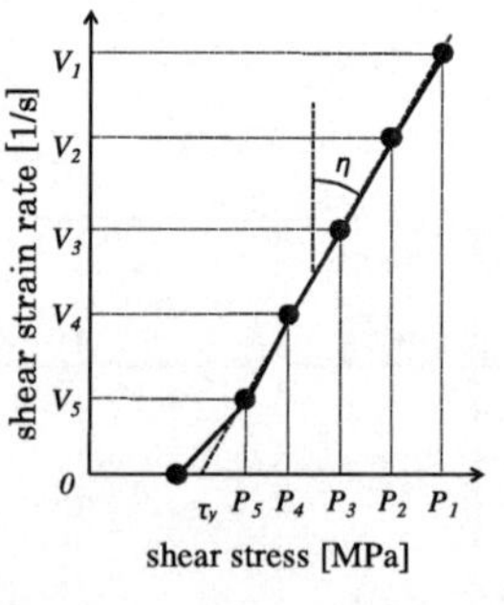

Fig. 3. Relation between shear stress and shear strain rare

As shown in Fig. 3, the intercept for the axis of shear stress gives the yield stress and the slope of the ascending linear line gives the plastic viscosity. Two times of the testing were performed for each case of the material combination.

2.4 Mechanical Test Method and Determination of Ductility Index

For determining ductility index of the hardened HFRCC, three-point bending tests were performed on notched beams with 40 x 40 x 160 (mm^3) of dimensions. Six specimens were tested for each case of the material combination. A notch, whose depth was 10mm, was introduced at the mid-span of the beams with a concrete cutter just before testing. Tests were carried out by an Instron-5567 universal testing machine at a loading speed of 0.4 mm/min with measurement of load-CMOD relation.

In this study, the ductility property of HFRCC was evaluated by the ductility index that is defined as the ratio between the area below the load-CMOD curve of the HFRCC and the area of an idealized rigid-perfectly plastic diagram with the same peak load as the specimen (see, e.g. Mihashi et al. [6]). Note that due to the limitation on the clip gage capacity, the load-CMOD curve stopped at CMOD of 5 mm.

3 Experimental Results and Discussion

3.1 Rheological Properties of Fresh State

Fig. 4 shows experimental results of the relation between t_e and the relative yield stress, where the relative value means the ratio between the value obtained for HFRCC and one for the corresponding paste. The yield stress and the plastic viscosity of the paste was 3.48 Pa and 2.76 /(Pa-sec), respectively. On the other hand, Fig. 5 shows experimental results of the relation between t_e and the relative value of plastic viscosity. According to these figures, the rheological properties decrease as t_e increases. Only by means of t_e, however, influence of S/B and/or type of fiber on the properties can't be properly quantified.

Detailed analyses of the experimental results showed that the contribution of t_e to the yield stress was changed by adding fibers and that the extent was dependent on the parameters of fibers such as volume contents of steel fiber (V_s) and polymeric fiber (V_p), and aspect ratio of steel fiber (L_s/d_s). Thus the modified excess paste thickness: t_e' was defined by Eq. (5) and Eq. (6).

$$t_e' = at_e \tag{5}$$

$$a = (V_s)^{n_1}(V_p)^{n_2}(L_s/d_s)^{n_3} \tag{6}$$

where $n_1 \sim n_3$ were determined by data fitting. Fig. 6 obviously shows that all of experimental data shown in Fig. 4 are better fitted to a function by introducing the modified excess paste thickness: t_e'.

In case of plastic viscosity, however, it was not succeeded to find any simple tendencies for parameters of fibers to influence the property. Then a multiple regression analysis was performed among six parameters of V_s, V_p, L_s, d_s, L_s/d_s and L_sd_s. As a result, it was clarified that V_s, V_p, and L_s are dominant parameters to influence the plastic viscosity.

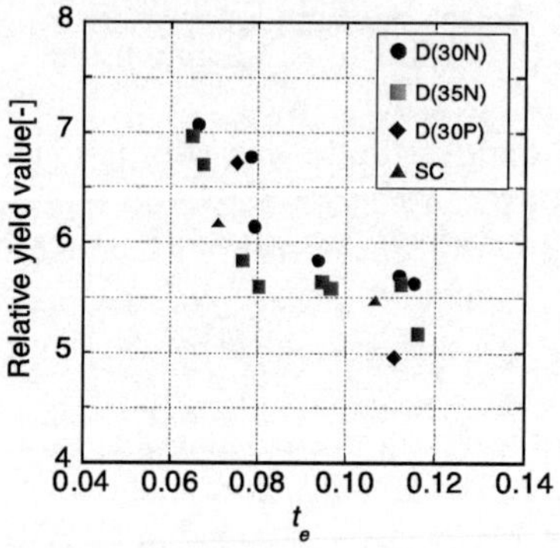

Fig. 4. Relation between t_e and relative yield stress

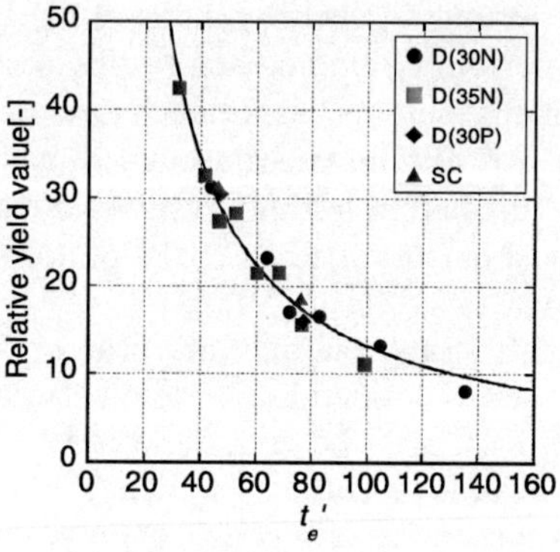

Fig. 5. Relation between t_e' and relative yield stress

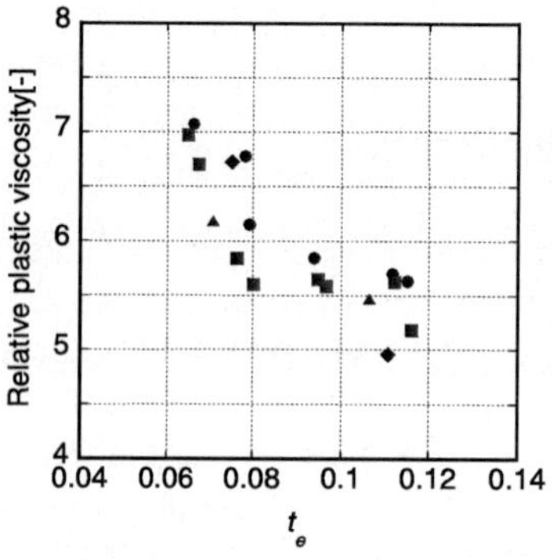

Fig. 6. Relation between t_e and relative plastic viscosity

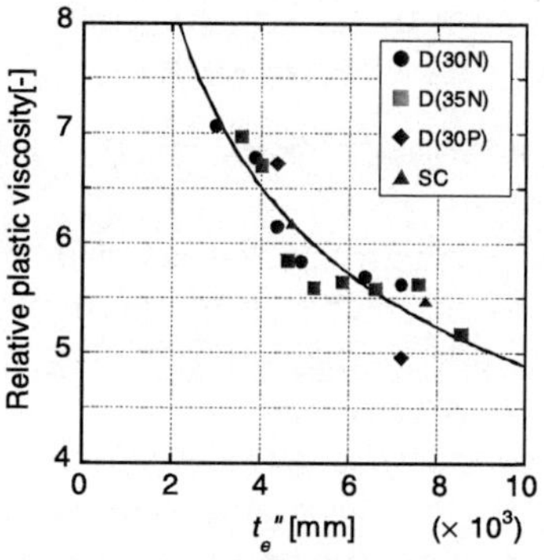

Fig. 7. Relation between t_e'' and relative plastic viscosity

Thus, the modified excess paste thickness: t_e'' was defined for the plastic viscosity by Eq. (7) and Eq. (8).

$$t_e'' = bt_e \tag{7}$$

$$b = (V_s)^{n_1}(V_p)^{n_2}(L_s)^{n_3} \tag{8}$$

Fig. 7 shows the relation between t_e'' and the plastic viscosity. Hence the regression curve can be used for quantifying the property.

3.2 Mechanical Properties of Hardened state

Fig. 8 shows experimental results of the relation between t_e and the ductility index. Scatter is so large that there are no obvious tendency to relate these two variables. Then, based on the detailed analyses of the experimental data, the modified excess paste thickness: t_e' given by Eq. (5) and Eq. (6) was introduced. Fig. 9 shows the relation between t_e' and the ductility index in which the ductility increased as t_e' increased. While many previous studies reported that larger amounts of fibers and higher aspect ratios increase the ductility of FRCC, Fig. 9 shows a unified relation can be obtained even for HFRCC in which even totally different types of fibers are mixed together.

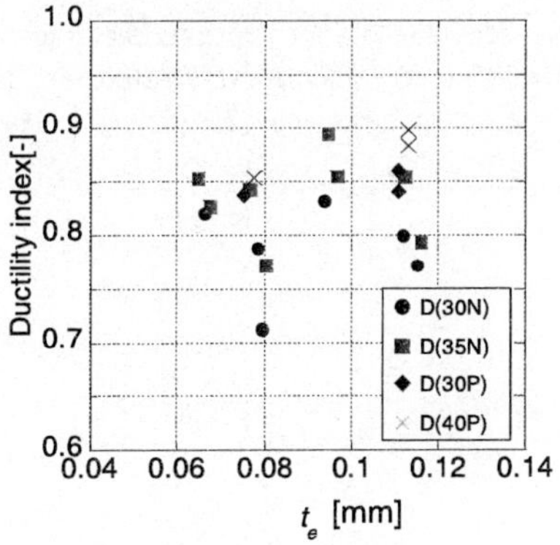

Fig. 8. Relation between t_e and ductility index

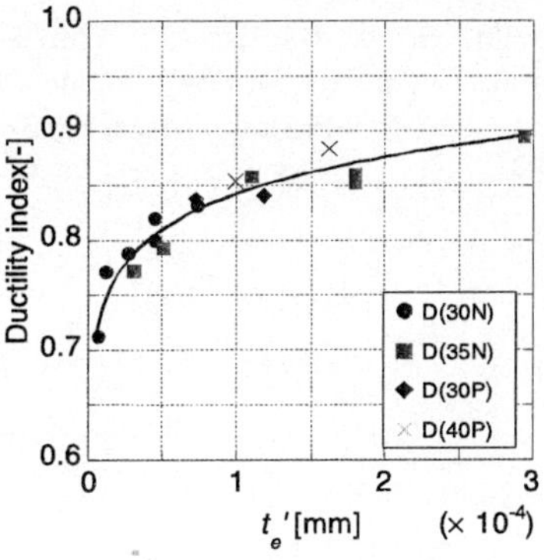

Fig. 9. Relation between t_e' and ductility index

4 Concluding Remarks

Experimental study on fresh properties and ductility of hybrid-type fiber reinforced cementitious composites (HFRCC) was carried out and the experimental results were analyzed. It was shown that the yield shear stress, plastic viscosity and ductility index of HFRCC can be quantified by the excess paste thickness modified with parameters related to volume contents of steel fiber (V_s) and polymeric fiber (V_p), and geometry of steel fiber. Finally the following concluding remarks were obtained.

1) The relative yield stress of HFRCC decreased but the ductility index increased as the excess paste thickness modified with V_s and V_p and the aspect ratio of steel fiber increased.
2) Plastic viscosity of HFRCC decreased as the excess paste thickness modified with V_s, V_p and the length of steel fiber increased.

References

[1] Kennedy, C.T.: The design of concrete mixes. ACI 36, 373–400 (1940)
[2] Ding, Y., Thomaseth, D., Niederegger, C., Thomas, A., Lukas, W.: The investigation on the workability and flexural toughness of fiber cocktail reinforced self-compacting high performance concrete. In: BEFIB 2004, pp. 465–475 (2004)
[3] Bui, V.K., Geiker, M.R., Shah, S.P.: Rheology of fiber-reinfoced cementitious materials. In: HPFRCC-4, pp. 221–231 (2003)
[4] Yang, E.H., Sahmaran, M., Yang, Y., Li, V.C.: Rheological control in production of engineered cementitious composite. ACI Material J. 106, 357–366 (2009)
[5] Yamamoto, Y., Homma, A., Kitsutaka, Y.: Study on the testing method of the rheological characteristic of high-fluidity concrete. J. Struct. Constr. Eng., AIJ 489, 9–16 (1996) (in Japanese)
[6] Mihashi, H., de Barros, J.P., Yamakoshi, S., Kawamata, A.: Optimizing fracture toughness of matrix for designing ductile fiber reinforced cementitious composites. In: Li, et al. (eds.) Fracture Mechanics of Concrete Structures, vol. 2, pp. 1021–1028 (2004)

An Investigation of Mechanical Properties of Jute Fiber-Reinforced Concrete

J. Kim[1], C. Park[1], Y. Choi[1], H. Lee[2], and G. Song[2]

[1] School of Architecture & Civil Engineering, Kyungpook National University, Korea
[2] Gyeongsangbuk-do Office Education, Sankyuk-dong, Pook-gu, Daegu, Korea

Abstract. An experimental investigation of mechanical properties of jute fiber-reinforced concrete (JFRC) is reported for making a suitable building material in terms of reinforcement. Two jute fiber reinforced concretes, so called jute fiber reinforced normal strength concrete (JFRNSC) and jute fiber reinforced high-fluidity concrete (JFRHFC), were tested in axial compression, flexure and splitting tensile strength. Jute fiber is economic and environmental-friendly fiber because it can be cultivated in nature. The slump value, compressive, flexural and splitting tensile strengths of specimens were investigated to four levels of jute fiber contents by volume fraction. The results showed that the slump value of JFRHFC decreased sharply as the increase of fiber content, which indicated the jute fiber in HFC should limited for maintain the HFC characteristics. Also the effect of jute fiber in HFC increased its compressive and splitting tensile strength significantly when compared to those of NSC while the flexural strength was not significantly different compared to NSC.

1 Introduction

One of the important properties of the hardened concrete is its strength. Since the unreinforced concrete has adequate strength for many structural applications, but is relatively brittle material. The addition of fibers in concrete has been used to overcome this problem. Factors affecting properties of fiber reinforced concrete are fiber type, geometry, forms, surface condition, mixing proportions, and curing method. Jute is a long, soft, shiny vegetable fiber that can be spun into coarse, strong threads. It is produced from plants in the genus Corchorus, which has been classified in the family Tiliaceae, or more recently in Malvaceae. Jute also is one of the most affordable natural fibers and is second only to cotton in amount produced and variety of uses of vegetable fibers. Jute fibers are composed primarily of the plant materials cellulose and lignin [1].

An investigation conducted on the mechanical properties of normal strength and high-fluidity concretes with jute fiber is reported in this paper. Different fiber

G.J. Parra-Montesinos, H.W. Reinhardt, and A.E. Naaman (Eds.): HPFRCC 6, pp. 75–82.

contents were used as reinforcement which were assumed randomly oriented and uniformly distributed in the matrix. Specimens with varying fiber contents were tested in splitting tension, flexure and axial compression. The results of this investigation have shown the feasibility of using jute fibers in developing a low-cost construction material when jute fibers are readily available.

2 Experimental Program

The experimental program was designed to evaluate mechanical properties of JFRNSC and JFRHFC. Although it is not possible to cover the entire actual conditions, laboratory tests may be used to predict or understand the behavior of the certain material in some desired respect.

2.1 Materials and Specimens

Fig. 1. Jute fibers

Type I/II Portland cement which meets the Korean standards KSL5201, 19 mm crushed maximum size of coarse aggregate with a specific gravity of 2.64 kg/dm^3, natural sand with a specific gravity of 2.56 and a fineness modulus of 2.42, a commercially produced jute fiber with a specific gravity of 1.03 and 9 mm length as shown in Fig. 1, were used in this work. The mix design of JFRNSC was as follows: ordinary Portland cement (type I/II) 410 kg/m^3, aggregate 1,629 kg/m^3, water-cement ratio kept of 0.50. The mix design of JFRHFC was as follows: ordinary Portland cement (type I/II) 400 kg/m^3, aggregate 1,629 kg/m^3, superplasticizer 8 kg, VMA 1 kg, water-cement ratio 0.50, and a 25% fly-ash replacement of the weight of the cement. Typical cylinder specimens of 100 mm in diameter and 200 mm in height were used for compressive and splitting tests. The size of specimen for flexural test was 100 x 100 x 400 mm. The fibers were uniformly dispersed in the mixer manually. Four different fiber contents by volume, 0, 0.25, 0.5 and 1.0%, were considered in the present work. In general, a relatively low content of fiber seems to have a small positive influence on the concrete. A relatively high content of fiber, however, may have a difficulty in mixing and the uniform distribution of fiber. Therefore the fiber contents used in this study seem to be reasonable for practical applications [2].

2.2 Testing Procedure

The slump tests were conducted with the regular slump test equipment as shown in Fig. 2. Compression tests were conducted on a total of 30 cylinder specimens of JFRNSC, JFRHFC and plain concrete at 28 days in accordance with the Korean standards KSF 2405 [3]. The load was applied at a rate of 0.02 mm/sec with a preload of about 200 N. The peak load and the load-axial displacement were recorded during the test by an acquisition system. Commonly, three test methods are used to measure tensile strength of concrete [5]: 1) the direct tension test, 2) the modulus of rupture or flexural test, 3) the split-cylinder test. The direct tension test can be attributed to the difficulty of insuring that the load is truly axial in direct tension when some eccentricities of the load may underestimate the tensile strength. The flexural (modulus of rupture) test is used to determine the tensile strength by using the maximum load obtained to cause failure. Although, the flexural tensile strength from a flexural test may show the true tensile strength of concrete, many researchers have indicated that the true tensile strength is approximately 65% - 70% of the flexural tensile strength [5-6]. The split-cylinder test is relatively simple and seems to provide a reliable test results to calculate the tensile strength of concrete under uniform stresses at the top and bottom across the diameter of the tested cylinder specimens [6]. Therefore, in this work the split-cylinder test and flexural test are used for investigation of tensile strength. Split-cylinder tests were carried out according to the Korean standards KSF 2423 [4]. Each cylinder specimen was placed on its side and loaded in compression along a diameter of specimen. The load was applied at a rate of 0.01 mm/sec with a preload of about 50 N. The flexural tests were carried out according to KSF 2408 standards [7] at the same ages of concrete used for the compression tests. The load was applied at a rate of 0.005 mm/sec with a preload of about 25 N.

Fig. 2. Slump test

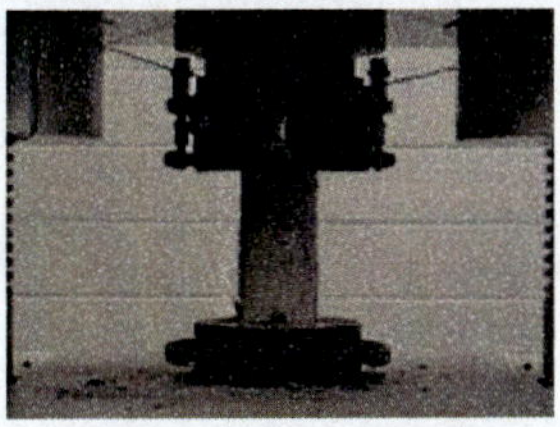

Fig. 3. Compressive test

Fig. 4. Splitting tensile test

3 Results and Discussion

The fiber length was kept constant at 9 mm while the volume fraction was varied from 0 to 1.0%. The volume fraction is one of the important parameter which affects the most mechanical properties of concrete.

3.1 Slump Value

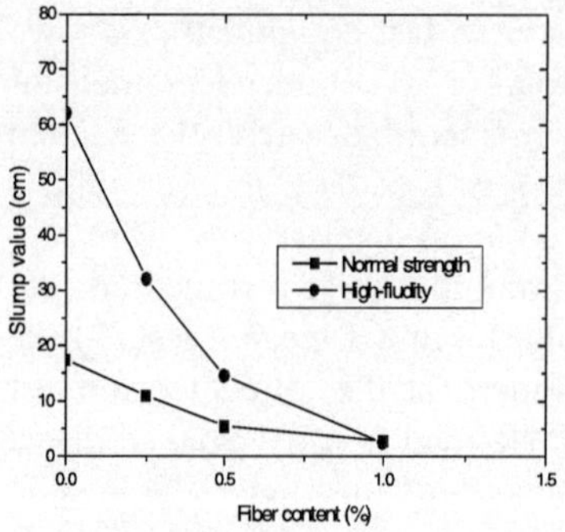

Fig. 5. Slump value curves

Fig. 5 shows the slump value to compare the effects of jute fiber with different fiber content. Test results show that the addition of jute fiber results in a more dramatic decrease in slump for the high-fluidity concrete when compared to the normal strength concrete. The mix of HFC should have sufficient flowability to maintain its own advantages. Therefore, the use of fiber in HFC must be carefully considered as reinforcement. In case of jute fiber, the expected reasonable volume fraction is considered by less than 0.25% to maintain HFC serviceability.

3.2 Compressive Strength

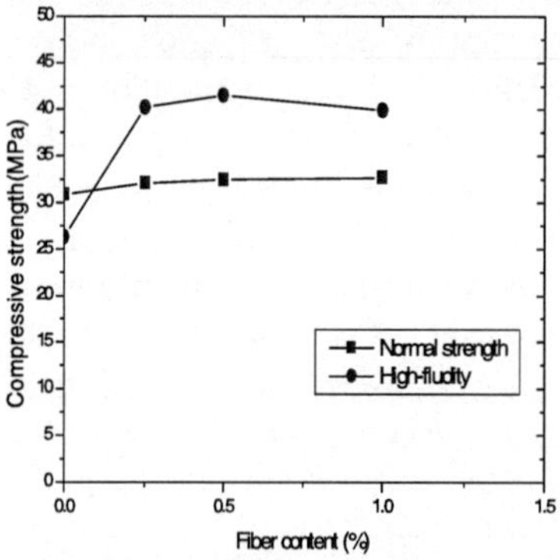

Fig. 6. Compressive strength curves

Compressive strength at 28 days is shown in Fig. 6. Each of the results was the average of 3 test specimens. There are significant differences between JFRNSC and JFRHFC. The addition of jute fiber up to 1% by volume in the normal strength concrete does not shows a significant increase, but HFC shows a significant increase of compressive strength by approximately 55% when compared to the one without fiber. These increase may come from the fibers interacting with the advancing cracks. Also the strains corresponding to peak load mostly increased, and the toughness also increased considerably. Furthermore, both the ascending and descending portions of the stress-strain curves were affected by the fibers.

3.3 Splitting Tensile Strength

The average splitting tensile strength at 28 days is shown in Fig. 7. The addition of jute fiber in the normal strength concrete and HFC increased the splitting tensile strength by approximately 6% and 30% when compared to the plain concrete, respectively. This result may be due to the reinforcing fiber's role to resist cracking and spalling of specimen across the failure plains. Also the average splitting tensile strength from tests observed from 9% to 12% for NSC, and from 10.5% to11% for HFC of its compressive strength. These ranges are similar to those found for SFRC, GFRC and PFRC [2, 8].

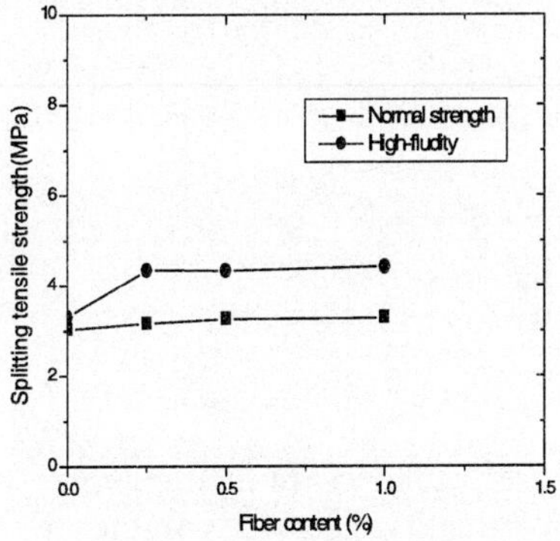

Fig. 7. Splitting-tensile strength curves

In order to estimate the effect of jute fiber on the splitting tensile strength, a regression analysis (curve fitting technique) was performed. From a large number of tests on concrete, a simple 0.5 power law model has become one of the most widely used analytical models for describing the relationship between the splitting tensile strength and compressive strength of concrete [9].

In this investigation, the tensile strength of concrete containing jute fiber is assumed proportional to the square root of their compressive strengths. The general form of the 0.5 power law equation is given as:

$$F_{(sp)_i} = a\sqrt{(f_{ck})_i} \tag{1}$$

$F_{(sp)_i}$ and $(f_{ck})_i$ are the splitting tensile and compressive strengths at the ith day, respectively. The intercept a represents the value of the constant at each age from the regression analysis.

Eqs. (2) and (3) have been proposed for the normal strength concrete and High-fluidity concrete containing jute fiber, respectively: The obtained equations are similar to those found for GFRC, PFRC and SFRC by Choi and Nataraja et al. [2,8]

$$F_{(sp)28} = 0.57\sqrt{(f_{ck})_{28}} \tag{2}$$

$$F_{(sp)28} = 0.67\sqrt{(f_{ck})_{28}} \tag{3}$$

These proposed equations are plotted in Fig. 8 with the experimental results. It can be seen that the regression lines of equations showed a relatively good relation for the splitting and compressive strength with R^2 =0.92 and 0.78, respectively. Most statisticians consider a R^2 =0.70 or higher for a reasonable regression [10]. Actually, it is noted that tensile strength of fiber reinforced concrete could be affected by many other factors such as the types of fiber, fiber content, length of fiber, directions of embedded fibers, aspect ratios of fiber. Furthermore, the number of tested specimens is important because more test data may provide better statistical validation for various factors.

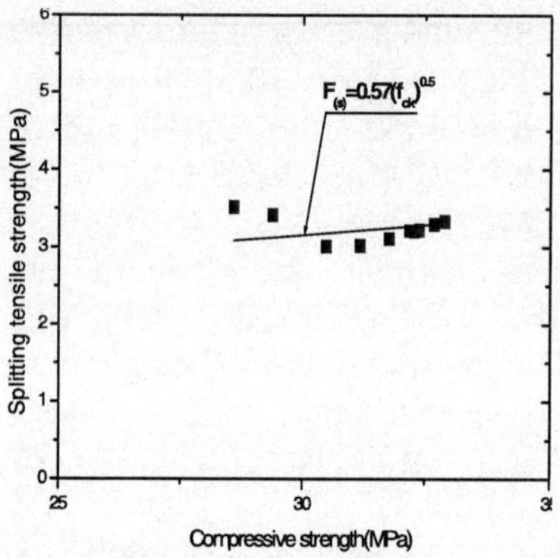

a) Normal strength concrete

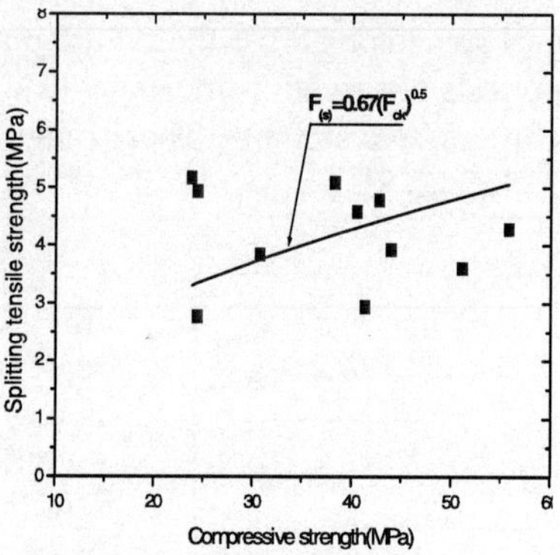

b) High-fluidity concrete

Fig. 8. Suggested equations with experimental results

The proposed equations, however, were given only as a function of their compressive strength from a relatively small number of experiments. Therefore, further research is needed to verify the proportional relationship, and to determine the effects of fiber types, the number of specimens, curing time and other factors.

3.4 Flexural Strength

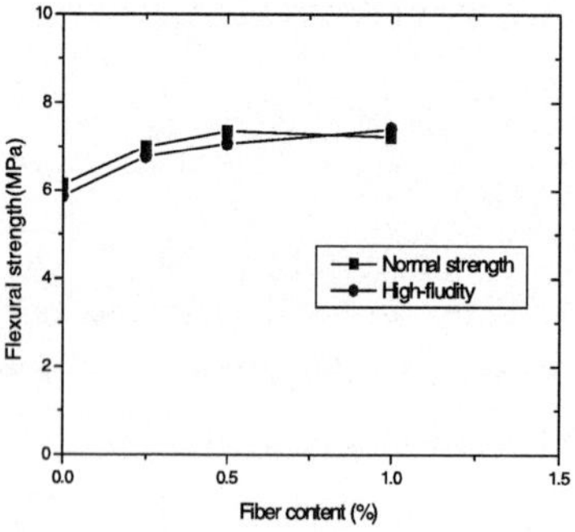

Fig. 9 Flexural strength curves

The modulus of rupture from the simple beam test at 28 days is shown in Fig. 9. It can be observed that the addition of jute fiber in the normal strength concrete and HFC increased the flexural strength by approximately 17% and 20% when compared to the plain concrete, respectively. Also the average flexural strength from tests ranged from 21% to 24% for JFRNSC, and from 16% to18% for JFRHFC of its compressive strength, respectively. Unlike compressive and splitting strengths, the increase in flexural strength was not significantly different for normal and HFC. Fig. 10 shows the load-deflection curves for flexural test of JFRNSC and JFRHFC, respectively. In addition to the increase in flexural strength, a significant increase in toughness is also observed. The obtained flexural toughness is determined as the area under the load-deflection curve up to a deflection of 0.6mm, except for the plain concrete. The calculated toughness values of JFRNSC and JFRHFC were improved about 400% and 200% when compared to the plain concrete, respectively.

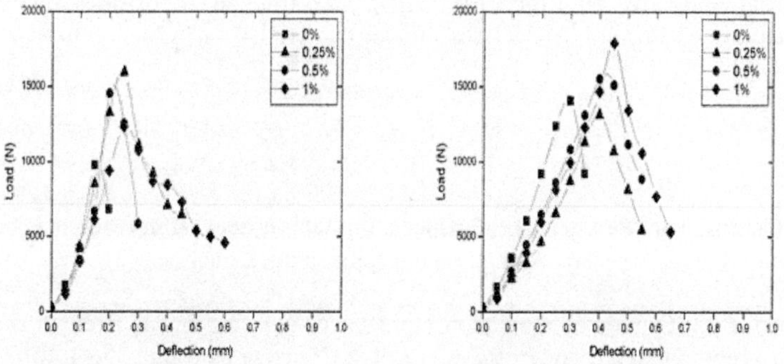

a) Normal strength concrete b) High-fluidity concrete

Fig. 10. Load-Deflection curves

4 Conclusion

This paper describes the influence of volume fraction on the compressive, flexural and splitting strength of jute fiber reinforced normal strength concrete and high-fluidity concrete. Based on the experimental and analytical results, the following conclusions are obtained:

(1) Jute fibers do affect the workability, the decreasing of slump value of jute fiber reinforced high-fluidity concrete (JFRHFC) is significantly higher than that of jute fiber reinforced normal strength concrete (JFRNSC)

(2) The compressive strength of the jute fiber reinforced high-fluidity concrete improved by 55% when compared to the one without fiber. The jute fiber reinforced normal strength concrete, however, does not show a significant increase of compressive strength.

(3) Additions of jute fibers in JFRNSC and JFRHFC increase the splitting tensile strength by approximately 6 and 30% that of plain concrete, respectively. Also the splitting tensile strengths approximately ranged from 9% to 12% of their compressive strengths.

(4) The addition of jute fiber in the normal strength concrete and HFC increased the flexural strength by approximately 17% and 20% when compared to the one without fiber, respectively. Also the average flexural strength from tests ranged from 21% to 24% for normal strength concrete, and from 16% to18% for HFC of its compressive strength, respectively

(5) Using a relatively small amount of experimental data, a simple 0.5 power law relationship between the tensile strength and the compressive strength of JFRC can be successfully used to estimate their splitting tensile strength.

(6) Further research is needed to verify the proportional relationship, and to determine the effects of the fiber types, fiber content, the number of specimens, curing time and other factors.

Acknowledgements. The work presented in this paper was funded by Center for Concrete Corea (05-CCT-D11), supported by Korea Institute of Construction and Transportation Technology Evaluation and Planning (KICTTEP) under the Ministry of Construction and Transportation (MOCT).

References

[1] Aziz, M.A., Paramasivan, P., Lee, S.L.: Prospects for natural fiber reinforced concrete in construction. Int. J. of Cem. Comp. and Lightweight Concrete 3(2), 123–132 (1981)

[2] Choi, Y., Yuan, R.L.: Experimental relationship between splitting tensile strength and compressive strength of GFRC and PFRC. Cement & Concrete Research 35(4), 1587–1591 (2005)

[3] KSF 2405, Standard Test Method for Compressive Strength of Cylindrical Concrete Specimens. Annual Book Koran Standards

[4] KSF 2423, Standard Test Method for Splitting Tensile Strength of Cylindrical Concrete Specimens. Annual Book Korean Standards

[5] Zain, M.F.M., Mahmud, H.B., Ilham, A., Faizal, M.: Predicting of Splitting Tensile Strength of High-Performance Concrete. Cement & Concrete Research 32(10), 1251–1258 (2002)

[6] Rocco, C., Guinea, G.V., Planas, J., Elices, M.: Review of the Splitting-test Standards from a Fracture Mechanics Point of View. Cement and Concrete Research 31(3), 73–82 (2001)

[7] KSF 2408, Standard Test Method for Flexural Strength of Concrete (Using Simple Beam with Third-Point Loading). Annual Book Korean Standards

[8] Nataraja, M.C., Dhang, N., Gupta, A.P.: Splitting Tensile Strength of SFRC. The Indian Concrete Journal 75(4), 287–290 (2001)

[9] Francis, A.O., Edwin, G.B., Deatherage, J.H.: Splitting Tensile Strength and Compressive Strength Relationship at Early Ages. ACI Material Journal 88(2), 115–121 (1991)

[10] Ostle, B., Turner, K.V., Hicks, C.R., Mcelrath, G.W.: Engineering Statistics: The Industrial Experience. Duxbury Press, Pacific Grove (1999)

[11] Akers, S.A.S., Studinka, J.B.: Ageing Behaviour of Cellulose Fibre Cement Composites in Natural Weathering and Accelerated Tests. Int. J. of Cem. Comp. and Lightweight Concrete 11(2), 93–97 (1989)

[12] Ferrara, L., Park, Y.-d., Shah, S.P.: A method for mix-design of fiber-reinforced self-compacting concrete. Cement & Concrete Research 37, 957–971 (2007)

[13] Corinaldesi, V., Moriconi, G.: Durable fiber reinforced self-compacting concrete. Cement & Concrete Research 34, 249–254 (2004)

[14] Grunewald, S., Walraven, J.C.: Parameter-study on the influence of steel fibers and coarse aggregate content on the fresh properties of self-compacting concrete. Cement & Concrete Research 31, 1793–1798 (2001)

Back-Calculation of Tensile Properties of Strain Softening and Hardening Cement Composites

M. Bakhshi, C. Barsby, and B. Mobasher

Ira A. Fulton School of Engineering, Arizona State University, USA

Abstract. Back-calculation procedures are used to obtain uniaxial tensile properties of cement composites from closed loop flexural tests. The proposed methodology uses a forward calculation of obtaining load deflection results based on tensile data in conjunction with crack localization rules. A closed form solution for moment–curvature and load deflection response is obtained using these material parameters. Test results are applied to a series of strain hardening composites to obtain material parameters for polymeric fiber reinforced concrete samples. Test results are compared with data reported for Average Residual tensile Strength (ARS) obtained from ASTM tests such as ASTM C1399 and C1609 and a correlation between the ARS and proposed post-peak strength are obtained. Results point out to the un-conservative nature of reporting the material properties data using the ASTM recommended test procedures.

1 Introduction

A formulation is presented to extend the available modeling technique for back-calculation of material properties from experimental data. The experimental data are fitted using the load deflection results of a nonlinear parameterized material model described using tensile parameters [1]. This method is attractive as an inverse analysis algorithm to back-calculate material parameters from flexural tests and may explain the differences between the tensile and flexural strengths of strain hardening and strain softening composites [2].

Fig. 1 presents the constitutive model for strain softening and hardening materials. The linear portion of an elastic-perfectly-plastic compressive stress-strain response terminates at yield point (ε_{cy}, σ_{cy}) and remains constant at compressive yield stress σ_{cy} until the ultimate compressive strain ε_{cu} as shown in Fig. 1(a). The tension model in Fig. 1(b) is described by a tri-linear response with an elastic range E, and post cracking modulus E_{cr} which is assigned a negative or positive value in order to simulate both strain softening or hardening materials. The third region is a constant stress range defined with stress σ_{cst} in the post crack region and terminates at ε_{tu}. Two strain measures define the first cracking and transition strains (ε_{cr}, ε_{trn}). Two intrinsic material parameters first cracking tensile strain

G.J. Parra-Montesinos, H.W. Reinhardt, and A.E. Naaman (Eds.): HPFRCC 6, pp. 83–90.

ε_{cr} and tensile modulus E are used to express seven normalized parameters as shown in Eqs. (1):

$$\omega = \frac{\varepsilon_{cy}}{\varepsilon_{cr}}; \alpha = \frac{\varepsilon_{trn}}{\varepsilon_{cr}}; \beta_{tu} = \frac{\varepsilon_{tu}}{\varepsilon_{cr}}; \lambda_{cu} = \frac{\varepsilon_{cu}}{\varepsilon_{cr}}; \gamma = \frac{E_c}{E}; \eta = \frac{E_{cr}}{E}; \mu = \frac{\sigma_{cst}}{E\varepsilon_{cr}} \tag{1}$$

Using these normalized parameters stress-strain & toughness G_f are expressed as:

$$\frac{\sigma_c(\lambda)}{E\varepsilon_{cr}} = \begin{cases} \gamma\lambda & 0 \le \lambda \le \omega \\ \gamma\omega & \omega < \lambda \le \lambda_{cu} \\ 0 & \lambda_{cu} < \lambda \end{cases} \qquad \frac{\sigma_t(\beta)}{E\varepsilon_{cr}} = \begin{cases} \beta & 0 \le \beta \le 1 \\ 1 + \eta(\beta - 1) & 1 < \beta \le \alpha \\ \mu & \alpha < \beta \le \beta_{tu} \\ 0 & \beta_{tu} \le \beta \end{cases} \tag{2}$$

$$G_f = E\varepsilon_{cr}^2 \left[\left(\frac{\alpha - \mu - \mu\alpha}{2} \right) + \mu\beta_{tu} \right] \tag{3}$$

In a flexural test a rectangular cross section with a width b and depth d is considered and the Kirchhoff hypothesis is applied. Maximum tensile and compressive strains β and λ are linearly related through the neutral axis parameter, k as Eq. 4.

$$\beta = \frac{\varepsilon_{tbot}}{\varepsilon_{cr}}; \quad \lambda = \frac{\varepsilon_{ctop}}{\varepsilon_{cr}}; \quad \lambda = \frac{k}{1-k}\beta \tag{4}$$

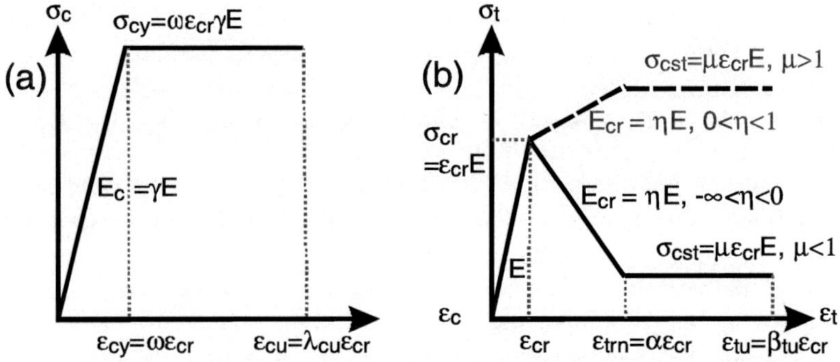

Fig. 1. Material models for FRC materials (a) compression (b) tension

By assuming linear strain distribution across the depth and ignoring shear deformations, stress distribution across the cross section at three stages of imposed tensile strain: $0<\beta \le 1$, $1<\beta \le \alpha$ and $\alpha < \beta \le \beta_{tu}$ is obtained in closed form [1]. Internal moment is obtained using the force components and their moment arm and the curvature is determined as the ratio of compressive strain ($\varepsilon_{ctop}=\lambda\varepsilon_{cr}$) to the depth of neutral axis kd. The moment and curvature M_i, ϕ_i at each stage i are expressed using the values at cracking M_{cr} and ϕ_{cr} and the parametric function m', ϕ' [2].

$$M_i = m' M_{cr}; \quad \phi_i = \phi_i ' \phi_{cr}; \quad (M_{cr} = \frac{1}{6} bd^2 E\varepsilon_{cr}, \; \phi_{cr} = \frac{2\varepsilon_{cr}}{d}) \tag{5}$$

By applying the moment-area method, mid-span deflection of a flexural beam is derived explicitly as presented in previous publications [3, 4].

2 Experimental Program

Four different polymeric macro type fibers were used as shown in Table 1. Their mechanical and physical properties cover a board range of fibrillated polypropylene, monofilament polypropylene/polyethylene blend, modified olefin, modified polypropylene blend, and high stiffness fibrillated polypropylene. Five different mixtures were prepared for testing under three-point bending. The w/cm ratio is 0.4 for all mixes and cement content is 475 kg/m^3. The fly ash and silica fume were added to the mixtures as supplementary cementitious materials with the dosage of 60 kg/m^3 and 15 kg/m^3, respectively. The sand-to-cement (s/cm) ratio is 1.9 for all mixtures using 1050 kg/m^3 fine aggregate and 400 kg/m^3 coarse aggregate. A constant fiber volume fraction of 6 kg/m^3 was used for all samples. Flexural test samples were 53 x 15 x 15 cm^3 beam specimens with an initial notch length of 2.5 cm and test span of 45 cm. Three-point bending closed loop tests were conducted using the Crack Mouth Opening Deformation (CMOD) across the notch as the controlled variable.

Table 1. Physical properties of polymeric fibers used

Fiber Type	Type A	Type B	Type C	Type D
Base	Monofilament Polypropylene/ Polyethylene blend	Modified Olefin	Modified Polypropylene Blend	Fibrillated polypropylene fiber
S. Gravity	0.92	0.9-0.92	0.91	0.91
E (GPa)	5	10	6.55	5.5
UTS (MPa)	600 -650	550	655	410
Length (mm)	50	50	37.5	50

3 Discussion of Results

Two replicate samples were tested for each test series. As the cracks initiate from the notch and extend up to the upper side of the beam, they are bridged by the fibers which increase the capacity of the cracked section, increasing the post peak strength by dissipating additional energy. Results of three-point bending tests at the age of 16 h are shown in Fig. 2, and summary results are in Table 2. Average flexural stiffness of Type B are 17, 12, 5 and 6% higher than A, C, B&D, and C&D, respectively. PolyF10-TypeB and C have the highest maximum flexural load and modulus of rupture (MOR). The flexural toughness of PolyF10-TypeB is 65, 29, 15 and 13% higher than Type A, C, B&D, and C&D, respectively.

Table 2. Summary of experimental analysis of Polymeric FRC materials

Mix ID	K, kN/mm	CMOD at Max Load, mm	MOR, MPa	Max CMOD mm	Flex. Tough., kN.mm	F'c, MPa
PolyF10-A	239 (15.4)	0.0457	2.43 (0.28)	3.83 (0.01)	9.6 (0.2)	29.0[a]
PolyF10-B	281 (5.6)	0.0431 (0.005)	2.76 (0.16)	3.83 (1.69)	15.8 (8.3)	27.1[a]
PolyF10-C	251 (27.8)	0.0457 (0.0076)	2.79 (0.01)	3.82 (1.00)	12.3 (3.2)	28.8[a]
PolyF10-B&D	269 (81.6)	0.0609 (0.0304)	2.48 (0.06)	3.82 (0.01)	13.8 (7.0)	28.7[a]
PolyF10-C&D	266 (29.1)	0.0508 (0.0018)	2.28 (0.11)	3.81 (0.00)	14.0 (1.8)	25.8[a]

[a] The results are for the age of 24 h, K = Stiffness

The load deflection response of a beam is obtained using the moment–curvature response, crack localization rules, and moment-area method. For a given cross section and material properties, the normalized tensile strain at the bottom fiber β is incrementally imposed to generate the moment–curvature response. The beam is segmented into finite sections. For a given load step, static equilibrium is used to calculate moment distribution along the beam and hence the curvature. The deflection at mid-span is calculated by numerical moment-area method of discrete curvature between the support and mid-span [4]. Back-calculated tensile parameters of all flexural samples are shown in Table 3.

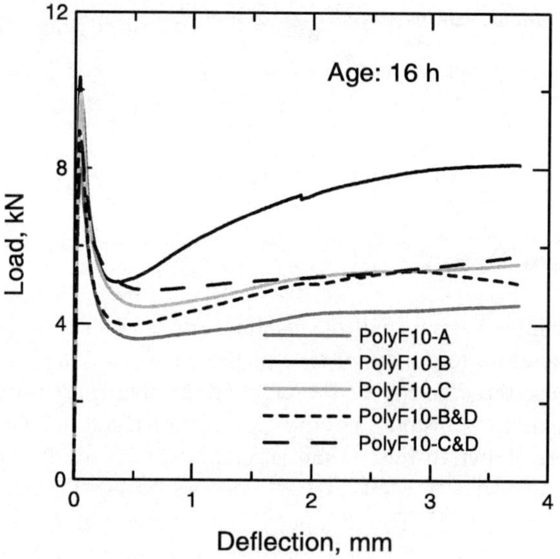

Fig. 2. Effect of different types of Fibers on the flexural response

4 Data Reduction by Average Residual Strength Method

The Averaged Residual Strength (ARS) proposed by Banthia and Dubey [5, 6] has been used by ASTM C1399 [7, 8]. The post-peak load is used to define an elastically equivalent flexural stress using the section modulus of an uncracked beam, Whereas, it is known that due to cracking, the neutral axis shifts significantly toward the compression zone [7, 9, 10]. The ARS value does not represent an equivalently elastic stress associated with the post crack tensile strength, or any tensile residual strength measure. Fig. 3(a-d) shows the stress distribution during the loading across the depth of the section. As the neutral axis moves toward the compression zone, a uniform tensile stress distribution is distributed over the tensile zone. Fig. 3(d) implies that the use of residual flexural strength as a residual tensile strength is un-conservative as it overestimates the tensile strength. Fig. 4 shows the tensile stress strain response and the fit obtained by backcalculation of samples PolyF10-B.

Table 3. Tensile parameters of flexural samples using back-calculation

Mix ID	Sample Code	ε_{cr} μstr	E GPa	μ	α	ε_{trn} μstr	ε_{tu} μstr
PolyF10-A	Avg.	58.5	20.684	0.285	14	793	52000
	Std. dev.	9.2	0.000	0.05	5.7	202	2121
PolyF10-B	Avg.	55	24.132	0.37	14	755	50000
	Std Dev	5.6	0.000	0.08	5.7	233	1414
PolyF10-C	Avg.	55.5	23.787	0.33	19	1032	43000
	Std. dev.	7.8	2.434	0	5.7	166	15556
PolyF10-B&D	Avg.	76	17.237	0.27	9	684	49500
	Std. dev.	10.8	11.032	0.06	5.7	192.4	1361
PolyF10-C&D	Avg.	66	17.237	0.39	16	1056	54000
	Std. dev.	6.7	0.000	0.02	5.7	210.6	1847

Direct correlation of residual strength from ASTM C1399 and the present method as shown in Fig. 5 indicate that the ARS method overestimates the residual tensile strength by as much as 2.2 times. The comparison between toughness of ARS and post-peak residual strength (Fig. 3d) shows the degree of error with the assumption that the neutral axis remains constant during loading.

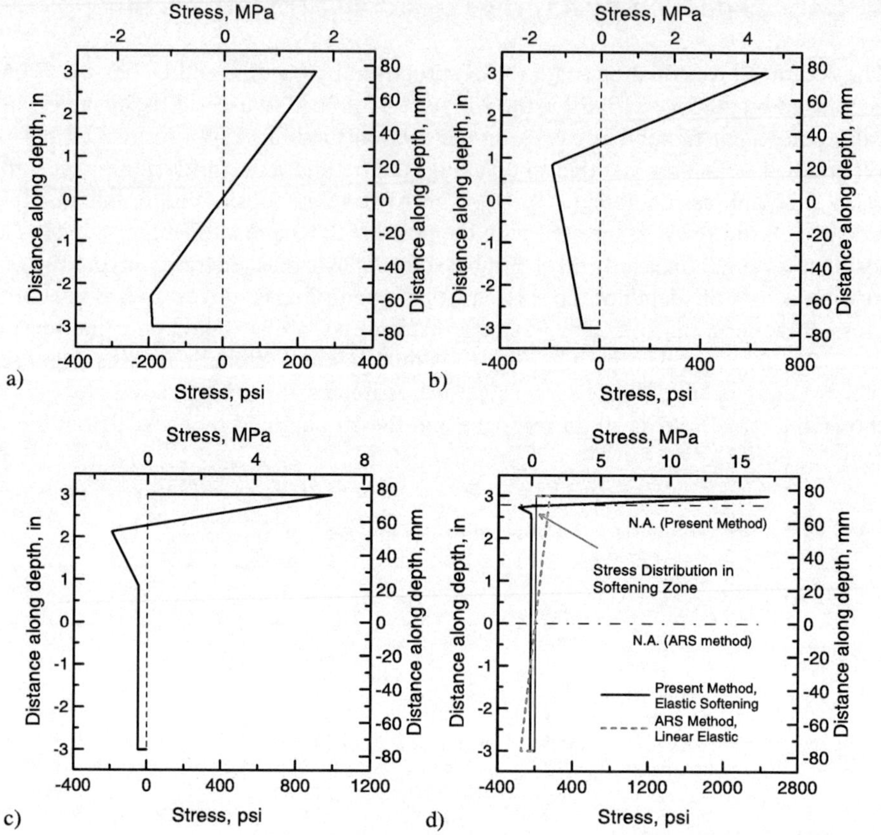

Fig. 3. Stress distribution through sample depth from initial to advanced stages of loading

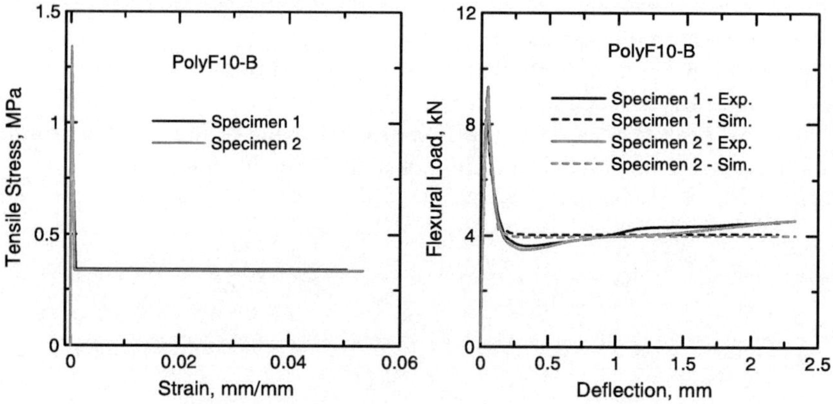

Fig. 4. Back-Calculated Tensile Response and Load-Deflection of PolyF10-TypeB

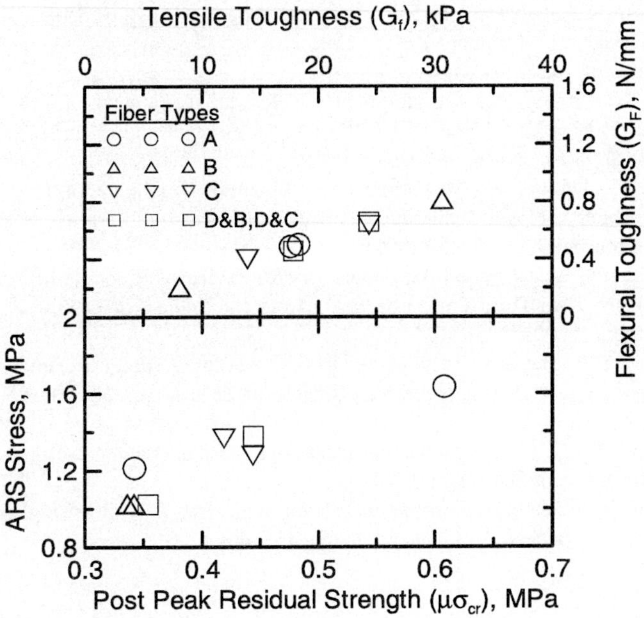

Fig. 5. Comparison of post peak strength with ARS and specific flexural toughness with back-calculated tensile toughness

5 Conclusion

Characterization of tensile-flexural strain softening and strain hardening of fiber cement composites shows that the presence of fiber significantly increases the ductility of the material. By applying the load deflection back-calculation technique one can generate strength data with a higher degree of accuracy than the current ASTM methods. Using a closed form set of governing parameters and variables applied through each stage of material response, the stress distribution that considers a shifting neutral axis cab be simulated which provides a more accurate representation of the residual strength of the fiber cement composites.

References

[1] Soranakom, C., Mobasher, B.: Closed-form moment-curvature expressions for homogenized fiber-reinforced concrete. ACI Material Journal 104(4), 351–359 (2007)

[2] Soranakom, C., Mobasher, B.: Closed form solutions for flexural response of fiber reinforced concrete beams. Journal of Engineering Mechanics 133(8), 933–941 (2007)

[3] Soranakom, C., Mobasher, B.: Flexural Design of Fiber Reinforced Concrete. ACI Materials Journal, 461–469 (September-October 2009)

[4] Soranakom, C., Mobasher, B.: Moment-Curvature Response of Strain Softening and Strain Hardening Cement Based Composites. Cement and Concrete Composites 30(6), 465–477 (2008)
[5] Banthia, N., Dubey, A.: Measurement of Flexural Toughness of Fiber-Reinforced Concrete Using a Novel Technique - Part 1: Assessment and Calibration. ACI Materials Journal 96(6), 651–656 (1999)
[6] Banthia, N., Dubey, A.: Measurement of Flexural Toughness of Fiber-Reinforced Concrete Using a Novel Technique - Part 2: Performance of various composites. ACI Materials Journal 97(1), 3–11 (2000)
[7] ASTM C1609, Test Method for Flexural Performance of Fiber-Reinforced Concrete (Using Beam With Third-Point Loading). American Society of Testing and Materials, Phila-delphia (1999)
[8] ASTM C 1399, Standard Tests Method for Obtaining Average Residual Strength of Fiber-Reinforced Concrete. American Society of Testing and Materials, Philadelphia (1999)
[9] Mobasher, B., Li, C.Y.: Mechanical properties of hybrid cement-based composites. ACI Mater J. 93(3), 284–292 (1996)
[10] Mobasher, B., Shah, S.P.: Test parameters for evaluating toughness of glass fiber reinforced concrete panels. ACI Mater J. 86(5), 448–458 (1989)

Basis of a Finite-Element Simulation Tool to Predict the Flexural Behavior of SFRC Prisms

T. Soetens[1], S. Matthys[1], L. Taerwe[1], and A. Van Gysel[2]

[1] Magnel Laboratory for Concrete Research, Department of Structural Engineering, Faculty of Engineering and Architecture, Ghent University, Ghent, Belgium
[2] Campus De Nayer, Materials and Structures Research Group, Department of Applied Engineering, Lessius University College Mechelen, Belgium

Abstract. The fiber pull-out response of a steel fiber, both straight and end-hooked, embedded in a concrete matrix has been widely investigated in the past. Attention was given to the influence of the type of fiber, the concrete strength and the inclination of the pull-out force. Hooked-end fibers need more energy to be pulled out of the concrete and thus are preferably used in steel fiber reinforced concrete (SFRC) for structural purposes.

Based on this single fiber pull-out behavior, numerous analytical and numerical methods were verified with experimental results of three- or four-point bending tests on small prisms. In this paper, a new approach to characterize the non-linear bending behavior of steel fiber reinforced concrete is proposed. Finite element (FE) analyses were conducted on a small prism-model, containing randomly generated small embedded reinforcement elements, which material properties are based on the pull-out behavior derived from earlier research activities. In this stage, attention was also given to the orientation of the fibers.

Eventually, a good correlation was found between the numerical and experimental results and the fine-tuned model was then used to perform a large number of numerical bending tests. In that way, the scatter of the parameters that describe the non-linear bending behavior of small steel fiber reinforced concrete prisms was estimated in a much faster way than by experimental testing. These parameters are usually needed to evaluate the structural behavior of SFRC elements according to current design guidelines for SFRC.

1 Introduction

In the past three decades, many researchers studied the single fiber pull-out response of a fiber embedded in a cementitious matrix. This research led to the development of several pull-out models to describe the pull-out response of both straight [1, 2] and hooked-end [3-8] fibers. Accurate predictions of the pull-out response served as a basis to develop computational models to characterize the behavior of SFRC subjected to pure bending or uni-axial tension. This characterization,

G.J. Parra-Montesinos, H.W. Reinhardt, and A.E. Naaman (Eds.): HPFRCC 6, pp. 91–98.

in particular the bending behavior of small prisms either by three- or four-point bending tests, is very important to derive parameters that are used in current design guidelines for SFRC [9].

In this paper a new approach to estimate flexural toughness parameters is explained. The pull-out response of a single fiber was used to generate different samples of a FE-model with randomly distributed fibers and results from both experimental and numerical analyses are compared and discussed.

2 Experimental Testing

2.1 Test Setup

Four small prisms with a span length l of 450 mm were subjected to a four-point bending test with the test setup in accordance with the Belgian standard NBN B15-238 (1992). The prisms have a square cross section with a height h of 150 mm and the total length is equal to 600 mm. 40 kg/m³ of DRAMIX® fibers, type ZC 60/.80 HC, were added to the concrete, which general characteristics are showed in Table 1. To achieve a load-deflection curve which can quantify the post-peak behavior of the prism, a displacement controlled test was necessary.

Table 1. Main properties of the used concrete

Concrete characteristics	
Mean compression strength f_{cm}	90.3 N/mm²
Mean tensile strength f_{ctm}	4.89 N/mm²
Modulus of elasticity E	45 800 N/mm²
Fracture Energy G_f	163 N/m
W/C ratio	0.30

2.2 Characterization of the Flexural Toughness

Flexural toughness can be quantified in two different ways as stated in the Belgian standard. According to the first method, the ratio D_n between the applied load F_n at a certain deflection l/n (i.e. $l/600$, $l/450$, $l/300$ and $l/150$) and the force F_r when cracking occurs, is calculated. The second method defines the flexural toughness as the surface B_n beneath the load-deflection curve until a prescribed deflection (i.e. $l/300$ and $l/150$) is reached. The conventional flexural strength is denoted as $f_{f,n}$ and is given by equation (1).

$$f_{f,n} = \frac{B_n n}{bh^2} \tag{1}$$

All values of D_n and $f_{f,n}$, derived from four experimental four-point bending tests, performed by Van Gysel, are summarized in Table 2. These values were derived from the load-deflection curves (solid black lines) displayed in Fig. 2.

Table 2. Parameters to characterize the flexural toughness of SFRC

Prism nr.	D_{600} *[-]*	D_{450} *[-]*	D_{300} *[-]*	D_{150} *[-]*	$f_{f,300}$ *[MPa]*	$f_{f,150}$ *[MPa]*
1	1.39	1.44	1.37	1.09	7.03	6.93
2 *	-	-	-	-	-	-
3	1.37	1.36	1.16	-	6.39	-
4	1.18	1.15	1.06	0.74	6.69	4.92
Mean	1.31	1.32	1.20	0.92	6.70	5.93

* failure at a central deflection of about 0.75 mm.

Regarding the values of Table 2, a general remark should be made: although all four-point bending were precisely executed as described in the Belgian standard, the results were not sufficient for all prisms to determine the flexural toughness indices. Without guarantee in advance of achieving useful test results, a significant number of tests is needed to determine the scatter of flexural toughness parameters of SFRC.

3 Numerical Analysis

To avoid time-consuming test procedures and to assure a useful load-deflection curve, a solution of the problem stated earlier can be found through numerical simulations. By the continuous improvement of computer hardware, computation times of complex numerical problems are reduced in a very fast way and new approaches to elaborate a finite element simulation tool for SFRC has become feasible: it takes about 3 hours[1] to perform a complete non-linear analysis with the software program Atena Science[2].

3.1 FE-Model

Due to the three dimensional distribution of fibers within a prescribed volume, an accurate simulation cannot be obtained by a much simpler two-dimensional numerical model and thus the whole prism was divided in tetrahedral (i.e. 3D solid) quadratic elements with a size of 15 mm, representing the coarseness of the concrete matrix. In the crack model, the option ‘rotating’ was chosen and a crack bandwidth of 15 mm was assumed. A simulated specimen without fibers was used to calibrate the input parameters for concrete. It followed that a small adaptation

[1] Duration of a single analysis with one central processing unit of 2,67 Ghz.

[2] AtenaWin v4 Statics module, developed by Cervenka Consulting.

of the input tensile strength and fracture energy, with respect to the values given in Table 1, was necessary.

3.2 Distribution of Fibers

When adding 40 kg/m³ of DRAMIX® ZC 60/.80 HC fibers to a prism with a volume of 150 x 150 x 600 mm³, about 2280 fibers should be drawn as a straight line. Considering this to be impossible to do by hand, a small software program was developed that reads all input data from the calculation file of Atena and adds an amount of fibers automatically, based on fiber properties and the shape of the volume (i.e. a rectangular prism). This volume is formed by a cross-section in a *yz*-plane that is extruded along the *x*-axis, in the longitudinal direction of the prism.

Although the main objective of the program is to add fibers in a volume in such way that the distribution of fibers approaches the real configuration, the presence of coarse aggregates was not taken into account at this stage. The orientation of a single fiber is represented by the angle θ with respect to the longitudinal axis x of the prism (i.e. the x-axis). Based on the experimental investigation of Laranjeira et al. [10], a Gaussian distribution was assumed for the orientation angle θ, which is defined by an orientation number η_θ, given by (2).

$$\eta_\theta = \frac{1}{N}\sum_{i=1}^{N}\cos\theta_i \tag{2}$$

The average θ_m and standard variation $\sigma(\theta)$, both in radians, of the normal distribution can be calculated by equations (3) and (4).

$$\theta_m = \arccos(\eta_\theta) \tag{3}$$

$$\sigma(\theta) = \frac{\pi}{2}\eta_\theta(1-\eta_\theta) \tag{4}$$

A second objective was to implement the pull-out model for hooked-end fibers developed by Van Gysel [8]. The pull-out response is mainly affected by the inclination of the pull-out force and the embedded length. So, for each fiber within the concrete volume a unique load-slip diagram exists, but implementing this would lead to a too complex model. To solve this problem, both the range of orientation angles and the embedded lengths are divided in discrete intervals.

In this way, the orientation angle θ is divided in six intervals from 0 to 90 degrees and each orientation angle interval is divided in another six sub-intervals going from 5 mm to half the fiber length, with an interval width of 5 mm. The discretization matrix for such a division is displayed in Table 3.

Table 3. Discretization matrix and corresponding material numbers

	Embedded length l_f					
Orientation interval [°]	5 mm	10 mm	15 mm	20 mm	25 mm	30 mm
[0;7.5]	1	2	3	4	5	6
[7.5;22.5]	7	8	9	10	11	12
[22.5;37.5]	13	14	15	16	17	18
[37.5;52.5]	19	20	21	22	23	24
[52.5;67.5]	25	26	27	28	29	30
[67.5;90]	31	32	33	34	35	36

Thus, when sampling a specimen that includes fibers, each fiber is linked to a material number between 1 and 36 within the FE-model, taking into account the normal distribution of the orientation angle θ and a uniform distribution of the embedded length l_f.

To model the pull-out response of a fiber, an equivalent stress-strain diagram was defined based on a load-slip diagram. The equivalent strain ε_{eq} is calculated as the ratio between the slip s and the total length of the fiber l_{tot} and the equivalent stress as the ratio between the pull-out-force F_u and the cross-section of the fiber A_f. As a result of this approach, cable elements were needed to assure that a constant stress would appear over the whole length of the fiber when a strain was locally induced due to crack formation. Fig. 1 shows a sampled FE-model with fibers.

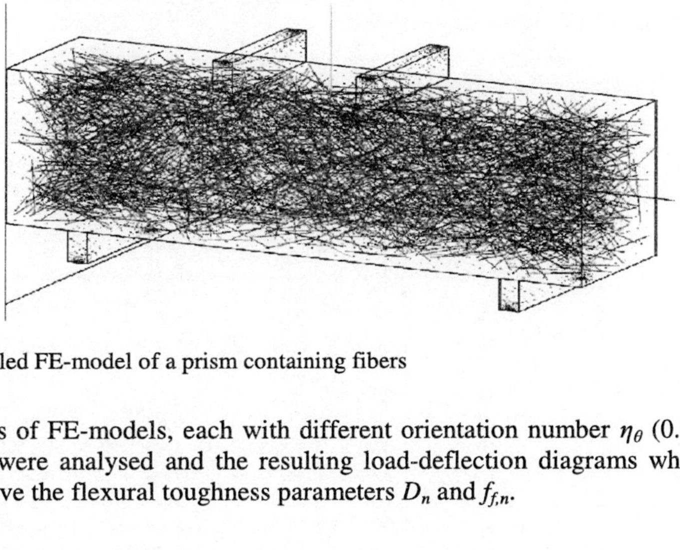

Fig. 1. Sampled FE-model of a prism containing fibers

Three series of FE-models, each with different orientation number η_θ (0.80, 0.75 and 0.70), were analysed and the resulting load-deflection diagrams where then used to derive the flexural toughness parameters D_n and $f_{f,n}$.

4 Results

Thirty different SFRC prisms were sampled with the software tool and analysed with Atena. All load-deflection curves are displayed in Fig. 2.

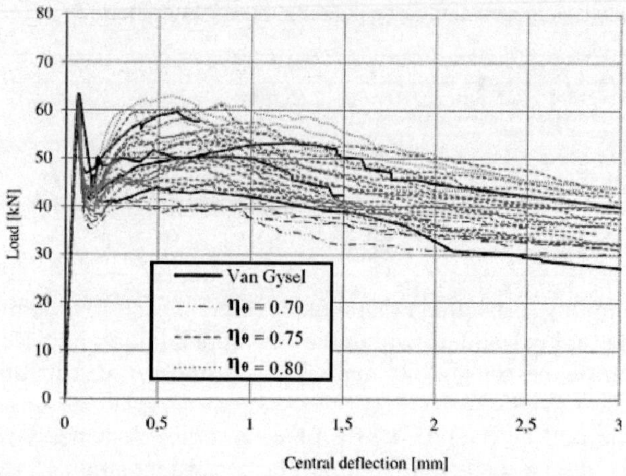

Fig. 2. Load deflection curves as a function of the orientation number η_θ

When an orientation number of 0.80 (5 simulations) was assumed, a systematic overestimating of residual strength occurred, while an orientation number of 0.70 (5 simulations) appears to be too conservative. A good correlation with the experimental results of the four-point bending tests by Van Gysel [8] was found with η_θ equal to 0.75. All results, according to an orientation number of 0.75 (20 simulations), are summarized in Table 4.

Table 4. Flexural toughness parameters for $\eta_\theta = 0.75$

Prism nr.	D_{600} [-]	D_{450} [-]	D_{300} [-]	D_{150} [-]	$f_{f,300}$ [MPa]	$f_{f,150}$ [MPa]
1	1.52	1.42	1.26	1.07	6.79	6.16
2	1.24	1.16	1.12	0.88	5.77	5.28
3	1.43	1.30	1.12	0.88	6.36	5.52
4	1.43	1.42	1.32	*	6.52	*
5	1.37	1.37	1.21	*	6.21	*
6	1.49	1.43	1.27	1.12	6.73	6.32
7	1.21	1.19	1.10	0.98	5.87	5.35
8	1.21	1.15	1.08	0.87	5.77	5.15
9	1.30	1.27	1.21	0.96	6.11	5.65
10	1.36	1.34	1.21	1.07	6.27	5.9
11	0.13	0.13	1.08	*	5.81	*
12	1.31	1.19	1.06	0.84	5.92	5.25
13	1.57	1.46	1.32	*	7.04	*
14	1.22	1.19	1.10	0.94	5.8	5.23
15	1.46	1.31	1.09	*	6.42	*
16	1.63	1.51	1.37	1.18	7.29	6.46
17	1.56	1.57	1.43	1.18	7.33	6.81
18	1.46	1.37	1.27	1.03	6.66	6.13
19	1.33	1.32	1.21	1.00	6.47	5.89
20	1.35	1.23	1.12	1.02	6.01	5.6
Average	1.33	1.27	1.20	1.00	6.36	5.78
Std. Dev.	0.31	0.29	0.11	0.11	0.50	0.51
C.O.V.	23%	23%	9%	11%	8%	9%

The mean values of the experimental results (Table 2) and the simulated tests (Table 3) are in good agreement.

5 Conclusions

Finite-element simulations with randomly distributed embedded cable elements, representing steel fibers, were able to predict the bending behavior of small steel fiber reinforced concrete prisms used to perform four-point bending tests according to the Belgian standard NBN B15-238 (1992).

Three different orientation numbers were compared to each other and the following conclusion can be made: the proper orientation number for a small prism subjected to bending should be in a range between 0.70 and 0.80.

Eventually, a good correlation was found for about twenty samples with an orientation number equal to 0.75 and the scatter of the flexural toughness parameter D_n and $f_{f,n}$ could be estimated accurately. For the values of D_{600} and D_{450}, a coefficient of variation equal to 23 % was found and for the values D_{300}, D_{150}, $f_{f,300}$ and $f_{f,150}$ the coefficient of variation was around 9 %. This scatter is inherent to the randomness of the fiber position and orientation.

References

[1] Naaman, A.E., Shah, S.P.: Pull-out Mechanism in Steel Fiber-Reinforced Concrete. Proc. ASCE, Journal of the Structural Division 102(ST 8), 1537–1548 (1976)

[2] Stang, H., Li, Z., Shah, S.P.: Pull-out Problem: Stress versus Fracture Mechanical Approach. Journal of Engineering Mechanics, ASCE 116(10), 2136–2150 (1990)

[3] Van Gysel, A.: A pullout model for hooked end steel fibres. In: Reinhardt, H.W., Naaman, A.E. (eds.) 3th RILEM Workshop on HPFRCC, pp. 351–359 (1999)

[4] Taerwe, L., Van Gysel, A.: Influence of steel fibres on design stress-strain curve for high-strength concrete. ASCE Journal of Engineering Mechanics 122(8), 695–704 (1996)

[5] Stroeven, P., de Haan, Y.M., Bouter, C., Shah, S.P.: Pull-out Tests of Steel Fibres. In: Swamy, R.N. (ed.) Testing and Test Methods of Fiber Cement Composites, pp. 345–353. The Construction Press, UK (1978)

[6] Chanvillard, G., Aïtcin, P.-C.: On the Modelling of the Pull-out Behaviour of Steel Fibers. In: Reinhardten, H.W., Naaman, A.E. (eds.) High Performance Fiber Reinforced Cement Composites, pp. 388–406. RILEM Proceedings 15, E & FN Spon, London (1992)

[7] Naaman, A.E., Najm, H.: Bond-Slip Mechanisms of Steel Fibers in Concrete. ACI Materials Journal 88(2), 135–145 (1991)

[8] Van Gysel, A.: Studie van het uittrekgedrag van staalvezels ingebed in een cementgebonden matrix met toepasing op staalvezelbeton onderworpen aan buiging (in Dutch), PhD thesis, Magnel Laboratory for concrete research, Ghent University, Belgium (2000)

[9] Jansson, A.: Analysis and design methods for fibre reinforced concrete: a state-of-the-art report. In: Dep. of Civil & Environmental Engineering, Div. of Structural Engineering/Concrete structures, vol. 16, p. 196. Chalmers University of Technology, Göteborg (2007)

[10] Laranjeira, F., Grünewald, S., Walraven, J., et al.: Characterization of the orientation profile of steel fiber reinforced concrete. Materials and Structures, RILEM, 1–19 (November 2010)

Multifunctional Carbon Black Engineered Cementitious Composites for the Protection of Critical Infrastructure

M. Li, V. Lin, J. Lynch, and V.C. Li

Department of Civil and Environmental Engineering, University of Michigan, USA

Abstract. A new generation of self-sensing structural materials that can provide spatially distributed sensing of material damage is being developed in the context of structural health monitoring and protection of deteriorating concrete infrastructure. Engineered Cementitious Composites (ECC) doped with carbon black (CB) nano-particles are proposed as highly damage-tolerant materials whose electrical properties can be correlated to strain and cracking. Specifically, this study investigated the effect of CB dosage on the CB-ECC rheological, mechanical and electrical properties. The results show the promise of CB-ECC as a new multifunctional material that combines damage tolerance with damage self-sensing capabilities; both contribute to infrastructure safety and serviceability.

1 Introduction

The deterioration of large-scale infrastructure is a critical national problem that, left unchecked, will continue to lead to catastrophes like the I-35W Bridge collapse under service loads [1]. Deterioration will also increase structural vulnerability to man-made and natural hazards, if not detected and repaired in a timely fashion. This problem is further compounded by the shrinking of economic and natural resources needed to maintain civil infrastructure at an alarming rate [2].

Two major contributors to this compelling problem are inadequate material performance and the limitations of current structural health monitoring (SHM) approaches. Concrete materials are susceptible to cracking and fracture failure due to their inherent brittleness, which also lead to other common deterioration mechanisms such as embedded steel corrosion. Transverse cracking in bridge decks is one of the dominant deterioration problems in many States in the U.S. [3], Freeze-thaw cycles of water in cracks and leakage of water to supporting structures can reduce the bridge service life. Cracks also accelerate chloride penetration into the concrete and initiate corrosion of reinforcements, especially in marine environments or regions where deicing salts are applied in winter. The U.S. spends about $8.3 billion annually to directly address corrosion in highway bridges [4]. Indirect

G.J. Parra-Montesinos, H.W. Reinhardt, and A.E. Naaman (Eds.): HPFRCC 6, pp. 99–106.

costs related to traffic delays and productivity losses exceed 10 times the direct cost of maintenance/repair/rehabilitation.

While detecting damage or deterioration in concrete structural components is critical, current infrastructure management practices have relied on bi-annual visual inspections that can be subjective and limited to accessible locations [5]. While great technological advances have been made in recent years on many fronts in the SHM field, there still remain very few implementations of SHM systems in operational structures [6]. The key flaws include: indirect damage sensing that needs physics-based models to correlate structural response measurements to damage state; point-based sensors (e.g. strain gages, thermal couples, accelerometers) that cannot accurately identify spatially distributed damage; and tethered monitoring systems that require wires, often kilometers long, to connect sensors to a central data repository, leading to expensive installation and maintenance.

This research aims to address these limitations through the realization of multifunctional materials with intrinsic damage tolerance as well as distributed damage self-sensing capacity; both characteristics are essential for enhancing infrastructure resilience, reducing maintenance needs and life cycle impacts, and improving user experience. The beauty of this self-sensing multifunctional material is twofold. First, it serves as a major material component for structures to significantly improve resistance to deterioration and damage (*e.g.* cracking, corrosion and cover spalling), by eliminating concrete brittleness and intrinsically controlling crack width to a small scale. Second, it offers the capability of distributed and direct sensing of damage wherever the material is located, by exhibiting changes in electrical properties due to mechanical (*e.g.,* strain and multiple cracking) stimulus.

Resistivity of cementitious materials has been explored as a sensing functionality for hydration monitoring, composition assessment, and currently as an emerging method for measuring mechanical state. Cementitious materials are semiconductors with bulk resistivity $10^3 \sim 10^8$ Ω-cm. Under an applied steady electric field, the ions in pore water are mobilized to create current. Piezoresistivity is defined as strain induced resistivity change. Seminal work by Chung [7] explored multi-point probing of plain cement matrix and cement pastes with conductive fibers, and revealed that the electrical properties of cementitious materials can be accurately correlated to their mechanical behavior. Electrical Impedance Spectroscopy (EIS) characterization of fiber reinforced cementitious composites (FRC) was first reported by Peled et al. [8], and illuminated the electrical properties of composite phases (matrix versus fiber). While previous work focused on establishing the piezoresistive behavior of FRCs with conductive fibers, this study targets at simultaneously achieving strain-hardening, crack width control and enhanced piezoresistive behavior within a single material platform. The strain-hardening response corresponds to an extended damage process (inelastic stage), which provides damage tolerance while allowing detection of microcrack damage long before unstable localized fracture occurs, so that early warning is possible. This is achieved through incorporating conductive carbon black nano-particles into the Engineered Cementitious Composites (ECC) system with non-conductive polymeric fibers to tailor composite rheological, mechanical and electrical properties.

2 Materials and Processing

The CB-ECC composite material design is achieved by ingredients selection, rheology control during processing, and microstructure tailoring. The mixture proportions of three CB-ECC designs are summarized in Table 1. The dry ingredients consisted of Type I Portland cement (C), fine silica sand (S) with average and maximum grain sizes of 110 μm and 200 μm, normal Class-F fly ash (FA), REC-15 poly-vinyl-alcohol (PVA) fibers, and carbon black nano-particles (CB). The PVA fiber was 12 mm long, 39 μm in diameter, and surface-coated with 1.2% by weight of a proprietary hydrophobic agent. CB with an average particle size of 30 nm and bulk resistivity of approximately 10^{-1} Ω-cm was incorporated at increasing ratios of weight to total cementitious ingredients (cement plus fly ash) of 0.25%, 0.5% and 1%.

Table 1. Mix compositions by weight ratio of cement

#	C	FA	S	CB (%)	W	SP	PVA (vol%)	Initial Resistivity (ohm-cm)	Gage Factors Elastic	Gage Factors Inelastic
CB-ECC -0.25%	1.0	1.2	0.8	0.55	0.60	0.013	2.0	-	-	-
								6.489×10^6 (1yr)	14.9 (1yr)	187.3 (1yr)
CB-ECC -0.5%	1.0	1.2	0.8	1.1	0.60	0.016	2.0	5.597×10^4 (28d)	44.4 (28d)	376.5 (28d)
								5.623×10^6 (1yr)	21.5 (1yr)	696.0 (1yr)
CB-ECC-1.0%	1.0	1.2	0.8	2.2	0.60	0.022	2.0	9.579×10^4 (28d)	23.8 (28d)	202.2 (28d)
								1.294×10^6 (1yr)	133.6 (1yr)	672.6 (1yr)

The mixtures were prepared in a Hobart mixer with dry ingredients mixed first. Water (W) and super-plasticizer (SP) were then added and mixed until homogeneous cement slurry with controlled rheology was formed. PVA fibers were then slowly added and mixed until uniform dispersion. The mixtures were cast into 152.4 mm × 76.2 mm × 12.7 mm tensile specimens, covered with plastic sheets, and demolded after 24 hours. They were first cured in plastic bags at 95 ± 5% RH, 20 ± 3°C for six additional days and then left to cure in laboratory air under at 20 ± 5% RH and 20 ± 3°C until the age of 28 day or 1 year for testing.

The sample specimens are shown in Fig. 1. Specimens with higher carbon content were darker colored. Under optical microcopy and high magnification scanning electron microscopy, the matrix phase appears homogeneously black with no CB aggolomeration found, indicating good CB dispersion (Fig. 2).

Introducing a large volume of 30 nm porous CB into the ECC system created significant processing challenges in terms of material workability and dispersion of PVA fibers. To address this, the amount of water and superplasticizer was optimized to achieve a homogenous slurry with plastic viscosity ranging from 7.67 Pa.s to 10.25 Pa.s and yield stress ranging from 5.47 Pa to 128 Pa, as measured by a rheometer, which allowed proper fiber dispersion and workability. To examine fiber dispersion, florescence microscopy was adopted to detect PVA fibers and their distribution at the specimen cross section. Through controlling plastic viscosity above the minimum requirement [9], good fiber dispersion was achieved, indicated by the fiber dispersion coefficients α that are in the range of 0.79-0.83 ($0 < \alpha \leq 1$; $\alpha = 1$stands for perfectly uniform fiber dispersion). The α was determined based on an image analysis method applied to the flurescence image [9, 10]. However, reduction of total fiber number at the cross section was found for CB-ECCs compared with ECC with no CB, meaning that the high viscosity led to larger entrapped air pores that reduced the net cross section area.

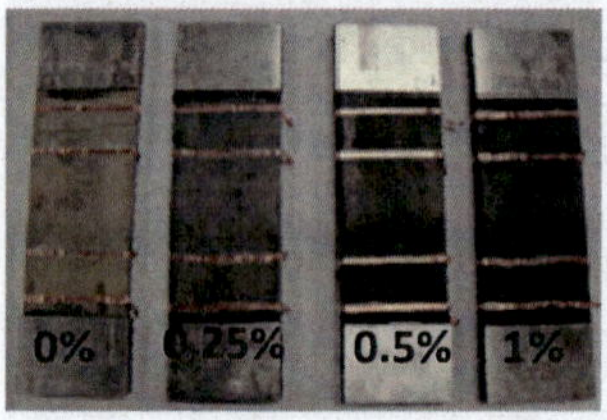

Fig. 1. CB-ECC Specimen

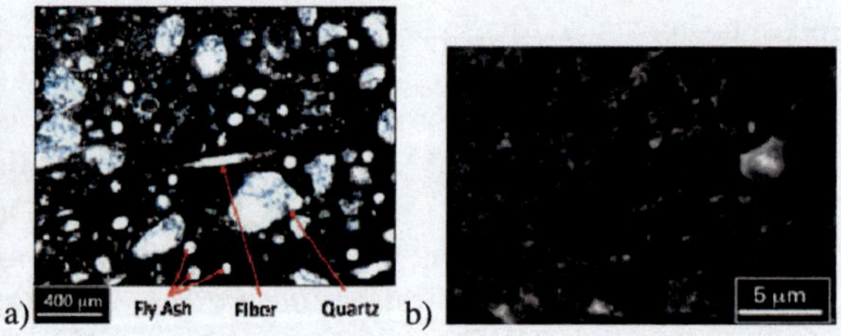

Fig. 2. CB-ECC-1.0% (a) Optical Microscopy (b) Scanning Electron Microscopy

3 Mechanical and Electrical Properties

The composite tensile stress-strain response was tested by an Instron machine. Two LVDTs were attached to the specimen with gage length of 100 mm (Fig. 3) to measure tensile strain. The load was applied under displacement control at a rate of 0.05 mm/min during the material elastic stage and 0.5 mm/min during the inelastic stage.

The composite resistivity was measured using the four-point electronic impedance spectroscopy (EIS) method (Fig. 3). EIS is an electrical testing technique that can be used to characterize the frequency dependent electrical properties of materials and their interfaces. Four electrical contacts, i.e. copper tapes, were applied onto the surface of each specimen with highly conductive silver colloidal paste. The outer electrodes passed current into the specimen, while the inner electrodes measured the in-situ voltage within the gage length. The resistivity measurements were carried out using the Solartron 1260 impedance/gain-phase analyzer, in which the excitation voltage was set at 100 mV.

It was observed that the bulk resistivity decreased as CB content increased, which was most prominent at the lower input current frequency (Fig. 4). This identified the effectiveness of CB in increasing bulk material conductivity, and the optimal range of excitation current frequency for sensing. Consequently, the input current was fixed at 100 Hz for the composite piezoresistive response testing.

Fig. 5 shows the mechanical and piezoresistive response of CB-ECCs at 28d and 1yr ages. The tensile stress-strain curve contains three stages: (1) elastic before multiple cracking occurs, (2) strain hardening accompanied by multiple cracking process, and (3) tension softening corresponding to failure localization at one of the cracks. When the final failure occurred outside the gage length for some of the specimens, the tension softening part of the stress-strain curve exhibited a strain reduction due to the stress relaxation of the segment within gage length.

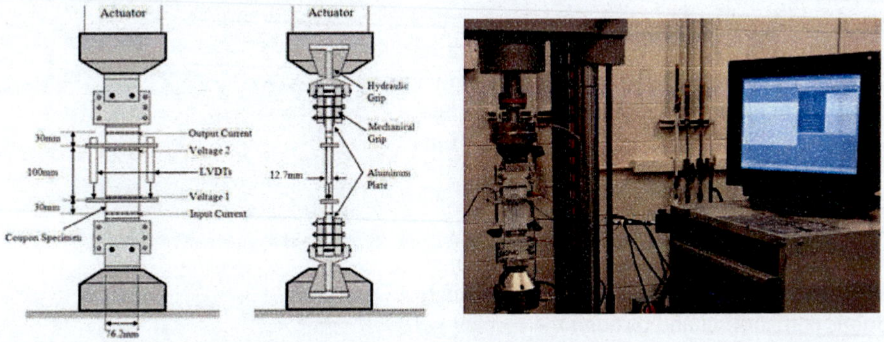

Fig. 3. 4-point EIS for piezoresistivity measurement under uniaxial tensile loading

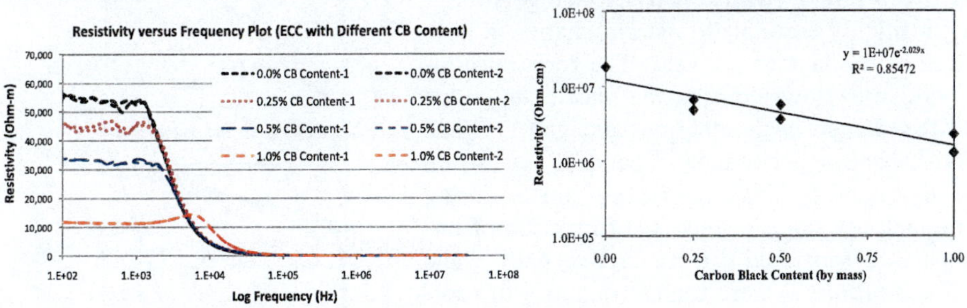

Fig. 4. 1-year-old CB-ECC Bulk Resistivity (a) vs. Frequency; (b) vs. Carbon Black Content

While all CB-ECCs showed strain hardening and multiple cracking behavior, their tensile strain capacity was in the rage of 0.8-1.8%, which were lower than typical ECCs with strain capacity of 3-5%. This reduction was due to two reasons: (1) the reduction in average crack width during multiple cracking process. It was found that the average crack width of CB-ECCs was 30-40 μm (during loading), while the average crack width of a standard ECC (ECC M45) is 60-70 μm. Single fiber pullout tests revealed enhancement of fiber/matrix interfacial bond due to the presence of CB nano-particles, leading to the reduction of average crack width. (2) The reduction of number of fibers, revealed by fluorescence microscopy at the final failure ("weakest") cross section, had led to a reduction of fiber bridging capacity. This is caused by the increase of plastic viscosity at fresh state when a large volume of porous and nano-size CB was incorporated. The detailed discussion on the experimental measurement of micro-parameters (e.g. interfacial bond), rheology and fiber dispersion, and their effects on the hardened CB-ECC composite tensile properties, will be included in a separated journal paper.

The tensile strain capacity of CB-ECCs, despite the reduction, is approximately 100 times larger than that of concrete or FRCs, and one-two order magnitude higher than shrinkage strain of cementitious materials. Shrinkage is considered to be the major cause of concrete deck cracking [3], which can be suppressed by the tensile ductility of CB-ECC [11]. Furthermore, the reduction in average crack width is favorable for resisting chloride penetration through the CB-ECC cover to reach the reinforcement and initiate corrosion [12]. It also promotes self-healing capacity [13, 14]. Through achieving tensile ductility as well as controlling crack width to a smaller scale, the CB-ECCs possess damage tolerance for structural components to remain "healthy" under non-severe service conditions.

The bulk resistivity of CB-ECCs was two-order magnitude higher at 1 year than 28 days. This increase in resistivity with hydration process was due to the self-desiccation, and the change in concentration ions and their mobility through a tortuous path around and through the cement gel.

The CB-ECCs exhibited prominent piezoresistive behavior, with resistivity change strongly positively correlated to strain change at both ages of 28 days and 1 year. The resistivity increase was slower during the elastic stage due to the change in the energy band gap between its valence and conductive band at low strain levels[20], and became significantly faster during the inelastic stage because the formation of microcracks separated the conductive path. Gage factors (Table 2) were used to quantify the correlation, defined as $\Delta R/R_G\varepsilon$, where ΔR is the change in resistivity caused by strain, R_G is the resistivity of unreformed gage, and ε is strain. The results confirmed that increasing CB content was effective in increasing gage factors of CB-ECC at elastic stage due to the increase in bulk material conductivity. However, the gage factor increasing trend was not clear for the inelastic stage based on the preliminary data. This suggested that multi-cracking pattern and process played a critical role on gage factors in the inelastic stage, at which the increase of CB-ECC resistivity under uniaxial tension should be due to the combined effects of matrix elastic straining, formation of new cracks, enlargement of previously-formed cracks, and crack propagation (e.g. an edge crack turns into a full crack). While the multiple cracking process depends on natural flaw distribution within the specimen, the randomness of flaws entrapped during CB-ECC processing can cause variation in the resistivity-strain curves.

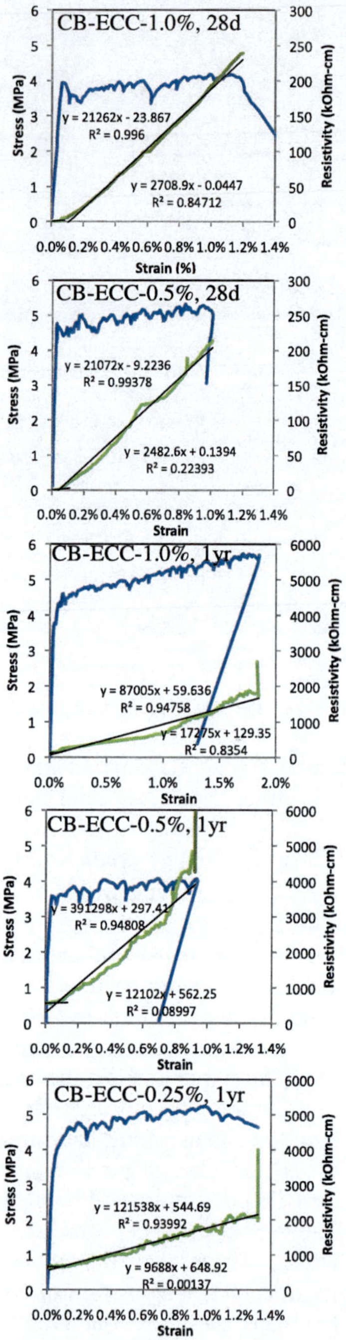

Fig. 5. Tensile stress-strain curve and resistivity-strain curve of CB-ECCs

To further understand the effect of cracking behavior on the resistivity and strain changes,

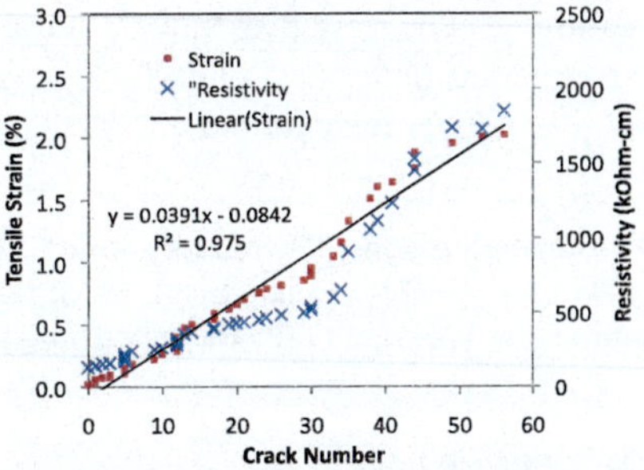

Fig. 6. Correlation between tensile strain, resistivity and crack number

78 images of the cracking pattern of the CB-ECC-1.0% specimen during unaxial tensile testing was taken at fixed time intervals. Each image was analyzed to count crack number at corresponding strain level, based on the assumption that each edge crack was considered as a single crack. The relation between crack number, tensile strain and resistivity is plotted in Fig. 6. Approximately linear relation was shown between tensile strain and crack number, with a slope suggesting 39 μm average crack width during multiple-cracking process, which corresponds well with the observation using a microscope during testing. The correlation between resistivity and crack number is non-linear, indicating a non-uniform multiple cracking process, e.g. unequal width of microcracks formed at lower strain vs. higher strain levels, full crack vs. edge crack, and networking of cracks. With more tests conducted in future to establish the correlation between resistivity with cracking pattern, such correlation can be used for damage sensing.

4 Conclusions

This study shows the promise of CB-ECC as a multifunctional infrastructure material with damage tolerance and self-sensing capacities, both are important for the protection of critical infrastructure. By incorporating CB nano-particles into the ECC system through rheology control, the CB-ECC achieved adequate workability, close-to-uniform PVA fiber and carbon black dispersion, reduced bulk resistivity by one magnitude, strain hardening behavior with tensile strain capacity of 0.8-1.8% that is essential for resisting shrinkage cracking, and reduced crack width to 30-40 μm during loading state that provides resistance to chloride penetration. All CB-ECCs exhibited prominent piezoresistive behavior, with increasing resistivity with increasing tensile strain and cracking number, indicating the potential of CB-ECC for strain and damage sensing.

CB-ECC's lower tensile strain capacity than typical ECCs was due to the larger entrapped air pores that reduced the fiber bridging capacity at the final failure cross section, and the tighter crack width as a result of CB's effect on improving fiber/matrix interfacial bond. To reduce larger pores, the optimization of material ingredients, proportion and rheology is suggested for future studies.

While the gage factors have been determined for both the elastic and inelastic stages of CB-ECCs, more experimental studies are necessary to accurately establish the correlation between resistivity change and strain as well as cracking pattern, and to fully understand the effect of the multiple cracking process on the variation in gage factors. This includes characterizing the resistivity change under a single crack opening scenarios with different stress-crack opening relationship, and the resistivity change under various multiple cracking processes caused by the randomness of flaw distribution. Future studies will also focus on the effect of environmental factors on gage factors, and optimizing CB-ECC composite design to achieve robust mechanical and piezoresistive behavior.

Acknowledgement. The authors would like to gratefully acknowledge the U.S. Department of Commerce, National Institute of Standards and Technology (NIST) Technology Innovation Program (TIP) for supporting this research work.

References

[1] Wardhana, K., Hadipriono, F.C.: Analysis of recent bridge failures in the United States. Journal of Performance of Constructed Facilities 17(3), 144–150 (2003)
[2] Anderson, J.: Cities debate privatizing public infrastructure. The New York Times (2008)
[3] Saadeghvaziri, M.A., Hadidi, R.: Transverse Cracking of Concrete Bridge Decks: Effects of Design Factors. ASCE Journal of Bridge Engineering 10(5), 511–519 (2005)
[4] Bavarian, B., Reiner, L.: Migrating Corrosion Inhibitors for Steel Rebar in Concrete. Materials Performance, 3–5 (2003)
[5] Rolander, D.D., Phares, B.M., Graybeal, B.A., Moore, M.E., Washer, G.A.: Highway bridge inspection: State-of-the-practice survey. Transportation Research Record 1749(1), 73–81 (2001)
[6] Derriso, M.M., Olson, S.E., DeSimio, M.P., Pratt, D.M.: Why are There Few Fielded SHM Systems for Aerospace Structures? In: 6th International Workshop on Structural Health Monitoring, Stanford, CA (2007)
[7] Chung, D.: Piezoresistive Cement-Based Materials for Strain Sensing. Journal of Intelligent Material Systems and Structures (13), 599 (2002)
[8] Peled, A., Torrents, J.M., Mason, T.O., Shah, S.P., Garboczi, E.J.: Electrical Impedance Spectra to Monitor Damage during Tensile Loading of Cement Composites. ACI Materials Journal 98(4), 313–322 (2001)
[9] Li, M.: Multi-Scale Design for Durable Repair of Concrete Structures, Ph.D Thesis, University of Michigan (2009)
[10] Lee, B.Y., Kim, J.K., Kim, J.S., Kim, Y.Y.: Quantitative Evaluation Technique of Polyvinyl Alcohol (PVA) Fiber Dispersion in Engineered Cementitious Composites. Cement and Concrete Composites 31(6), 408–417 (2009)
[11] Li, M., Li, V.C.: Influence of Material Ductility on the Performance of Concrete Repair. ACI Materials Journal 106(5), 419–428 (2009)
[12] Sahmaran, M., Li, M., Li, V.C.: Transport Properties of Engineered Cementitious Composites Under Chloride Exposure. ACI Material Journal 104(6), 604–611 (2007)
[13] Li, M., Li, V.C.: Cracking and Healing of Engineered Cementitious Composites under Chloride Environment. ACI Materials Journal (November 2010) (accepted)
[14] Kan, L.-L., Shi, H.S., Sakulich, A.R., Li, V.C.: Self-Healing Characterization of Engineered Cementitious Composites. ACI Materials Journal 107(6), 617–624 (2010)

Effects of Fiber Dispersion and Flaw Size Distribution on the Composite Properties of PVA-ECC

R. Ranade, M.D. Stults, B. Lee, and V.C. Li

Department of Civil and Environmental Engineering,
University of Michigan, Ann Arbor, USA

Abstract. Over the last decade, Engineered Cementitious Composites (ECC) containing Poly-Vinyl Alcohol fibers (PVA-ECC) have been extensively researched and used in a wide variety of structural applications utilizing the composite's high tensile ductility and durability. Fiber and flaw size distributions of PVA-ECC, which greatly affect its composite properties, have been studied in this research using fluorescence imaging and optical microscopy. Statistical analysis revealed a double-Gaussian best-fit distribution showing possible non-conservative preferential alignment of fibers in dogbone specimens along the longitudinal axis of the specimen. Maximum flaw sizes at various sections ranged from 0.6 to 6.3 mm with a combination of lognormal and Gaussian distributions best-fitting the observed data. The effects of the above statistical distributions on composite stress-strain behavior are studied using micromechanics and scale-linking models. The predicted composite properties are then compared with the experimental data of the direct uniaxial tension tests on PVA-ECC dogbone specimens.

1 Introduction

Engineered Cementitious Composites (ECC) with Poly-Vinyl Alcohol (PVA) fibers are materials with the highest tensile ductility of 2-6% [1] in the class of High Performance Fiber Reinforced Cementitious Composites (HPFRCC). ECC are micromechanically designed materials [2] and their composite performance is greatly affected by flaw size distribution, and fiber number, dispersion, and orientation distribution. For instance, the load carrying capacity of the composite can drop by up to 20% [3], when the fiber distribution is changed from 1D uniform alignment to 3D random distribution. Similarly, the ductility of ECC can be improved by more than 100% [4] through incorporation of artificial flaws of appropriate size range. The fiber and flaw size distributions, therefore, greatly affect the composite properties of ECC, especially tensile strength and ductility. The design of ECC utilizes scale-linking models to predict the composite behavior based on the changes in material microstructure [5]. These models make theoretical assumptions for fiber and

G.J. Parra-Montesinos, H.W. Reinhardt, and A.E. Naaman (Eds.): HPFRCC 6, pp. 107–114.

flaw size distributions, as the observational data for these distributions is very limited. Recent studies in authors' laboratory have shown deviation of predicted tensile stress-strain curves from the test results (especially with dogbone shaped specimens), which questions the validity of theoretical fiber and flaw size distribution assumptions. The research presented here, along with a series of similar subsequent studies, attempts to quantify these distributions through detailed observations and analysis. Fluorescence imaging of fibers [6] has emerged as a possible technology for quality control of HPFRCCs in terms homogeneity of fiber dispersion, and it has been used in this research to determine the fiber distribution. There are, however, certain limitations to this technique, which have also been discussed here. Observations of fiber distribution and flaw size distribution at numerous cross sections of dogbone specimens, along with their effects on composite tensile behavior of PVA-ECC, are presented in following sections.

2 Experimental Investigation

The material investigated in this research is an Engineered Cementitious Composite (ECC) using short randomly distributed Poly-Vinyl Alcohol (PVA) fibers in a cementitious matrix containing high volumes of Class F Fly Ash. The mix proportions of ECC used in this research are given in Table 1. A detailed description of materials can be found in Yang et al. (2009) [7].

Table 1. Mix proportions of PVA-ECC by weight of cement

Type I Cement	Class F Fly Ash	F-110 Silica Sand	Tap Water	HRWRA	PVA Fiber*
1	2.8	1.4	0.988	0.016	0.080

* Length (L_f) = 12 mm, diameter (d_f) = 40 μm, Denstiy (ρ_f) = 1.3 g/cm^3, Strength (σ_f) = 1 GPa, Modulus (E_f) = 22 GPa, oil coating = 1.2% by weight of fiber, and fiber volume compared to mix volume (V_f) = 2%.

The specimens for tension testing are dogbone shaped specimens recommended by Japan Society of Civil Engineers (JSCE) [8] for standardized testing of HPFRCCs. The specimen geometry is shown in Fig. 1a. For this research, 3 such dogbones were prepared and tested. ECC mix in its fresh plastic state is poured at the middle of the specimen mold, which is constantly vibrated using a vibration table and as a result, the material slowly flows towards the ends. Each mold is vibrated at moderate energy for about 3 minutes before stowing away for curing.

In the setup for direct uniaxial tension tests, the dogbone specimens are gripped through wedge action on its slanting edges. Sufficient degrees of freedom are provided in the grips to ensure the application of almost pure uniaxial tension along the longitudinal axis. The tensile tests on dogbones were carried out at 0.005 mm/s (0.0002 in/s) using a displacement controlled test system with maximum load capacity of 10 kN (2.2 kips). The strain is computed from the extension of the specimen measured by two LVDTs mounted diagonally opposite to each other.

The cross-sections for microscopic observation were prepared after performing uniaxial tensile tests on all the three dogbone specimens. Twenty cross-sections of the material within the gage length were cut perpendicular to the loading direction for each dogbone specimen for the maximum flaw size measurement at each section, with area 30 x 12.7 mm^2 (1.18 x 0.5 $inch^2$) and about 10 mm (0.4 in) thick (Fig. 1b). For each dogbone specimen, three sections were selected: one section very close to the final failure plane (denoted by "F"), one section very close to a crack with average crack width in the mid-gage ("M") area, and one section from the gripping ends ("G"). The cross-sections of these 9 sections (3 sections per specimen) were further ground for observation under a fluorescence microscope for fiber distribution. PVA fibers are known to fluoresce and emit green light in the range of 440 to 460 nm when excited by ultra-violet incident light of 370 to 390 nm wavelength. The field of view of the fluorescence microscope is about 2 x 2 mm^2 (0.08 x 0.08 in^2) and each cross section is sampled by 9 images taken at the locations (small squares) shown in Fig. 1b.

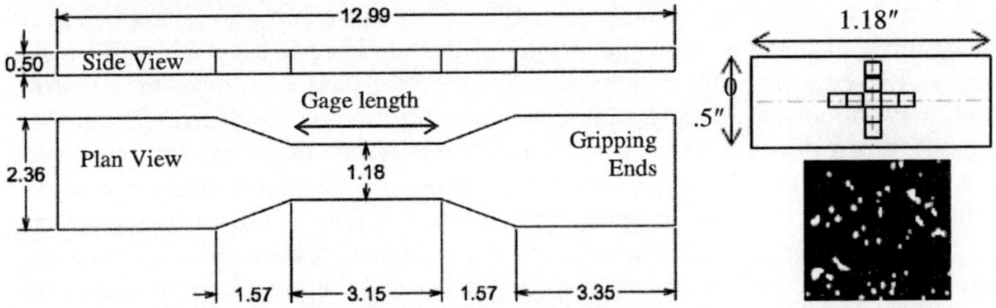

Fig. 1. (a) Dogbone specimen geometry for tensile testing of HPFRCC (JSCE) (all dimensions in inches) (b) Image sample locations (small squares) for fiber orientation distribution analysis and sample Fluorescence microscope image

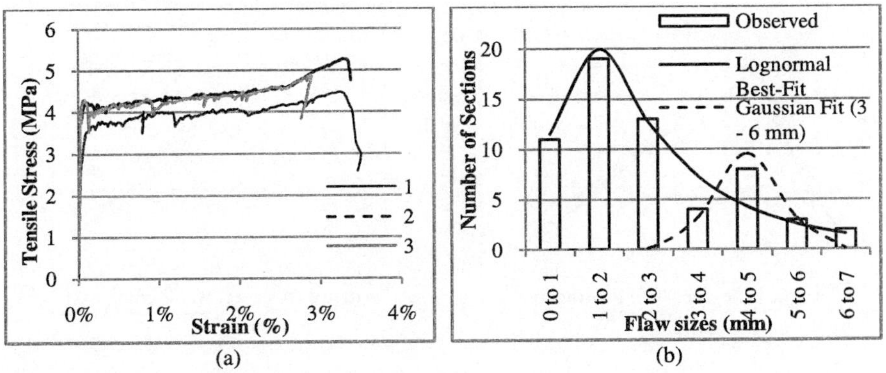

Fig. 2. (a) Measured tensile stress-strain curves of PVA-ECC of 3 specimens (b) Observed distribution of maximum flaw sizes for 60 sections (20x3), Lognormal overall Best-Fit for all sections, and Gaussian fit for flaws in 3-6 mm range

3 Results

The results of the direct uniaxial tension tests on three PVA-ECC dogbone specimens at 28 days after casting are shown in Fig. 2a. The three specimens show consistent stress-strain behavior with average peak tensile stress of 4.8 MPa and average tensile strain capacity (up to peak stress) of 3.1%. The average elastic secant modulus is 18 GPa and the average first crack strength of the matrix is 3.9 MPa. A total of 60 specimen cross-sections (20 sections per specimen) were observed under an optical microscope. Maximum flaw size was recorded for each section and the data is shown as a histogram in Fig. 2b along with an overall best-fit lognormal distribution with parameters $\mu = 0.70$ mm and $\sigma = 0.77$ mm.

A MATLAB digital image processing program developed by Lee et al. (2009) [9], utilizing advanced mathematical techniques such as watershed segmentation and morphological reconstruction, was used to extract useful information about fiber number, dispersion, and orientation from the images captured by fluorescence microscope (one such image is shown in Fig. 1b). Table 2 below shows the number of fibers observed at all the 9 sections and the fiber dispersion coefficient $\alpha \in [0,1]$, which is a measure of homogeneity of fiber dispersion. $\alpha = 1$ implies fully homogenous dispersion and $\alpha = 0$ implies all fibers clumped together with no dispersion. PVA Fibers with cylindrical shape are projected as ellipses on a cutting plane, with minor axis equal to the fiber diameter (d) and major axis (l) equal to $d/\cos(\theta)$, where θ is the angle w.r.t. loading axis. Thus, by measuring the aspect ratio (major axis/minor axis or l/d) of the ellipse, the fiber orientation can be determined by $\cos^{-1}(d/l)$. The computed orientation distribution for the mid-gage section "M" (with average crack width) of each specimen is shown in Fig. 3a, along with the average distribution and best fit double-Gaussian distribution. The mean, standard deviation, and weight of the first Gaussian are 25°, 13°, and 0.6, and that for the second Gaussian are 54°, 16°, and 0.4, respectively. The average (over 3 dogbones) distributions for M, F, and G sections are shown in Fig. 3b.

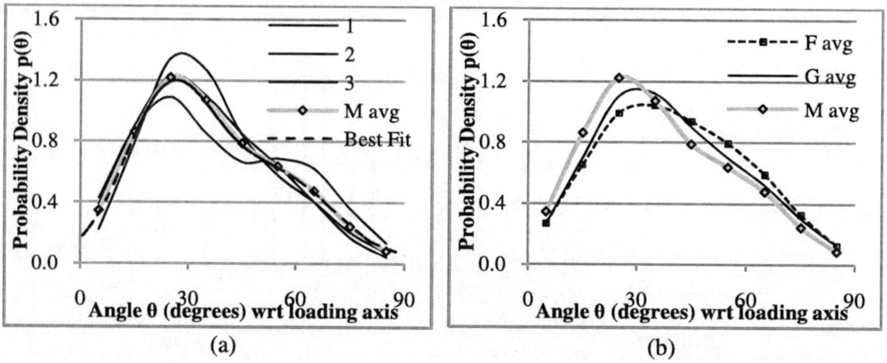

Fig. 3. (a) Observed, average, and best fit fiber orientation distributions at mid (M) sections of the 3 specimens (1,2,3) (b) Observed average fiber orientation distributions at Final failure (F), Grip (G), and Mid (M) sections

Table 2. Observed number of fibers per unit area and fiber dispersion coefficient for Final failure (F), Grip (G), and Mid-gage (M) sections of the 3 specimens and their average

Spec No.	Number of PVA Fibers per cm^2			Fiber Dispersion Coefficient		
	F	G	M	F	G	M
1	1289	1330	1390	0.363	0.365	0.372
2	1361	1216	1376	0.360	0.369	0.363
3	1025	1197	1235	0.361	0.362	0.369
Avg.	1225	1247	1334	0.361	0.365	0.368

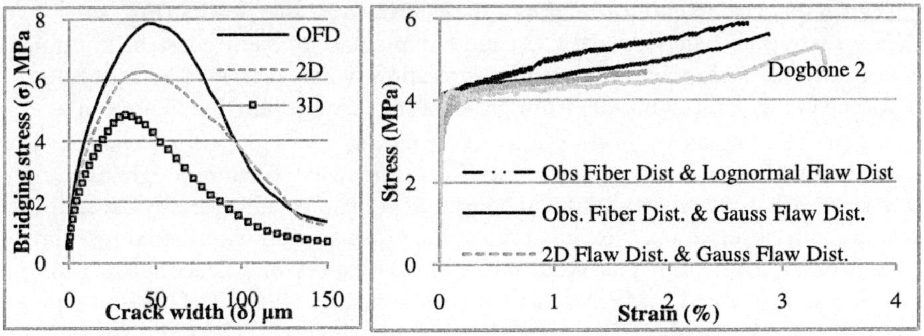

(a) (b)

Fig. 4. (a) σ-δ curves for various fiber distributions at $V_f = 2\%$ (OFD – Best-Fit to Observed Fiber Distribution, 2D and 3D distributions) (b) Composite stress-strain curve corresponding to various Fiber Distributions and Flaw Size Distributions

4 Analytical Investigation

In this section, the tensile stress-strain behavior of ECC at composite scale is explained based on material microstructure (fiber, matrix, and interfacial properties) and material processing (effective fiber number, fiber orientation distribution, and flaw size distribution) using micromechanics and scale-linking, which form the basis of the ECC design methodology. Single fiber pullout test results of previous studies [7] with the same fiber-matrix combinations have been used in this research to determine the interfacial properties. Analytical models of fiber debonding and pullout developed by Li and co-workers [1, 4] are used to express the pullout load (P) on a single fiber as a function of relative displacement (δ), fiber embedment length (l_e), and fiber orientation (θ). A statistical scale linking model (Eq.1), utilizing the best-fit fiber orientation distribution p(θ) (Fig. 3a), is used to compute the average bridging stress vs. crack width (σ-δ) relation (Fig. 4a-OFD) for multiple fibers bridging across a single crack. In Eq. 1, $p(l_e)$ is the probability distribution for embedment length: $2/L_f$, where L_f is fiber length.

$$\sigma = \frac{V_f}{A_f}\int_0^{\frac{\pi}{2}}\int_0^{(L_f/2)\cos\theta} P(\delta, l_e, \theta)p(l_e)p(\theta)dl_e d\theta \qquad (1)$$

The conditions for steady state multiple cracking are checked using the above σ-δ relation. Various cross sections are bridged by varying number of fibers (Table 2) due to inhomogeneity of fiber dispersion in the ECC mix. As a result, the final failure occurs at the crack with least number of effective fibers. One of the conditions of multiple cracking, therefore, requires the stresses at which cracks initiate to be smaller than the maximum bridging capacity (peak of σ-δ curve) of the least bridged crack $(\sigma_0)_{min}$. From Table 2, the ratio of minimum number of fibers (Specimen 3F) to average number of fibers at mid (M) sections is 1025/1334 = 0.77. The peak bridging stress (OFD - Fig. 4a) for average number of fibers ($V_f = 2\%$) is 7.7 MPa and, therefore, $(\sigma_0)_{min}$ is taken as 0.77 times 7.7, i.e. 5.9 MPa, which is higher than the first cracking strength of the matrix equal to 3.9 MPa (Fig. 2a).

The stress-strain behavior of ECC can be explained by the process of multiple cracking, where each crack obeys the σ-δ relation which serves as a constitutive law for the crack. Cracks initiate from pre-existing flaws (often due to air voids) at stresses (σ) determined by Irwin's fracture criterion, $\sigma = K_m/(Q\sqrt{\pi a})$, where a is the maximum flaw size at a cross section, K_m is the matrix fracture toughness, and Q is a geometrical parameter assumed constant here and is calibrated by using first crack strength observation, which occurs at the cross-section with the largest flaw size (6.3 mm - average of observed maximum flaw sizes in 3 specimens). Using $a_{max} = 6.3$ mm, $K_m = 0.5$ MPa$\sqrt{\text{m}}$, and σ (first crack) = 3.9 MPa, Q is computed equal to 0.91. The smallest flaw size that can be triggered before the stress exceeds $(\sigma_0)_{min} = 5.9$ MPa (K_m and Q same), $a_{min} = 2.7$ mm. For saturated cracking, the crack spacing between cracks is computed equal to about 1.1 mm; therefore, a set of 70 randomly distributed flaws (for a gage length of 80 mm) following the best-fit Gaussian distribution is generated using MATLAB between a_{min} and a_{max}. Corresponding initiation stresses are calculated for every flaw size. The stress drops every time a new crack forms because the stiffness across the crack drops suddenly, from combined stiffness of matrix and fibers to only that of fibers, at the same displacement. The stress is gained back slowly as the crack opens, contributing strain. The stress continues to rise until it reaches the initiation stress level for the next smaller flaw – the process is repeated until the stress is below $(\sigma_0)_{min}$; and beyond that, the damage is localized and the least bridged crack opens showing the softening corresponding to its σ-δ curve.

The effects of 3 different fiber distributions, viz. the observed (best-fit) double Gaussian distribution, 2D uniform distribution $p(\theta) = 2/\pi$, and 3D distribution $p(\theta) = \sin(\theta)$, on σ-δ relation are shown in Fig. 4a. These in turn affect the stress-strain relation as shown in Fig. 4b. The stress-strain relation due to 3D fiber distribution is not plotted because first crack stress is larger than $(\sigma_0)_{min}$ (assuming the same ratio of 0.77 between an average bridged crack and the least bridged crack), which violates the condition of multiple cracking. The 2 different flaw size distributions considered, viz. the observed lognormal distribution which best fits the overall data (0 to 7 mm flaw sizes) and Gaussian distribution which only fits the data in 3 to 6 mm flaw size range are shown in Fig. 2b, and their effects on stress-strain curves are shown in Fig. 4b.

5 Discussion and Conclusions

Fiber Orientation: The distribution observed using fluorescence microscopy (Fig. 3a) is close to a double Gaussian distribution, instead of theoretically expected 2D uniform distribution, $p(\theta) = 2/\pi$. Two possible reasons are (a) rotation and alignment of fibers between 30° and 45° towards smaller angles during the casting process when the fresh plastic material flows along the longitudinal axis of the specimen (b) limited freedom of rotation of fibers in the third dimension (thickness), which increases the number of fibers at smaller angles. This causes the peak of the first Gaussian distribution to be higher than the second. The fiber detection procedure also has a few limitations which may contribute towards the observed Gaussian peaks. The function $\cos(\theta)$ changes by only about 6% between 0° and 20°, which is sometimes higher than the stigmatism error in microscope images causing all fibers in 0° to 20° to be counted with angles greater than 20°. One possible way of correcting this error is to take the average number of fibers between 0° and 25°. Very highly inclined fibers (between 80° and 90°) tend to bend due to their low transverse stiffness, which reduces the aspect ratio of their projection on the section plane and are, therefore, detected as less inclined fibers. Also, the watershed algorithm incorrectly detects the highly inclined fibers as two separate fibers instead of one due to randomly varying intensity of image pixels.

Fiber number, dispersion, and bridging efficiency: It can be seen in Table 2 that the sections cut near the final failure (F – sections) have lesser number of fibers than M and G sections, which validates the theoretical assumption that the final failure occurs at a plane with lesser (more strictly least) number of fibers. Average F, G, and M sections are compared in Fig. 3b, and average bridging efficiency is computed to be the maximum (0.76) in M-sections as compared to the minimum in F-sections (0.71) due to higher alignment of fibers in M-sections. There is moderate level of segregation of fibers ($\alpha \approx 0.36$) but the non-homogeneity of dispersion is almost the same at all sections.

Effect of fiber distribution on σ-δ relation and composite stress-strain behavior: Preferential orientation of fibers between 20° and 80° increases the bridging capacity of the fibers because (a) the bridging efficiency of effective fibers is improved, and (b) the number of highly inclined fibers prone to early breakage due to snubbing is reduced. The increase in bridging capacity enhances ductility by increasing the range of flaw sizes, which can act as potential sites for crack initiation, as higher peak bridging capacity $(\sigma_0)_{min}$ reduces the critical minimum flaw size a_{min}. This can be seen in the analytical results of Fig. 4b, where the ductility improves by about 50% when the observed fiber distribution (a_{min} = 2.7 mm) is used instead of the theoretical 2D distribution (a_{min} = 4.3 mm).

Best-fit flaw size distribution and its effect on stress-strain behavior: The Gaussian distribution with mean at 4.5 mm (standard deviation of 0.6 mm) is used to locally fit the distribution from 3 to 6 mm size range. This is the most important range for crack initiation because a_{min} = 2.7 mm and a_{max} = 6.3 mm for observed best-fit fiber distribution. The change in distribution from lognormal (which fits the overall flaw size distribution from 0 to 7 mm) to Gaussian significantly affects the shape of the stress-strain curve, although it does not change the ductility substantially (Fig. 4b) because area under the two distributions (which represents the number of flaws) between a_{min} and a_{max} is almost the same. The secondary hardening or increase in the slope of the stress-strain curve near the last 0.5% strain, often observed in high volume fly ash PVA-ECC, can be explained by the falling arm of this Gaussian distribution causing fewer flaws to be triggered (area under the Gaussian distribution) for the same increase in stress, which reduces the increment in strain and shows as a slight hardening (Fig. 3a and Fig. 4b).

Conclusions: This research shows the possibility of preferential alignment of PVA fibers in ECC when cast in dogbone specimens resulting in a double Gaussian distribution with higher peak near 25° and a lower peak at 54° angles w.r.t. loading axis, which affects the composite stress-strain behavior non-conservatively; although, the data for this conclusion is very limited and continued work with more samples is further required. Flaw size distribution affects the shape of the stress-strain curve of high volume fly ash (HVFA) ECC; although its effect on ductility is less significant. Lognormal distribution seems to best-fit overall flaw size range, whereas a Gaussian distribution with mean of 4.5 mm and standard deviation of 0.6 mm seems to best-fit 3-6 mm flaw size range. Average maximum flaw size in HVFA-PVA-ECC was found to be about 6.3 mm.

References

[1] Li, V.C., Wang, S., Wu, C.: Tensile Strain-Hardening Behavior of PVA-ECC. ACI Materials Journal 98(6), 483–492 (2001)
[2] Li, V.C., Ogawa, C., Wu, A., Saito, T.: Interface Tailoring for Strain-Hardening PVA-ECC. ACI Materials Journal 99(2), 463–472 (2002)
[3] Cox, H.L.: The elasticity and strength of paper and other fibrous materials. British Journal of Applied Physics 122(1), 10–18 (1952)
[4] Wang, S., Li, V.C.: Tailoring of pre-existing flaws in ECC matrix for saturated strain hardening. In: Proceedings of FRAMCOS-5, pp. 1005–1012 (2004)
[5] Li, V.C., Wang, S.: Microstructure variability and macroscopic composite properties of high performance fiber reinforced cementitious composites. Journal of Probabilistic Engineering Mechanics 21(3), 201–206 (2006)
[6] Torigoe, S., Horikoshi, T., Ogawa, A.: Study on evaluation method for PVA fiber distribution in engineered cementitious composite. Journal of Advanced Concrete Technology 1(3), 265–268 (2003)
[7] Yang, E.H., Yang, Y., Li, V.C.: Use of high volumes of fly ash to improve ECC mechanical properties and material greenness. ACI Materials Journal 104(6), 620–628 (2007)
[8] JSCE. Recommendations for design and construction of high performance fiber reinforced cement composites with multiple fine cracks. Japan Soc. of Civil Engineers, Tokyo (2008)
[9] Lee, B.Y., Kim, J.K., Kim, J.S., Kim, Y.Y.: Quantitative evaluation technique of Polyvinyl Alcohol (PVA) fiber dispersion in engineered cementitious composites. Cement and Concrete Composites 31(6), 408–417 (2009)

Part 2
Bond and Pull-Out Mechanisms

Groups of Physical Parameters Influencing the Three Stages Pull-Out Behavior of Glass Multi-filament Yarns Embedded in Micro-concrete

H. Aljewifi, B. Fiorio, and J.-L. Gallias

Université de Cergy-Pontoise, Cergy-Pontoise, France

Abstract. Five different continuous glass yarns have been embedded in cement based micro concrete. The pull-out behavior of the yarns has been characterized by the way of a classical pull-out test when the physical state of the impregnation has been characterized by different dedicated methodology (differential mercury intrusion porosimetry, flow test, SEM characterization). By studying the correlation between the mechanical behavior and the physical properties of the impregnated yarns, groups of physical parameters have been identified as associated to each of the three stages of the pull-out behavior. The overall result of this work is a better understanding of the relationship between, on one hand, the physical state and the impregnation of the yarns and, on the other hand, the pull-out behavior.

1 Introduction

Glass fiber reinforced cement composites have been used for many nonstructural applications since the early 1970s. An important step in the development of this kind of reinforcement was the introduction of the first alkali-resistant (AR) glass fiber by Pilkington about three decades ago [1]. As glass fibers are not sensitive to corrosion, unlike steel, the concrete cover can be reduced in order to minimize the thickness of the structural members. This reduction is particularly high in the case of textile reinforced concrete (TRC).

The properties, performances and applications of TRC have been described by many authors [2]. Current studies have confirmed that complex processes of interaction occur inside the composite between the yarns and the cementitious matrix, and inside the yarns between the glass filaments and the penetrating cement paste or between the filaments themselves [3]. These interactions are largely influenced by the yarns impregnation, and govern the mechanical properties of the TRC.

The purpose of this paper is to present results of measurements of different physical parameters related to the yarn impregnation made for different yarns and different impregnation conditions and to study the relationship between these parameters and the mechanical pull-out behavior.

G.J. Parra-Montesinos, H.W. Reinhardt, and A.E. Naaman (Eds.): HPFRCC 6, pp. 117–125.

Especially, attention was paid to the relationship between the physical state of the impregnation and the three stages of the pull-out behavior.

2 Glass Yarn Embedded in Matrix

2.1 Yarn Impregnation

Five different kinds of multi-filament yarns have been used in this work (the basic properties of the yarns are given in Table 1).

Table 1. Physical and mechanical characteristics of the multi-filament yarns

	SG1	SG2	SG3	OC1	OC2
Type of glass	AR	AR	AR	E	E
Type of roving	Assembled	Assembled	Assembled	Assembled	direct
Filament diameter (μm)	14	14	14	12	17
Finesses (tex = g/km)	2450	2450	2500	2400	2400
Glass density (kg/m^3)	2680	2680	2680	2530	2530
Max.strength (N)	714.4	856,6	830.4	1116.3	900.5
Yarn stiffness (N/mm)	565	449	455	548	559

It should be noticed that the geometrical properties of the yarns are much more numerous than given in Table 1, due to the multi-scale assembly of the filaments. Moreover glass filaments were coated by sizing which influenced the yarns processing but also the adhesion with the matrix. Yarns were centered in a cylindrical shape cementitious matrix using a specific mold [4]. The matrix was a micro-concrete made from CEM I cement (0.48 water on cement ratio; 1.4 sand on cement ratio with 0-1.25 mm sand; a superplasticizer was used with a dry extract dosage of 0.125 % of the cement mass). Properties of the matrix were as follow: 2.135 g/cm^3 density, 31500 MPa Young's modulus, 55 MPa compressive strength.

Before casting, yarns were applied one of the three following pre-treatments: water pre-wetting (W); air drying (D); mechanically induced pre-impregnation by a cement slurry (PI; composition of the slurry was identical to those of the cement paste of the matrix). The pre-treatment had an influence on the resulting impregnation of the yarn by the matrix [4].

2.2 Physical Characterization of the Impregnation

Impregnation of the yarns was characterized by the mean of three different techniques: SEM observations of longitudinal section of impregnated yarns, flow rate measurements, differential mercury intrusion porosimetry.

The SEM observation allows the average penetration depth of the cementitious matrix into the yarn to be determined. An impregnation index i_y was defined as the ratio of the impregnated area to the apparent area of the cross-section of the yarn [4]. It was determined considering a cylindrical shape of the yarn with constant penetration depth of the matrix.

The values of the impregnation index significantly vary with the pre-treatment applied to the yarn. W pre-treatment, by water- saturating the porosity of the yarn, prevents the yarn from matrix penetration and gives low values of impregnation index. D pre-treatment leads to a moderate increase of the impregnation index for SG2, SG3 and OC1 yarns when the increase was almost unobserved for SG1 and OC2 yarns. This can be explained by a filter effect of the glass filaments, that stops the in-depth penetration of the water carried cement particles during the wetting of the yarn. The wettability of the yarn, which depends on the type of sizing, is also an influential parameter. PI pre-treatment, by forcing the introduction of matrix inside the yarn, allow the full impregnation (100%) to be obtained.

Flow rate measurements were achieved on 2 cm thickness sample of micro concrete in which a pre-treated yarn was embedded. The test consists in measuring the steady state flow rate of water along the yarn, under a 107.5 kPa/cm pressure gradient [4].

Mercury intrusion porosimetry (MIP) was used to determine the total pore volume of samples of embedded yarns and surrounding cementitious matrix. Samples were obtained by coring the center of the cylindrical specimens detailed in 2.1. By comparison with reference matrix samples for capillary diameter in the range 0.3 to 30 µm, the void volume inside the impregnated yarn was determined (see [4] for details). It is called "yarn pore volume" in the following.

Fig. 1, left, shows the relationship between the impregnation index and the flow rate measurement.

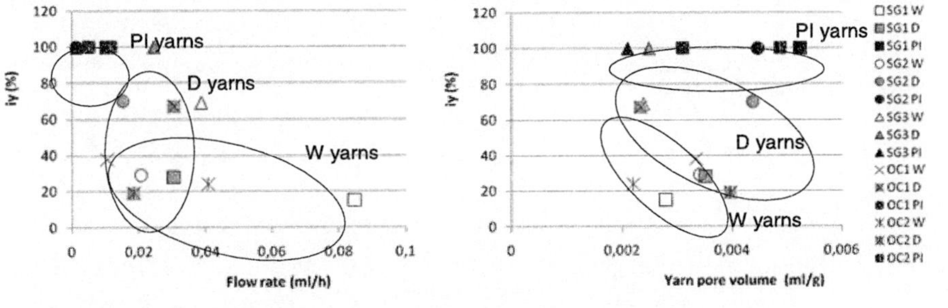

Fig. 1. Impregnation index, i_y vs flow rate (left) and yarn pore volume (right)

Flow rate decreases when the impregnation index increases, which is the consequence of the matrix filling of the inter-filaments voids. For low impregnated yarns, water is in direct contact with glass filaments and the flow rate depends on the wettability of the yarn when for high impregnation there is no influence of the yarn (flow occurs in the porosity of the matrix).

Fig. 1, right, gives the relationship between the impregnation index and the yarn pore volume. This volume is generally more important when the impregnation index is high, due to the disorder in the filaments organization induced by the introduction of the matrix. This is particularly the case for PI pre-treatment for which a mechanical action is required.

3 Mechanical Pull-Out Behavior of the Embedded Yarns

3.1 Testing Method

The load-displacement (P-δ) curves for pull-out tests were obtained from tension test performed with a universal testing device (30 kN capability) as described in [5]. The pull-out tests were made by applying a tensile load on the free end of a yarn, which extremity was embedded in a micro-concrete cylinder (6 embedded lengths have been used in the range 1 to 25 cm). The free length of the yarn was 10 cm. Loading was applied at constant strain rate (0.01 mm/min). Three tests were performed for each configuration (one set of yarn type, pre-treatment type, and embedded length).

Three different stages are usually observed for the pull-out behavior of yarns (St1 to 3 in Fig. 2). The first stage (St1) corresponds to the progressive loading of the filaments. It is characterized by delayed activation of the filaments associated to the progressive failure of some of them. In the second stage of the behavior (St2), the maximum strength is reached and a majority of filaments progressively failed, leading to a progressive fall of the load when the pull-out displacement is increased. The third stage (St3) of the pull-out behavior corresponds to the residual friction of the intermixed broken filaments. It is characterized by slowly decreasing level of friction.

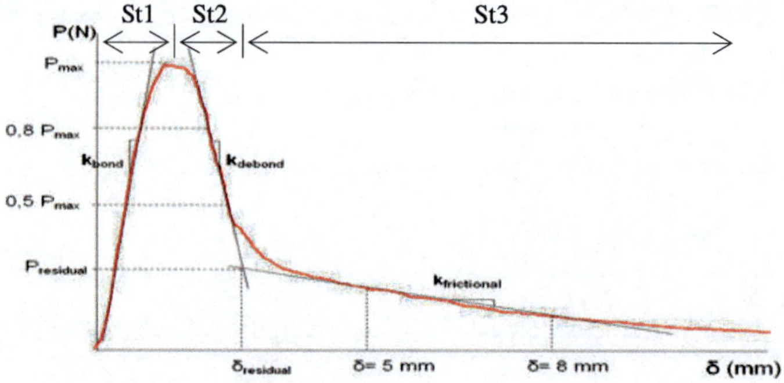

Fig. 2. Parameters of the pull-out behavior

Fig. 2 shows the parameters retained as characteristic for the pull-out behavior: maximum pull-out strength P_{max}; stiffness k_{bond} and k_{debond} of the first and second stages of the pull-out behavior (determined in the range 0.5-0.8 P_{max}); stiffness $k_{frictional}$ of the third stage (residual load; determined in the δ range 5-8 mm); and second to third stages transition load $P_{residual}$.

3.2 *Contribution of the Embedded Length to the Pull-Out Behavior*

The evolution of the maximum pull-out strength P_{max} for increasing embedded length is given in Fig. 3 (example of SG1 yarns). The pull-out strength increases when the embedded length increases as long as the embedded length is lower than a threshold length L_{min} (defined in Fig. 3 by the slope change of the line).

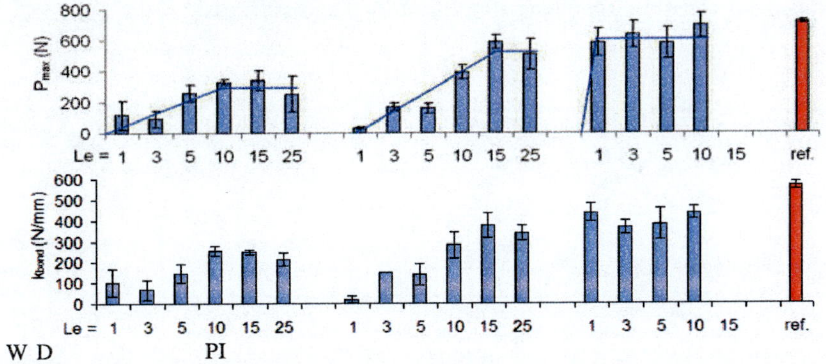

Fig. 3. Values of P_{max} and k_{bond} versus the embedded length in cm (SG1 yarn)

This length corresponds to the length needed to have all the filaments of the yarn in contact with the matrix, considering the twisted shape of the yarn. For embedded length lower than L_{min}, the number of anchored filaments and therefore the pull-out strength increases when the embedded length increases. Over L_{min}, the pull-out strength remains constant as a change of the embedded length does not modify the anchoring conditions of the filaments. L_{min} is influenced by the yarn pre-treatment. In particular, PI pre-treatment gives very low L_{min} (under 1 cm). The same analysis can be done with the values of the stiffness k_{bond}. In the following section, we retain the average values of the behavior parameters obtained for embedded length over L_{min} as characteristic of the pull-out behavior.

4 Relationship between Physical and Mechanical Parameters

4.1 Pre-peak Stage of the Pull-Out Behavior

The pre-peak stage of the pull-out behavior is characterized by the maximum pull-out load ($P_{max,\ lim}$) and the first stage stiffness $k_{bond,lim}$. Index "lim" refers to the average of the values obtained for embedded length higher than L_{min}.

As shown in Fig. 4, the pull-out load is an increasing function of the impregnation index. Differences are observed from one yarn to the other. In particular, OC yarns give slightly lower values of the pull-out load and seem to be less sensitive to the impregnation index. These tendencies are also observed for the relationship between the stiffness and the impregnation index but in this case stiffnesses of OC and SG yarns are very similar. The increase of the impregnation index leads to a better penetration of the matrix in the yarn, which tends to reduce the free length of the filaments. As a consequence, filaments work more simultaneously, leading to enhanced pull-out load and stiffness.

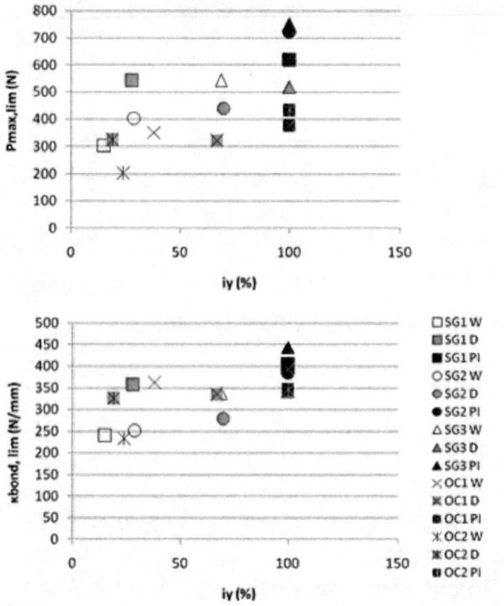

Fig. 4. $P_{max,lim}$ (left) and $k_{bond,lim}$ (right) vs impregnation index i_y

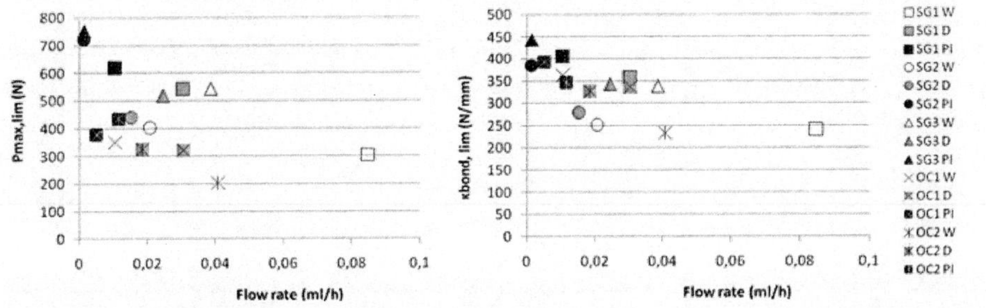

Fig. 5. $P_{max,lim}$ (left) and $k_{bond,lim}$ (right) vs flow rate

As the penetration of the matrix in the yarn creates obstacles to the flow of water, the correlation between i_y and the pre-peak behavior is also observed with the flow rate (Fig. 5): a decrease of the flow rate is associated to an increase of pull-out load and stiffness. The correlation with the flow rate is particularly good in the case of k_{bond} where it seems to be almost uninfluenced by the type of yarn.

Concerning the relationship to the yarn pore volume, no general correlation is observed (Fig. 6), neither for the pull-out strength nor for the stiffness.

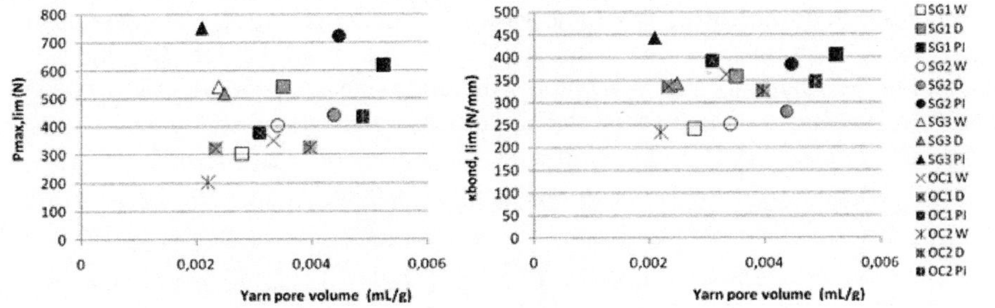

Fig. 6. $P_{max,lim}$ (left) and $k_{bond,lim}$ (right) vs yarn pore volume

4.2 Post-peak Stage of the Pull-Out Behavior

The post-peak stage of the pull-out behavior is characterized by the second stage stiffness $k_{debond,lim}$. Fig. 7 gives the values of $k_{debond,lim}$ versus flow rate and impregnation index. As there is no specific correlation, relationship of $k_{debond,lim}$ to the yarn pore volume is not given.

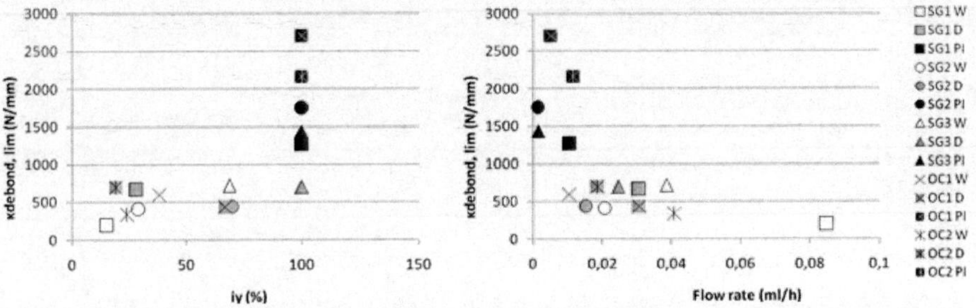

Fig. 7. $k_{debond,lim}$ vs impregnation index (left) and flow rate (right)

Whatever the type of yarn is, values of $k_{debond,lim}$ is low for W and D pre-treatment, when they vary significantly with the type of yarn for PI pre-treatment (Fig. 7). In the case of PI yarns, $k_{debond,lim}$ is very close to the measurements made on yarns in direct tension. Concerning the relationship to the flow rate, it is similar to those observed for $k_{bond,lim}$ with enhanced variation of the stiffness (Fig. 5).

4.3 Residual Stage of the Pull-Out Behavior

The residual stiffness and load that characterized this stage of the pull-out behavior are both increasing when the yarn pore volume increases. As explained in 2.2, this can be linked to the penetration of the matrix in the yarn that induces disorder in the filaments arrangement and therefore enhances the friction between filaments (possibly with cement particles acting as a third body).

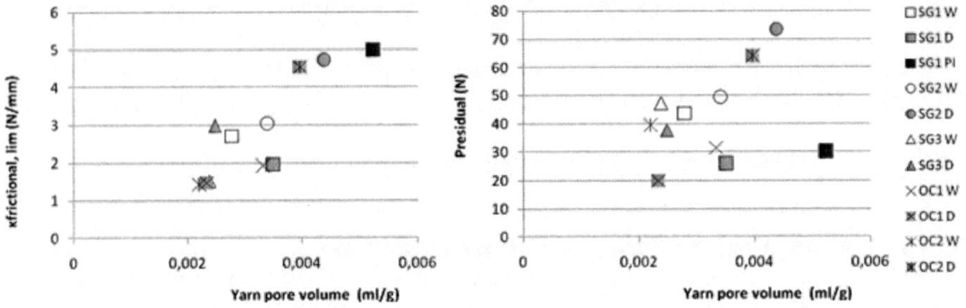

Fig. 8. Relationship between $k_{frictional}$ (left), residual strength (right) and yarn pore volume

5 Conclusion

The here-presented results appear as complex because the specific structure of impregnated yarns generates multi-influenced behavior. Despite of this difficulty, the results give a general trend associated to the pull-out behavior. In particular, it appears that the pre- and post-peak behavior are linked to the capability of water to flow along the impregnated yarn, when the residual behavior can be correlated to the specific pore volume of the yarn. These results make it possible to connect the micro-structure of the impregnated yarns to its pull-out behavior.

References

[1] Jesse, F.: The present and the future of textile reinforced concrete. In: FRPRCS 5 Conference, Cambridge, July 16-18, vol. 2 (593) (2001)

[2] Brameshuber, W., Brockmann, T., Banholzer, B.: Textile reinforced ultra high performance concrete. In: Proceedings: International Symposium on UHPC, Kassel, September 13-15, pp. 511–532 (2004)

[3] Konrad, M., Chudoba, R., Butenweg, C., Bruckermann, O.: Textile Reinforced Concrete Part II: Multi-Level Modeling Concept. In: Proceedings Internationales Kolloquium über Anwendungen der Informatik und der Mathematik in Architektur und Bauwesen, Weimar (2003)

[4] Aljewifi, H., Fiorio, B., Gallias, J.L.: Characterization of the impregnation by a cementitious matrix of five glass multi-filament yarns. European Journal of Environmental and Civil Engineering, EJECE 14(5), 529–544 (2010)

[5] Aljewifi, H., Fiorio, B., Gallias, J.L.: Pull-out behaviour of a glass multi-filaments yarn embedded in a cementitious matrix. In: EUROC 2010, Austria, March 15-18 (2010)

Tailor-Made Steel Fiber Reinforced Ultra High Performance Concrete – Single Fiber Pull-Out, Bending Capacity and Fracture Toughness

T. Stengel, X. Lin, P. Schießl, C. Gehlen

Centre for Building Materials cbm, Technische Universität München TUM, Germany

Abstract. Micro steel fibers used in UHPC account for a significant portion of the overall cost and the environmental footprint of the material. It is therefore aimed to replace these fibers by bond optimized macro steel fibers with diameters between 0.3 to 0.5 mm. Single fiber pull-out and beam tests were performed to characterize fiber bonding and fiber i.e. crack bridging efficiency and model the resulting composite behavior. Different fiber diameters, surface roughness, fiber inclinations and concrete compositions were considered. This contribution presents the results of one concrete composition in detail and summarizes the main findings of the overall study. Furthermore, the derivation of input parameters for modelling bending beam behavior is demonstrated.

1 Introduction

With its outstanding mechanical properties ultra high performance concrete (UHPC) opens up a field of new construction methods and applications for concrete. However, the extreme brittleness of UHPC, characterized by explosive failure under critical load, restricts the exploitation of the otherwise advantageous properties of this material. To improve ductility, typically 2 to 3% by volume of micro steel fibers 0.15 to 0.20 mm in diameter with a length of 10 mm are added to UHPC. Micro steel fibers (MSF) account for a considerable portion of the overall material cost as well as of the environmental footprint of UHPC [1, 2]. It is therefore the aim of the present study to replace the MSF by steel fibers thicker in diameter with optimized bonding properties. To optimize bonding properties the effect of UHPC composition, steel fiber geometry and surface roughness as well as coatings of fibers was investigated with the help of single fiber pull-out tests. To evaluate the crack bridging efficiency of different steel fibers, inclined fiber pull-out tests (15°, 30°, 45°, 60°) were also carried out. Bending capacity and fracture toughness of the UHPC containing different fiber mixes were then evaluated by three point bending tests and analyzed employing the two parameter model (TPM). Ten different mix compositions together with nine different types of fibers

G.J. Parra-Montesinos, H.W. Reinhardt, and A.E. Naaman (Eds.): HPFRCC 6, pp. 127–135.

were included in the investigations (undulated 0.50/0.40, helix-type, smooth 0.48/0.40/0.29/0.20/0.16, with/without brass coating, 3 different surface roughnesses). The present contribution summarizes the results of single fiber pull-out tests of one of the mixes and one fiber diameter and illustrates the derivation of input parameters for modelling the bending beam behavior with the TPM.

2 Composition and Mechanical Properties of Plain UHPC

The compressive strength of the heat-treated UHPCs ranged from 180 to 260 MPa depending on the mix composition. OPC was used in all cases at an amount of 220 to 850 kg/m³ and w/c-ratios from 0.89 to 0.22. Besides silica fume, cement and fine quartz sand (max. grain size 0.5 mm) six different mineral additions, namely three lime stone powders of different fineness, two quartz powders of different fineness and grounded granulated blast furnace slag were used. The mix compositions varied in order to investigate the effect of mineral type as well as the effect of mix fineness in combination with different fiber surface roughnesses. Here only results for one of the mixes combined with smooth untreated fibers are presented. The mix composition of the unreinforced UHPC is given in Table 1.

Table 1. Composition of the plain UHPC

Material	Amount [kg/m³]	Material	Amount [kg/m³]
Cement CEM I 42,5 N	680	Quartz sand	990
Silica fume	138	Water	166
Quartz powder	360	Superplasticizer	35

The compressive and splitting tensile (Brazilian test) strengths were measured using cylinders with a diameter of 50 mm and aspect ratios of 1 and 2, respectively. The Young`s modulus (compression) was determined using cylinders with same aspect ratio as the Brazilian test specimen. Mechanical properties and fiber pull-out were tested at an age of 7 days. The mean compressive and splitting tensile strength was 212.5 and 11.6 MPa, respectively and the Young`s modulus was 50,000 MPa.

3 Single Fiber Pull-Out Tests

The fibers were cut and cleaned with isopropyl alcohol until there were no impurities (e.g. fat) on the surface of fibers. The fibers were mounted in appropriate formworks which were then filled with the self-compacting UHPC (without any vibration) and stored in the humidity cabinet for 24 h. Afterwards, the specimens were demoulded and cured in the humidity cabinet for four days. Following this, the specimens were heat-treated at 90 °C for 48 h before being prepared for the pull-out test. For testing, each cylindrical sample (30 mm in diameter) was first glued to the middle of a steel plate with epoxy resin. After the resin had hardened, the steel plate was mounted in the testing machine (spindle driven testing machine with 500 N load cell). The free-fiber length, i.e. the distance between the matrix surface and the chuck holding the fiber, was adjusted to about 8 mm. The fiber was fixed and pulled out at a constant displacement rate of 0.03 mm/s. The pull-out load was recorded during the entire pull-out procedure. Fig. 1 shows single fiber pull-out behavior of aligned fibers 0.16 (brass coated) and 0.48 mm in diameter with an embedded length of 5 to 5.5 mm aligned in the direction of the load. The linear branch is not shown in the diagram. Displacement due to elastic strain of the free fiber length was taken into consideration (E = 210 GPa).

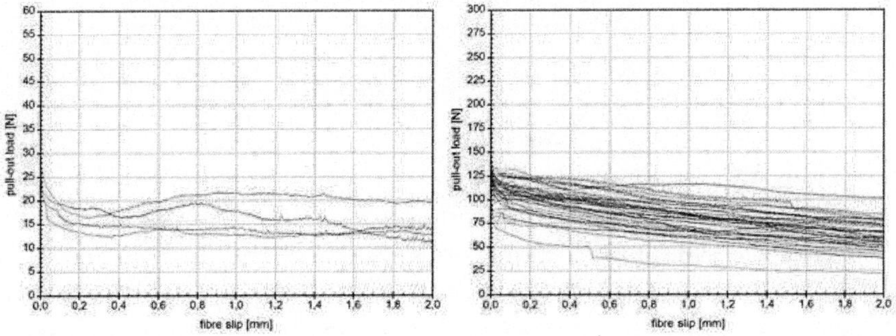

Fig. 1. Pull-out curves of aligned 0.16 mm micro steel fibers (left) and 0.48 mm fibers (right)

The bonding behavior of aligned steel fibers can be affected significantly (factor 3 to 4) by using roughened fiber surfaces and appropriate mix composition, e.g. finer constituents. Compressive strength of the matrix (180 to 260 MPa) does not affect bond behavior of aligned fibers significantly. Considering inclined fiber pull-out it is apparent that the compressive strength of the matrix slightly affects bond behavior; in this case mix composition i.e. fineness doesn't play an important part.

If the pull-out behavior of a single aligned fiber is taken as an input parameter for modelling bending capacity and fracture toughness of the fiber reinforced composite, it should be kept in mind that the

- orientation of fibers i.e. inclination will cause different pull-out behavior,
- embedded length may vary over a fiber bridged crack in the composite,
- fibers may interact at higher volume fractions of fibers.

In this study only the two first aspects are considered. Two efficiency factors, namely η_θ for fiber inclination and η_l for embedded fiber length were derived based on experiments and simple geometric considerations. Note that the factors are not the same as those obtained by Swamy [3]. For simplification it is assumed that both factors are independent of each other and can be superposed by multiplication. The efficiency factor η_θ is derived from results of aligned and inclined fiber pull-out tests (0, 15, 30, 45 and 60°), where the angles are related to fiber inclination between 0 and 90° with respect to the crack surface in the composite as shown in Fig. 2. As indicated, the quadrant $0° \leq \theta \leq 90°$ is divided into 12 equal sectors each covering 7.5°. Hence, fibers which are aligned 0° to 7.5° with respect to the crack surface are allocated to the aligned pull-out test results; fibers which are inclined 7.5° to 22.5° ($\Delta\theta = 2 \times 7.5°$) with respect to the crack surface are allocated to the 15° pull-out test results and so on. The probability of every inclination range is then calculated based on the surface of a spherical cap. This gives for 0° to 7.5° 1-cos(0.131) = 0.009 and for 7.5° to 22.5° cos(0.131)-cos(0.393) = 0.068 and so on.

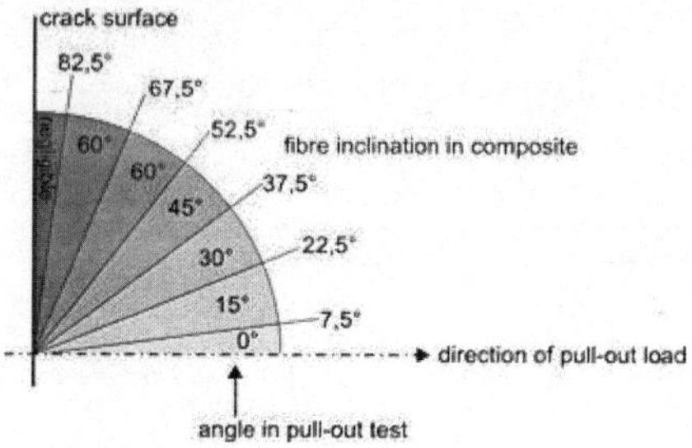

Fig. 2. Allocation of angle in pull-out test to fiber inclination in composite

The effect of the inclination of fibers on the peak pull-out load was almost the same for 0.29 and 0.48 mm fibers and is given in Table 2 as standardized peak load P'^{θ}_{i}.

Table 2. Effect of inclination on peak pull-out load of single fibers

Angle in pull-out test	0°	15°	30°	45°	60°
$P'^{\theta}_{i}=P_{\max\,\theta}/P_{\max\,0°}$	1.00	1.35	1.65	1.80	1.30

Therefore η_θ can be written as: $\eta_\theta = 0.009 + 0.068 \cdot P'^{\theta}_{15°} + ...$

This yields $\eta_\theta = 1.269$ for the present study. The mean peak pull-out load of a single fiber with embedded length $l_e = l_f/2$ can then be derived by: $\langle P^{\theta}_{\max} \rangle = \eta_\theta \cdot P_{\max 0°}$. As mentioned earlier, the embedded length of fibers affects bond behavior i.e. peak bond load. The effective embedded length of fibers crossing a crack in a composite varies from $0 \leq l_e \leq l_f/2$. Hence, when modelling the fiber bridging over a crack with the help of single fiber pull-out behavior, an efficiency factor η_l for the embedded length has to be taken into consideration. This is carried out in a simplified way similar as for the inclination efficiency. If one considers a layer made of fiber reinforced material divided by a crack with almost zero crack opening and crossed by fibers, the length of fibers in the left l_l being in contact with the crack is between 0 and l_f (i.e. $0 \leq l_l / l_f \leq 1$) and vice versa on the right, see Fig. 3. It is assumed that the number of fibers with a certain embedded length is always the same on both sides, i.e. the embedded length is equally distributed. If one now assumes ranges of embedded length l_e as given in Fig. 3 and

relate them to a peak load one can obtain a simplified efficiency factor for embedded length. The total range of embedded length $(0...l_f/2)$ is divided into 20 subranges of approximately $\Delta l_e / l_f = 0.04...0.05$. Every subrange contributes with $p_{ij} = 1/20$ to the total range, i.e. $p_{16}(0...0.04) + p_{61}(0...0.04) = 2/20$, $p_{15}(0.05...0.14) + p_{51}(0.05...0.14) = 4/20$, $p_{11}(0.45...0.55) = 2/20$ and so on.
Hence, η_l can be written as: $\eta_l = 1/10 \cdot P'^{l}_{11} + 1/5 \cdot P'^{l}_{12} + ... + 1/5 \cdot P'^{l}_{15}$
where $P'^{l}_{ij} = P^{l}_{ij} / P(l_e = l_f/2)$.

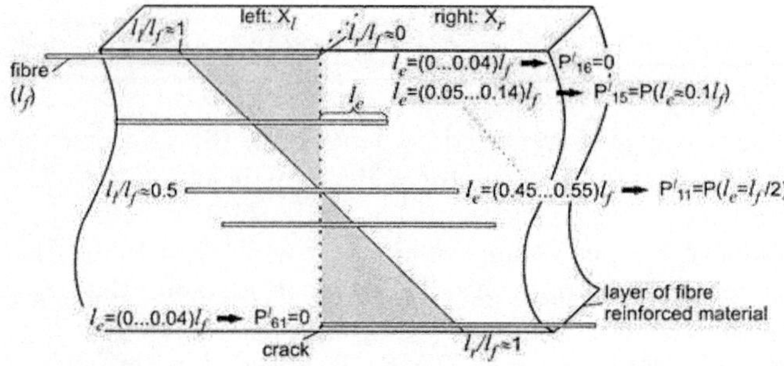

Fig. 3. Different embedded length of fibers crossing a crack and resulting peak bond loads

For a given fiber length l_f a simplified relation between peak pull-out load P^{l}_{ij} and embedded length is obtained from fiber pull-out tests applying an empirical fit curve similar to the equation derived analytically by Gao [4]: $P_l(\varsigma) = \sigma_0 \cdot r_f^{\,2} \cdot \pi + (\bar{\sigma} - \sigma_0) \cdot (1 - \exp(-\lambda \cdot \sqrt{\varsigma \cdot l_f})) \cdot r_f^{\,2} \cdot \pi$ where $\zeta = l_e / l_f$ $(0 \leq \zeta \leq 0.5)$, $\sigma_0 = 124$ MPa, $\bar{\sigma} = 1905$ MPa and $\lambda = 0.025$. Table 3 shows the normalized peak pull-out load P'^{l}_{i} for different normalized embedded lengths $\zeta = l_e / l_f$.

Table 3. Normalized peak pull-out load P'^{l}_{i} for different normalized embedded lengths $\zeta = l_e / l_f$

ζ	0.5	0.4	0.3	0.2	0.1
$P'^{l}_{i} = P_{max\ l} / P_{max\ 0.5lf}$	1.00	0.90	0.78	0.64	0.45

Therefore $\eta_l = 0.654$. The mean peak pull-out load of a single fiber taking into consideration both efficiency factors is $\langle P^{l\theta}_{max} \rangle = \eta_\theta \cdot \eta_l \cdot P_{max}(l_e = l_f/2; \theta = 0°)$ which yields a mean $P^{l\theta}_{max} \approx 91$ N for the 0.48 mm fibers used this study (mean $P_{max}(l_e = l_f/2,\ \theta = 0°) = 110$ N, 38 specimen). To implement the pull-out behavior (fiber slip = crack opening displacement) a linear relationship was applied in this study: $\frac{P(w)}{\langle P^{l\theta}_{max} \rangle} = 1 - \frac{2 \cdot w}{l_f} ... 0 \leq w \leq l_f/2$ and $\frac{P(w)}{\langle P^{l\theta}_{max} \rangle} = 0 ... w > l_f/2$. To model the bending behavior of notched beams the so-called crack closing pressure, which depends on the crack opening displacement (COD) has to be known. The closing pressure $p_{cl}(w)$ can be obtained by the mean pull-out load slip curve of a single fiber based on a simplified relation derived by Swamy [3] and applied in a similar way to fiber reinforced mortar by e.g. Wecharatana [5]:

$$p_{cl}(w) = p_{\max} \cdot \left(1 - \frac{2 \cdot w}{l_f}\right) \quad \text{and} \quad p_{\max} = \eta^l \cdot \eta^\theta \cdot \tau_{\max}(l_e = l_f/2; \theta = 0°) \cdot V_f \cdot \frac{l_f}{d_f} = \left\langle P^{l\theta}{}_{\max} \right\rangle \cdot V_f \cdot \frac{1}{d_f^{\,2} \cdot \pi}$$

The mean closing pressure p_{max} for different fibers used in this study are given in Table 4.

Table 4. Closing pressure p_{max} of steel fibers used in this study ($V_f = 0.01$, fiber length: 10 mm)

d_f [mm]	0.16[1)]	0.20[1)]	0.20	0.29	0.40	0.48
p_{max} [MPa]	2.79	1.59	2.71	1.93	1.90	1.26

[1)] brass coated

This is in the same range as reported by Jenq et al. [6] who calculated closing pressure by analysis of steel fiber reinforced bending beams at high deflection, i.e. crack opening displacement. They derived p_{max} = 1.16 MPa up to p_{max} = 2.32 MPa for a fiber content of 1% per volume in a mortar with w/c = 0.40 (fiber diameter: 0.25 mm, fiber length: 25.4 mm). Taking into consideration the results of fiber pull-out tests with an embedded length of 10 mm (i.e. fiber length: 20 mm) and fibers 0.29 mm in diameter p_{max} = 3.52 MPa is obtained within this study. This seems to be reasonable for UHPC when compared to [6].

4 Fracture Toughness and Bending Capacity

A typical load-CMOD plot of an unreinforced notched UHPC beam subjected to three-point bending is shown in Fig. 4, see for example [8] for test setup. The peak load P_{max} was 6.947 kN and the corresponding $CMOD_t$ was 58.6 μm.

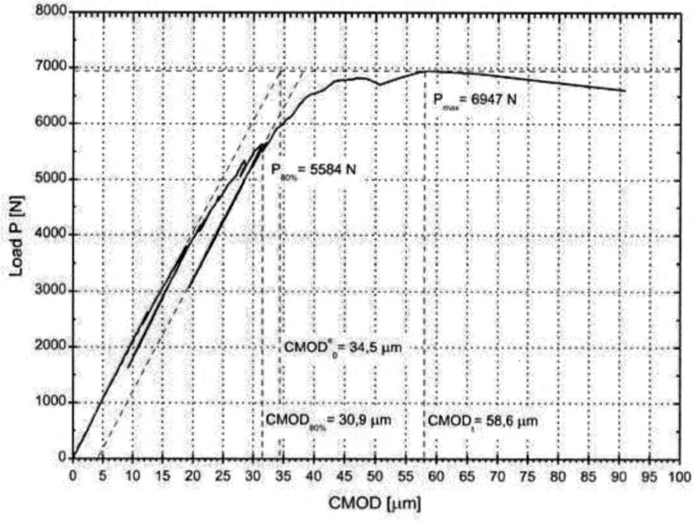

Fig. 4. Typical load-CMOD plot of unreinforced UHPC in three-point bending

The compliance of unloading $C_{80\%}$ at 80% of P_{max} was the same as the initial compliance C_i. From the plot it can be seen that there is non linear prepeak crack growth. At load levels up to about 55% of P_{max} there is a linear relationship between CMOD and load P. The elastic $CMOD^e_0$ at peak load was 34.5 μm in the case of no stable (prepeak) crack growth. Despite the use of a stiff servo controlled closed-loop testing machine it was not possible to distinguish the elastic $CMOD^e_s$ due to nonlinear effects (e.g. microcracking within the FPZ and slow crack growth) and inelastic $CMOD^*$ at peak load. In order to apply the two parameter model (TPM) according to Jenq and Shah [7] assumptions had to be made for the portion of $CMOD^e_s$ and $CMOD^*$ respectively. This was carried out based on the fact that

$$CMOD_t = CMOD^e{}_0 + CMOD^* + CMOD^e{}_s \qquad \text{and} \qquad CMOD^* \approx \frac{CMOD_t}{\alpha^*}$$

The inelastic $CMOD^*$ in cementitious composites is mainly attributed to friction at rough crack surfaces, i.e. particle interlock or fiber bridging [7]. In accordance with this, Jenq and Shah [7] showed that α^* is slightly increasing i.e. $CMOD^*$ decreasing with decreasing maximum grain size (from $\alpha^* = 8.5$ for mortar with w/c = 0.45 to $\alpha^* = 10$ for cement paste with fine sand with w/c = 0.45).

Comparison of two mortars with same maximum grain size but with different w/c-ratios (0.65 and 0.45), i.e. strength, indicates a much more significant effect of strength on the factor α^*; α^* increased by 70% to 8.5 for the mortar with w/c = 0.45. Assuming a simplified linear relation for w/c-ratio a factor $\alpha^* \approx 11.5$ is derived for w/c = 0.28. Hence the inelastic $CMOD^*$ becomes approximately 5.1 μm ($CMOD^*_{80\%} \approx 4$ μm) and the nonlinear elastic $CMOD^e_s$ app. 19 μm, i.e. $CMOD^e = 53.5$ μm. With that the unloading compliance C_u is estimated to be (58.6 μm – 5.1 μm) / 6.947 kN which yields 7.7×10^{-3} μm/N. From equations (2) and (3) in [8] together with the Young`s modulus of the UHPC, the effective critical crack length $\underline{a} = a_0 + l_{ec}$ is then obtained ($\underline{a}$ = 67.59 mm) so that the critical stress intensity factor for mode I of the plain UHPC matrix can be calculated by

$$K^s{}_{Ic} = \frac{3 \cdot P_{max} \cdot l_s}{2 \cdot b \cdot h^2} \cdot \sqrt{\pi \cdot \underline{a}} \cdot F(\varsigma)$$

where l_s is the span between the two supports, b is the beam width, h is the beam depth and $F(\zeta)$ is a compliance function according to [8]. The mean effective crack length for the four beams tested was 63.62 mm with a coefficient of variation (CV) of 4.5%. The mean fracture toughness K^S_{Ic} was shown to be 1.79 MPaxm$^{1/2}$ with CV equal to 23.8%. Considering a simplified method assuming zero nonlinear inelastic behavior before peak load (LEFM, see Freese [9]) a mean K_{Ic} fracture toughness around 1.33 MPa m$^{1/2}$ is obtained. The mean critical crack tip opening displacement $CTOD_c$ calculated by equation (6) in [8] is then 14.41 μm (CV = 34.3%). The mean critical effective crack extension l_{ec} was found to be 13.62 mm (CV = 21%). The factor β necessary for application of the model [6, 7] was derived using the same assumptions as for α^* [7], hence $\beta = 1.18$. Combining fracture toughness of plain UHPC and fiber closing pressure one can easily compute the load-CMOD and load-deflection curves of three point bending beams using the model given in [6]. This will be the next step within the study and shown elsewhere.

5 Summary

The bonding of aligned steel fibers can be optimized by the use of appropriate fiber surface roughness and mix composition of UHPC. The compressive strength of UHPC has almost no effect on the bonding of aligned fibers. When considering inclined fiber pull-out, compressive strength (180 to 260 MPa) slightly affects bonding behavior. The approach presented here for fiber efficiency in fiber reinforced composites yields crack closing pressures in good agreement with literature.

Despite the very brittle failure of unreinforced UHPC, an appreciable amount of nonlinear prepeak crack growth was observed in three point bending beam tests. This crack growth can be accounted for with the help of a nonlinear fracture mechanics model, as e.g. the two parameter model according to Jenq and Shah [7]. Applying this model, a mode I fracture toughness of 1.79 $MPa \times m^{1/2}$ was obtained for the UHPC presented in this study. Based on the aforementioned crack closing pressure and the mode I fracture toughness the bending behavior of fiber reinforced UHPC can easily be modelled. Hence, optimization of fiber reinforced UHPC can be performed based on bond behavior of single fibers and fracture toughness of plain UHPC.

The authors gratefully acknowledge the financial support of the German Federal Ministry of Education and Research (WING Nanotecture program).

References

[1] Stengel, T.: Sustainability aspects of traffic bridges made from UHPC. In: Proc. AFGC / fib Int. Workshop on UHPFRC Designing and Building with UHPFRC, France (2009)
[2] Stengel, T., Schießl, P.: Sustainable Construction with UHPC. In: Proc. 2nd Int. Symp. UHPC, Germany (2008); ISBN 978-3-89958-376-2
[3] Swamy, R.N., Mangat, P.S., Rao, C.V.S.K.: The mechanics of fiber reinforcement of cement matrices. ACI SP-44, 1–28 (1974)
[4] Gao, Y.-C., Mai, Y.-W., Cotterell, B.: Fracture of fiber-reinforced materials. ZAMP 39(4), 550–572 (1988)
[5] Wecharatana, M., Shah, S.P.: A model for predicting fracture resistance of fiber reinforced concrete. Cement and Concrete Research 13, 819–829 (1983)
[6] Jenq, Y.S., Shah, S.P.: Crack propagation in fiber-reinforced concrete. J. of Struct. Eng. 112(1) (1986)
[7] Jenq, Y.S., Shah, S.P.: A fracture toughness criterion for concrete. Engineering Fracture Mechanics 21(5), 1055–1069 (1985)

[8] Shah, S.P., Carpinteri, A.: Fracture Mechanics Test Methods for Concrete. RILEM Report 5, pp. 4–6. Chapman and Hall, Boca Raton (1991)
[9] Freese, C.E., Baratta, F.I.: Single edge-crack stress intensity factor solution. Engineering Fracture Mechanics 73(5), 616–625 (2006)

Study on Size Effect in Bond Splitting Behavior of ECC

K. Asano and T. Kanakubo

University of Tsukuba, Tsukuba, Ibaraki, Japan

Abstract. This paper describes the test results of the pullout test in order to obtain the local bond behavior between Engineered Cementitious Composites (ECC) and steel reinforcing bar. There is a possibility that the size of cover thickness of ECC around reinforcing bar affects the orientation of fiber, so bridging performance of fiber is influenced by specimen size. To evaluate size effect, the main parameters of specimens are reinforcing bar diameter and cover thickness. The shape of specimen is a similarity shape based on reinforcing bar diameter. From the test results, bond strength tends to increasing with cover thickness. Moreover, bond strength decreases with increasing reinforcing bar diameter. To evaluate size effect, cylindrical volume around reinforcing bar is defined as highly-stressed volume. As for specimen of each cover thickness, the normalized bond strength decreases with increasing of highly-stressed volume.

1 Introduction

High Performance Fiber-Reinforced Cementitious Composites (HPFRCC), which show a strain hardening branch and multiple cracking under uniaxial tensile stress, have been focused by lots of researchers because of its unique mechanical performance. Engineered Cementitious Composites (ECC) exhibit a maximum tensile strain of several percent owing to the synergetic effect of high-performance fiber and specifically designed mortar matrix [1]. Unprecedented high-performance structural members can be expected when ECC is applied to seismic components [2].

It has been cleared that cementitious materials such as concrete show scale effect on their mechanical properties due to size of aggregates, existence of air void, and so on. In addition, it is considered that the fiber in fiber-reinforced cementitious composite causes scale effect which is mainly influenced by fiber orientation. For example, the small size of specimens such as plate type shows higher tensile strength and deformation capacity because of two-dimensional fiber orientation [3]. If we use some test pieces to check the mechanical properties of ECC, it

G.J. Parra-Montesinos, H.W. Reinhardt, and A.E. Naaman (Eds.): HPFRCC 6, pp. 137–144.

is necessary to have information about the relationships between properties obtained by test pieces and those in actual structures.

This paper describes the test results of the pullout test to obtain the local bond behavior between ECC and steel reinforcing bar. There is a possibility that the size of cover thickness of ECC around reinforcing bar affects the orientation of fiber as shown in Fig. 1, so bridging performance of fiber at the splitting crack is influenced by specimen size. To evaluate the size effect, similar specimens using several diameters of reinforcing bars with same ratio of cover thickness to bar diameter are tested. The test results are mainly discussed by bond strength and slippage.

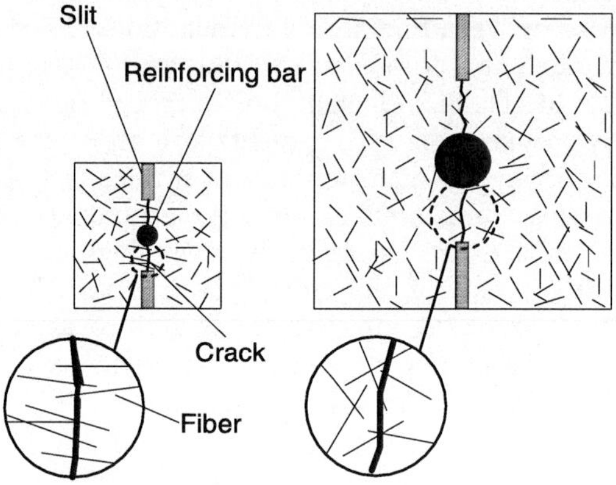

Fig. 1. Fiber orientation according to size difference

2 Employed Materials

Table 1 shows the characteristics of PVA (Polyvinyl Alcohol) fiber used in this study. The binders are ordinary Portland cement and fly ash. Fine aggregate is silica sand. The volume fraction of PVA fiber is 2.0%. Specimen is cast continuously to avoid the discontinuity of fiber. Table 2 shows the characteristics of ECC used in this study. Tensile strength and ultimate strain is calculated by 4-point bending test according to JCI-S-003-2007.

Table 3 shows the characteristics of reinforcing bar. The data of D22 was not obtained by data error. In this experiment, the bar does not yield.

Table 1. Characteristics of PVA fiber

Length (mm)	Diameter (mm)	Tensile strength (MPa)	Elastic modulus (GPa)
12.0	0.04	1690	40.6

Table 2. Characteristics of ECC

Fiber volume fraction (%)	Tensile strength* (MPa)	Ultimate strain* (%)	Compressive strength (MPa)	Elastic modulus (GPa)
2.0	4.24	1.25	47.9	16.2

* By 4-point bending test (JCI-S-003-2007)

Table 3. Characteristics of reinforcing bar

Type of bar	Yield strength (MPa)	Tensile strength (MPa)	Elastic modulus (GPa)
D10	378	533	192
D13	369	543	193
D16	442	665	190
D22			190*

* value of D16

3 Specimens

The shape of specimen is shown in Fig. 2. The slits are set in the two side of specimen to cause splitting cracks around reinforcing bar. The parameters of specimens are reinforcing bar diameter (d_b=10, 13, 16, 22 mm) and cover thickness (C). Cover thickness is adjusted by the size of slit. Three specimens are tested for each parameter, and total number is 48. Specimen list is shown in Table 4. Sectional size of specimen is set to square by 14 times of reinforcing bar diameter, and bond length is 4 times of the bar diameter. At the both ends of reinforcing bar of 1.5 times of the bar diameter, the bar is covered by Teflon sheet to insulate the bond to ECC. To investigate the size effect of bond behavior between ECC and reinforcing bar, size of specimen is proportional to the bar diameter.

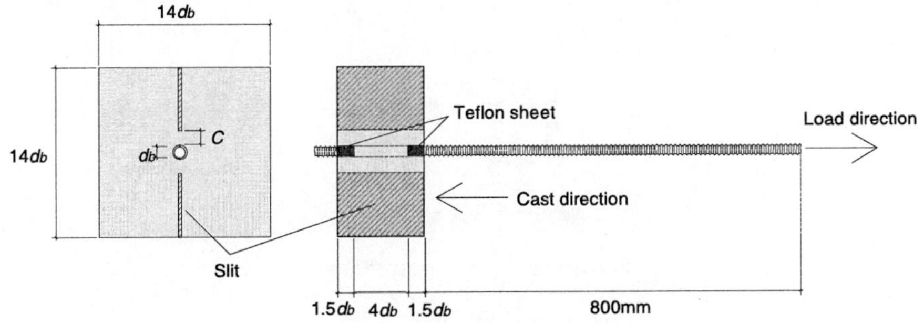

Fig. 2. Shape of specimen

Table 4. Specimen list

Name of specimen		Diameter d_b (mm)	Sectional size (mm)	C/d_b	Cover thickness C (mm)
D10-05-	1～3	10	140 × 140	0.5	5
D10-10-	1～3			1.0	10
D10-15-	1～3			1.5	15
D10-20-	1～3			2.0	20
D13-05-	1～3	13	182 × 182	0.5	6.5
D13-10-	1～3			1.0	13
D13-15-	1～3			1.5	19.5
D13-20-	1～3			2.0	26
D16-05-	1～3	16	224 × 224	0.5	8
D16-10-	1～3			1.0	16
D16-15-	1～3			1.5	24
D16-20-	1～3			2.0	32
D22-05-	1～3	22	308 × 308	0.5	11
D22-10-	1～3			1.0	22
D22-15-	1～3			1.5	33
D22-20-	1～3			2.0	44

4 Loading and Measurements

Fig. 3 shows the method of loading. The loading is conducted by monotonic pullout test. Teflon sheet is set between specimen and support plate to not restrict lateral displacement of ECC block. Measurement values are pullout load and free end slippage.

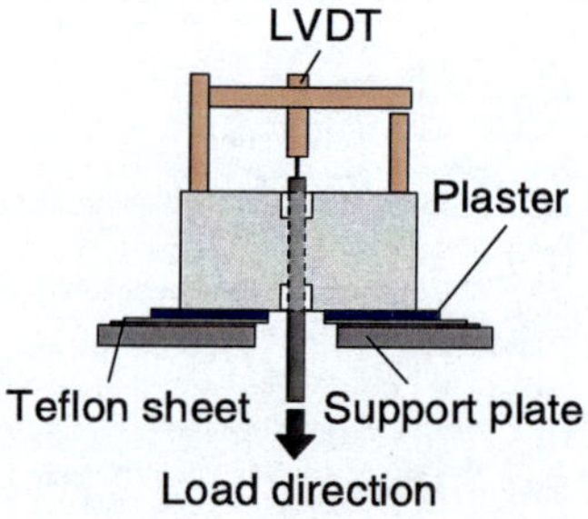

Fig. 3. Method of pullout test

5 Test Results

Fig. 4 shows the examples of sketch of specimens after loading. In case of the small cover thickness, splitting cracks were recognized between reinforcing bar and the slit at the load end. The reinforcing bar diameter increases, the tendency that the splitting cracks took place through both slits is observed. At the free end, the observed crack is fewer than the load end.

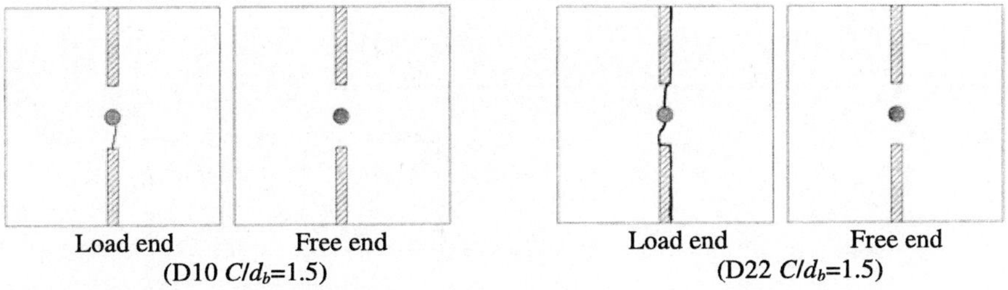

Fig. 4. Splitting crack after loading

Fig. 5 shows the relationship between bond stress and load end slippage (τ-s relationship). Bond stress is the average value of bonding area.

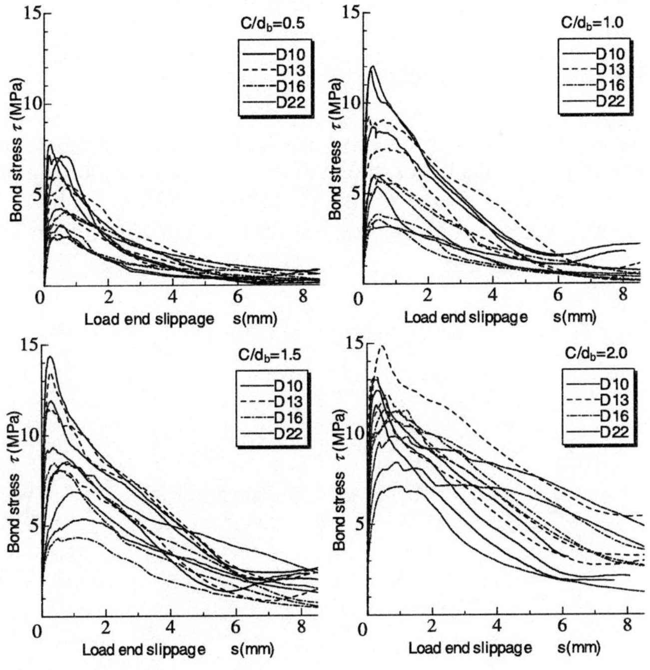

Fig. 5. Bond stress and load end slippage curves

The load end slippage is calculated by free end slippage adding elongation of reinforcing bar under the uniform bond stress assumption. After the maximum bond stress, the bond stress decreases gradually without sudden failure. The slope of decrement of bond stress becomes larger as the maximum bond stress increases.

Fig. 6 shows the relationship between the averages of bond strength and cover thickness. Bond strength (τ_{max}) tends to increasing with cover thickness. The size of reinforcing bar seems to not so effect on the increasing ratio. Fig. 7 shows the relationship between bond strength and reinforcing bar diameter. In all case, bond strength decreases with increasing reinforcing bar diameter.

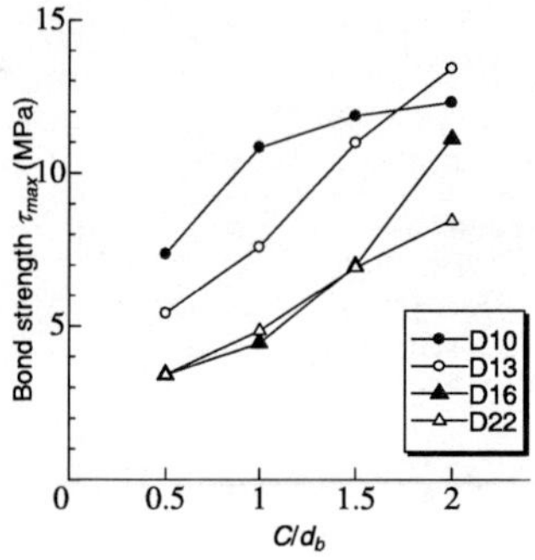

Fig. 6. τ_{max}-C/d_b relation

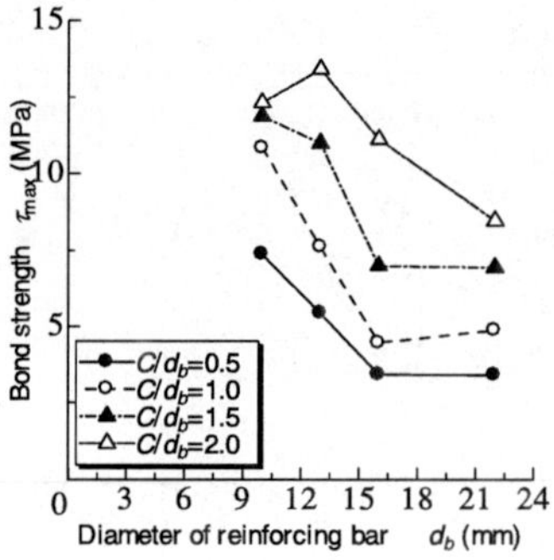

Fig. 7. τ_{max}-d_b relation

6 Size Effect on Bond Strength

It is known that the size effect of HPFRCC is expressed by the power of the highly-stressed volume [4]. In this paper, the evaluation of size effect for bond strength is conducted quantitatively by the result of pullout test. The highly-stressed volume in pullout test is defined as the volume of a cylindrical column as shown in Fig. 8.

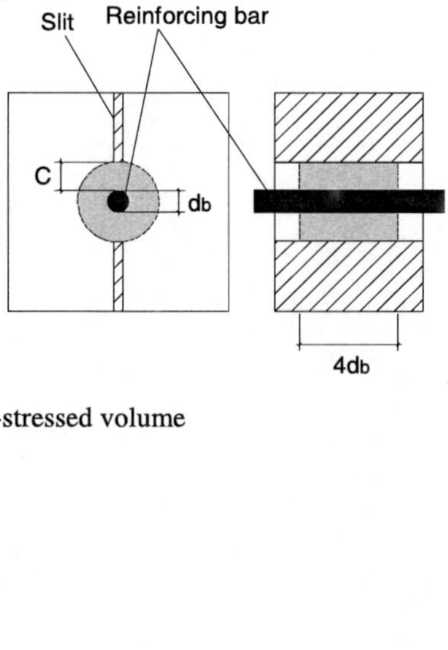

Fig. 8. Definition of highly-stressed volume

The experimental bond strength is standardized by the calculated value by Eq. (1) proposed by the previous study [5]. In the previous study, the specimens with D13 reinforcing bar were tested. The shape of specimens and loading method is same as this experiment.

Fig. 9 shows the relation between the normalized bond strength and the highly-stressed volume. As for specimen of each cover thickness, the normalized bond strength decreases with increasing of highly-stressed volume. However, the united evaluation to all specimens is difficult. It may be considered that other indexes which represent stress condition of ECC around the reinforcing bar will be required.

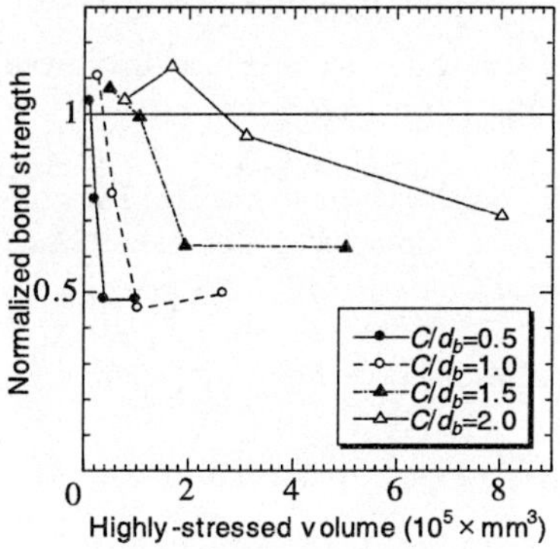

Fig. 9. Normalized bond strength-highly stressed volume relation

$$\frac{\tau_{max}}{\sigma_t} = \frac{2}{d_b}\left\{\left(r_i - \frac{d_b}{2}\right) + \frac{r_i\left(r_u^2 - r_i^2\right)}{r_u^2 + r_i^2}\right\}\cot\alpha \quad (1)$$

τ_{max} : bond strength

σ_t : tensile strength of ECC

d_b : diameter of reinforcing bar

r_i : radius of inner cracked zone $= 0.113\, r_u + 0.520\, d_b$

r_u : radius of modeled cylinder $= d_b / 2 + C$

C : cover thickness

α : angle of inner force to axial direction ($\cot\alpha = 1.46$)

7 Conclusions

In this study, the pullout bond test, in which the cover thickness and the diameter of reinforcing bar are selected as a parameter, is conducted to evaluate the size effect.

The findings obtained in this study are shown as follows.

1. In case of the small cover thickness, splitting cracks were recognized between reinforcing bar and the slit.
2. The size effect is clearly recognized on the bond strength evaluated by the ratio of cover thickness to diameter of reinforcing bars.
3. From the result of evaluation of bond strength using highly-stressed volume, the normalized bond strength decreases with increasing of highly-stressed volume.

Acknowledgement. The authors acknowledge the kind support of Okumura Corporation and Techno Material Co. for producing the specimens.

References

[1] Li, V.C.: From Micromechanics to Structural Engineering, The Design of Cementitious Composites for Civil Engineering Applications. Journal of Structural Mechanics and Earthquake Engineering, JSCE 10(2), 37–48 (1993)

[2] Kanda, T., Tomoe, S., Nagai, S., Maruta, M., Kanakubo, T., Shimizu, K.: Full Scale Processing Investigation for ECC Pre-Cast Structural Element. Journal of Asian Architecture and Building Engineering, Architectural Institute of Japan 5(2), 333–340 (2006)

[3] Kanakubo, T.: Tensile Characteristics Evaluation Method for Ductile Fiber-Reinforced Cementitious Composites, Japan Concrete Institute. Journal of Advanced Concrete Technology 4(1), 3–17 (2006)

[4] Yamada, K., Mihashi, H., Itagaki, N., Ishiyama, S.: Stochastic Study on Size Effect in Tensile Strength of Short-Fiber Reinforced Cementitious Composites. Journal of Struct Constr. Engng. (540), 7–12 (2001)

[5] Shigemizu, N., Yano, M., Hosoya, H., Kanakubo, T.: Study on Bond Splitting of SHCC Beam. In: Summaries of Technical Papers of Annual Meeting, Architectural Institute of Japan, pp. 115–118 (2010) (in Japanese)

Experimental and Nonlinear Finite Element Analysis of Fiber-Cementitious Matrix Bond-Slip Mechanism

C.S. Chin and R.Y. Xiao

Department of Urban Engineering, London South Bank University, UK

Abstract. The fiber-matrix bond-slip process is highly complex that involves cohesion, interfacial debonding, deformation with geometric nonlinearity, material plasticity as well as sliding frictional contact. It is also the governing failure mechanism during the post-peak phase of loading that significantly affects the ductility and toughness of fiber reinforced concrete (FRC). Therefore, a detailed analysis of the embedded mechanisms is essential to fully enumerate the composite behavior of FRC. This paper has been to experimentally and numerically investigate the complete bond-slip response of deformed metallic fiber pullout from cementitious matrix. Extensive validations using the experimental results obtained from recent experimental investigations and technical literature have been conducted and will be presented in this paper.

1 Introduction

Fiber-bridging-pullout is one of the main failure mechanisms of FRC composite under applied loads particularly during the post-cracking phase which causes the high energy absorption capacity of the material [1, 2]. The fiber-cementitious matrix interaction is normally studied by pullout test. To date, there is no standard testing method available. A variety of testing methods have been proposed by using either single-sided [3–7] or double-sided [8–11] configurations. Some of which have also been performed by adopting multiple fibers in a double-sided specimen [3, 8, 12, 13]. The most common variables researched up till now include: a) Fiber type [1, 3-7, 9, 10, 13]; b) Type of cementitious matrix [1, 3, 4, 6–13]; c) Fiber inclination angle [10, 13]; d) Fiber pullout rate [9]. In this paper, fiber-matrix bond-slip behavior has been measured on a specially designed double-sided specimen. The investigated parameters include the geometrical and cross-sectional dimension of the fiber. Nonlinear finite element analysis will also be presented where the numerical results have been validated by using experimental results.

G.J. Parra-Montesinos, H.W. Reinhardt, and A.E. Naaman (Eds.) HPFRCC 6, pp. 145–152.

2 Experimental Investigation

The fiber pullout specimen was designed to enable complete fiber embedment with an artificial pre-crack located at the centre of the specimen. A steel tube was embedded at each side of the specimen during the casting in order to establish connection to the loading grips by using the friction shafts. The floor-mounted closed-loop servo hydraulic testing machine was set-up at a loading rate of 0.10 mm/s and retrieves data concerning the load and slips every 200 ms; this was carried on until complete fiber pullout (or fiber fracture in some cases).

Five different types of fibers with different physical and mechanical properties have been used. These include: a) flat-crimped steel fiber (FC), b) flat-end steel fiber (FE); c) hooked-end steel fiber (HE); d) round-crimped steel fiber (RC); and e) double-anchored steel fiber (DA). The physical and material properties of all fibers employed are illustrated in Table 1. They can be summarized into two main categories where FC and RC are fully deformed fibers while FE, HE, and DA are end deformed fibers. The mortar mix with a water/cement ratio of 0.40 was adopted. The casting procedures were divided into two stages. One portion of the specimen was initially cast, with the fiber located precisely at the specified position and allowed to cure in room temperature. Extreme care was taken when positioning and aligning the fiber symmetrically during the first stage of casting. After 24 hours, a 0.10 mm thick plastic separator (with a hole slightly smaller than the corresponding fiber) was penetrated through the fiber and attached to the artificial cracking plane to prevent any bond established between the two half-specimens. An identical mortar mixture was then poured to the other portion of the mould and the assembly was further cured for 28 days at room temperature until tested [14, 15]. The overall testing results are summarized in Table 2.

The FC steel fiber has recorded the greatest maximum pullout load (the maximum resistance the fiber can offer to crack-widening [10]) followed by RC, FE, DA and HE steel fibers. For all fibers investigated, total dissipated pullout energy of between 1.4 and 10.8 Nm were obtained; the highest value has corresponded to FE steel fiber followed by RC, DA, FC and HE steel fibers.

It has been shown in Fig. 1 that the bond-slip curves behaved differently depending on the fiber types. The variations are caused by the geometrical dimension and tensile strength of the fiber. The curve can generally be divided into pre-peak and post-peak branches. The ascending portion up to the maximum pullout load represents the fiber-matrix debonding and initial fiber deformation caused by pullout load which is mainly affected by the fiber tensile strength and the total contact surface area of the fiber bonded in the mortar matrix. The descending part of the curve (corresponds to energy absorption capacity of the material) is significantly influenced by the mechanical anchorage and the geometry of the fiber. It is also attributable to the fiber straightening and loads transferred to the mortar by the frictional as well as shear stress developed between the fiber and matrix interface.

Table 1. Fibers, specimen dimension and testing apparatus

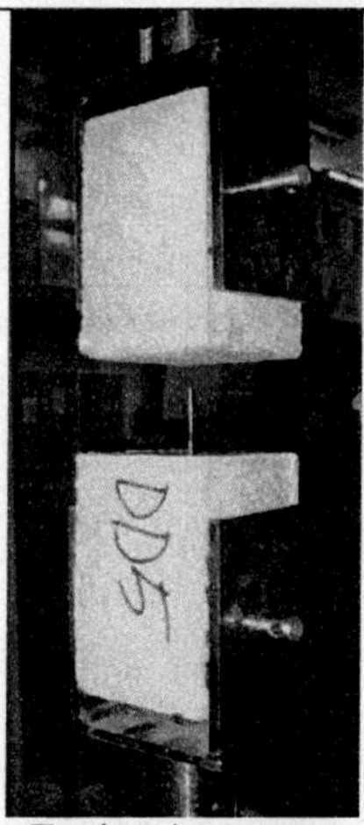

Testing Apparatus

Unit: mm

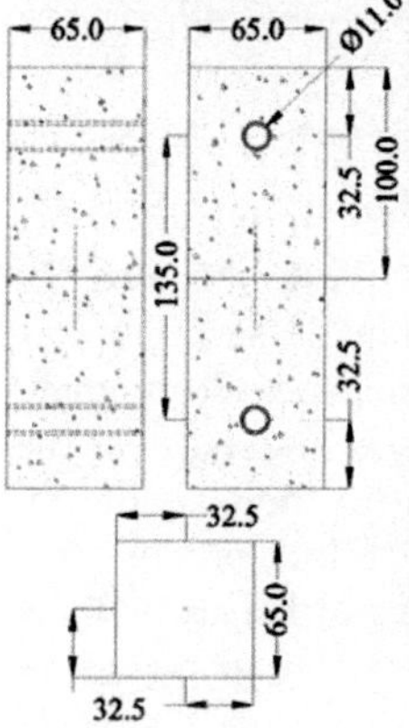

Specimen Dimension

Illustration	Fiber Length mm	Equivalent Diameter mm	Tensile Strength MPa	Elastic Modulus GPa
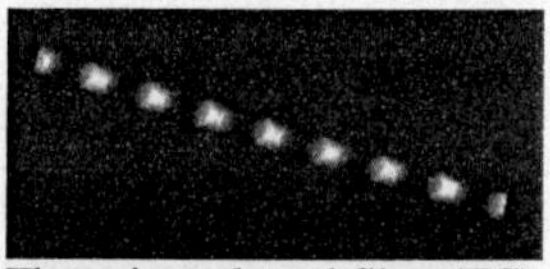 Flat-crimped steel fiber (FC)	50	1.14	825	200
 Flat-end steel fiber (FE)	50	1.00	1050	200
 Hooked-end steel fiber (HE)	50	1.00	1050	200
 Round-crimped steel fiber (RC)	60	1.00	1500	200
 Double-anchored steel fiber (DA)	50	1.00	1050	200

Table 2. Overall testing results

Label	Fiber Type	f_{cu} MPa	f_{sp} MPa	P_{max} N	E_t Nm
SFC	FC	45.6	2.1	1185.6	3.6
SFE	FE	44.5	1.9	585.8	10.8
SHE	HE	42.4	2.0	274.3	1.4
SRC	RC	42.4	2.0	933.0	7.3
SDA	DA	44,5	2.0	470.5	5.8

Note: 1) f_{cu} = concrete cube strength; 2) f_{sp} = cylinder splitting strength; 3) E_t = total dissipated pullout energy.

In general, the test results have revealed that less deformed fiber provides relatively higher dissipated pullout energy but with lower maximum pullout load while heavily deformed fiber will has much higher maximum pullout load but with less dissipated pullout energy (due to fiber fracture).

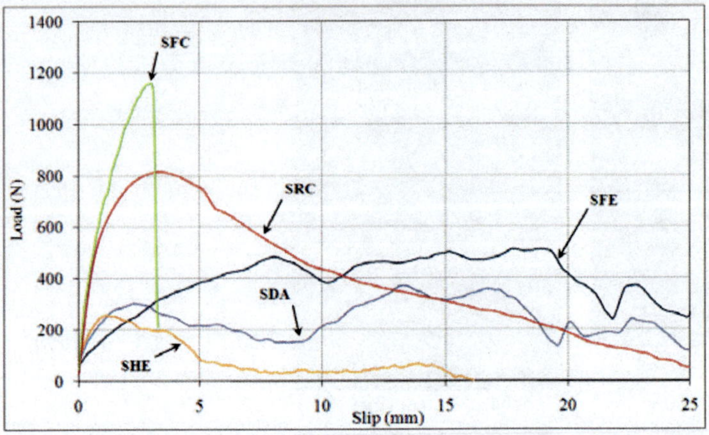

Fig. 1. Average bond-slip curve for each fiber types

Fully deformed fibers such as RC and FC tend to have greater initial gradients when compared to end-deformed steel fibers (e.g. FE and HE). However, heavily deformed FC fiber registered fiber fracture failure when its tensile strength was exceeded. For RC steel fiber, an equal possibility of fiber fracture and complete fiber pullout was observed. Minor matrix spalling has also been noticed near the fiber exit point during the pullout process. This is rarely seen for the case of end deformed fibers where they tend to straighten up during the pullout process followed by complete fiber withdrawal. Some of the enlarged ends of FE steel fibers were fractured followed by complete pullout.

3 Nonlinear Finite Element Analysis

3D solid elements were used to model the fiber and cementitious matrix. ANSYS [16] was selected as a modelling platform. The typical finite element mesh is illustrated in Fig. 2. Fiber pullout is a highly nonlinear process that requires in-depth understating of all involved mechanisms. For the fiber itself, cross-sectional and geometrical properties, nonlinear material behavior and its interaction with cementitious material are all the governing factors to accurately describe the bond-slip response. The plasticity behavior of steel fiber was introduced via von Mises yielding criteria. The failure criterion for cementitious matrix due to multiaxial state of stress used was the modified Willam and Warnke constitutive model [17].

The inter-connecting bond between the fiber and cementitious matrix (represented by contact elements associated with the corresponding deformable bodies) was described by the Coulomb friction model where two contacting surfaces would carry shear stresses up to a certain magnitude across their interface. Once the shear stress has been exceeded, the two surfaces will slide relative to each other. The development of the numerical model has been reported at different stages by the authors [18–20].

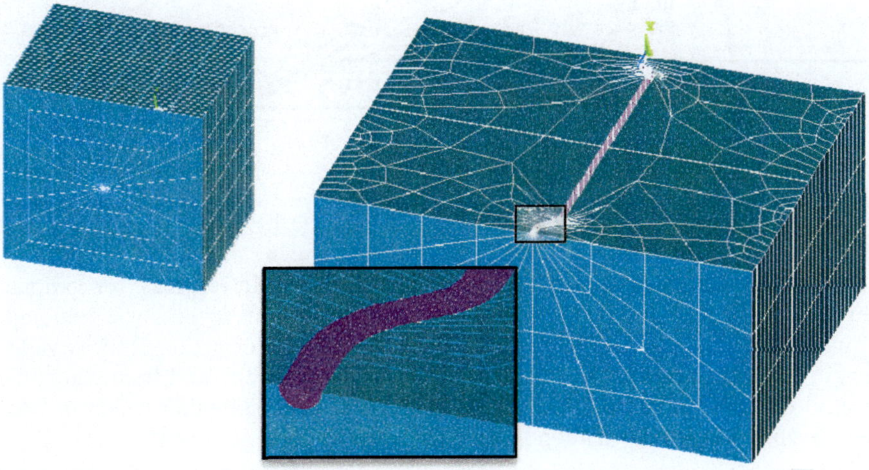

Fig. 2. Typical finite element mesh

Experimental data obtained from various literatures and current study [4, 5, 10, 13, 14] have been used for validation (Table 3). They consist of various steel fibers performed on single or double sided pullout specimens. Other variables include the fiber embedment length (l_{em}), fiber diameter (d_f) and tensile strength (σ_f).

Table 3. Summary of specimens used for validation

Label	Specimen Type Dimension	Fiber Type	Fiber Properties l_{em} mm	 d_f mm	 σ_f MPa
ST1 [4]	Single-sided 25.4 x 101.6 x 177.8 mm	Straight steel fiber	25.40	0.75	1000
ST2 [5]	Single-sided 150 x 150 x 150 mm		30.00	0.80	1000
HE1 [14]	Double-sided 2 x 65 x 65 x 100 mm	Hooked-end steel fiber	50.00	1.00	1050
HE2 [10]	Double-sided 2 x ϕ63.5 x 63.5 mm		60.00	0.80	1115
CR1 [13]	Double-sided 2 x ϕ100 x 100 mm	Corrugated steel fiber	50.00	0.85	1200
CR2 [10]	Double-sided 2 x ϕ63.5 x 63.5 mm		40.00	1.00	1037

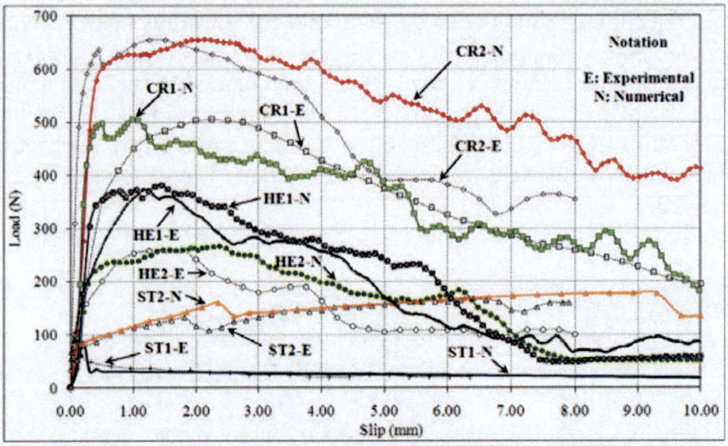

Fig. 3. The comparisons of bond-slip curves between experimental and numerical results

In general, the numerical model captures well the bond-slip response for various fiber types (Fig. 3). From the numerical analysis, the pullout process was initiated by simultaneous interfacial debonding and fiber elastic deformation until its yield strength was exceeded. It was then undergo material plasticity and frictional sliding until the maximum load was reached. The post-peak softening response was due to further plastic deformation and the frictional anchorage slips (Fig. 4).

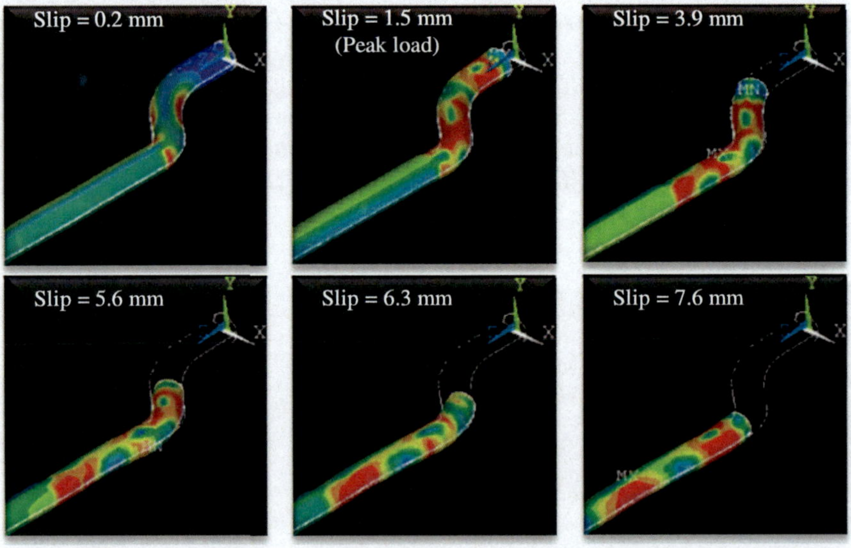

Fig. 4. Deformation of HE steel fiber of at various slip levels (HE1-N model)

4 Conclusions

The fiber pullout testing apparatus have shown its capability of accurately measuring the fiber-matrix bond-slip response for different types of fibers. The testing results have revealed that the bond-slip behavior can be significantly affected by the geometrical and material properties of the fiber. The proposed numerical model has shown to perform well where good correlations were obtained between the experimental and numerical results. Optimization of fiber with new mechanical deforming shape, length, cross section area, or material type can be carried out by using the numerical model presented herein.

References

[1] Groth, P.: Cracking in concrete - Crack Prevention with Air -Cooling and Crack Distribution with Steel Fibre Reinforcement. Luleå University of Technology, Division of Structural Engineering, Licentiate Thesis, 37L (November 1996)

[2] Li, V.C., Wang, Y., Backer, S.: Effect of Inclining Angle, Bundling and Surface Treatment on Synthetic Fibre Pull-Out From A Cement Matrix. Composites 21(2), 132–140 (1990)

[3] Burakiewicz, A.: Testing of Fibre Bond Strength In Cement Matrix. In: RILEM International Symposium on Testing and Test Methods of Fibre Cement Composites, pp. 355–365 (1978)

[4] Naaman, A.E., Najm, H.: Bond-Slip Mechanisms of Steel Fibers in Concrete. ACI Materials Journal 88-M17, 135–145 (1991)

[5] Weiler, B., Grosse, C.: Pullout Behaviour of Fibers In Steel Fiber Reinforced Concrete. Otto Graf Journal (Annual Journal on Research and Testing of Materials, Otto-Graf-Institute, Research and Testing Establishment for building and Construction [FMPA] – part of the University of Stuttgart, Germany) 7, 116–127 (1996)

[6] Shannag, M.J., Brincker, R., Hansen, W.: Pullout Behavior of Steel Fibers from Cement-Based Composites. Cement and Concrete Research 27(6), 925–936 (1997)

[7] Alwan, J.M., Naaman, A.E., Guerrero, P.: Effect of Mechanical Clamping On The Pull-Out Response of Hooked Steel Fibers Embedded In Cementitious Matrices. Concrete Science and Engineering 1, 15–25 (1999)

[8] Mandel, J.A., Wei, S., Said, S.: Studies of the Properties of the Fiber-Matrix Interface in Steel Fiber Reinforced Mortar. ACI Materials Journal 84-M12, 101–109 (1987)

[9] Banthia, N.: A Study of Some Factors Affecting the Fiber-Matrix Bond in Steel Fiber Reinforced Concrete. Canadian Journal of Civil Engineering 17, 610–620 (1990)

[10] Banthia, N., Trottier, J.-F.: Concrete Reinforced with Deformed Steel Fibers, Part 1: Bond-Slip Mechanisms. ACI Materials Journal 91-M43, 435–446 (1994)

[11] Banthia, N., Yan, C.: Bond-Slip Characteristics of Steel Fibers in High Reactivity Metakaolin (HRM) Modified Cement-Based Matrices. Cement and Concrete Research 26(5), 657–662 (1996)

[12] Wei, S., Mandel, J.A., Said, S.: Study of the Interface Strength in Steel Fiber-Reinforced Cement-based Composites. ACI Journal 83-53, 597–605 (1986)

[13] Chanvillard, G., Aïtcin, P.: Pull-Out Behavior of Corrugated Steel Fibers: Qualitative and Statistical Analysis. Advanced Cement Based Materials 4(1), 28–41 (1996)

[14] Chin, C. S.: Experimental and Computational Analysis of Fibre Reinforced Cementitious Composites. PhD Thesis, Civil & Computational Engineering Centre, Department of Civil Engineering, School of Engineering. University of Wales Swansea (2006)
[15] Chin, C.S., Xiao, R.Y.: Study on the Fibre-Mortar Bond-Slip Mechanism. In: 29th Cement & Concrete Science Conference, Leeds, UK, pp. 86–89 (2009)
[16] ANSYS, ANSYS Help Documentation, ANSYS Inc. (2010)
[17] Willam, K.J., Warnke, E.D.: Constitutive Model for the Triaxial Behavior of Concrete. In: Proceedings of the International Association for Bridge and Structural Engineering, ISMES, Bergamo, Italy, vol. 19, p. 174 (1975)
[18] Xiao, R.Y., Chin, C.S.: Nonlinear Finite Element Modelling of Fibre Pullout from Cementitious Matrix. In: 9th USA Congress on Computational Mechanics, Berkeley, USA, pp. 23–26 (2007) (invited paper)
[19] Chin, C.S., Xiao, R.Y.: Computational Analysis of the Mechanism of Metallic Fibre Pullout from Cementitious Matrix. In: 9th World Congress on Computational Mechanics and 4th Asian Pacific Congress on Computational Mechanics (WCCM/APCOM 2010), Sydney, Australia, p. 275 (2010)
[20] Chin, C. S., Xiao, R. Y.: Nonlinear Finite Element Modelling of the Fibre-Bridging-Pullout Mechanism Under Static Load. In: NAFEMS (National Agency for Finite Element Methods and Standards) World Congress 2011 - A World of Engineering Simulation, Boston, USA (2011) (paper accepted)

Part 3
Durability

Self-healing of Engineered Cementitious Composites in the Natural Environment

E.N. Herbert and V.C. Li

Department of Civil and Environmental Engineering, University of Michigan, USA

Abstract. This research investigates the self-healing behavior of Engineered Cementitious Composites (ECC) in the natural environment. ECC specimens were damaged to 0.5% tensile strain and allowed to heal outdoors, under random and sometimes extreme environmental conditions. Resonant frequency measurements and uniaxial tensile tests were used to quantify the rate and robustness of self-healing, while photo documentation was used to obtain visual evidence of self-healing products. It was found that there was a significant recovery of resonant frequency and stiffness in the damaged specimens after they were exposed to the natural environment. Specimens were able to recover up to 90% of their original, pre-damaged, resonant frequency values and up to 31% and 68% of their initial stiffness after one and three months of exposure. Photo documentation also showed self-healing in cracks up to 20 μm in width. This suggests that ECC is not only capable of self-healing in controlled laboratory conditions, but also in the natural environment.

1 Introduction

Cracks are unavoidable during the lifetime of a concrete structure. Structures directly exposed to the natural environment are susceptible to cracking not only from factors such as excessive loading and restrained shrinkage, but also from harsh environmental conditions. Cracking lowers the durability of concrete structures by creating pathways for harmful agents to penetrate the structure and potentially attack the reinforcing steel or the surrounding concrete. Cracks may also weaken the structure by negatively impacting the mechanical properties of the concrete. Therefore, the development of a concrete that can heal itself and regain this loss of performance due to cracking is highly desirable.

Studies have shown that cracked cementitious materials have the ability to heal themselves over time when exposed to water. It has been found that there is a gradual reduction in the permeability of damaged cementitious materials as water is allowed to flow through the cracks. This decrease in permeability is due to

G.J. Parra-Montesinos, H.W. Reinhardt, and A.E. Naaman (Eds.): HPFRCC 6, pp. 155–162.

diminishing crack widths as cracks are filled with healing products. In some extreme cases with small crack widths, cracks may heal completely, thus increasing the durability of the damaged material [1-3]. However, this is rare since most cementitious materials are brittle and incapable of achieving crack widths small enough to undergo self-healing.

ECC is a fiber reinforced cementitious composite that has been optimized through the use of micromechanics to achieve high tensile ductility and tight crack widths. ECC has a tensile strain capacity of 3-5% and develops extremely small microcracks (< 60 μm) under loading [4-6]. These tight crack widths are an intrinsic material property of ECC and promote robust self-healing behavior that is not easily attainable in brittle concrete with uncontrolled crack widths. It has been found that self-healing can occur in ECC under controlled laboratory conditions [7-8], and this study builds on that research by allowing ECC to heal outdoors, in the natural environment, under random and sometimes extreme environmental conditions.

2 Experimental Investigation

2.1 Mix Proportion and Raw Materials

The ECC mix proportion used in this study is given in Table 1. Type I ordinary Portland cement, Class F normal fly ash conforming to ASTM C618 requirements, fine silica sand with an average particle size of 110 μm, a polycarboxylate-based high range water reducer (ADVA® Cast 530), and polyvinyl alcohol (PVA) fibers were used to prepare the ECC material. The PVA fibers account for 2% of the total mix volume and are 12 mm in length with an average diameter of 39 μm. The fibers have a tensile strength of 1600 MPa, a density of 1300 kg/m^3, an elastic modulus of 42.8 GPa, and a maximum elongation of 6%. Also, the PVA fiber manufacturer coated the surface of the fibers with an oiling agent (1.2% by weight) to reduce the fiber/matrix interfacial chemical bond caused by the strong hydrophilic nature of the PVA fiber [4, 9].

Table 1. Mix proportion of ECC

Component	Cement	Fly Ash	Sand	Water	HRWRA[a]	Fiber
Weight %	27	33	22	16	0.4	1.3

[a] HRWRA: High Range Water Reducing Admixture

2.2 Specimen Preparation

A batch of ECC was prepared using a force-based Hobart mixer with a 20L capacity. Coupon specimens measuring 300 x 76 x 12.5 mm were then cast and covered with wet plastic sheets. Specimens were demolded after 24 hours and cut down to 200 mm in length to minimize bending stresses caused by misalignment during tensile loading. Specimens were then air cured at laboratory temperature and humidity (20 ± 1°C, 50 ± 5%) until testing. The day prior to testing, aluminum plates were glued to the ends of the coupons to facilitate gripping within the load frame.

All specimens were preloaded to 0.5% tensile strain at an age of 3 days. Uniaxial tensile loading was applied using a load frame (MTS Model 810) with a 25 kN capacity under displacement control and a loading speed of 0.0025 mm/s. Two Linear Variable Displacement Transducers (LVDTs) were attached to the specimens during loading to measure tensile deformation. When the tensile strain reached 0.5%, the tensile load was released and the samples were unloaded and removed from the load frame.

2.3 Natural Environment Exposure

After preloading, specimens were placed outdoors in a location where they would be fully exposed to all environmental conditions. Since this experiment took place in Michigan between February and May, the samples were exposed to rain, snow, and temperatures ranging from -14 to 28°C.

2.4 Self-healing Evaluation Methods

2.4.1 Photo Documentation

In order to obtain visual evidence of self-healing, images of cracks in the preloaded specimens were taken once a week. Specimens were chosen for photo documentation based on crack width data in hopes of determining the largest crack width capable of self-healing in the natural environment.

2.4.2 Resonant Frequency

Resonant frequency (RF) measurements based on ASTM C215 for the longitudinal mode were used to monitor the rate of self-healing in the preloaded specimens. Although the RF measurement technique is commonly used to evaluate concrete damage after exposure to freezing and thawing cycles, it has proven to be a useful method for determining the rate and extent of self-healing in ECC [8]. RF measurements were taken before and after preloading the specimens to quantify the amount of damage, and then twice a week throughout the duration of natural environment exposure to evaluate the rate of self-healing.

2.4.3 Uniaxial Tensile Test (Reloading)

To assess the robustness of the self-healing products, mechanical properties were measured under uniaxial tensile loading once the specimens were allowed to heal in the natural environment. After damage was induced during preloading, the specimens were placed outdoors and allowed to heal under various natural environment conditions. Specimens were then reloaded using uniaxial tensile tests and these mechanical properties (primarily stiffness) were then compared to those measured during preloading to determine the level of recovery. Two sets of specimens were used for reloading: one set was reloaded after 1 month of natural environment exposure, and the other was reloaded after 3 months of exposure. These reloading tests were conducted using the same method as described in Section 2.2 for preloading, but the specimens were reloaded until failure.

3 Experimental Results and Discussion

3.1 Photo Documentation

Fig. 1 shows a 10 μm crack before and after self-healing was allowed to occur. This crack healed gradually over time as various precipitation events occurred and, after 1 month of natural environment exposure, the crack appeared to be completely healed. This gradual type of healing was found to be typical of cracks that were 10 μm or less in width.

a) b)

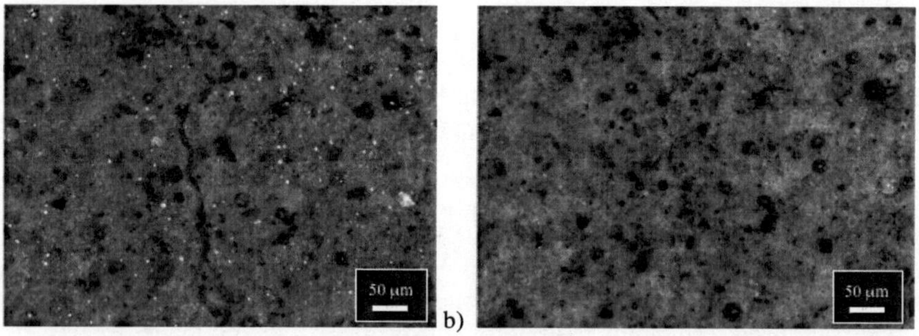

Fig. 1. 10 μm crack (a) after preloading to 0.5% and (b) after 1 month of natural environment exposure (48.3 cm of precipitation)

Fig. 2 shows another 10 μm crack that experienced a slightly different form of healing than the crack in Fig. 1. Fig. 2 (a) and (b) show the crack before environmental exposure and after the crack appears to be completely healed. However, in Fig. 2 (c), the crack becomes much more visible and, after 2 more weeks of environmental exposure, appears to be completely healed again in (d). Based on these images, it seems as though healing products are being washed out of cracks during heavy rainfalls. Although other studies [8, 10] have found that healing products are most likely a combination of calcium carbonate and C-S-H gel, these would not easily dissolve during large precipitation events. Therefore, it is believed that the healing products being washed out are calcium hydroxides (CH) which are formed during cement hydration. CH would normally be consumed during pozzolanic reactions, but these reaction kinetics are slow and a precipitation event could occur before the reaction is complete and dissolve the CH between the crack faces. This type of washout was not seen in all cracks, but more detailed studies will be required to confirm the chemical composition of the healing products.

Although most self-healing was visible in cracks that were 10 μm or less in width, healing products were seen in cracks as large as 20 μm. It has been found that crack widths must be below 150 μm for self-healing to occur under controlled laboratory conditions [8], however, no significant amount of healing products were observed in cracks larger than 20 μm when healing was allowed to occur in the natural environment.

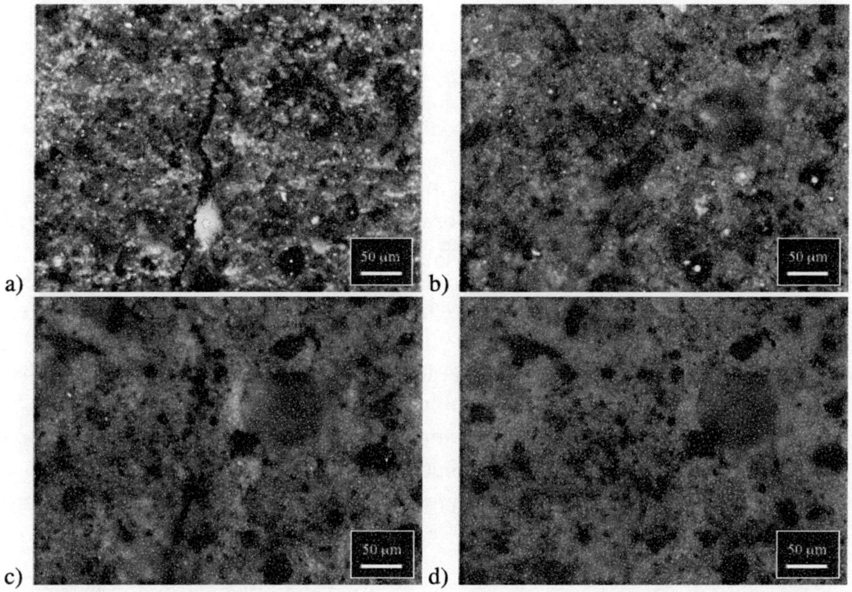

Fig. 2. 10 μm crack (a) after preloading to 0.5%, (b) after 2 weeks of natural environment exposure (22.2 cm of precipitation), (c) after 1.5 months (50.9 cm of precipitation) with possible washout of healing products, and (d) after 2 months (54.3 cm of precipitation)

3.2 Resonant Frequency

In addition to self-healing, continued hydration creates an increase in RF values. Therefore, the RF data in Figs. 3 and 4 has been normalized to account for the effects of continued hydration and these graphs represent the true RF recovery from self-healing. Due to the formation of cracks within a sample during loading, there was a large drop in RF values after samples were preloaded. However, it is clear that the RF values recover greatly even after only 1 month of natural environment exposure (Fig. 3). As shown in Fig. 4, the rate of self-healing drastically decreases after 1 month of natural environment exposure, but most samples were able to recover up to 90% of their original, pre-damaged, RF values.

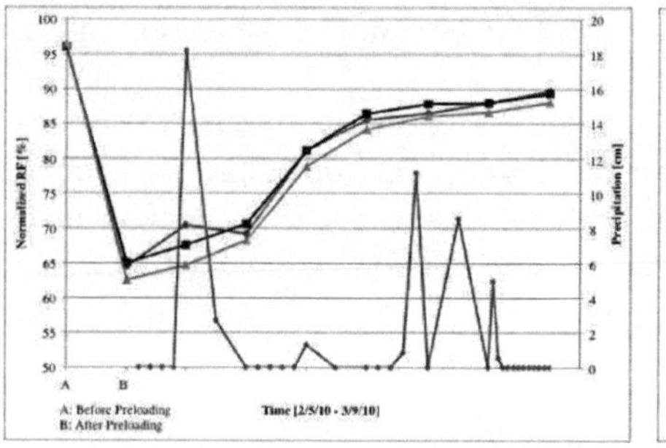

Fig. 3. Precipitation data and true RF recovery of samples reloaded after 1 month of natural environment exposure

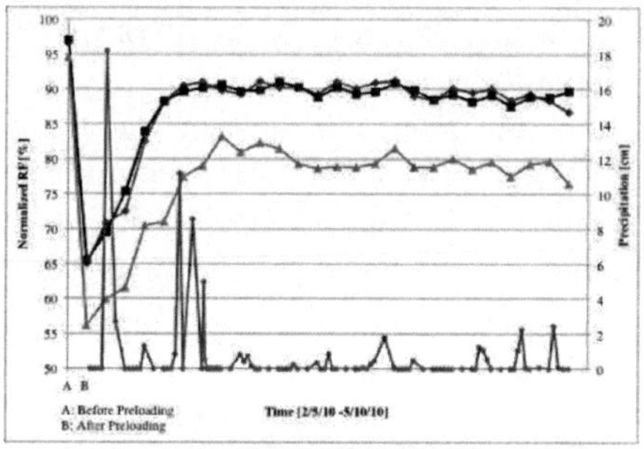

Fig. 4. Precipitation data and true RF recovery of samples reloaded after 3 months of natural environment exposure

Although most samples recovered to within 90% of their initial RF value, it has been seen that samples healed under laboratory conditions can recover up to 100% of their original value. Also, the rate of RF recovery seen in laboratory conditions is much more rapid than the rates seen in the natural environment [7-8, 10]. This is due to the fact that self-healing in the laboratory is achieved using consistent wetting and drying cycles, while self-healing in the natural environment is more sporadic and the duration of the precipitation events are usually much shorter than the duration of the wetting cycles used in laboratory conditions.

The RF data in Fig. 4 is fairly consistent after 1 month of natural environment exposure, but there are noticable fluctuations. These fluctuations correspond to precipitation events where the RF values decrease after one rainfall event and increase after the next. This is consistent with the observations made during photo documentation discussed in Section 3.1. The decrease in RF could be caused by the washout of healing products during heavy rainfalls, and the increase in RF would then be due to the reformation of healing products after the next precipitation event.

3.3 Uniaxial Tensile Test (Reloading)

Fig. 5 shows the preloading and reloading curves of 2 typical ECC samples that were reloaded at 1 and 3 months of natural environment exposure. For comparison, each graph also includes a sample that was reloaded without any self-healing. It can be seen that the samples exposed to the natural environment recover some of their initial stiffness due to self-healing. Samples were able to recover up to 31% and 68% of their initial stiffness after 1 and 3 months of exposure. Although this is a significant regain in stiffness, samples healed under laboratory conditions were able to recover up to 90% of their initial stiffness values [8]. Like RF recovery, the greater stiffness recovery in laboratory samples is due to the fact that self-healing in the laboratory is achieved through consistent wet-dry cycles, while healing in the natural environment is more irregular since the amount of precipitation and temperatures are constantly changing.

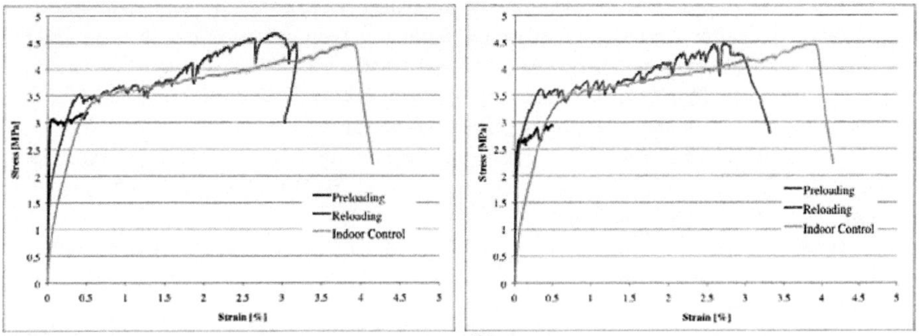

(a) (b)

Fig. 5. Preloading and reloading curves for (a) samples reloaded after 1 month of natural environment exposure and (b) samples reloaded after 3 months of natural environment exposure.

4 Conclusions

(1) Self-healing can occur in damaged ECC samples exposed to the natural environment, under random and sometimes extreme environmental conditions.

(2) Through photo documentation, it was observed that the majority of healing products formed in cracks that were less than 10 μm in width. However, healing products were seen in cracks up to 20 μm in width.
(3) Washout of healing products is a possibility due to the slow kinetics of the pozzolanic reaction.
(4) Samples recovered up to 90% of their initial RF values.
(5) Samples recoverd up to 31% and 65% of their initial stiffness after 1 and 3 months of natural environment exposure.
(6) Although self-healing in the natural environment is promising, it is not as robust as the self-healing seen under controlled laboratory conditions.

References

[1] Edvardsen, C.: Water Permeability and Autogenous Healing of Cracks in Concrete. ACI Materials Journal 96, 448–455 (1999)
[2] Granger, S., Loukili, A., Pijaudier-Cabot, G., Chanvillard, G.: Experimental Characterization of the Self-Healing of Cracks in an Ultra High Performance Cementitious Material: Mechanical Tests and Acoustic Emission Analysis. Cement and Concrete Research 37(4), 519–527 (2007)
[3] Wang, K., Jansen, D., Shah, S., Karr, A.: Permeability Study of Cracked Concrete. Cement and Concrete Research 27(3), 381–393 (1997)
[4] Li, V.C., Wang, S., Wu, C.: Tensile Strain-Hardening Behavior of PVA-ECC. ACI Materials Journal 98, 483–492 (2001)
[5] Li, V.C.: On Engineered Cementitious Composites (ECC): A Review of the Material and Its Applications. Journal of Advanced Concrete Technology 1(3), 215–230 (2003)
[6] Li, V.C.: From Micromechanics to Structural Engineernig – The Design of Cementitious Composites for Civil Engineering Applications. JSCE Journal of Structural Mechanics and Earthquake Engineering 10(2), 37–48 (1993)
[7] Li, V.C., Yang, E.H.: Self Healing in Concrete Materials. In: Zwaag, S.v.d. (ed.) Self Healing Materials: An Alternative Approach to 20 Centuries of Matierals Science, pp. 161–193. Springer, Heidelberg (2007)
[8] Yang, Y.Z., Lepech, M., Yang, E.H., Li, V.C.: Autogenous healing of Engineered Cementitious Composites under wet-dry cycles. Cement and Concrete Research 39, 382–390 (2009)
[9] Li, V.C., Wu, C., Wang, S., Ogawa, A., Saito, T.: Interface Tailoring for Strain-Hardening PVA-ECC. ACI Materials Journal 99, 463–472 (2002)
[10] Kan, L.L., Shi, H.S., Sakulich, A.R., Li, V.C.: Self-Healing Characterization of Engineered Cementitious Composites (ECC). ACI Materials Journal 107(6), 617–624 (2010)

Resistance to Corrosion Induced Cracking in Self Consolidating Hybrid Fiber Reinforced Concrete

G. Jen and C.P. Ostertag

University of California, Berkeley, USA

Abstract. Corrosion of reinforcing steel is a familiar form of environmental attack on reinforced concrete structures. Common design methodology attempts to combat this threat by means of a specified cover thickness which increases the diffusion length over which chlorides must travel and subsequently the time before corrosion can initiate. However, once active corrosion begins and expansion stresses at the rebar surface crack the cover concrete, this does little to prolong the life expectancy of the structure with respect to corrosion resistance. In an effort to enhance structural durability, we have developed a self-consolidating hybrid fiber reinforced concrete (SC-HyFRC) which improves the cover concrete's crack propagation resistance to expansion forces caused by rust formation at the rebar surface. By exposing reinforced concrete samples in an aqueous environment to a constant amplified current we are able to accelerate the rate of corrosion at the rebar surface and quickly develop an equivalent amount of rust formation to that which would be expected from multi-year exposure under marine environmental conditions. Electrochemical measurements taken post-acceleration provide us with corrosion rates which exhibit the improved performance of the self-consolidated hybrid fiber reinforced concrete relative to a standard self-consolidated concrete mix without fibers. In contrast to the non-fiber reinforced samples, hybrid fiber reinforced specimens did not exhibit any cracking after accelerated corrosion.

1 Introduction

The need for remedial actions to counteract corrosion damage in reinforced concrete structures is a constant problem in the inventory of US infrastructure, requiring billions of dollars annually. American Concrete Institute Committee 222 provides design guidelines including necessary precautions to be taken when undertaking new construction with respect to cover concrete properties and delaying corrosion initiation [1]. During the corrosion propagation phase, the recommended remedial actions often involve removal of said cover concrete due to its loss of protection caused by cracking, delamination and spalling. We propose the use of a hybrid fiber reinforced concrete with improved cracking resistance to control

G.J. Parra-Montesinos, H.W. Reinhardt, and A.E. Naaman (Eds.): HPFRCC 6, pp. 163–170.

corrosion rates at damage levels in which conventional concrete would require removal and repair.

One of our research programs explores the effect of fiber hybridization on durability. Various HyFRC mix designs were developed to resists deterioration of reinforced concrete structures exposed to mechanical and harsh environmental loading conditions as well as seismic loading conditions [2-5]. The rationale behind the use of a HyFRC based mix can be referred to as Deterioration Reduction through Micro- and Macrocrack Control, or the DRMC approach. PVA microfibers and hooked end steel macrofibers are used simultaneously to control the micro and macrocracks, respectively. Multi-scale crack control is essential for enhancing the durability and corrosion resistance of reinforced concrete structures [6-8]. Since cracking due to expansive deterioration processes such as corrosion initiates as microcracks in close vicinity to the steel reinforcing bar, these cracks need to be influenced at onset before they become macrocracks. Reducing crack growth and width, limits both the egress of the corrosion products away from the reaction site and the ingress of reactants to the corrosion site [7].

Despite the enhanced properties of high performance fiber reinforced composites, their use has been hampered by the inability of the fibers to pass through the congested areas of reinforcements in heavily reinforced concrete structures. Hence we developed a SC-HyFRC that does not require vibration during placement and is able to flow under its own weight, completely filling the formwork and achieving full compaction, even in the presence of congested reinforcements [5].

This paper provides quantitative and qualitative measurements on the effect of a self-consolidating hybrid fiber reinforced concrete on corrosion propagation rates in reinforced concrete. Recent corrosion propagation measurements with sodium chloride added to the mixing water to depassivate the steel rebar indicated levels of active corrosion ($i_{corr} \approx 1\ \mu A/cm^2$) but no significant difference in corrosion propagation rates between HyFRC and control specimens over a 70 week test period, if no cracks were present. However, accelerated corrosion tests revealed promising results. Accelerated corrosion testing provides an opportunity to quickly determine whether fiber reinforcement can influence the cracking behavior of cover concrete and alter the apparent rate of corrosion of steel reinforcement.

Potentiostatic tests performed by Li et al. [9] on the corrosion resistance of Engineered Cementitious Composite (ECC) reinforced elements provided inspiration for the galvanostatic testing protocol used in this study. While potentiostatic testing involves applying a constant potential, the galvanostatic test applies a constant current which allows the rate of mass loss to be tuned to match and corroborate prior data measurements.

2 Corrosion Tests

2.1 Test variables

Three concrete mixes are used to test for variations in cracking behavior to corrosion-induced damage. An accelerated testing system is adopted to achieve equivalent bar loss as would be experienced under long term exposure. Under Faraday's Law shown in Equation 1, mass loss (*M*) is a function of both current (*I*) and time (*t*). Manipulation of the system current reduces the time parameter proportionally to achieve equivalent charge transfer and mass loss.

$$M = \frac{ItA_w}{nF} \tag{1}$$

Where A_w = atomic weight of iron (55.85g), n = valency (2) and F = Faraday's constant (96,500 coulombs/equivalent mass).

Test samples are subjected to either 0, 24 or 48 hours of acceleration at a current equivalent to 1000 times higher than previously tested specimens were exposed to. Assuming unchanged corrosion rates, charge transfer equivalence would be reached by the undisturbed samples (i.e. 0 hours of acceleration) after 2.74 or 5.48 years for the 24 and 48 hours tests respectively.

2.2 Material Properties

Concrete mixes included a self-consolidating concrete (SCC) control mix and a self-consolidating hybrid fiber reinforced concrete (SC-HyFRC) mix. All mixes have an identical water-binder ratio of 0.45 using a Quickcrete Type I/II cement, and type C fly ash as a partial cement replacement. Aggregate is comprised of pea gravel with a maximum aggregate size of 10 mm and Vulcan sand with a measured fineness modulus of 3.2. Chemical admixtures used to produce self-consolidating behavior are supplied by BASF and consisted of Glenium 3030NS Superplasticizer (SP) and Rheomac 362 Viscosity Modifying Admixture (VMA). The hybrid fiber mixture, with a total volume fraction (V_f) of 0.015, contained Dramix ZP305 steel fibers (V_f = 0.013) by Baekert and Kuralon RECS15 polyvinyl alcohol (PVA) fibers (V_f = 0.002) by Nycon. Mix proportioning is shown in Table 1.

Representative axial compression and flexure response of a SC-HyFRC mix in comparison to a non-reinforced control sample are shown in Fig. 1a and 1b respectively. The SC-HYFRC revealed extended softening behavior in compression.

The flexural response is characterized by deflection hardening behavior due to stabilization and control of micro- and macrocracks bridged by PVA microfibers and steel macrofiber, respectively. The SC-HyFRC in table 1 was utilized in 1:4.5 scale reinforced concrete bridge columns to enhance their damage resistance when subjected to cyclic loading. The SC-HyFRC bridge columns were successful in resisting damage due to spalling of concrete up to drift ratios of 4% due to their enhanced crack growth resistance [5].

Table 1. Concrete mix composition (kg/m^3)

	Cement	Fly Ash	Water	Coarse Agg.	Fine Agg.	SP	VMA	Fiber V_f	
								RECS15	ZP305
SCC	397	131	237	497	1006	0.42	0.40	-	-
SC-HyFRC	397	131	237	418	1044	0.93	2.21	0.002	0.013

Where Superplasticizer (SP) and Viscosity Modifying Admixture (VMA) are given in weight percentage of binder materials.

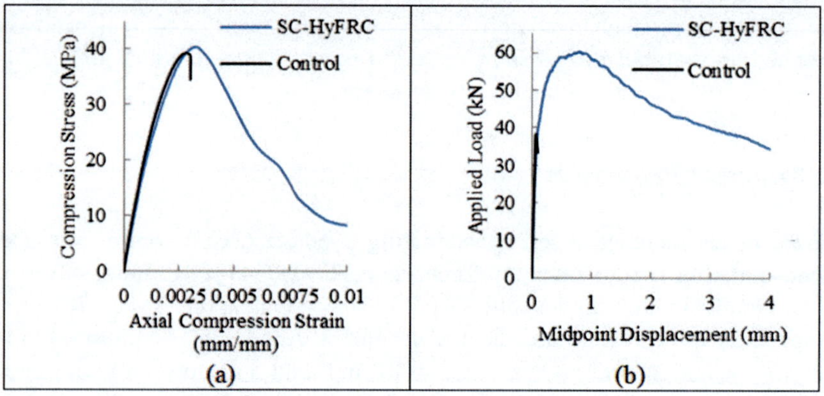

Fig. 1. Mechanical properties of SC-HyFRC specimens relative to non-fiber reinforced control samples; (a) axial compression test response, (b) third-point flexure test response

2.3 Specimen Configuration

Test samples cast from each of the 3 concrete mixes consisted of 101.6 mm by 203.2 mm cylinders, with a 254 mm long #8 reinforcing bar embedded along the central axis. The surface of the reinforcing bar was insulated along the lower 32 mm and upper 83 mm by a layered coating of Scotchkote electrical coating, vinyl electrical tape and a heat shrink tube sealed at either bar end. Minimum cover distance to the 140 mm length of exposed bar surface was 38 mm. The top end of each bar was drilled and threaded to accommodate a 6-32 x 3/8 stainless screw.

After casting and prior to subjecting the specimens to the corrosion propagation test protocol, samples were wet cured for 7 days, followed by a minimum of 21 days of dry curing.

2.4 Accelerated Corrosion Testing

Accelerated corrosion testing cycles were performed with a Gamry G750 Potentiostat, using the DC105 software package. Connection of the working electrode leads to the reinforcing bar was made through an insulated wire and ring terminal on the stainless steel screw. Counter electrode leads connected to a stainless steel mesh encircling the perimeter of the concrete cylinder. A saturated calomel electrode (SCE) was used for a reference electrode. Samples were immersed in 3.5 wt. % NaCl solution up to a height roughly one half inch from the top of the concrete cylinder. A continuous current of 111.5 mA (1 mA/cm^2) was applied for either 24 or 48 hours, after which samples were allowed to equilibrate in solution for 1 day before linear polarization resistance test monitoring commenced.

Polarization resistance and Tafel scans were performed on a testing schedule which included monitoring measurements at 1, 3, 7, 14, 28, 56 and 98 days following accelerated testing during which samples remained immersed in a salt water bath at room temperature. Linear polarization resistance measurements were conducted by sweeping potential ± 10mV from the corrosion potential. Tafel scan measurements required a potential sweep ± 75mV from the corrosion potential on companion samples having undergone the same exposure condition.

3 Test Results and Discussion

Corrosion propagation rates of the self-compacting specimens are given in Table 2. Samples in Table 2 are designated by concrete mix type (Self-Consolidating Concrete (SCC) and Self-Consolidating Hybrid Fiber Reinforced Concrete (SC-HyFRC)) and accelerated test duration (no acceleration (0), 24 hours (1) and 48 hours (2)). The corrosion rates were calculated using Equation 2 and normalized by the area of exposed reinforcing steel bar. Values of the Stern-Geary constant, *B*, are computed from the anodic and cathodic slopes of the Tafel Scan. Polarization resistance Rp was determined experimentally. The computed corrosion rates reveal different levels of corrosion. From a practical point of view corrosion rates less than 0.2 $\mu A/cm^2$ are considered negligible, whereas values above 0.5 $\mu A/cm^2$ are moderate and above 1 $\mu A/cm^2$ indicate high levels of corrosion as suggested by Andrade and Alonso [10]. A current density of 1 $\mu A/cm^2$ is equivalent to a rate of corrosion penetration into the steel of about 12 μm/year.

Using the criteria described above, both non-accelerated samples (SCC0 and SC-HyFRC0) are below 0.2 μA/cm^2 and hence are not yet corroding during the 98 day test period. The low corrosion rate is due to a lack of cracking (as will be discussed later) and hence low chloride penetration rates to the still passivated reinforcing bar surface. A quantitative improvement with the inclusion of fibers to the concrete matrix is evident after exposing the specimens to accelerated corrosion. Exposure to amplified current caused active corrosion in all samples but the SC-HyFRC exhibiting the lowest corrosion rates after both 24 and 48 hours. The largest difference in corrosion rate was observed between SSC2 and SC-HyFRC2. While the corrosion rate of SCC2 remained nearly constant during the 98 days of immersion in the salt water bath following the galvanostatic test, the corrosion rate of SC-HyFRC2 decreased and after the 98 day test period was an order of magnitude lower than its non-fiber reinforced counterpart (SCC2).

Table 2. Computed rates of corrosion (μA/cm^2) at sample monitoring intervals after accelerated testing

	1 day	7 days	28 days	98 days
SCC0	0.01	0.01	0.20	0.02
SCC1	6.2	3.3	0.68	1.3
SCC2	64	24	23	22
SC-HyFRC0	0.01	0.06	0.09	0.09
SC-HyFRC1	2.2	0.55	0.28	0.61
SC-HyFRC2	14	11	5.9	2.9

$$I_{corr} = \frac{B}{R_p} \tag{2}$$

Where B is the Stern-Geary constant and R_p is the polarization resistance.

A visual comparison of the SCC and SC-HyFRC cylinders provides insight into behavior and damage levels experienced between the different specimens. Surface appearance allows for a comparison of cracking patterns, and determination of the impact that fiber inclusion may have on cracking resistance of cover concrete. Representative photographs of cylinder surfaces taken after testing and 98 days of immersion in a salt water bath are shown in Fig. 2, with the section of surface shown being either representative of the entire perimeter or revealing only the most severe cracking. For those samples exhibiting cracking, a trace of the surface cracks and their crack width is provided as well.

As expected from the corrosion rate results in Table 1, none of the non-accelerated samples (SCC0 and SC-HyFRC0) exhibit any cracking behavior. Following the first level of galvanostatic testing, one of the self-consolidating control samples exhibits a crack at roughly mid height with a build-up of rust evident in the crack width (Fig. 2a), however, none of the fiber reinforced samples

(SC-HyFRC1) exhibit any cracking (Fig. 2b). At the highest level of current exposure, rust formation on the steel rebar surface was sufficient to cause a crack parallel to the length of the rebar and along the full cylinder height to form in both SCC2 samples. At this level of damage, remedial action in the field would likely require removal of the cover concrete and replacement with a patch. In contrast to the single extended crack formation of the non-fiber reinforced samples, hybrid fiber reinforced specimens (SC-HyFRC2) exhibit no such crack formations but isolated rust spots where fibers are located at the surface of the specimen.

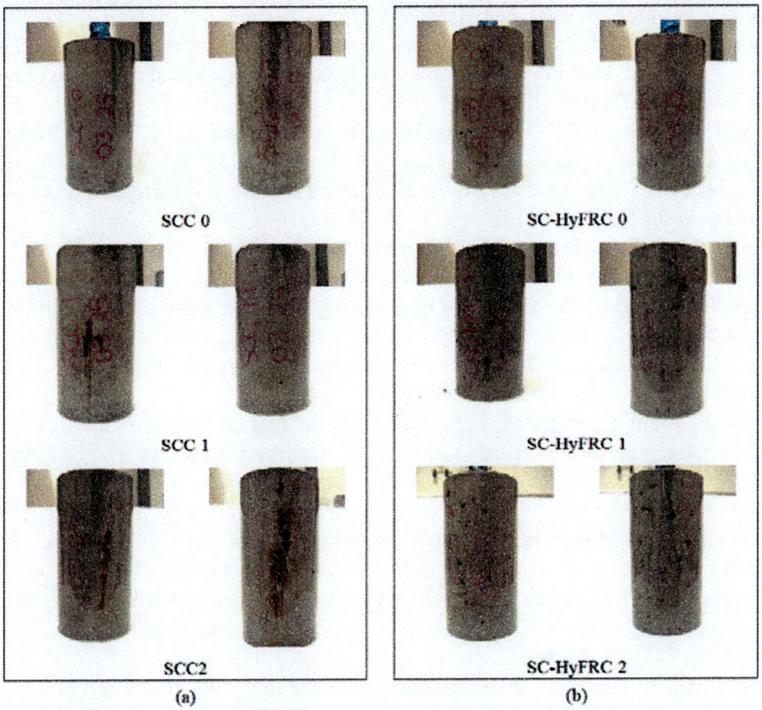

Fig. 2. Observed cracking patterns in self-consolidated specimens after 98 days of salt water immersion following galvano-static testing; a) SCC0, SCC1, SCC2 specimens, b) SC-HyFRC0, SC-HyFRC1, SC-HyFRC2 specimens

4 Summary

The influence of hybrid fiber reinforcement in resisting cracking of cover concrete induced by corrosion is demonstrated photographically. After 48 hours of continuous exposure to an amplified current of 1 mA/cm^2, samples incorporating a self-consolidating hybrid fiber reinforced composite resist the formation of a single large crack evident in their non-reinforced counterparts. Furthermore, in comparing the computed rates of corrosion occurring on the reinforcing bar surface after 98 days of storage in a salt solution, fiber inclusion is responsible for an order of magnitude reduction in corrosion rate.

Acknowledgements. The laboratory assistance of Mr. Lev Stepanov and Ms. Ekaterina Kostyukova and material donations from Bekaert and BASF are gratefully acknowledged.

References

[1] ACI 222R-01 Protection of Metals in Concrete against Corrosion. American Concrete Institute, Farmington Hills, Michigan (2010)
[2] Blunt, J., Ostertag, C.P.: A Performance Based Approach for the Design of a Deflection Hardened Hybrid Fiber Reinforced Concrete. ASCE Journal of Engineering Mechanics (135), 978–986 (2009)
[3] Blunt, J., Ostertag, C.P.: Deflection Hardening and Workability of Hybrid Fiber Reinforced Composites (HyFRC). ACI Materials Journal 106(3), 265–272 (2009)
[4] Ostertag, C.P., Blunt, J.: Use of Fiber Reinforced Concrete in Bridge Approach Slabs, Caltrans report No. CA09-0632, pp. 1–74 (2008)
[5] Kumar, P., Jen, G., Trono, W., Lallemant, D., Panagiotou, M., Ostertag, C.P.: Self Compacting Hybrid Fiber Reinforced Concrete Composites for Bridge Columns. PEER 2010/01. Pacific Earthquake Engineering Research Center, University of California, Berkeley (2010)
[6] Yi, C.K., Ostertag, C.P.: Mechanical Approach in Mitigating Alkali-silica Reaction. Cement and Concrete Research 35(1), 67–75 (2005)
[7] Grubb, J., Blunt, J., Ostertag, C.P., Devine, T.: Effect of steel microfibers on corrosion of steel reinforcing bars. Cement and Concrete Research 37(7), 1115–1126 (2007)
[8] Ostertag, C.P., Blunt, J.: Effect of crack control in hybrid fiber reinforced concrete composites on corrosion rate of steel reinforcing bars. In: FRAMCOS7, pp. 894–900 (2010)
[9] Sahmaran, M., Li, V., Andrade, C.: Corrosion Resistance Performance of Steel-Reinforced Engineered Cementitious Composite Beams. ACI Materials Journal 105(3), 243–250 (2008)
[10] Andrade, C., Alonso, C.: Corrosion rate monitoring in the laboratory and on site. Construction and Building Materials 10(5), 315–328 (1996)

Basic Creep under Compression and Direct Tension Loads of Self-compacting-steel Fibers Reinforced Concrete

E. Marangon, R.D. Toledo Filho, and E.M.R Fairbairn

Department of Civil Engineering, COPPE/UFRJ, Rio de Janeiro, Brazil

Abstract. Most previous research carried out in self-compacting steel fiber reinforced concrete (SCSFRC) have focused on the short-term mechanical properties, with few works dedicated to its creep behavior. Moreover, the study of creep plays an important role for the evaluation of the long-term service-state of concrete structures. In this work, it is presented an experimental investigation that includes tensile and compressive creep tests on SCSFRC. The self-compacting mixes were reinforced with volume fractions of 0%, 1% and 1.25% of steel fibers (35 mm and 65 mm). The results indicated that tensile creep is smaller than compressive creep and that fibers do not significantly change compressive creep.

1 Introduction

In the last few years researches on self-consolidating steel fiber reinforced concrete have greatly increased. It is well know that fibers improve the short-term mechanical characteristics of concrete, such as, compressive strength, flexural strength, toughness and cracking control. However, few studies are dedicated to the creep behavior of such composites [1].

Most of the research related to creep of concrete concerns compressive creep. This is due to the fact that compressive creep tests are easier to be realized. Creep under tension is, however, very important because concrete structural members are often submitted to tensile stresses fields that can originate cracking compromising the durability and the integrity of structures. As an example, if creep is not precisely considered, the thermal stress at the early ages can be overestimated for massive structures such as dams. On the other hand tensile creep strains leading to cracking can occur under certain conditions for the long-term behavior of nuclear power plants [2]. Moreover, fibers are mainly used to improve the tensile behavior of concrete.

Tensile creep tests are not used frequently due to the intrinsic complexity of testing concrete under tension. Additionally, strain rates are very low, which can lead to inaccurate measures [3], [4].

G.J. Parra-Montesinos, H.W. Reinhardt, and A.E. Naaman (Eds.): HPFRCC 6, pp. 171–178.

In the literature, there is no consensus about the role of steel fibers in concrete creep. According to [5] the addition of steel fibers can reduce creep. On the other hand, [6] and [7] showed that creep is higher for concrete reinforced with steel fibers. Studies carried out by [8] have reported a negative influence of steel fibers on the tensile creep behavior of concrete because the fibers can possibly create voids in the cement paste structure as a consequence of the reduction of the workability. Another hypothesis is related to the creation of a porous region around the fibers, similar to the transition zone between aggregates and cement paste.

Considering the lack of consistent information in the literature, this work brings a contribution about the creep behavior of self-compacting concrete reinforced with steel fibers.

2 Methodology and Materials

The compressive creep tests were carried out on cylindrical specimens 300 mm long and 150 mm diameter, as suggested by NBR8224/83 [9] standards. The tensile test was carried out using prismatic specimens of dimensions 100 x 50 x 400 mm. For each concrete, four samples were cast. Two of them were used for long-term creep measurement and the other two were used to measure shrinkage and temperature strains of unloaded specimens (control samples). Strains were measured with strain gages embedded within the concrete mass. Thermometers were embedded in the concrete to monitor the temperature variations in the bulk of concrete specimens. After casting the samples were covered with damp cloths until demoulding. After 48 hours the specimens were demoulded and sealed with plastic film plus aluminum ribbon to prevent water loss to the external environment. For the creep test, two samples were loaded with constant load equivalent to 40% of the resistance obtained from the compressive and tensile test. Tests were carried out in the creep laboratory room with controlled relative humidity (50 ± 3%) and temperature (20 °C ± 1). The creep test setup are shown in Figs. 1e and 1f.

The concrete was produced using slag cement (from Brazilian standards CPIII-40 with up to 52 % of slag), fly ash, and silica fume, having specific weights of 3.02 g/cm^3, 2.35 g/cm^3, and 2.28 g/cm^3, respectively. Coarse aggregates of two sizes were used: 19 mm (2.68 g/cm^3) and 9.5 mm (2.64 g/cm^3). Two sands had been used, a natural sand of river with specific weight 2.44 g/cm ^{3} and the other one a quartz of 325 mesh and specific weight of 2.68 g/cm^3.The superplasticizer (SP) was an ether carboxilate with a total amount of solid particles of 31.2%. Two types of hooked end steel fibers with aspect ratios of 65 and 80 were used. The fiber volume fractions were 1% and 1.25%.

Four different self-compacted concrete mixtures were produced: the matrix (or reference) named C2; SCSFRC with fiber content of 1% (in volume) and aspect ratio of 80 named C2.1%80; SCSFRC with fiber content of 1.25 % (in volume) and aspect ratio of 65 named C2.1,25%65; SCSFRC with fiber content of 1.25 % (in volume) and aspect ratio of 80 named C2.1, 25%80. The proportions of the mixtures are summarized in Table 1. All concretes presented self-compacting characteristics [10].

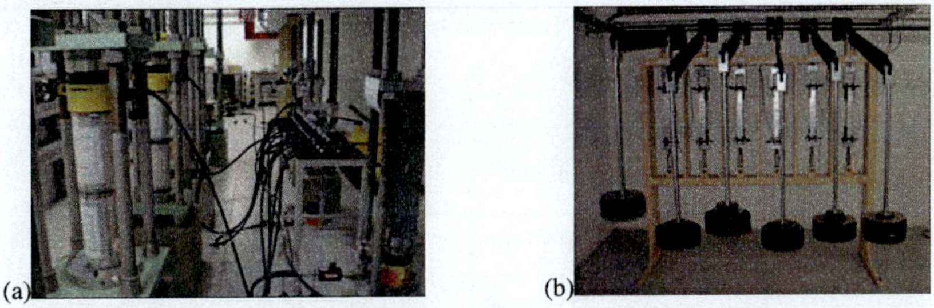

Fig. 1. Creep test setup (a); Compressive creep (b) tensile creep

Table 1. Proportions of concrete mixtures (kg/m³)

Mixtures	C2	C2.1%80	C2.1,25%65	C2.1,25%80
Aggregate 01	354	323	317	317
Aggregate 02	152	136	136	136
Sand	949	949	949	949
Silica 325	70	70	70	70
Cement	360	360	360	360
Fly Ash	185	185	185	185
Silica Fume	48	48	48	48
SP	17	28	46	46
Viscosity modifying agent	0.36	0.36	0.36	0.36
Water	158	159	150	150
Steel Fibers	0	78.0	97.5	97.5

2.1 Compressive and Tensile Strength Tests

Cylindrical samples 200 mm long and 100 mm diameter were tested under compressive loads. For the tensile test, the samples had the dimensions: 100 mm x 100 mm x 400 mm. After casting, the specimens were covered with a damp cloth to prevent water loss for 48 hours. Then the samples were demoulded and cured in a moist curing room (100% RH) during the 28 days.

The short-term uniaxial tensile and compressive tests for determining strength and elastic modulus were carried out in a servo-controlled testing machine with a capacity of 1000 kN. The samples were tested under compression at a loading rate of 0.02 mm/min. For the tensile tests it was used a loading rate of 0.1 mm/min. Three samples were tested for each mixture and its mean values were taken as representative.

3 Results

3.1 Compressive and Tensile Short-Term Properties

The mean and the coefficient of variation (CV) of the compressive strength (f_c), modulus of elasticity (E_c) and axial strain (ε_a) for the self-compacting mixtures are given in Table 2. Furthermore, Table 3 displays the mean and CV obtained for the tensile strength (f_t), the axial strain (ε_p) and the ratio between the maximum tensile stress and the cracking stress (ft_{ci}/ft_{c0}).

Table 2. Compressive strength, axial strain and modulus of elasticity

Mixtures	Compressive Strenght (MPa)		Axial Strain (με)		Modulus of Elasticity (GPa)	
	f_c	CV	ε_a	CV	E	CV
C2	70.20	2.76	2740	0.43	35.65	0.90
C2.1%80	70.32	0.25	2810	2.69	35.26	2.31
C2.1,25%65	70.34	1.21	3080	0.86	34.55	1.48
C2.1,25%80	76.06	3.48	3180	3.85	35.13	4.42

Table 3. Tensile strength and axial strain

Mixtures	Tensile Strength (MPa)				
	f_t (MPa)	CV (%)	ε_P (με)	CV (%)	ft_{ci}/ft_{c0}
C2	4.63	8.41	139	8.41	1.00
C2.1%80	4.65	8.93	192	24.39	1.01
C2.1,25%65	4.84	4.21	177	2.00	1.04
C2.1,25%80	5.02	6.31	814	39.15	1.08

3.2 Basic Creep in Compression

Fig. 2 shows the compressive creep curves for the reference and fiber reinforced concrete. The age of loading was 28 days and creep strains were measured until the age of 156 days. Then the sample was unloaded allowing computing the Young´s modulus at 156 days. The values of the Young´s modulus obtained with the creep test rig ($\bar{E}_c$) for loading and unloading and of the maximum compressive creep strains are given in Table 4. The obtained results indicate that the presence of the steel fibers, for the aspect ratios and volume fraction studied, did not change significantly the compressive creep behavior of the self-compacting concrete. The influence of fibers on compressive creep can also be shown using the creep coefficient (creep strain/elastic strain) which is displayed in Fig. 3.

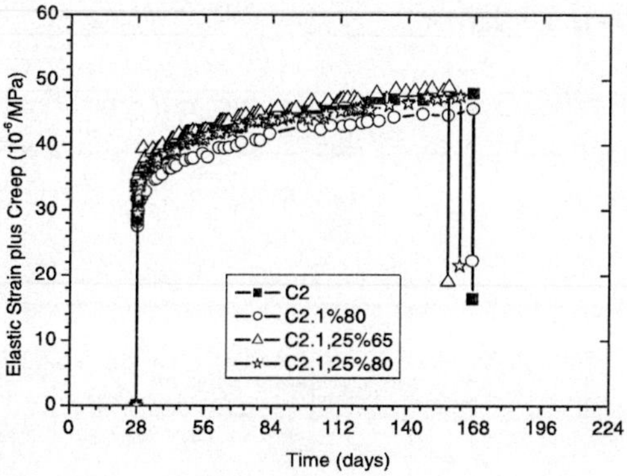

Fig. 2. Compressive creep curves for the studied mixes

Table 4. Young´s modulus and creep under compressive loading and unloading

Mixtures	Loading		Unloading
	$\bar{E}_{c\,28}$ GPa	Max. creep* $\cdot 10^{-6}$/MPa	$\bar{E}_{c\,156}$ GPa
C2	32.4	17.19	38.0
C2.1%80	36.2	17.35	41.4
C2.1,25%65	32.8	16.81	36.6
C2.1,25%80	33.8	17.48	40.8

* Creep values measured after 128 days of loading

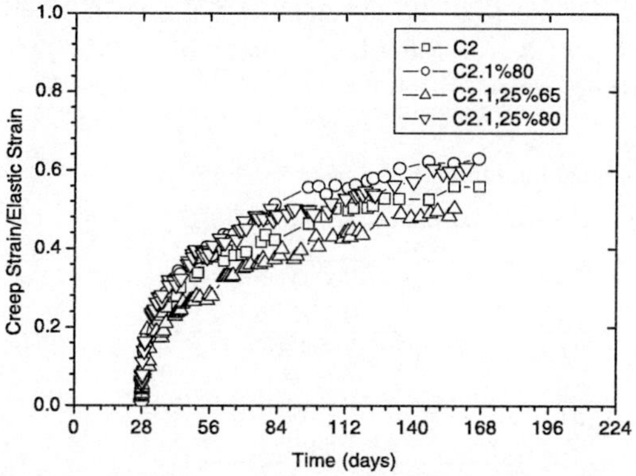

Fig. 3. Compressive creep coefficient curves for the studied mixes

The obtained results indicate that the compressive creep coefficient of the studied mixes (after 128 under loads) ranged between 0.52 and 0.62 with the SCC matrix presented a compressive creep coefficient of about 0.58. The mixtures C2.1%80 and C2.1.25%80 presented a slight increase (smaller than 5-8%) in the creep coefficient whereas the mixture C2.1,25%65 reduced the compressive creep coefficient by about 11%.

3.3 Basic Creep in Tension

Fig. 4 shows the tensile creep curves for the reference and fiber reinforced concretes. The age of loading was 28 days and creep strains were measured until the age of 120 days.

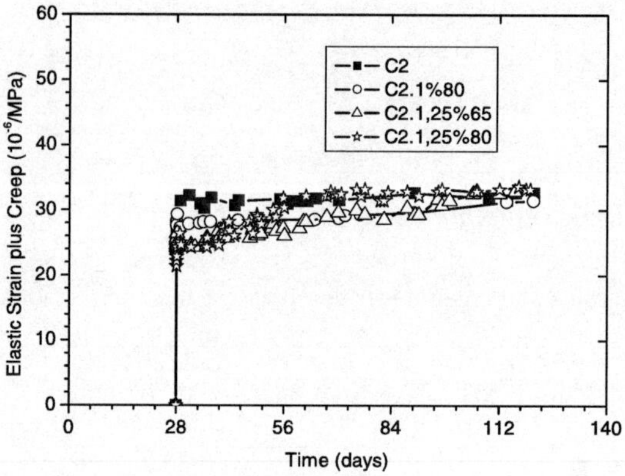

Fig. 4. Tensile Creep curves for the studied mixes

The values of the elastic modulus (Ēc) and of the maximum creep strain are given in Table 5. It was observed that the greater the volume of fibers the greater the elastic modulus. Otherwise, for the fiber content of 1.25% the elastic modulus increased with the aspect ratio.

Table 5. Elastic modulus and creep under tension loading

Mixtures	Loading	
	$\bar{E}_{c\,28}$ GPa	Max. creep* $\cdot 10^{-6}$/MPa
C2	37.4	6.89
C2.1%80	40.5	5.71
C2.1,25%65	41.6	7.50
C2.1,25%80	42.5	11.78

* Creep values measured after 92 days of loading

Fig. 5 shows the curves corresponding to the evolution of the creep coefficient for tensile loading. In what concerns tensile creep, the effect of fibers was not evident. For a volume fraction of 1% the creep of the SCC matrix was reduced whereas for volume fraction of 1.25% the tensile creep increased. Otherwise for the C2.1,25 concretes creep increased for increasing aspect ratios

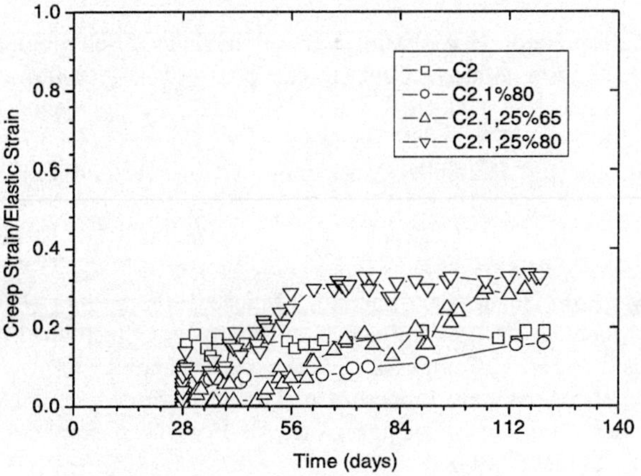

Fig. 5. Tensile- Creep coefficient: comparison for the studied concretes

4 Concluding Remarks

For the self-compacting fiber reinforced concrete studied in this paper the introduction of different types of fibers did not change significantly the compressive creep behavior. This tendency was not verified in reference [7] where the use of fibers increased the creep strains. However, in this case only the matrix was self-compacting what can explain the differences with the results presented in the present research.

In what concerns tensile creep, the effect of fibers was not evident. For 1% of fiber content creep decreased and for 1.25% creep increased. Otherwise for the C2.1,25 concretes creep increased for increasing aspect ratios. More studies are, therefore, necessary to allow a complete understanding of this behavior.

Tensile creep was smaller than compressive creep for all the concretes studied in this paper. This conclusion agrees with the results of French researchers of IRSN and LCPC [2].

Acknowledgements. The authors would like to thank the Federal University of Rio de Janeiro, CNPQ, CAPES, FAPERJ, ANP, FURNAS Centrais Elétricas and ANEEL for their support.

References

[1] Wüstholz, T., Reinhardt, H.W.: Deformation Behaviour of Self-Compacting Concrete under Tensile Loading. Materials and Structures 40, 965–977 (2006)
[2] Reviron, N., Nahas, G., Tailhan, J.-L., Le Maou, F., Benboudjema, F., Millard, A.: Experimental Study of Uniaxial Tensile Creep of Concrete. In: CONCREEP 2008, Creep, Shrinkage and Durability Mechanics of Concrete and Concrete Structures, vol. 1, pp. 453–457 (2008)
[3] Neville, A.M., Dilger, W.H., Brooks, J.J.: Creep of Plain and Structural Concrete. Construction Press (1983)
[4] Neville, A.M.: Propriedades Do Concreto, 2 Edição, Editora Pini (1997)
[5] Zhang, J.: Modeling of The Influence of Fibers on Creep of Fiber Reinforced Cementitious Composite. Composites Science and Technology 63, 1877–1884 (2003)
[6] Balaguru, P., Ramakrishnan: Properties of Fibre Reinforced Concrete: Workability, Behavior under Long-Term Loading, And Air-Void Characteristics. ACI Material Journal 85(3), 189–196 (1988)
[7] Velasco, R.V., Toledo Filho, R.D., Fairbairn, E.M.R., Silvoso, M.M.: Basic Creep of Steel Fibers Reinforced Composites. In: 8th International Conference on Creep, Shrinkage and Durability of Concrete And Concrete Structures 2008, Ise-Shima, vol. 1, pp. 735–739 (2008)
[8] Bissonnette, B., Pigeon, M., Vaysburd, A.M.: Tensile Creep of Concrete: Study of Its Sensitiviy to Basic Parameters. Aci Materials Journal 104-4, 360–368 (2007)
[9] Associação Brasileira De Normas Técnicas. Concreto Endurecido – Determinação Da Fluência: NBR 8224. Rio De Janeiro (1983)
[10] Efnarc. Specification and Guidelines for Self-Compacting Concrete. Applications to reproduce all or part of this document should be made to EFNARC at the address shown on the title page. Reino Unido (2002), http://www.efnarc.org
[11] Velasco, R.V.: Concretos Auto-Adensáveis Reforçados Com Elevadas Frações Volumétricas De Fibras De Aço: Propriedades Reológicas, Físicas, Mecânicas E Térmicas. Tese De Doutorado, COPPE/Universidade Federal Do Rio De Janeiro, Rio De Janeiro, p. 349 (2008)

Hot and Residual Behavior of Steel Fiber-Reinforced Structural Shotcrete Exposed to High Temperature

P. Bamonte, P.G. Gambarova, and A. Nafarieh

Politecnico di Milano, Milan, Italy

Abstract. Sprayed concrete (*shotcrete*) has been known for almost a century as a reliable and effective material for rock stabilization, fire proofing of metallic structures and jacketing of R/C members. Shotcrete structural utilization, however, has been very limited so far, but the situation is changing, as proposals have been lately put forward for the use of shotcrete in rather demanding structures, like the linings of blasted-off tunnels, where fire resistance is a must.

The lack of information on shotcrete's behavior at high temperature and past cooling is the starting point of this research project aimed to investigate the thermo-mechanical behavior of a number of shotcrete mixes (with alkaline or alkali-free accelerating agents, with or without steel fibers), in order to check whether – and to what extent – shotcrete's temperature-triggered mechanical decay is similar to that of ordinary concrete.

Here the compressive behavior of two alkali-free shotcrete mixes is examined at high temperature (*hot tests*, T = 20-600°C) and past cooling (*residual tests*, T = 20-750°C), and the thermal diffusivity is evaluated as well (T = 150-900°C). The two mixes (v_f = 0, 0.4%, expected strength f_c = 40 MPa) behave similarly to ordinary concrete, and exhibit a very low thermal diffusivity at any temperature.

1 Introduction and Nature of Problem

Shotcrete is predominantly used for supporting, stabilizing and water-proofing rock layers in tunnels and other underground constructions [1] (Fig. 1), and for jacketing R/C members, as it adheres to the surfaces via bond mechanisms that become effective in a very short time and are accompanied by a very rapid setting of the material itself. Shotcrete is often sprayed on rock surfaces immediately after the excavation of the bore to stabilize its arch-shaped top (*first-phase lining*). Hence, completing the lining (*final lining*) with structural shotcrete instead of reinforced concrete is very appealing.

Shotcrete's structural use in tunnels, however, is rather demanding in terms of fire resistance, toughness and permeability. As fires in tunnels are probably the

G.J. Parra-Montesinos, H.W. Reinhardt, and A.E. Naaman (Eds.): HPFRCC 6, pp. 179–186.

most dangerous – even if highly unlikely – accidents, the concerns about shotcrete's fire resistance should be dispelled, to have it accepted by designers and owners as a structural material. Unfortunately, no information is so far available on shotcrete's behavior at high temperature and past cooling. On the other hand, the larger amount of cement and fine aggregates generally found in shotcrete, and the relatively-high content of accelerating agents may alter its high-temperature behavior with respect to ordinary concrete [2-4].

Fig. 1. Structural use of shotcrete (left) in a cylindrical shield for the protection of an oil tank (France, Atlantic Coast); and use of sprayed mortar (right) for water-proofing the lining of an urban tunnel (Italy, Naples), both from Realtà Mapei, V. 96, N.9, 2009.

To have first-hand information on shotcrete's high-temperature behavior, a research project is currently in progress at the Politecnico di Milano on three shotcretes sprayed by means of the "wet" process (Mix C1, sodium silicates as accelerating agent and no fibers [5] expected strength in compression 15-20 MPa; Mixes C2F and C2, sulfo aluminates as accelerating agent, with/without steel fibers, f_c = 40-45 MPa). Here reference is made to the two alkali-free mixes.

2 Specimens, Test Set-Up and Experimental Campaign

2.1 Mix-Design and Compressive Strength of the Virgin Materials

The shotcretes investigated in this project (Mixes C2 and C2F, Table 1) are used in the construction of the first-phase lining of the service tunnels, that are required for the assemblage and dismantling of the tunnel boring machines operating in the excavation of the 54 km-long Brenner Base Tunnel, which will be a direct connection across the Alps, between Italy and Central Europe through Austria.

The expected compressive strength (C 28/35 - S5, f_c = 40 MPa) was confirmed by the mean compressive strength measured on cores in Milan (close to 50 MPa for Mix C2 and to 45 MPa for Mix C2F, at 60 days). To this end, an alkali-free accelerating agent containing sulfo-aluminates was used. (As it is well known, using alkali-free accelerating agents makes the mechanical properties of the shotcrete very close to – or even better than – those of the base material). The aggregate was in dry – surface conditions before the mixing operations and contained roughly 2.2% of water (36 kg/m^3). The steel fibers were Dramix hooked fibers (length L = 32 mm; diameter ∅ = 0.8 mm; L/∅ = 40).

Table 1. Mix design and properties of shotcretes C2 and C2F

Concrete (Portland cement)	C2	C2F
Sand (0-4 mm) [kg/m^3]	1324	1314
Aggregate : granite (4-8 mm) [kg/m^3]	336	336
Cement "c" [kg/m^3]	450	450
Added water "w" [kg/m^3] (w/c)	198 (0.44)	198 (0.44)
Superflux "sp" [kg/m^3] (sp/c)	4 (0.9%)	4 (0.9%)
Sulfo Aluminates "sa" [kg/m^3] (sa/c)	29 (6.4%)	29 (6.4%)
Fibers [kg/m^3] (v_f)	0	30 (0.4%)
Total mass [kg/m^3]	2341	2361
Target strength f_c at 28 days	40 MPa	40 MPa
Strength prior to testing (60 days)	50 MPa	45 MPa

The mean compressive strength measured by BBT at 28 days on short cores (∅/h = 1) and corrected for the slenderness (∅/h = ½) was 38/41 MPa (C2/C2F), in good agreement with that measured in Milan on slender cores at 90 days (50/45 MPa). Finally, the cubic compressive strength of the base concrete, measured in Milan at 90 days, was 51/53 MPa, which means roughly 40 MPa (cylindrical strength) at 28 days. Hence, there are no measurable differences between the base concrete and the shotcrete, contrary to what occurs - and is well known - when accelerating agents containing alkali silicates or aluminates are used.

2.2 Casting and Curing of the Specimens

The material for the specimens was collected directly in the building site of Mauls Nord, and the casting was performed according to typical on-site procedures.

The moulds filled with shotcrete were brought to the laboratory of the Department of Structural Engineering in Milan, and a few days later the formworks were removed. The blocks were cured in moist conditions for 30 days (T = 20°C; RH = 90%) and then were cored; the end sections of the cores (∅,h = 80,160 mm) were ground and polished according to ASTM C39-81 Standard.

The cores were kept in the above-mentioned moist conditions for 30/45 days prior to be thermally cycled and tested (residual tests, fall of 2009). The remaining cores were stored in a cellar (T = 15-25°C; R.H = 60-85%) for one year, prior to being heated and tested (hot tests, fall of 2010; f_c^{20} after one year = 48/47 MPa for Mix C2/C2F, practically coincident with the values at 60 days).

For each mix, 35 cores were extracted, 2 for measuring the thermal diffusivity, 18 for the residual tests (six thermal levels from 20 to 750°C), and 15 for the hot tests (five thermal levels from 20 to 600°C); in each sub-case, three nominally-identical cores were tested.

2.3 Thermal Cycles

Prior to the mechanical tests, the specimens were subjected to a thermal cycle (Fig.2a), in order to induce various levels of mechanical damage. The reference temperatures were 20, 105, 200, 400, 600°C for both residual and hot tests + 750°C for hot tests only. Each cycle was carried out in an electric furnace (T_{max} = 1000°C; size of the chamber 300 × 500 × 600 mm), and consisted of a heating phase, a rest at the maximum temperature (*reference temperature*) and a cooling phase, according to a well established procedure [6]. In the residual tests, the rest period was fixed (2 hours). In the hot tests, the rest period was variable (from 2 hours to 4.5 hours, see the solid circles in Fig. 2a) and the maximum temperature was set at a value 5% higher than the reference temperature, in order to have at least 2/3 of the specimen at or above 90% of the reference temperature during the 10-15 minutes requested by (a) the extraction from the furnace, (b) the placement of the insulated specimen between the press platens, and (c) the test itself.

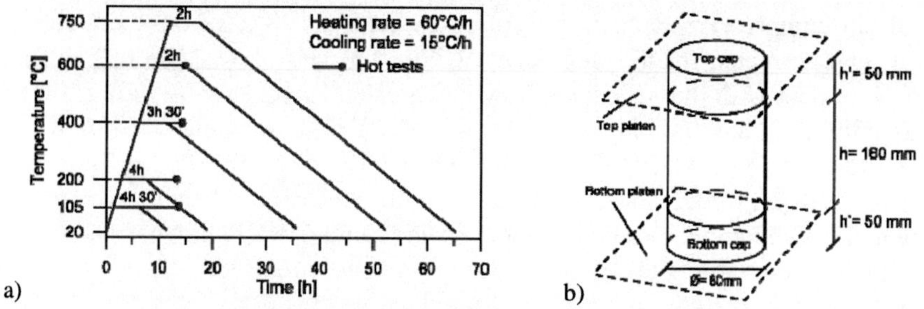

Fig. 2. (a) Thermal cycles: for the residual tests (heating + 2-hour rest + cooling), and for the hot tests (heating + variable rest and testing = solid circles); and (b) sketch of a specimen prior to be tested at high temperature (the insulating shield is not shown).

2.4 Test Set-Up and Instrumentation

The residual tests in compression were carried out by means of an electromechanical SCHENCK press (max. capacity = 1000 kN). All tests were displacement-controlled and the control signal was provided by the LVDT located inside the press. The shortening of the specimen was monitored via 3 LVDTs connected by means of an elastic ring and placed at 120° astride the mid-height section (base length = 50 mm). The displacement rate of the press platens was 0.005 mm/s for both the loading and unloading processes; hence, in each test the mean loading rate up to the peak load was roughly 0.6 MPa/s (virgin specimens).

The same loading rate was adopted in the force-controlled tests at high temperature, for which an electro-hydraulic CONTROLS press was used (max. capacity = 2500 kN). The shortening of the specimen was measured by means of two LVDTs in direct contact with the press platens (base length = 260 mm, Fig. 2b). The relationship between the "apparent" elastic modulus evaluated over the 260 mm-base length and the "local" modulus evaluated on the specimen over the 50 mm-base length was calibrated by testing a few virgin specimens. In this way, the values of the secant moduli were all referred to the base length of 50 mm, and the possible effects of specimen-to-cap and cap-to-platen contacts, as well as of caps deformability were taken care of.

In Fig. 3a a specimen instrumented for evaluating the thermal diffusivity is shown prior to being heated to 900°C. The same set-up was used to heat the specimens to be tested in residual conditions. (The steel pipe guarantees the homogeneity of the thermal field around the specimen). In Fig. 3b two insulated specimens to be tested at high temperature are shown prior to being heated. In order to keep the average temperature of the specimens as close as possible to the reference temperature past the extraction from the furnace and during the test, all specimens to be tested in hot conditions were laterally insulated by means of a 15 mm-thick hollow shield made of high-grade alumina silica and alumina fibers (thermal diffusivity from 1/4 to 1/2 of that of shotcrete). During the loading process, two 50 mm-thick HPC caps were placed between the end sections of the specimen and the platens, to insulate the specimen and to protect the press.

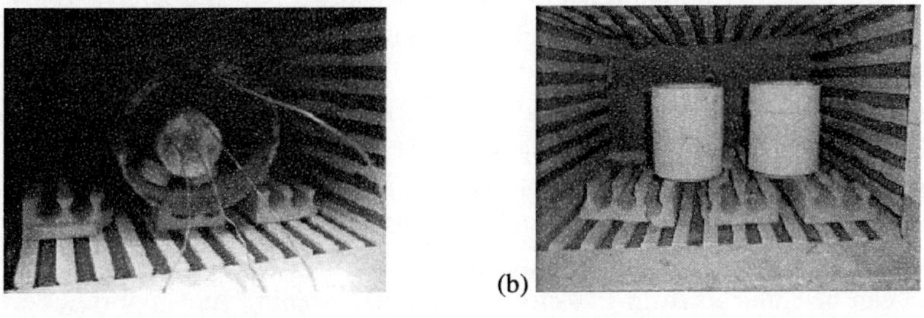
(a) (b)

Fig. 3. (a) Instrumented specimen for evaluating the thermal diffusivity and for thermally-cycling the specimens to be tested in residual conditions; and (b) two specimens placed inside the insulating shields prior to be heated and tested at high temperature.

3 Results

3.1 Thermal Diffusivity

The thermal diffusivity D (Fig. 4) was measured on instrumented cylindrical specimens (one per mix), in accordance with the theoretical solution of slowly-heated long cylinders (Ø/h ≥ 2) subjected to a constant heating rate.

Each cylinder was instrumented with three embedded thermocouples (Fig. 3a), the first placed close to the heated surface, the second along the axis and the third in between, all in the mid section.

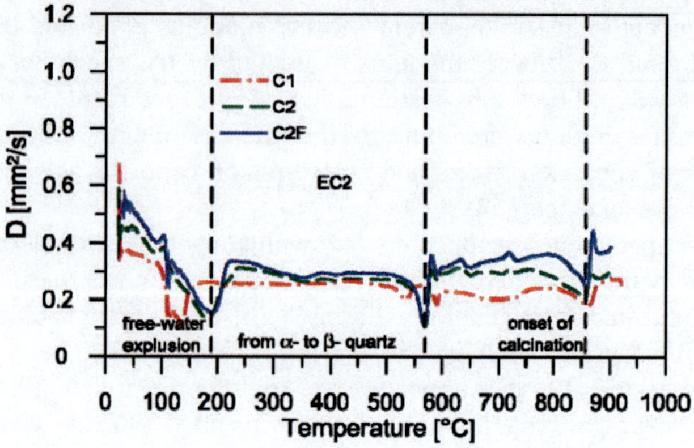

Fig. 4. Plots of the thermal diffusivity as a function of the temperature

3.2 Stress-Strain Curves in Compression and Normalized Quantities

The mean residual and hot curves are shown in Figs. 5a-d. Each mean curve was obtained by (a) normalizing each single curve ($\sigma_c/\sigma_{c,max}$ and $\varepsilon_c/\varepsilon_{c1}$), (b) working out the mean normalized curve, and (c) reconstructing the mean stress/strain curve on the basis of the mean values of $\sigma_{c,max}$ and ε_{c1}.

As should be expected, the hot curves exhibit higher peak values and (slightly) lower stiffness values in the loading branches (Fig. 5e).

As shown in Fig. 5f, the strain at the peak stress tends to increase with the temperature, but there is a common trend up to 400-500°C.

In Fig 6a the greyish envelope represents the provisions by EC2. A similar cloud can be inferred from EC4 for the residual strength (not reported here). According to EC4, the residual strength is always lower (by 5% at 100°C and by 10% above 200°C) than the hot strength, but the differences frequently found in performing hot/residual tests are much higher (more than 25-30%).

In Fig. 6b, the greyish envelope encompasses the values of the elastic secant modulus of many tests examined by Phan and Carino [7], and again the results concerning the shotcretes agree well with those pertaining to ordinary concrete.

4 Comments and Conclusions

- <u>Thermal diffusivity (Fig. 4):</u> in the range 200-600°C the two alkali-free shotcretes have a rather low diffusivity; a possible explanation may be found in the larger porosity resulting from the spraying process; fibers have no effect on thermal diffusivity; above 600°C there are no sizable differences between shotcrete and ordinary concrete.

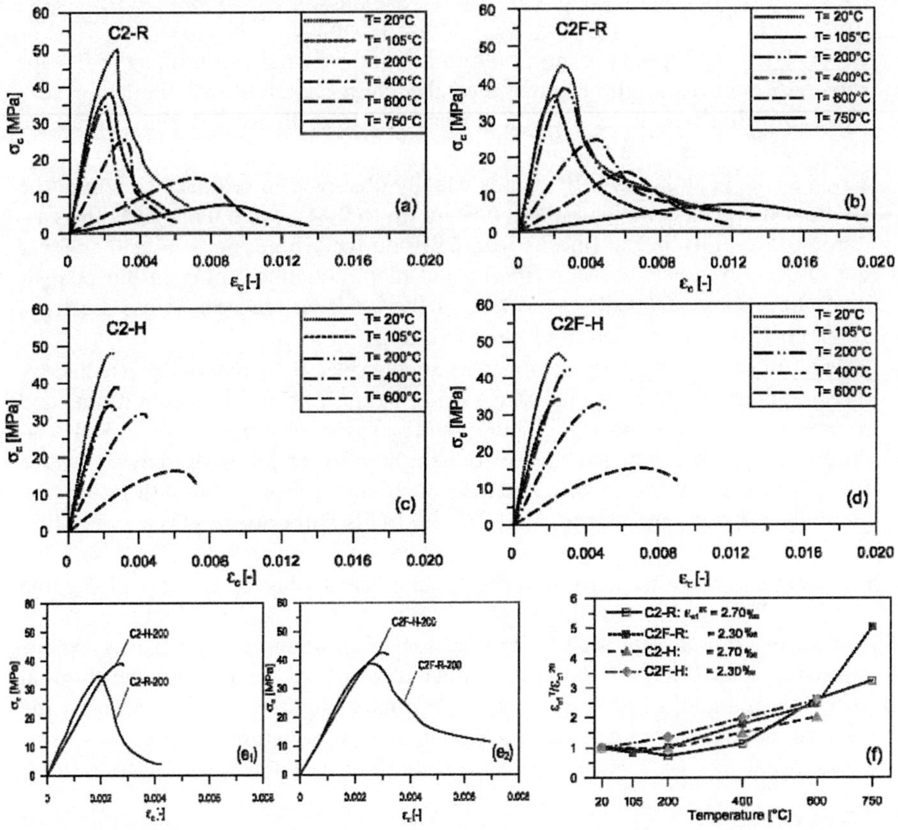

Fig. 5. Stress-strain relationship in residual and hot conditions for Mixes C2 (a,c) and C2F (b,d); typical residual and hot curves for T = 200°C (e); and normalized strain at the peak stress (f)

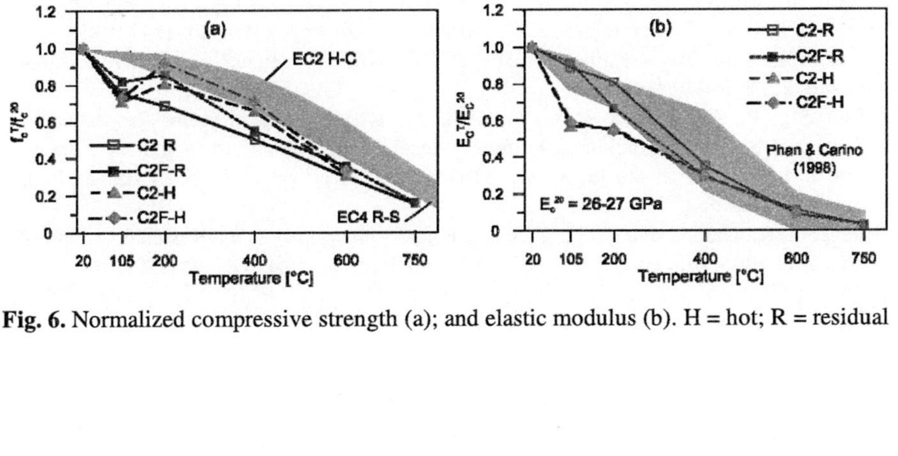

Fig. 6. Normalized compressive strength (a); and elastic modulus (b). H = hot; R = residual

- Stress-strain curves (Figs. 5a-d): because of the limited size of the aggregate particles (8 mm), the microstructure tends to be very homogeneous, which explains the remarkable linearity of the loading branches (at all temperatures) and the very limited toughness of Mix C2 (no fibers) at least up to 400°C; steel fibers give Mix C2F some toughness, to the detriment of the strength.
- Compressive strength (Fig. 6a): both in hot and residual conditions the fibrous shotcrete behaves slightly better than the plain shotcrete, and both are very close to ordinary concrete; as often observed in ordinary concrete, the range 100-200°C seems to be critical.
- Strain at the peak stress (Fig. 5f): as usually observed in ordinary concrete, the strain at the peak remains rather constant up to 200°C, and then starts increasing, more rapidly in the fibrous Mix C2F; on the whole, up to 600°C there is not much difference between fibrous and plain concrete, either in hot or residual conditions; at 600°C the values of the strain at the peak stress are from 100% to 150% larger than in virgin conditions.
- Elastic modulus (Fig. 6b): as in ordinary concrete, E_c is markedly affected by the temperature, more in Mix C2F than in Mix C2 in residual conditions, and in pretty much the same way in hot conditions; the differences in residual conditions might be a consequence of the slightly larger porosity induced by the fibers and responsible for the decrease of the compressive strength in the virgin material; the temperature range 100-200°C is still critical.

Summing up, the thermally-induced mechanical decay of a good alkali-free shotcrete is close to that of ordinary concrete, both in hot and residual conditions, either with or without steel fibers, for the usual fiber amounts ($v_f \leq 0.5\%$). As for the insulation capability of shotcrete (either alkaline or alkali-free), the thermal diffusivity is markedly lower than that of ordinary concrete, which confirms the efficacy of shotcrete as an insulating material, even at high temperature.

References

[1] Thomas, A.H.: Sprayed-Concrete Lined Tunnels. Taylor and Francis, London (2009)

[2] Malmgren, L.: Strength, Ductility and Stiffness of Fibre-Reinforced Shotcrete. Magazine of Concrete Research 59(4), 287–296 (2007)

[3] Jolin, M., Beaupré, D., Pigeon, M., Lamontagne, A.: Use of Set Accelerating Admixtures in Dry-Mix Shotcrete. ASCE – J. of Mat. in Civil Eng. 9(4), 180–184 (1997)

[4] Prudêncio, L.R.: Accelerating Admixtures for Shotcrete. Cement and Concrete Composites 20, 213–219 (1998)

[5] Bamonte, P., Gambarova, P.G., Nafarieh, A.: On High-Temperature Properties of Structural Shotcrete Containing Different Accelerating Agents. In: Kodur, V., Franssen, J.M. (eds.) Proc. 6th Int. Conf. "Structures in Fire", pp. 848–855. DEStech Pub. Inc, East Lansing (2010)

[6] Felicetti, R., Gambarova, P.G.: Effects of High Temperature on the Residual Compressive Strength of Siliceous HSCs. ACI Materials Journal 95(4), 395–406 (1998)

[7] Phan, L.T., Carino, N.J.: Review of Mechanical Properties of HSC at Elevated Temperature. ASCE – Journal of Materials in Civil Engineering 10(1), 58–64 (1998)

Part 4
Structural Elements: Design, Detailing, Shear, Tension Stiffening

Optimization of HPFRCC-Structures with Innovative Computational Methods

S. Grünewald[1], M. Flint[2], H. Han[3], J. Coenders[2,4], and J.C. Walraven[1]

[1] Section of Concrete Structures, Delft University of Technology, Delft/The Netherlands
[2] Arup Amsterdam/The Netherlands
[3] Royal BAM Group NV, Bunnik/The Netherlands
[4] BEMNext Lab, Delft University of Technology, Delft/The Netherlands

Abstract. A High Performance Fiber Reinforced Cementitious Composite (HPFRCC) is a tensile-hardening cementitious material with a post-cracking tensile strength equal to or higher than that of an un-cracked cross-section. As a consequence, the tensile strength can be taken into account for the structural design, which allows sophisticated computational methods facilitating the optimization process and designing innovative structures. This paper presents two approaches (Ant Colony Optimization and Automated design of Special Nodes) to optimize concrete structures by applying computational algorithms. The technical feasibility of the models was determined by a Finite-Element analysis. Case-studies demonstrate the potential of the proposed approaches. Designers can use such optimization tools in order to take into account the freedom of shape of HPFRCC.

1 Introduction

Recent material developments, enhanced production methods and new mould techniques yield in an improved structural performance. HPFRCC can be produced with a low to a very high strength as the examples of Engineered Cementitious Composites/ECC [1] and Ultra High Performance concrete/UHPC [2] show; new types of application with a slender and an enhanced aesthetical appearance can be developed with HPFRCC. Dependent on the application, fibers are able to replace rebars. They eliminate the needs for a concrete cover and to produce complicated rebar cages. Developing topologies for new materials can be a process of trial-and-error, such as happened in the past for many materials. Computational search algorithms can provide automated approaches to what topologies would be optimal for HPFRCC concerning material efficiency, structural stiffness, mould surface, pre-stress levels and other considerations. With improved and special material characteristics the opportunity evolves of finding new and optimized topologies for concrete structures.

G.J. Parra-Montesinos, H.W. Reinhardt, and A.E. Naaman (Eds.): HPFRCC 6, pp. 189–196.

2 Ant Colony Optimization

Ant Colony Optimization (ACO) was applied by Flint [3] to find an optimized structural topology for UHPC. The ACO-algorithm was introduced by Dorigo et al. [4] and it is a computational search algorithm that is capable of efficiently finding optimal solutions in a virtual search space that is based on the cooperating ability of ants to find the shortest route from their nest to a food source. Luh et al. [5] and Mitsui [6] proposed ACO for topology finding in structural design. Real ants find the quickest way by randomly choosing paths and lay pheromone on the paths they choose. The pheromone on a shorter and thus better path is reinforced more often as the ants that follow this path make more crossings in the same period of time. The trail therefore becomes stronger which motivates more and more ants to follow this strong trail of pheromone. Fig. 1 explains how the proposed ant system based approach is applied for topology optimization. A single-element (consisting of two nodes) can either contain material or it can be a void. A colony of ants can choose to follow both paths when walking from A to B. For all ants that chose path I, we say those ants chose to put material in the element, while all ants that chose path II chose to leave a void. A system will only work if elements actually contain material.

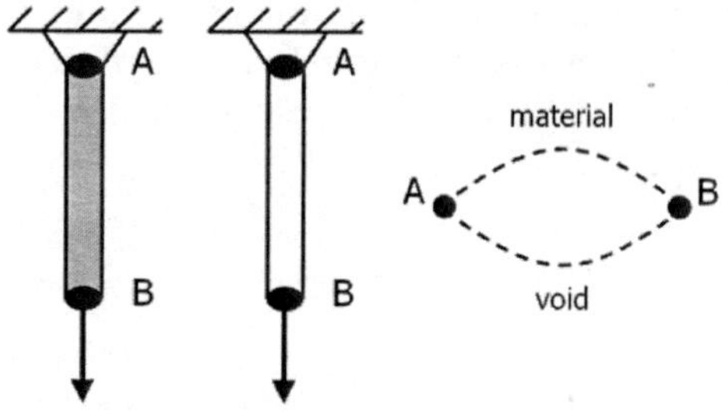

Fig. 1. Single-element optimization problem

When all ants have walked the path the structure they made is analyzed with a finite element application. The ants that have used the path that is associated with placing material in the element have generated a structure that is stable and can carry the load and as a result, the ants place pheromone on the path they chose. Ants make a probabilistic decision based on the amount of pheromone they find on a trail. Opposed to the first iteration, more ants now choose to follow path I. After following the path and hence creating a well-performing system, the amount of pheromone is increased further. As a consequence, all ants will quickly converge to this one path. Fig. 2 illustrates how computer software exploits the described process for structures containing more elements (2D- and 3D-structures). The design space is meshed into numbered elements. Through mapping a trail is constructed that the ants need to follow. They need to choose one out of two paths for every element, assigning material to an element or leaving a void.

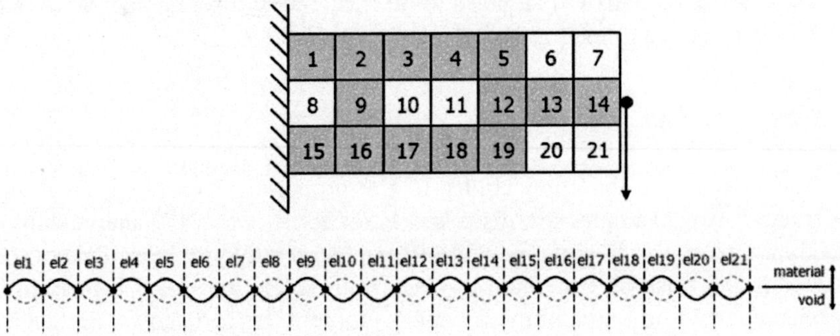

Fig. 2. Optimal solution and its array representation

Based on an initial pheromone distribution, ants choose their way and thus constructing solutions. Solutions are stored as arrays consisting of zeroes and ones. When all the ants have completed their trails, the structures they have generated accordingly are tested with a finite element application. The pheromone on the paths they chose is updated according to the performance of the structure they constructed. Well-performing solutions are more likely to be chosen again; room is left for the exploration of other possible solutions which allows the algorithm to find its way out of a local optimum. The user chooses how to mesh a structure with elements: using a coarse mesh reduces the number of possible solutions and an optimized solution is obtained with less iterations. A fine mesh increases the run time and the chance that the algorithm gets 'stuck' in a solution that is not the global optimum, but does give more detail to the developed solutions.

Flint implemented the ACO-algorithm in Fortran 77 and linked it to the finite element package DIANA 9.2. The algorithm is run with k iterations for a colony of j ants and a structure of i elements. The iteration steps are: initializing a structure, to put material or to leave voids, to calculate the performance, to select the best structure and to update the pheromone. An ant 'produces' one complete possible solution in each iteration and the number of ants of a colony is equal to the amount of structures that is generated within one iteration. The pheromone on a path represents the probability that this path is chosen by an ant as a part of its trail; the pheromone distribution is different for each iteration. The structure's performance is determined with a mathematical function, which can be defined by the user. It was decided to base the performance of the generated structures on its estimated costs. In order to calculate the costs the following input is required for the algorithm: volume of the concrete, mould surface area of the structure, the level of post-tensioning and maximum allowable stresses and displacements. Additional criteria can be added; the space available for ducts and gantries was used in the research. Before the performance is determined, unconnected elements are removed from the ants' solutions. The performance of the structure is set to almost zero when the stress or displacement criterion is not met. Otherwise, a rough approximation of the structure's cost is obtained; its inverse is a performance indicator. The 'holes'-criterion adds virtual costs to the structure if a set number of required holes for ducts and gantries is not met.

2.1 Case-Studies

In order to determine the applicability of the proposed algorithm several cases were studied; two examples are discussed hereafter. Michell [7] analytically developed minimum weight structures for different structural problems. One of these structures is the truss on two fixed-in-place hinges. The proposed algorithm has been benchmarked on this minimum weight problem. Fig. 3a shows the resulting structure after 250 iterations (32 by 48 elements; applied force: 600 kN); only half the problem is analyzed and the image is mirrored. An imaginary isotropic material with a yield stress of 50 MPa was assumed. Fig. 3b shows the solution of the problem obtained with BESO [8], a popular and well-proven computational optimization algorithm.

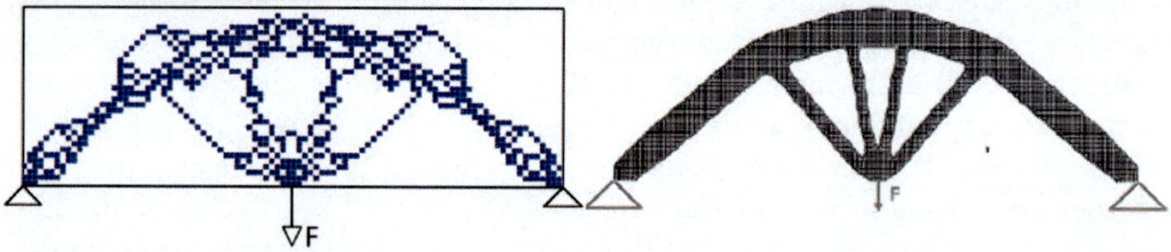

Fig. 3. Benchmark calculations a) Pheromone distribution and optimal structure after 250 iterations, b) Michell truss: BESO solution [9]

The random process has created some holes and extensions in the compression arch in Fig. 3a. These small discrepancies can appear because there is no incentive for the routine to minimize surface; it only optimizes towards minimum weight. If surface area is taken into account, it can be expected that the members become more homogeneous. The Michell truss benchmark shows that the routine converges towards a proper optimum and the structure found highly resembles the BESO solution. The tool also has been applied to the case of the design of a roof girder with a span of 105 meters. The aim of the case-study was to design an UHPC-variant to the current steel structure, which comprises a roof supported by steel trusses. Material characteristics of the UHPC-concrete BSI [10] were assumed to determine design values according to AFGC/SETRA [11]. Fig. 4 presents the results of a run with a coarse mesh.

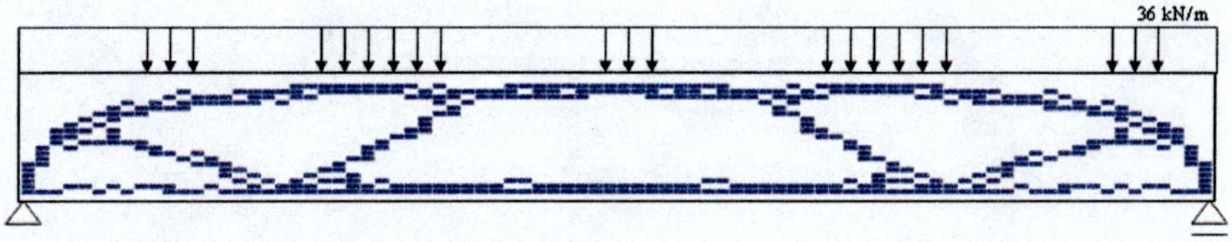

Fig. 4. Optimal structure as found after 500 iterations on coarse problem approximation

The resulting structure makes sense from a structural engineering point of view, because it results in a truss-like structural topology. Diagonals carry the shear forces near the supports. Random fluctuations cause small-scale discrepancies. The middle of the span is free of diagonals because of the symmetric approach; there is no incentive to create a diagonal here. The structure demonstrates that a truss-like structure can be optimal for an element spanning the 105 meter span of the hall roof in UHPC. Local stability has not been taken into account by the algorithm and needs to be considered for the global design of the Stadium. In a more detailed process the structural members were designed and would be post-tensioned to different levels according to the level of the stress of the individual members. The estimated material costs of the UHPC-structure were in the same order as the costs of the steel truss in the current steel design.

3 Automated Design of Special Nodes

Han [12] has developed the computational structural design tool called '*VisionNode*' which uses computational optimization to create Special Nodes. A Special Node is a structural element that connects incoming members of a frame structure, has a complex 3D-geometry and has geometrical distortions in the form of holes. User-defined material characteristics, geometrical boundary conditions and transferred force distributions are used as input parameters for the design tool. The application of *VisionNode* enables freedom of design while obtaining a certain degree of material efficiency which is achieved by reducing the material volume and satisfying the structural capacity of the Special Node. Fig. 5 shows three examples of Special Nodes obtained with *VisionNode*. Material characteristics of a tensile-hardening fiber-reinforced UHPC ($f_{ck,cube}$=200 MPa) were assumed for the optimization of the nodes; the assumed tensile strengths were 11, 15 and 30 MPa, respectively.

VisionNode was tested on the connection of the welded steel girders of the 'Yas-Hotel' in Abu Dhabi (Fig. 6) to provide realistic boundary conditions. The results of a structural study on the Yas-Hotel were girder dimensions and forces transferred by the incoming members. These results were used as input parameters for *VisionNode*.

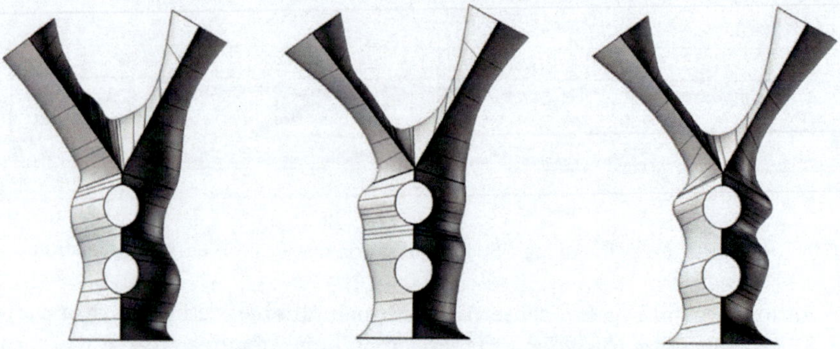

Fig. 5. Special Node with the same preliminary design but different tensile strengths

Fig. 6. Location of the Special Node [13]

After defining the boundary conditions and generating the initial structure, the optimization process (three steps: run the geometric modeler, run the finite-element analysis and the evaluation of the objective function) takes place which is followed by generating the output. *VisionNode* creates Special Nodes through computational optimization which is performed by a genetic algorithm. Through the interactive graphical-user-interface of *VisionNode*, users can create a Special Node based on their design preferences and design modifications. By explicitly specifying where material should be used and where it shouldn't be used by the geometric-modeler, the user has the means to influence the design of the node and to create a Special Node with the desired characteristics. Within a defined area of the node the user specifies a) material characteristics, b) coordinates and geometry of the incoming structural members, c) forces transferred from the incoming members and d) where the node must and must not contain material (thus specifying the locations of the holes). During the optimization the programs DAKOTA, Rhinoceros and Ansys communicate with each other and continuously exchange data. The goal is to minimize the objective function (the volume of the node) and to generate a new and optimized geometry; the objective-function is evaluated after every iteration. The design-variables are the coordinates on the XY-plane from which the geometry of the Special Node is created. The algorithm is incorporated in *VisionNode* and continuously exchanges data between a geometric-modeler, which creates the geometry of the Special Node, and a Finite-Element-Model, that determines the technical feasibility. For every geometry, the stresses in the node are calculated and are also uploaded to the optimization algorithm. Hereby, the structural capacity of the material is fulfilled taking into account the boundary conditions set by the user. A solution is considered feasible when it satisfies the Mohr-Coulomb-Failure-Check (Table 1); this value is calculated by *VisionNode* after extracting the largest principal stress S1, which is the tensile stress, and the largest principal stress S3, which is the compressive stress, from the FEM. By dividing these extracted values by the material capacities, the value of the Mohr-Coulomb-Failure-Check is obtained.

Table 1. Calculation results of three special nodes (Fig. 5) with differences in tensile strengths

Vision-Node solution	A	B	C
Tensile strength [MPa]	11,0	15,0	30,0
Volume of Special Node (mm^3)	1464257380	1368151684	1184695894
Volume percentage [%] relative to A	100%	93%	81%
Maximum tensile stress (MPa)	7.27	9.98	16.60
Maximum compressive strength (MPa)	-37.67	-33.87	-45.64
Mohr-Coulomb-Failure-Check	1.000082	0.966	0.961

VisionNode has the ability to create 2D- and 3D-Special Nodes in FEM which are based on the same geometrical data. The first kind is a 2D-geometry of the Special Node which consists of an extruded area, thus creating a solid. The second kind is a 3D-geometry of the Special Node that does not involve extruding but involves so-called lofting of surfaces. Every Special Node is initially created in two dimensions in the form of NURBS-curves that are extracted from the design-variables. The FEM-software Ansys creates a surface-area with depth out of the NURBS-curves. In order to create a 3D-geometry Rhinoceros was applied using its loft-surface functionality. The goal of switching between 2D- and 3D-geometries is to save computation time: 2D-geometries are calculated for the most part of the optimization process and after that, when a solution nearly converged, 3D-geometries are created to obtain the final design. The reason for this approach is that a 3D-FEM calculation takes considerably longer than a 2D-FEM calculation. Material efficiency is achieved by reducing the volume of the Special Node. As a result, a Special Node in HPFRCC is designed that satisfies all structural checks and has a minimal volume.

4 Conclusions

The combination of computational tools and innovative materials like HPFRCC is promising as it may lead to surprising concepts. Two innovative computational methods and their potential to optimize structures with HPFRCC are presented in this paper. The two algorithms have been developed, implemented, tested and applied. They allow finding solutions that otherwise may not have been considered. As a second step, the solutions should be interpreted and developed by a competent engineer. The tools can be used for common design problems but they are especially useful for complex design problems or where innovative solutions are required. The studies show that attractive and competitive structures can be produced with tensile hardening cementitious materials taking into account recent developments of production and mould techniques and computational methods.

References

[1] Li, V.C.: On Engineered Cementitious Composites (ECC) - A Review of the Material and its Applications. J. Advanced Concrete Technology 1(3), 215–230 (2003)

[2] Resplendino, J.: Introduction: What is a UHPFRC? In: Designing and Building with UHPFRC, pp. 3–14. John Wiley & Sons, Chichester (2011); ISBN: 978-1-84821-271-8

[3] Flint, M.G.: Ant Colony Optimization for Ultra High Performance Concrete Structures. In: Designing and Building with UHPFRC, pp. 601–617. John Wiley & Sons, Chichester (2011); ISBN: 978-1-84821-271-8

[4] Dorigo, M., Maniezzo, V., Colorni, A.: Ant System: Optimization by a Colony of Cooperating Agents. IEEE Transactions on Systems, Man, and Cybernetics–Part B: Cybernetics 26(1), 29–41 (1996)

[5] Luh, G.-C., Wu, C.-Y., Lin, C.-Y.: Multi-modal topological optimization of structure using ACO algorithms. In: Proceedings of the 8th Joint Conference on Information Sciences (JCIS 2005), Salt Lake City, UT, USA, pp. 479–482 (2005)

[6] Mitsui, K.: Optimal Shape and Topology of Structure Searched by Ants' Foraging Behavior, Report of the Research Institute of Industrial Technology, (83), p. 1–11. Nihon University (2006)

[7] Michell, A.G.M.: The Limits of Economy of Material in Frame-Structures. Philosophical Magazine, Ser. 6 8(47), 589–597 (1904)

[8] Huang, X., Xie, Y.M., Burry, M.C.: A new Algorithm for Bi-directional Evolutionary Structural Optimization. JSME International Journal, Series C 49(4), 1091–1099 (2006)

[9] Xie, Y.M., Steven, G.P.: Evolutionary Structural Optimization. Springer, London (1997)

[10] EIFFAGE, Caractéristiques Mécaniques et Physiques (2004), `http://www.bsieiffage.com` (retrieved November 26, 2007) (in French)

[11] AFGC/SETRA, Ultra High Performance Fibre-Reinforced Concretes; Interim Recommendations, Association Française de Génie Civil, France (2002)

[12] Han, H.: Special Nodes in Ultra High Performance Concrete, MSc thesis, Delft University of Technology, The Netherlands (2010)

[13] Yas-Hotel Press Release (2009), http://www.asymptote-architecture.com (retrieved August 2009)

Structural Applications of Hybrid Fiber Engineered Cementitious Composites - A Review

M. Maalej

Department of Civil & Environmental Engineering,
College of Engineering University of Sharjah, Sharjah, UAE

Abstract. This paper reviews some of the recent progress in the structural applications of hybrid fiber Engineered Cementitious Composite (ECC) materials. A summary of the design and characteristics of such materials is presented followed by a review of recent applications of hybrid fiber ECC. The reviewed applications include the use of hybrid fiber ECC for designing impact & blast resistant protective panels, strengthening of RC beams, and enhancing corrosion durability of RC beams. The review demonstrates that hybrid fiber ECC can significantly enhance the performance of structures incorporating these materials.

1 Introduction

In recent years, it has been demonstrated that Engineered Cementitious Composites (ECCs) can be designed to exhibit pronounced tensile strain-hardening by adding to the cement-based matrix a relatively low volume (typically $\leq 2\%$) of short randomly-distributed fibers of specific type and property. The resulting ECC composites are characterized by their high tensile strain capacity, fracture energy and notch insensitivity, making them ideal materials for various structural applications, including impact- and blast-resistant structures. Under uniaxial tension, sequentially developed parallel cracks contribute to the inelastic strain at increasing stress level. The ultimate tensile strength and strain capacity can be as high as 5 MPa and 4%, respectively (see Fig. 1). The latter is two orders of magnitude higher than that of normal or ordinary fiber reinforced concrete.

It has been demonstrated that, for a number of structural application, the use of more than one type of fiber as reinforcement will result in a *hybrid fiber composite* that is better able to meet the material performance requirements for the given application. For instance, to use ECC in protective structures, the material is required to possess sufficient strength to resist penetration, while at the same time it is required to absorb a large amount of energy, thereby minimizing fragmentation and reducing the velocities of the fragments (i.e. need material of sufficiently high strain capacity). For using ECC in corrosion-resistant structures, a low crack width is required to reduce the ingress of aggressive substances reaching the steel

G.J. Parra-Montesinos, H.W. Reinhardt, and A.E. Naaman (Eds.): HPFRCC 6, pp. 197–204.

reinforcement, while at the same time a high strain capacity is also required to prevent delamination and concrete cover spalling. Further, given that mono-fiber ECCs containing high modulus fibers (e.g. steel and carbon fibers) normally exhibit high ultimate strength, low crack width and low strain capacity, while those containing low modulus fibers (e.g. polyvinyl alcohol and polyethylene fibers) exhibit opposite behaviors, it becomes clear that a hybrid-fiber ECC with proper volume ratio of high and low modulus fibers can be designed to achieve an optimal balance between ultimate strength, crack width and strain capacity, and therefore better meet the functional requirement for these applications.

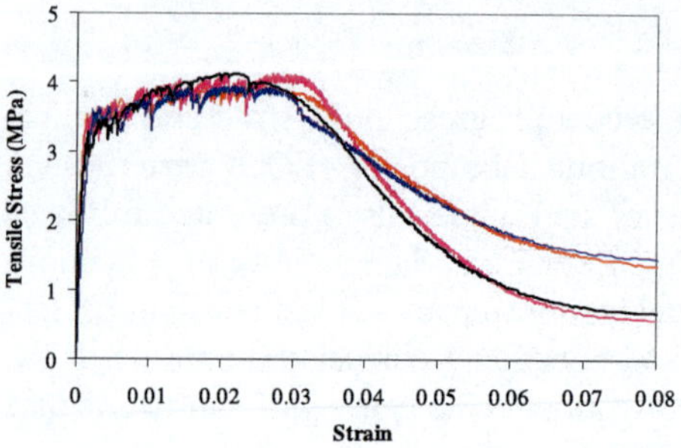

Fig. 1. Uniaxial tensile stress-strain curves of ECC from coupon specimen tests

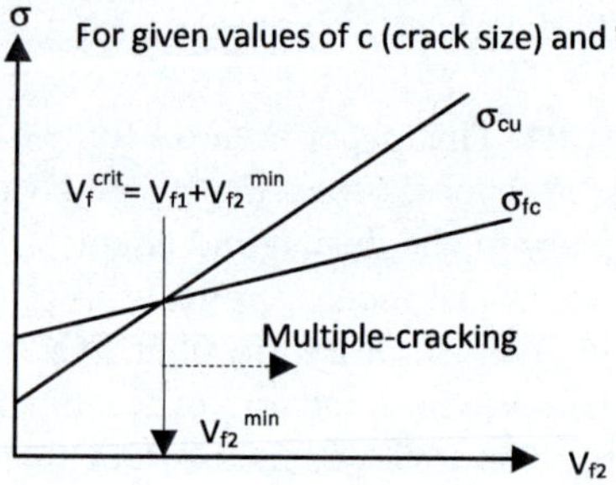

Fig. 2. Critical fiber volume fraction concept for hybrid fiber ECC

In sections to follow, the performance of hybrid fiber ECC in a number of structural applications is reviewed with the objective of assessing the potential of these materials in providing better functionality, in particular, in applications involving enhancing the impact- and blast-resistance of structures, improving the strength and ductility of FRP-strengthened beams and enhancing the corrosion durability of RC members.

2 The Critical Fiber Volume Fraction Concept

A micromechanical model for the design of hybrid fiber ECC was develop by Ahmed et al. [1] on the basis of fracture mechanics and deformation mechanism taking into account the effects of hybrid fibers. This micromechanical model for hybrid-fiber ECC is an extension of an earlier model proposed by Li and Leung [2] for mono-fiber composites.

The above model assumes that strain-hardening behavior could be achieved in a hybrid-fiber composite when the ultimate bridging strength (σ_{cu}) imposed by the fiber/matrix interaction exceeds the first crack strength (σ_{fc}) at which new matrix cracks can propagate (multiple cracking). For a given initial size of matrix crack (c) and a given volume fraction of one type of fiber (V_{f1}), the above condition ($\sigma_{cu} \geq \sigma_{fc}$) leads to the estimation of a minimum volume fraction of the other type of fiber (V_{f2}^{min}) necessary for the composite to exhibit strain-hardening behavior (Fig. 2), and thus a critical fiber volume fraction concept for hybrid-fiber composites dependent on the matrix, fibers and interface properties.

3 Blast/Impact Resistant Panels

In view of the high ductility and fracture energy of ECC, Zhang et al. [3] proposed the use of ECC materials for the construction of blast-resistant protective panels. The performance of ECC in these applications can be evaluated experimentally using drop weight impact tests to simulate the impact by large mass at relatively low strain rate and low impact velocity. When a structure is impacted by a blast wave or a large projectile at low impact velocity, the global response of the target is likely to dominate where tensile and flexural stresses must be sustained at large deformation without crack localization (unlike the case of impact by a small projectile at high velocity where the load application is spatially concentrated and the local response effects dominate).

Zhang et al. [3] conducted a series of drop weight impact tests to evaluate the damage and failure mode of hybrid-fiber ECC panels caused by large projectiles or fragments. One of the specific aims of the experiment was to quantify the extent to which hybrid-fiber ECC improves the resistance of blast panels against impact loading. For this purpose, drop weight tests were conducted on full-scale hybrid-fiber ECC blast/shelter panels (2 m x 1 m x 0.05~0.1 m) to study their response and performance under impact loading. Conventional steel reinforced concrete (RC) and steel fiber reinforced concrete (FRC) blast panels were also tested to identify the advantages of using ECC in this application. The impact resistance of blast panels of different materials was evaluated in terms of the extent of damage, energy absorption capacity and residual resistance against multiple impacts. The drop weight impact test results showed that the hybrid-fiber ECC panels exhibit lesser damage, significantly-improved impact resistance against multiple impacts and improved ductility and energy absorption capacity compared to both RC and FRC counterparts. The response of the ECC panels to drop weight impact was characterized as follows:

- Much smaller indent depth and crater size on the impact face. The indent depths of the crater for all the panels under multiple impacts are summarized in Fig. 3. Fig. 4 shows the damage on the impact face of the RC panel after the first three impacts. Similar indentation depth as the 1st impact on RC panel was induced on the ECC panels after ten impacts, as shown in Fig. 5.
- Much smaller exit crater on the distal face after perforation with the structure remaining largely intact.
- Ductile failure process characterized by large deformation limit prior to serious scabbing and total failure.

- Large improvement in the cracking behavior with a larger volume of material involved in energy absorption for better resistance.

Zhang et al. [3] have also adopted a Single Degree of Freedom (SDOF) concept to analyze the global behavior, especially the energy absorption capacity of RC and ECC panels. From the analysis, the higher ultimate resistance and maximum allowable deflection of the ECC panels appeared to be the main parameters contributing to the higher energy absorption capacity and hence the enhanced global resistance of the ECC panels over their RC counterparts. The results of this study and those of previous studies on the strain rate effect of hybrid-ECC and its impact resistance against high velocity impact from small-mass projectile provided reliable evidence for the advantages of using ECC materials in protective structures.

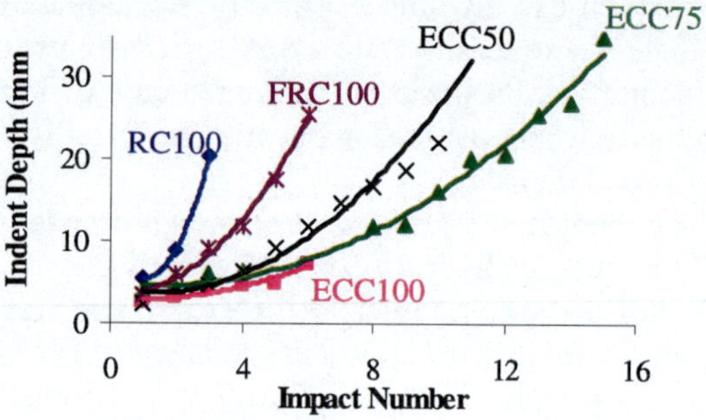

Fig. 3. Indent depth against number of impacts

Fig. 4. Damage development in RC100 panel on impact face at 1st impact (left), 2nd impact (middle), and 3rd impact (right)

Fig. 5. Damage development of ECC 100 panel on impact face at 1st impact (left) and at 10th impact (right)

4 Effective FRP-Strengthening of RC Beams

The applications of Fiber Reinforced Polymer (FRP) Composites to concrete structures have been studied intensively over the past few years in view of the many advantages that FRPs possess. While FRPs have been shown to be effective in strengthening RC beams, strength increases have generally been associated with reductions in the beams' deflection capacities due to premature debonding. Debonding failure modes occur mainly due to interfacial shear and normal stress concentrations at FRP-cut off points and at flexural cracks along the RC beam.

Maalej & Leong [4] suggested that if the quasi-brittle concrete material which surrounds the main flexural reinforcement is replaced with a ductile layer of ECC, then it would be possible to delay the debonding failure mode and hence increase the deflection capacity of the strengthened beam.

When ECC is introduced in a RC member, more but thinner cracks are expected to form on the beam tensile face rather than fewer but wider cracks in the case of an ordinary concrete beam. More frequent but finer cracks are expected to reduce crack-induced stress concentration and result in a more efficient stress distribution in the FRP layer. The objective of the above-study was to establish both experimentally and numerically the structural performance of FRP-strengthened RC beams incorporating a ductile ECC layer around the main flexural reinforcement. The load-carrying and deflection capacities as well as the maximum FRP strain at failure were used as criteria to evaluate the performance.

The experimental program included two series of RC beams where one series consisted of two ordinary RC beams (beam A1 and A2) and another series consisted of two ECC layered beams (ECC-1 and ECC-2). In each series, one specimen was strengthened using externally-bonded carbon fiber reinforced polymer (CFRP) sheets while the second was kept as a control[1] in order to compare its load-deflection behavior under third-point loading with the strengthened specimen.

It can be seen from Fig. 6 that the CFRP strengthened beams with an ECC layer (e.g. beam ECC-2) depicted higher load-carrying capacity compared to their ordinary reinforced concrete counterparts (e.g. beam A2). If expressed in term of strengthening ratio (SR, defined as the strength of beam with CFRP reinforcement divided by the strength of control beam), beam ECC-2 had a strengthening ratio of about 1.43, compared to 1.28 for beam A2. Also, it can be seen that beam ECC-2 showed a significantly higher deflection capacity (29.6mm) at peak load compared to beam A2 (21.9mm).

On the cracking behavior, both ECC-1 and ECC-2 showed a considerable number of fine cracks compared to the ordinary RC beams (beam A1 and A2) as revealed in Fig. 7. The crack spacings were consequently much smaller in the former beams, particularly in ECC-2. These multiple but fine cracks played a

[1] Specimens A1 and ECC1 are control specimens without any CFRP reinforcement.

major role in reducing crack-induced stress concentration resulting in more efficient stress distribution in the CFRP sheet and a better stress transfer between the CFRP and the concrete beam. This delayed intermediate crack-induced interfacial debonding and resulted in higher strengthening ratio and higher deflection capacity and, therefore, a more effective use of the CFRP material.

The above experimental results showed that the ECC had indeed delayed debonding of the CFRP and resulted in effective use of the CFRP material. With the use of ECC as a ductile layer, RC beams can effectively be strengthened while minimizing loss in deflection capacity. These positive results warrant further studies on the use of ECC in combination with CFRP to repair and strengthen deteriorating RC structures, particularly those where deteriorated concrete has to be replaced with a new repair material.

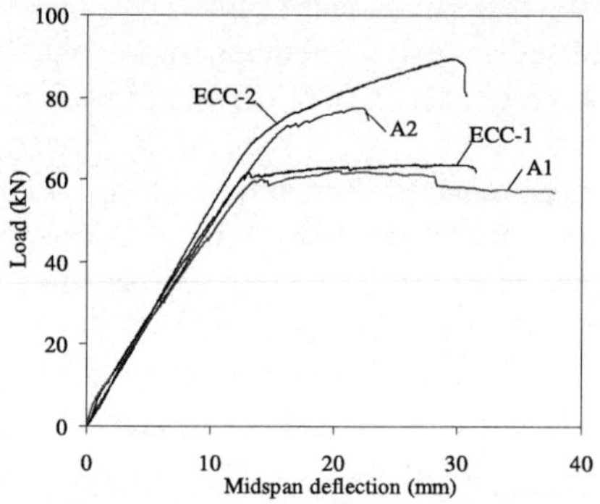

Fig. 6. Load-deflection curves of beams A1, A2, ECC-1, and ECC-2

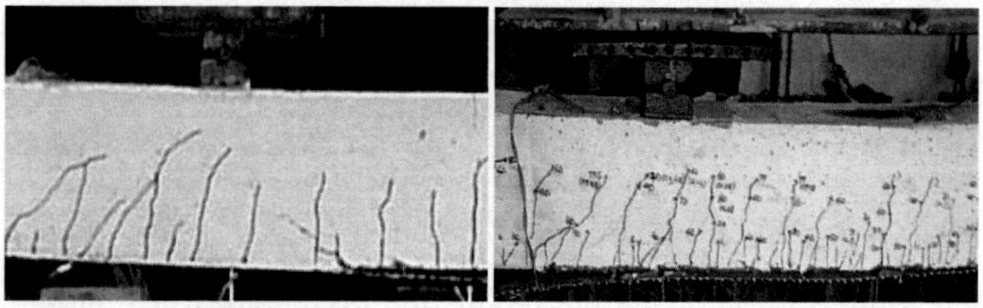

Fig. 7. Cracking patterns in beams A2 (left) and ECC-2 (right) around the loading point

5 Corrosion-Resistant RC Beams

As an alternate method of improving the corrosion resistance of RC beams exposed to aggressive substances, Maalej and Li [5] proposed a new design for RC flexural members where part of the concrete which surrounds the main flexural reinforcement is replaced with a strain hardening ECC material. This alternate design with layered ECC has been referred to as Functionally-Grade Concrete (FGC). It was suggested that the ECC material in FGC beams could provide two levels of protection. First, it could prevent the migration of aggressive substances into the concrete, therefore, preventing reinforcement corrosion. Second, in the extreme case when corrosion initiates, accelerated corrosion due to longitudinal cracks would be reduced (if not eliminated), and spalling and delamination problems common to many of today's RC structures would be prevented. This is expected due to the high strain capacity and fracture resistance of the ECC material.

Fig. 8. Corrosion status of steel in OPCC-2 beam (Top) and FGC-3 beam (bottom)

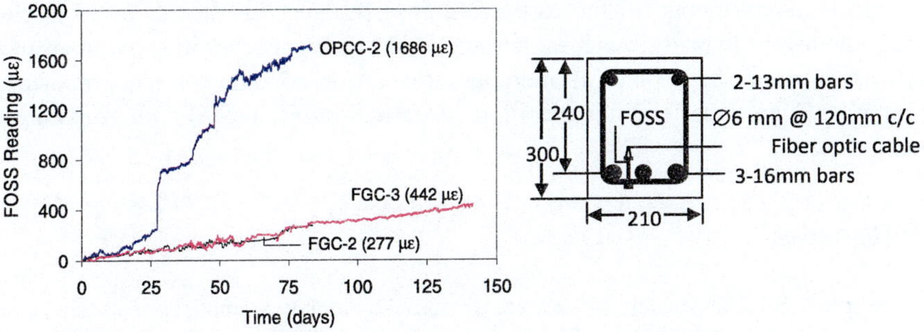

Fig. 9. Corrosion-induced tensile strain in concrete as measured by FOSS gauge

Maalej et al. [6] adopted the above concept to prepare a series of FGC beams where the main longitudinal reinforcements were surrounded by a hybrid fiber reinforced mortar material exhibiting strain hardening and multiple cracking under third-point flexural loading. The hybrid fiber reinforcement consisted of high modulus (steel) and low modulus (PVA) fibers used with respective volume fractions of 1% and 1.5%. The objective of the study was to evaluate the effectiveness of the FGC concept in retarding the corrosion of steel reinforcement in RC beams and reducing the tendency of the concrete cover to delaminate as measured by a concrete embeddable fiber optic strain sensor (FOSS). The effects of steel loss and corrosion damage on the flexural response of RC beams were also be evaluated. The experimental program included 2 control RC beams (OPCC beams) made or ordinary Portland cement concrete and 3 FGC beams.

At any given time, the FGC beams were found to exhibit lower level of steel loss than the OPCC beam (see Fig. 8). It was also observed that an FGC beam takes about 70% longer time to achieve the same level of induced steel loss compared to an OPCC beam. The better performance of the FGC beams over the OPCC beam was also evident from the absence of any corrosion-induced cracks or damage and the lowest tendency for the concrete cover to delaminate as measured by a concrete embeddable FOSS (see Fig. 9). For the same level of steel loss, an FGC specimen exhibited higher residual load and deflection capacities compared to its OPCC counterpart. While a corroded OPCC beam was found to experience widening of corrosion-induced cracks, delamination and spalling during loading, no such behavior was observed in the FGC beams. The FGC concept was found in the above study to be very effective in preventing corrosion-induced damage in RC beams and minimizing the loss in the beam's load and deflection capacities.

6 Conclusions

This paper reviewed some of the recent progress in the structural applications of hybrid fiber Engineered Cementitious Composites (ECC). In all cases, the ability to achieve a balance between high strength and high ductility of the ECC material through the use of hybrid fibers was very important in meeting the material performance requirements for the respective applications. For the reviewed applications, the use of hybrid fiber ECC had resulted in a significant improvement in the performance of the structural members (in terms of either impact resistance, strengthening and deflection capacities, or corrosion resistance) over conventional counterparts.

References

[1] Ahmed, S.F.U., Maalej, M., Paramasivam, P.: Analytical model for tensile strain-hardening and multiple-cracking behaviour of hybrid fiber engineered cementitious composites. ASCE J. of Materials in Civil Engineering 19(7), 527–539 (2007)

[2] Li, V.C., Leung, C.K.Y.: Steady State and Multiple Cracking of Short Random Fiber Composites. ASCE J. of Engineering Mechanics 118(11), 2246–2264 (1992)

[3] Zhang, J., Maalej, M., Quek, S.T.: Performance of Hybrid-Fiber ECC Blast/Shelter Panels Subjected to Drop Weight Impact. ASCE Journal of Materials in Civil Engineering 19(10), 855–863 (2007)

[4] Maalej, M., Leong, K.S.: Engineered Cementitious Composites (ECC) for Effective FRP-Strengthening of RC Beams. Composites Science and Technology 65(7-8), 1120–1128 (2005)

[5] Maalej, M., Li, V.C.: Introduction of Strain Hardening Engineered Cementitious Composites in the Design of Reinforced Concrete Flexural Members for Improved Durability. ACI Structural Journal 92(2), 167–176 (1995)

[6] Maalej, M., Ahmed, S.F.U., Paramasivam, P.: Corrosion Durability and Structural Response of Functionally-Graded Concrete Beams. Journal of Advanced Concrete Technology 1(3), 307–316 (2003)

D-Zones in HPFRC

M. Colombo and M. di Prisco

Politecnico di Milano, Milano, Italy

Abstract. In this work, the behavior of High Performance Fiber Reinforced Concrete (HPFRC) in the D-regions was experimentally investigated.

It refers to the bottle-shaped stress field which develops large transverse stresses: compression in the bottle neck and tension far away with a peak along the vertical symmetry axis (bursting). This is the case of a member subjected to a compressive axial force applied over a small bearing area, the typical case of a tunnel segment loaded by the TBM jack during the excavation advance. The distribution of the high local stresses under the bearing plate also causes transverse tensile stresses along the horizontal upper border which may crack the concrete (spalling).

The experimental tests were performed on square plates (270 x 270 x 60 mm). A uniform vertical displacement was imposed on top region characterized by a variable width. Several ratios β between the specimen size and the bearing plate width are considered (1, 1.5, 2 and 3).

1 Introduction

The design of detailing in discontinuity regions plays a crucial role both for the structure behavior and for its safety [1, 2]. The behavior of ordinary reinforced concrete structures was investigated by means of truss analogy starting from Ritter (1899) and Mörsch (1912), to arrive at Schlaich and Schäfer [2] who in 1991 proposed the 'strut and tie' model as a general method for design. With reference to compressed zones, in this approach three different strut models were presented: prismatic, fan and bottle shape. The last strut model can be regarded as the most complex one for structural design, because its bearing capacity should depend on the diffusion degree that can be quantified by the geometrical ratio between the strut-end widths. According to the authors [2], a not negligible compressive strength reduction has to be taken into account for geometrical ratios varying in the range 1.5 – 5 if any transverse reinforcement is introduced. The strength reduction valley can be significantly reduced by increasing the mechanical ratio of transverse reinforcement up to about ω_{st} = 0.10, a value for which no strength reduction has to be considered. Although this analysis was introduced in Model Code 90 [3], from a careful analysis of the physical problem, some doubts on the

G.J. Parra-Montesinos, H.W. Reinhardt, and A.E. Naaman (Eds.): HPFRCC 6, pp. 205–212.

procedure used to define the strength reduction may arise. First of all, the strength capacity does not take into account any stress redistribution in the 2D structure and assumes the local reaching of failure biaxial domain for the material as the structure collapse. The same assumption seems to be considered also when steel reinforcement is introduced. Again, the peak of transverse stresses, computed by referring to an elastic stress distribution that neglects crack propagation, is set equal to the spread reinforcement yielding strength, computed by substituting the geometrical reinforcement ratio multiplied by the yielding steel strength to the tensile strength of the plain concrete.

The research here presented is aimed to investigate the behavior of a High Performance Fiber Reinforced Concrete when subjected to concentrated loads. This material is going to be used in the design of the external layer of a precast tunnel lining where the concentrated load condition is quite important in the construction phase of the whole tunnel when TBM jacks act on the lining in order to allow the excavation to advance.

2 Material

The material used in the experimental investigation is a High Performance Steel Fiber Reinforced Concrete characterized by the mix – design summarized in Table 1. The fibers were straight with a length equal to 13 mm and a diameter of 0.16 mm; the steel used was high carbon and the fiber content was equal to 100 kg/m^3 (1.28% by volume).

The mechanical properties of the concrete considered are summarized in Table 2, where the residual tensile strengths defined according to four point bending tests as suggested by Italian Recommendation CNR DT 204 [4].

Table 1. Mix design of HPFRC used in the experimental investigation

Component	Conetnt
Cement Type I 52.5	600 kg/m^3
Water	200 l/m^3
Sand 0-2mm	983 kg/m^3
Slag	500 kg/m^3
Superplasticizer	33 l/m^3
Fibers	100 kg/m^3

Table 2. Mechanical properties of HPFRC [4]

f_{cc}	110 MPa
f_{If}	11.05 MPa
f_{eq1}	14.09 MPa
f_{eq2}	14.00 MPa

3 Experimental Program

The experimental investigation was carried out on square 270x270x60 mm plates. In order to investigate the influence of the loading area size, different ratios β (=b/a) between the specimen width (b = 270 mm) and the bearing zone width (a = variable) were considered (β = 1, 1.5, 2 and 3).

Three nominally identical tests were performed and each specimen was casted in a different day with the same mix design. (16, 23 and 29 December 2010).

For each casting date a 1400 x 400 x 60 mm slab was cast orienting the concrete flow, and therefore the fibers, parallel to the long side of the slab. Four specimens were sawn from the centre of each slab in order to neglect any wall effect. During the test the specimen was placed on the machine, orienting the fiber direction at right angle with the load one.

The tests were displacement controlled considering the actuator displacement as feedback parameter and using a stroke rate equal to 2 μm/s.

The load was applied by means of a steel plate 4 mm thick and characterized by a variable depth, "a", in the different tests carried out. The material and the thickness used for the bearing plate were chosen in order to ensure an almost constant stress distribution under the loading plate and, without introducing a too large deformation that could prevent the displacement controlled procedure.

Some stearic acid was placed between the bearing plates and the specimen in order to remove friction.

Three LVDTs were used on one side of the specimen to measure crack opening displacements along the depth (Fig. 1) in each test. Two LVDTs were placed respectively at the top and at the bottom of the specimen for all the cases, while the third one was placed in the middle for the β = 1 cases and at 160 mm from the bottom in the other cases. The selected points were located in the points in which the peak of the horizontal tensile stresses was computed according to a linear elastic analysis.

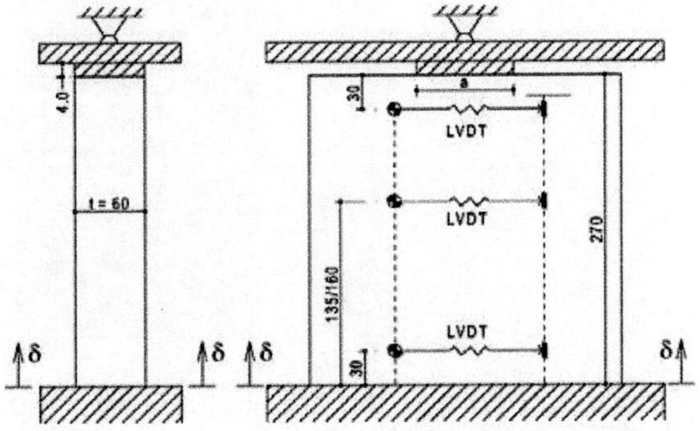

Fig. 1. Experimental set-up

The instrumented side was the free surface during casting while the opposite side was used to take pictures of the specimen in order to perform crack measurements by means of image crack detection techniques.

4 Test Results

The experimental results are represented in Figs. 1, 2 and 3 by means of the load (P) vs actuator displacement (δ) curves; the curves related to all the specimen tested together with the average curve are plotted. In the case of the tests β = 1.5 and 2 for the specimens cast on 29/12, the original curves was shifted in order to neglect the initial nonlinearity of the curves due to the boundary conditions releted to the imposed displacement procedure. This shift was performed by adopting as initial stiffness the one detected in the linear phase of the curve itself. The original curve is plotted in gray in the same figure.

Some pictures of the crack pattern after failure are shown with reference to the instrumented side and both the crosswise sides.

All the cases investigated show a good repeatability of the initial stiffness while the peak load shows very small scattering for the cases of β = 1 and 3. In any case the standard deviation is lower than 20% (Table 3).

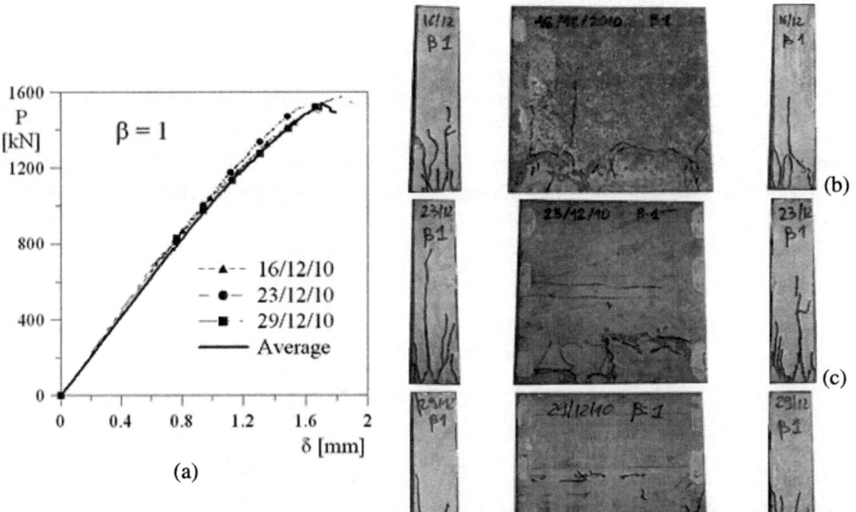

Fig. 1. Experimental results for β=1: (a) load vs. displacement curve and crack pattern of specimens 16/12/10 (b), 23/12/10 (c) and 29/12/10 (d)

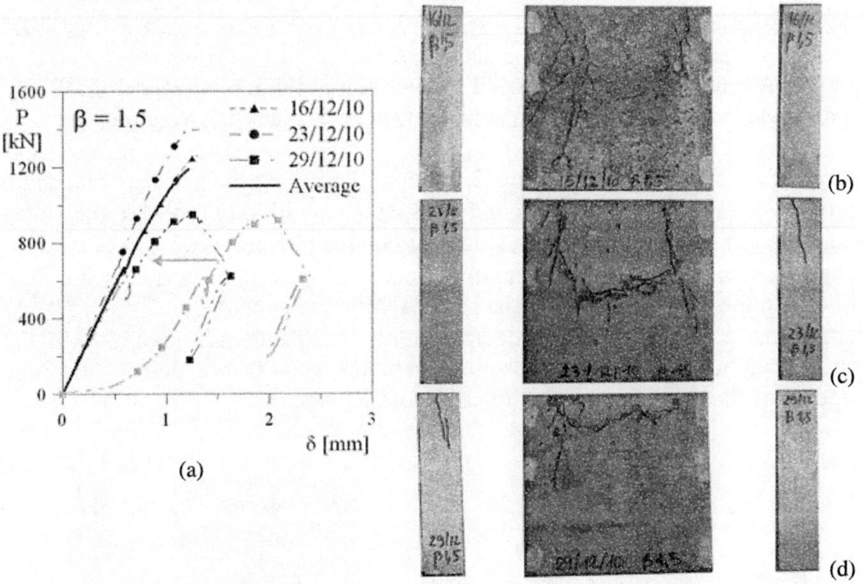

Fig. 2. Experimental results for β=1.5: (a) load vs. displacement curve and crack pattern of specimens 16/12/10 (b), 23/12/10 (c) and 29/12/10 (d)

Looking at the crack patterns in the cases of β = 2 and 3, the formation of a wedge under the bearing area and of an almost vertical crack in the bottom part of the specimen, where a distributed load pressure acts, can be observed. In the case of uniaxial compression (β = 1) the failure seems to be related to vertical cracks in the specimen plane.

Fig. 5a compares the results of all the tests performed by means of the average load vs. displacement curves. The results are also compared in Fig. 5b that shows the behavior of the peak pressure acting under the bearing plate normalized with respect to the pressure of the case of pure compression (β = 1).

The results of the experimental investigation here presented are compared to those obtained from nominally identical experimental tests carried out on two Steel Fiber Reinforced Concrete. The reinforcement content of the materials were 25 [5] and 50 kg/m^3 [6]; the s used were steel hooked end, 60 mm long and with a diameter of 0.6 mm

The same comparison is reported in Table 3 in terms of peak load. It is worth noting that the cubic average compressive strengths of the materials were 41 MPa [5] and 63.7 MPa [6] respectively for the 25 and 50 kg/m^3. A quite significant increase of the peak load is achieved with respect to traditional SFRC.

The behavior of the normalized pressure with respect to the load ration seems to be quite similar for all the materials. In the same Fig. 6b the behavior suggested by Model Code for traditional reinforcement in the case of a mechanical reinforcement ratio ω = 0.1 is also shown.

5 Conclusions

The experimental investigation carried out on 60 mm thick plate made of HPFRC to investigate the bottle shaped concrete struts suggests the following remarks.

- Any significant strength reduction is observed changing the geometrical ratio β between the bearing plate width and the specimen width. In any case the valley suggested by Model for medium values of β cannot be observed.
- The behavior of the peak pressure under the bearing plate for growing β is very similar to the one of traditional steel fiber reinforced concrete.
- The increase of the material properties allow the specimen to exhibit a higher resistance, but the failure mechanism remains very brittle despite the high amount of fiber and the bending ductility of the material.

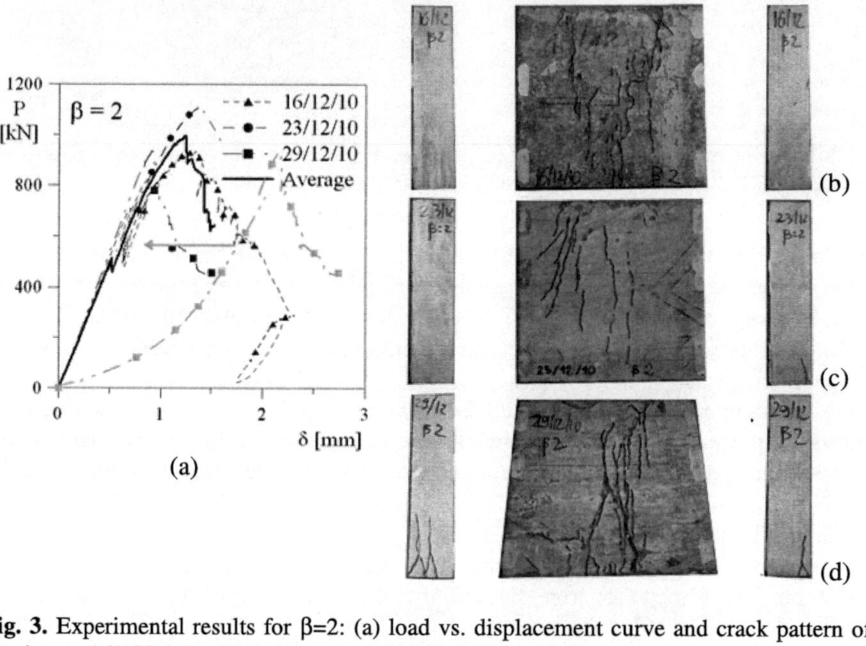

Fig. 3. Experimental results for β=2: (a) load vs. displacement curve and crack pattern of specimens 16/12/10 (b), 23/12/10 (c) and 29/12/10 (d)

Table 3. Peak loads of HPFRC and of two different SFRC in kN [5, 6]

β	HPFRC				SFRC	SFRC
	16/12/10	23/12/10	29/12/10	Average (std%)	25 kg/m^3	50 kg/m^3
1	1444.47	1519.50	1575.28	1513.08 (4.34%)	452.13	807.93
1.5	1259.68	1405.47	950.53	1205.23 (19.28%)	357.66	-
2	935.26	1117.96	936.70	996.64 (10.54%)	272.23	582.93
3	771.23	766.03	756.84	764.70 (0.95%)	238.87	362.11

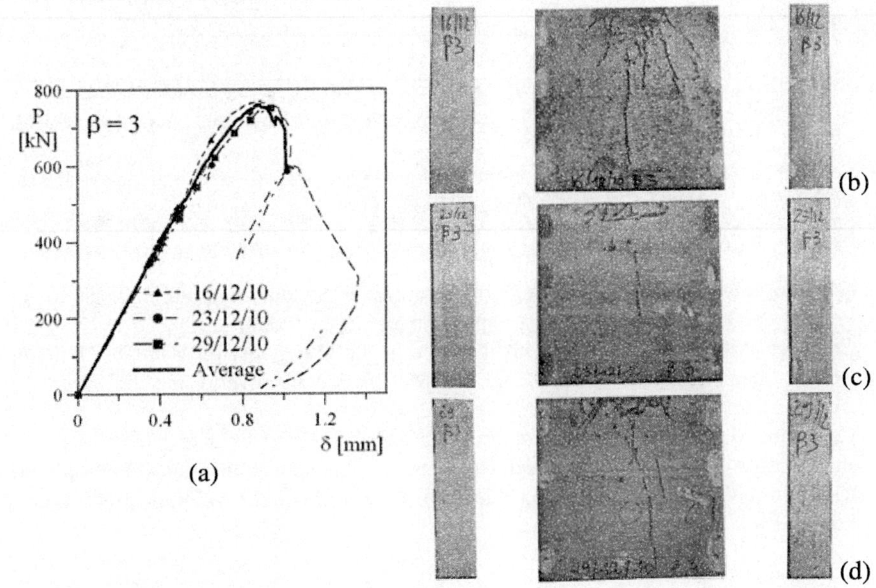

Fig. 4. Experimental results for β=3: (a) load vs. displacement curve and crack pattern of specimens 16/12/10 (b), 23/12/10 (c) and 29/12/10 (d)

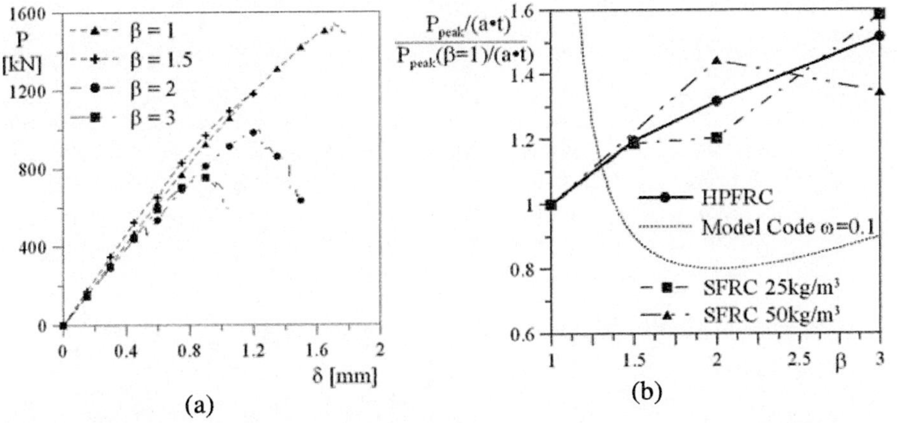

Fig. 5. Experimental results: Average load vs. displacement curves (a) and normalized maximum pressure vs. βb). t is the specimen thickness (t = 60 mm).

Acknowledgements. The authors would like to acknowledge the students R. Bazzi, E. Moroni, L. Pavone, G. Robustellini, L. Vidali and S. Rusconi for their help in the experimental investigation.

References

[1] Biondini, F., Bontempi, F., Malerba, P.G.: Stress path adapting Strut-and-Tie models in cracked and uncracked R.C. elements. Structural Eng. and Mechanics 12(6), 685–698 (2001)

[2] Schlaich, J., Schäfer, K.: Design and detailing of structural concrete using strut and tie models. The Structural Engineer 69(1), 113–125 (1991)

[3] CEB – FIP Model Code 90

[4] Caverzan, A., Colombo, M., di Prisco, M.: High Performance steel fibre reinforced cocnrete: Residual behaviour at high temperature. In: Sonoda, Banthia, Fujikake (eds.) Proc. of PROTECT 2009, Hayama, Japan, August 19-21 (2009)

[5] Colombo, M., di Prisco, M.: Steel-fiber role in the D-regions. In: Banthia, Uomoto, Bentur, Shah (eds.) Proc. Of Int. Conf. ConMat 2005, Vancouver, Canada, August 22-24 (2005)

[6] Colombo, M., di Prisco, M., Lamperti, M.: SFRC D-Regions: size effects in bottle shape struts. In: Fibre Reinforced Concrete: design and applications, Proc of 7th RILEM Int. Conf BEFIB 2008, pp. 115–124. BAGNEAUX Rilem Publications S.a.r.l, France, PRO 60 (2008)

Effect of Fiber Reinforced Concrete in Members with Highly Complex Stress Fields

S.-H. Chao[1], T. Pareek[1], and D.R. Sahoo[2]

[1] Department of Civil Engineering, University of Texas at Arlington, USA
[2] Indian Institute of Technology at Delhi, India

Abstract. Reinforced concrete (RC) members with significant geometric discontinuities and complex stress distributions under loading require considerable analyses and usually complicated reinforcement detailing. RC members with large openings are one of the examples. These large openings may interrupt the load transfer by direct concrete struts and cause substantial decrease in strength and unpredictable failure modes. The reinforcement detailing of these concrete members based on strut-and-tie models (STMs) is generally complicated and very often, these models cannot predict the failure mechanism due to localized damages. The actual stress fields in such members are typically very different from that predicted by STMs, as evidenced by many experimental investigations. This study investigates the performance of steel fiber reinforced concrete (SFRC) deep beams with multiple large openings under monotonically increased load. A simple design approach based on elastic finite element analysis (FEA) was used for the reinforcement detailing of the SFRC specimen. Experimental results indicated that, although the complex reinforcement detailing as per STM was not used, the SFRC specimen with 1.5% volume fraction of fibers reached much higher strength than the design load and exhibited ductile mode of failure.

1 Introduction

Sufficient plastic redistribution of internal forces is essential for a structure to sustain expected and unexpected loads, and to fail in a ductile manner if over-loaded. In RC members, due to the brittle nature of concrete, this redistribution primarily relies on the steel reinforcing bars and their layouts, in which bars are placed at locations where the concrete is overly stressed beyond its cracking strength. For typical concrete members with simple and regular geometries, those locations can be easily predicted by classical elastic theory. It is well known, however, that the stress pattern is highly non-linear and deviates considerably from the classical elastic theory for RC members with significant geometric discontinuities. One such example is RC deep beams with large web openings in a building. These openings in the web region of deep beams are provided for essential services and

G.J. Parra-Montesinos, H.W. Reinhardt, and A.E. Naaman (Eds.): HPFRCC 6, pp. 213–220.

accessibility. Large openings, if located between the loading point and support, will disrupt the flow of force transfer, and usually significantly reduce the load-carrying capacity [1]. In-span hinges in RC box-girder bridges are another example. Those hinges are typically used to accommodate the longitudinal expansion and contraction of the structure and to allow independent vibrations of two adjacent bridge frames. In-span hinges have very complex internal stress distribution due to the geometries of the seat and discrete bearing locations, as well as the geometric discontinuities when utility openings exist [2]. These members with significant geometric discontinuities and complex stress distributions under loading require considerable analyses and usually complicated reinforcement detailing. The reinforcement detailing of these concrete members based on STMs [3], is generally complicated and very often, these models cannot predict the failure mechanism due to localized damages. Also, the actual stress fields in such members are typically very different from that predicated by STMs, as indicated by many experimental investigations [4-7]. It should be noted that the concept of STMs were originally developed based on plastic truss analogy, in which the structure is assumed sufficiently ductile. However, due to the fact that concrete has a limited capacity to sustain plastic deformation, along with the complex stress field after cracking, those members deigned based on STM generally have limited post-peak ductility [8].

This study investigates the performance of SFRC deep beams with two large openings under monotonically increased loads. Complex reinforcement detailing as per STM was not used. The use of SFRC is based on the assumption that, if breakdown of the most stressed locations are prevented by reinforcing bars, the greater plastic deformation capacity of SFRC will allow considerable internal force redistribution, thus increasing the ultimate load-carrying capacity.

2 Experimental Program

The specimens tested in this experimental study was a deep beam and had the same but 1/4 scale geometry as the analytical model originally considered by Schlaich et al. [9]. Earlier, the specimens with the same geometry and dimensions were also tested by other researchers [7]. The overall dimension of the specimens was 1880 mm (74 in.) long, 1170 mm (46 in.) deep, and 112 mm (4.4 in.) thick. The specimens had two 381 mm (15 in.) square opening near the left lower corner and the right upper corner of the beams, respectively. The position and size of the openings in the specimens were selected to interfere with the direct load paths that could potentially form between the loading point and the supports [7]. Two deep beam specimens were tested in this study; one specimen was RC and the other one was SFRC. The properties of the concrete mixtures and steel fibers are listed in Tables 1 and 2, respectively. SFRC specimen contained end-hooked steel fibers of 1.5% by volume.

The detailing of reinforcing bars in the RC specimen was originally designed as per STM [7], as shown in Fig. 1. Test specimens were designed for an ultimate load-carrying capacity of 151.7 kN (34.1 kips). Detailed information of the strut geometry, forces, and the efficiency factors were given elsewhere [10]. Steel reinforcing bars in the RC specimen was provided in two layers using No. 10M (10 mm) bars with a nominal bar area of 71 mm^2 (0.11 in.^{2}). The clear cover of concrete of 25 mm (1 in.) to the edge of the steel reinforcing bars was provided in all specimens. All reinforcing bars were provided with standard hooks to provide anchorage in order to avoid their pullout. Secondary reinforcing bars were not used in the RC specimen as per the prior study by other researchers [7]. It is noted that although the secondary reinforcement was expected to increase the ductility of the concrete thus allowing a truss-like plastic mechanism to form, it was hardly realized as evidenced by prior experiment results [4, 7, 8]. Experimental studies with the same specimen geometries and dimensions [7, 11] showed that RC specimen designed based on the STM (Fig. 1) suffered severe cracking and crushing of concrete near the supports. To avoid these premature local failures, steel cages formed by four No. 10M longitudinal reinforcement at corners and No. 10M transverse stirrups at a spacing of 100 mm (4 in.) were used as boundary elements near the supports of the specimen. Fig. 2 shows the reinforcement layout of the RC specimen, which was the same as that used by Breña and Morrison [7], except the additional support cages.

Table 1. Mix proportion by weight for the three mixtures

Mix type	Cement Type 1	Fly ash (Class C)	Sand[1]	Coarse Aggregate [2]	Water	Steel Fiber	f'_c [3], MPa
RC	1.0	0.50	1.7	1.0	0.60	0	46.3
SFRC	1.0	0.50	1.7	1.0	0.58	0.247	44.2

Note: [1] ASTM natural river sand (Fineness Modulus = 2.57); [2] ASTM C 33 Size Number 8 (4.75 mm to 9.5 mm), 95% of mass finer than 9.5 mm (3/8 in.), and nominal maximum size = 9.5 mm (3/8 in.); [3] average of six cylinders (100 mm x 200 mm) tested on the same day when the specimens were tested

Table 2. Properties of steel fibers

Fiber type	Length (L) mm (in.)[4]	Diameter (D) mm (in.)[4]	Aspect ratio (L/D)	Tensile strength MPa (ksi) [5]
Steel Hooked Fiber	47.5 (1.87)	0.76 (0.031)	62.5	1100 (160)

Note: [4] measured; [5] provided by manufacturer

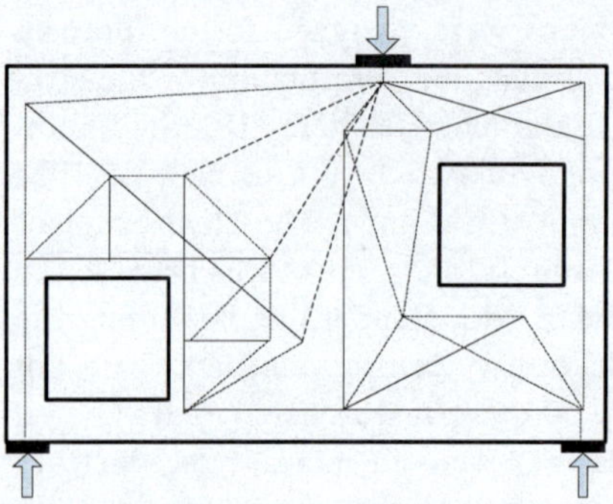

Fig. 1. Design STM for RC specimen: solid lines represent ties and dashed lines represent struts [7]

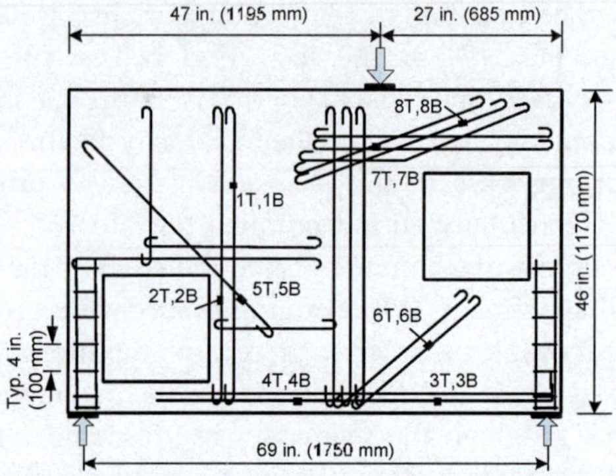

Fig. 2. Reinforcement layout (two layers for each bar location), and locations of strain gauges in RC specimen

Design of the SFRC specimen was based on a simple procedure in which a two dimensional elastic finite element analysis [12] was first carried out by applying the ultimate design load and the principal stress distributions are shown in Fig. 3.

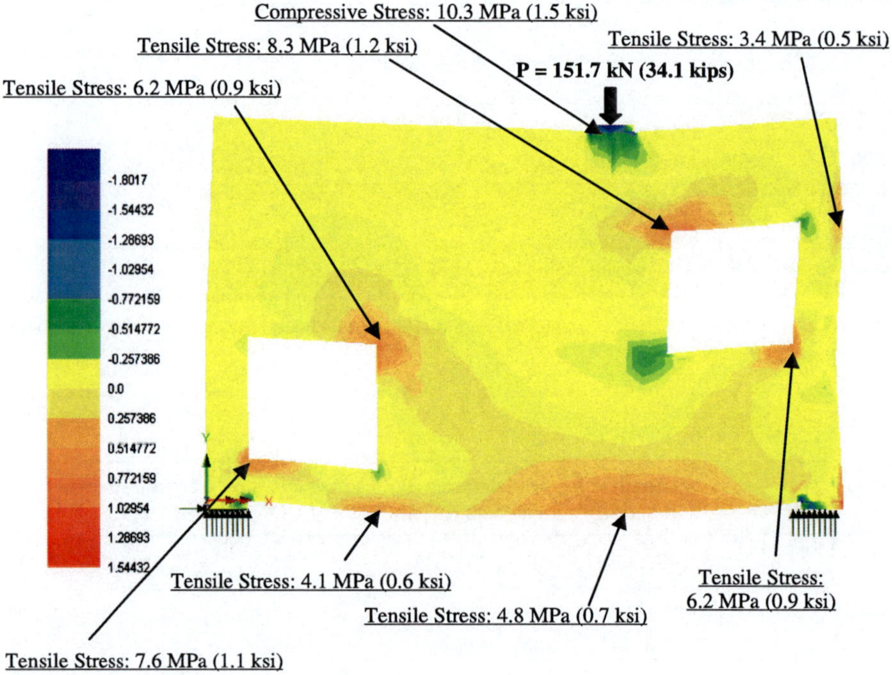

Fig. 3. Elastic finite element analysis: contour of principal stresses at design load of 151.7 kN (34.1 kips)

The critical locations where the principal tensile stresses are close to or greater than the peak flexural strength, f_P, of the SFRC material were identified, as indicated in Fig. 3. The numbers of critical locations are expected to decrease for members with less complex geometries. f_P was determined according to ASTM C1609 [13], and the average value was approximately 5.9 MPa (0.86 ksi) for the SFRC used in this study (Fig. 4). Steel reinforcing bars were then placed in those critical locations in two layers by using No. 10M (10 mm) bars. The detailed reinforcement layout is shown in Fig. 5. As can be seen, the detailing is much simpler than that for the RC specimen designed according to STM (Fig. 2).

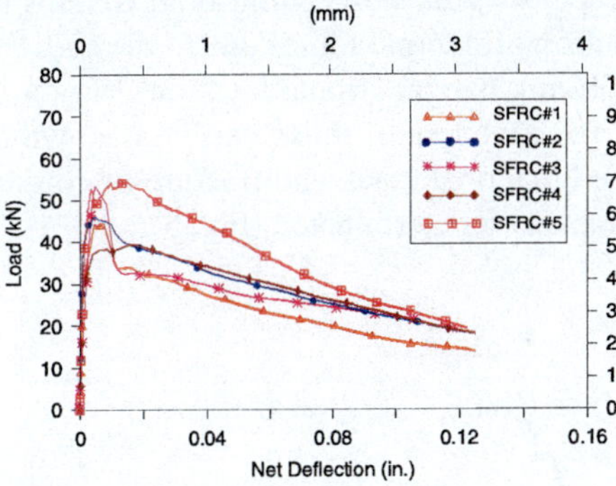

Fig. 4. ASTM C1609 test results for the SFRC material used in this study

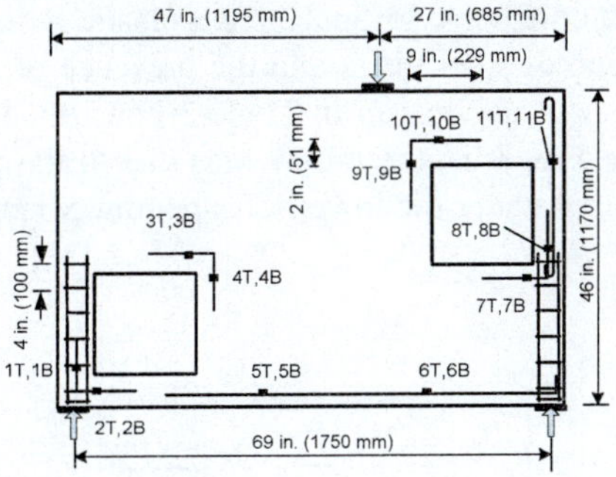

Fig. 5. Reinforcement layout (two layers for each bar location), and locations of strain gauges in SFRC specimen

3 Experimental Results

Both specimens were loaded monotonically by using a 1780 kN (400 kips) Tinius-Olsen universal testing machine. The loading was gradually increased at an interval of 22.5 kN (5 kips) and was continued till the failure of specimens was observed. Uniaxial electrical strain-gauges were attached to the surface of steel reinforcing bars at specified locations to measure the state of strains. Figs. 2 and 5 show the locations and numbering of the strain gauges used in the RC and SFRC specimens, respectively. As shown in Fig. 6, four linear variable differential transformers (LVDTs) were used on the surface of specimens to measure the deformation of concrete struts formed in the specimens during the testing. A pair of string potentiometer (front and back of the specimens) was used below the load point to measure the displacement at approximate mid-height of the upper right opening [7].

The load-displacement responses of the two specimens are compared in Fig. 7. It should be noted that the slightly greater displacement of the SFRC specimen at the ascending branch was attributed to the local deformation near the opening. The maximum load carried by the RC specimen was 445 kN (100 kips) before the testing was stopped due to unexpected and explosive local fracture of concrete near the upper right corner of the specimen, as shown in Fig. 8(a). This premature failure led to abrupt decrease of the load-carrying capacity. It is noted that no reinforcement was placed at the fractured region according to the STM shown in Fig. 1. As can be seen from Table 3, measured steel strains at various bar locations indicated that a fully yielded truss mechanism was not reached when the failure occurred.

Response of the SFRC specimen showed stably increased load up to a peak strength of 356 kN (80 kips), which was much higher than the design load of 151.7 kN (34.1 kips). A deflection-hardening response was noticed after the peak strength was reached, followed by a gradual post-peak descending branch. This ductile behavior indicated a plastic redistribution of internal forces due to the addition of steel fibers and the presence of reinforcing bars at critical locations. Strain gauge data given in Table 3 confirmed the effectiveness of those bars. As shown in Fig. 8(b), the failure was due to the opened diagonal crack and fracture of concrete under the lower left opening, where the bars were terminated (Fig. 5).

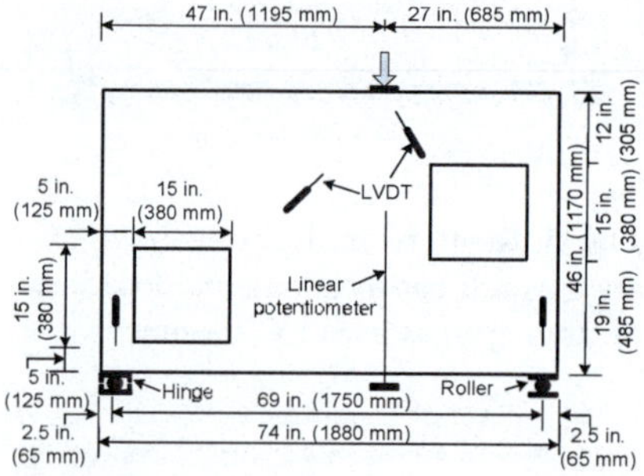

Fig. 6. Details of test specimen and location of LVDTs

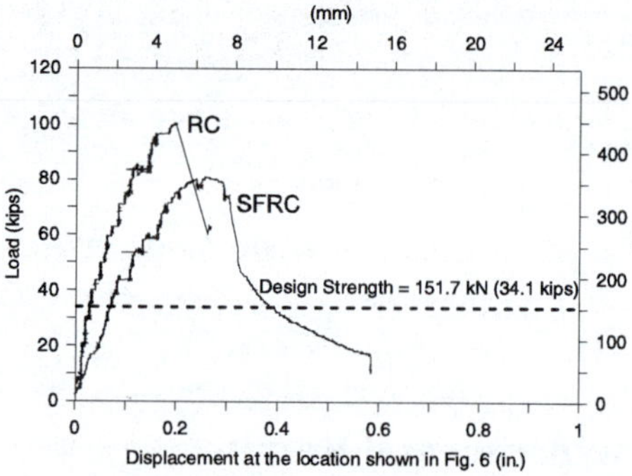

Fig. 7. Load versus displacement response of test specimens

(a) RC specimen (b) SFRC specimen

Fig. 8. Mode of failure of specimens

Table 3. Recorded strain values at various bar locations

Strain Gauge Label	Micro-Strains of Reinforcing Bars at Ultimate Load	
	SFRC	RC
1T	1539	1503
1B	*	1641
2T	2373§	2083§
2B	2987§	1851
3T	3715§	3028§
3B	*	7562§
4T	*	2509§
4B	2804§	2457§
5T	386	*
5B	380	*
6T	6339§	*
6B	3343§	*
7T	1137	1388
7B	1186	1446
8T	1524	1477
8B	1190	1517
9T	394	€
9B	1007	€
10T	928	€
10B	1699	€
11T	2439§	€
11B	2360§	€

* Damaged strain gauges; € Strain gauges not used; § bar yielded; T: top layer of the bars, B: bottom layer of the bars; all strain values were in tension except Strain Gauge 1T in SFRC specimen

4 Summary and Conclusions

This paper investigated the effectiveness of using steel fibers in reinforced concrete members with significant geometric discontinuities. The effectiveness was evaluated by the load-carrying capacity and ductility. The study deep beam specimens had two large openings which induced complex stress fields upon loading. The RC specimen was designed according to strut-and-tie model and the SFRC specimen with 1.5% steel hooked fibers by volume was designed based on two dimensional elastic finite element analysis (FEA). Reinforcing bars in the SFRC specimens were required only at a few critical locations as suggested by FEA. Those bars served as "ductile links" to prevent the breakdown of the highly stressed regions before the fully plastic redistribution of internal forces through steel fibers. The construction of RC specimen can be time-consuming and labor-intensive due to complicated detailing of reinforcing bars in contrast to that of the SFRC specimen. Both specimens were tested under monotonically increased load until failure.

RC specimen failed due to an unexpected local brittle fracture, which could hardly be predicted by the strut-and-tie model. On the other hand, even though there was much less quantity of steel reinforcing bars used, the SFRC specimen reached approximately 2.4 times the design load and exhibited ductile mode of failure. This study clearly shows that using SFRC and simple reinforcement detailing can ensure sufficient load-carrying capacity and ample ductility of reinforced concrete members with significant geometric discontinuities.

References

[1] Ray, S.P.: Deep Beams with Web Openings. In: Kong, F.K. (ed.) Reinforced Concrete Deep Beams, p. 288. Van Nostrand Reinhold, New York (1990)

[2] Hube, M. A., Mosalam K. M.: Experimental and Computational Evaluation of Current and Innovative In-Span Hinge Details in Reinforced Concrete Box-Girder Bridges—Part 1: Experimental Findings and Pre-Test Analysis. Pacific Earthquake Engineering Research Center, PEER 2008/103 (January 2009)

[3] ACI Committee 318, Building Code Requirements for Reinforced Concrete and Commen-tary (ACI318-08/ACI318R-08). American Concrete Institute, Detroit (2008)

[4] Maxwell, B.S., Breen, J.E.: Experimental Evaluation of Strut-and Tie Model Applied to Deep Beam with Opening. ACI Structural Journal 97(1), 142–149 (2000)

[5] Chen, B.S., Hagenberger, M.J., Breen, J.E.: Evaluation of Strut-and-Tie Modeling Applied to Dapped Beam with Opening. ACI Structural Journal 99(4), 445–450 (2002)

[6] Park, J.W., Kuchma, D.: Strut-and-Tie Model Analysis for Strength Prediction of Deep Beams. ACI Structural Journal 104(6), 657–666 (2007)

[7] Breña, S.F., Morrison, M.C.: Factors Affecting Strength of Elements Designed Using Strut-and-Tie Models. ACI Structural Journal 104(3), 267–277 (2007)

[8] Kuchma, D., Yindeesuk, S., Nagle, T., Hart, J., Lee, H.H.: Experimental Validation of Strut-and-Tie Method for complex Regions. ACI Structural Journal 105(5), 578–589 (2008)

[9] Schlaich, J., Schäfer, K., Jennewein, M.: Toward a Consistent Design of Structural Concrete. PCI Journal 32(3), 75–150 (1987)

[10] Breña, S.F., Morrison, M.C.: Authors' Closure on the Discussion on Factors Affecting Strength of Elements Designed Using Strut-and-Tie Models. ACI Structural Journal 105(2), 236 (2008)

[11] Flores, C. A.: Performance of large scale steel fiber reinforced concrete deep beam with single opening under monotonic loading, Master Thesis, Department of Civil Engineering. The University of Texas at Arlington (2009)

[12] LUSAS, Version 14.5 (2010)

[13] ASTM C1609/C1609M-10, Standard Test method for Flexural Performance of Fiber-Reinforced Concrete (Using Beam with Third-Point Loading), ASTM International, West Conshohocken, PA, 9 pages (2010)

Towards a Design Model for Steel Fiber Reinforced Concrete in Bending

G.P.A.G. van Zijl and P.B.K. Mbewe

Department of Civil Engineering, University of Stellenbosch, South Africa

Abstract. This paper outlines a bending model for steel fiber reinforced concrete (SFRC) with strain softening, deflection hardening behavior. Models for the determination of flexural capacity of SFRC rectangular sections based on equivalent stress blocks for both compression and tension are developed using strain compatibility and force equilibrium principles. The equivalent stress blocks are derived from an elastic perfect plastic model for compression, and an elastic-constant post-peak response for tension. An experimental program was performed to characterise the material model parameters and to verify the proposed bending model.

1 Introduction

Steel fiber reinforced concrete (SFRC) is a composite cement-based material with enhanced post-cracking tensile resistance due to the fibers bridging the crack surfaces [1]. This behavior is influenced mainly by the amount of fibers crossing a crack effectively and the bond and strength properties of the fibers used. By proper design, steel fibers may be used to substitute conventional reinforcement [1-3].

The use of SFRC in structural systems requires thorough research on the material behavior, analysis models, verification and structural reliability [4]. Current intense research effort towards developing reliable design guidelines for SFRC [2, 3, 5-8] is justified, as no codified design approach exists for the material. Bending models based on a drop down model for tension, i.e. plastic post-peak tensile response, but at a level lower than the cracking strength, and a bilinear model for compression has been proposed recently [2, 3] where closed form equations are used. ACI [7] outlines the possibility of using a rectangular stress block for tensile stresses as for compression. A closed form model has been proposed using average post cracking tensile strength smeared over the whole beam section part under tensile strain, in combination with either the ACI compression stress block or a triangular compression block [9].

This paper elaborates a flexural model for strain softening SFRC using equivalent rectangular stress blocks for compression and tension. Modifications are made as compared to the recommendations of ACI, Rilem and Eurocode while using a

G.J. Parra-Montesinos, H.W. Reinhardt, and A.E. Naaman (Eds.): HPFRCC 6, pp. 221–229.

drop down tensile model and Rilem's tensile model [5] as bases for derivation of the equivalent tensile stress block. This is done to (i) facilitate evaluation of the influence of the assumed tensile behavior on modelling assumptions, (ii) evaluate the simplified, closed form models, and (iii) provide the eventual user a programmable model while retaining insight in strain limits.

2 Design Stress Blocks for SFRC

Simplified beam section bending stress distributions are derived here, based on force equilibrium and equivalent moment resistance with the bilinear compressive stress-strain model (Fig. 1). The following expressions can be derived:

$$\lambda_c = \frac{2\left(\omega^2 - 3\omega + 3\right)}{3(2-\omega)}, \eta_c = \frac{3(2-\omega)^2}{4\left(\omega^2 - 3\omega + 3\right)}, \beta_c = \lambda_c \eta_c = \frac{2-\omega}{2}, \text{ with } \omega = \varepsilon_{cy} / \varepsilon_{cp} \quad (1)$$

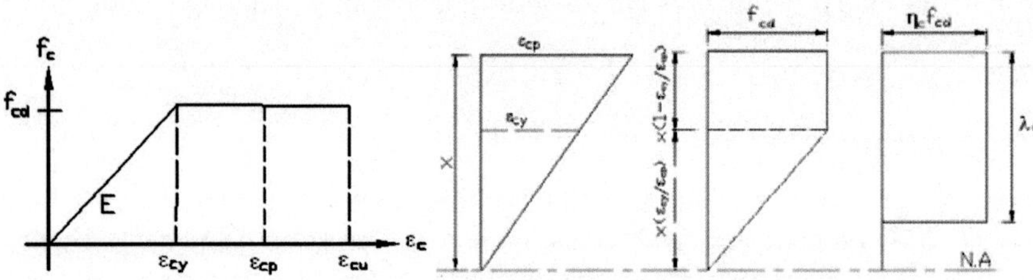

Fig. 1. Definition of rectangular stress block from bilinear compressive stress-strain model

Using appropriate strain limits, the design conversion factors λ_{cd} and η_{cd} are similar to the conversion factors given in Eurocode [10].

A drop-down constant tensile stress model as proposed in [11] and enhanced by [2] is used (see Fig. 2). Generally, two stress parameters should be known if this model is to be used. These are ultimate tensile strength (peak tensile strength) and an assumed constant post peak tensile strength. It is against this background that equivalent post–peak tensile parameters, $f_{eq,2}$ and $f_{eq,3}$ defined in [5] can be used. Since post cracking tensile strength of SFRC depends on the amount and properties of steel fibers and fiber-matrix interaction, the grade of concrete will not be used as a basis for converting the stress distribution into a rectangular stress block. The post peak tensile behavior for a particular grade of concrete will have to be determined through testing. It is assumed that for a specified sample, with consistency in mix design, the ratios of peak tensile strength to post peak tensile strength, and that of cracking strain to ultimate strain can be defined.

Factors λ_t and η_t are used to convert the drop down constant to a rectangular stress block as shown in Fig. 2. These values depend on the ratio of post peak tensile strength to peak tensile strength, μ_i, and $\omega' = \varepsilon_{ty}/\varepsilon_{tp}$, which parameters are assumed to fully define the post peak material model properties.

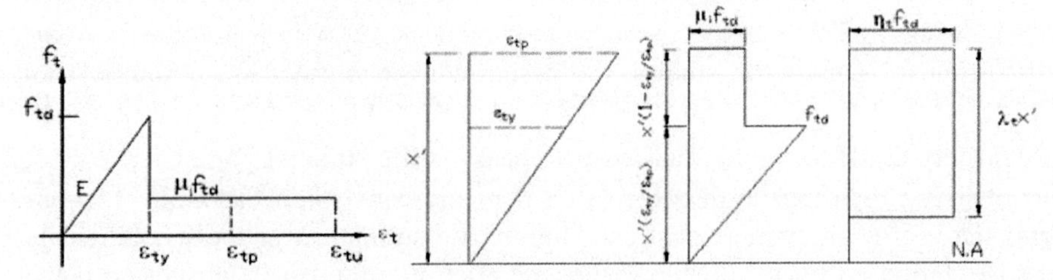

Fig. 2. Definition of tensile rectangular stress block from drop down constant stress-strain model

The conversion factors, λ_t, η_t and β_t can be found to be defined as follows:

$$\lambda_t = \frac{2\left(\omega'^2(3\mu_i - 2) + 3\omega'(1 - 2\mu_i) + 3\mu_i\right)}{3(\omega'(1 - 2\mu_i) + 2\mu_i)}, \quad \eta_t = \frac{3(\omega'(1 - 2\mu_i) + 2\mu_i)^2}{4\left(\omega'^2(3\mu_i - 2) + 3\omega'(1 - 2\mu_i) + 3\mu_i\right)},$$

$$\beta_t = \lambda_t \eta_t = \frac{\omega'(1 - 2\mu_i) + 2\mu_i}{2} \tag{2}$$

Since different SFRC grades may have different ratios of post peak tensile strength to peak tensile strength, the values of λ_t and η_t can be provided for different values of μ_i. A parametric study is conducted to observe the variability of parameters λ_t and η_t within a practical range of ω'. Fig. 3 shows the values of λ_t and η_t for selected values of μ_i as function of the ratio of ω'.

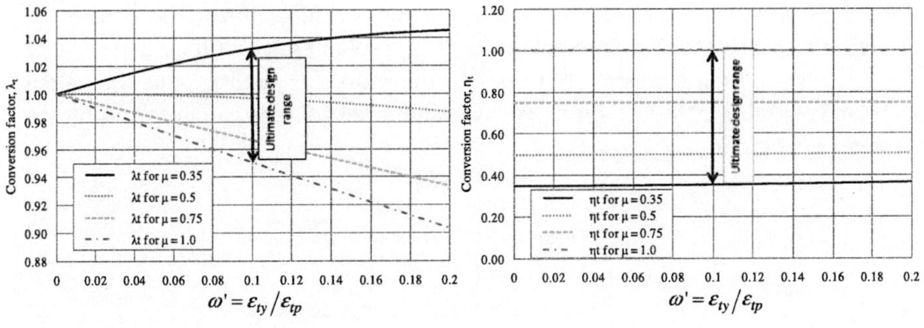

Fig. 3. Effects of normalised post peak strength μ_i on conversion factors λ_t and η_t

The value μ_i = 0.35 represents the minimum ratio that achieves deflection hardening in strain softening SFRC [2], while μ_i = 1.0 is the maximum value that can be used by the proposed flexural model and represents strain hardening material with perfect plastic post peak tensile behavior.

For the ultimate design of the section, the range of conversion factors has been shown in Fig. 3. This range is based on the practical values in which the ratio ω' falls. From the parametric studies, the design conversion factor, λ_{td} ranges from 0.95 for μ_i = 1.0, to 1.05 for μ_i = 0.35 for $\omega' = 0.1$ while the factor η_{td} is approximately equal to the normalised post peak tensile strength parameter, μ_i. So, for relatively high tensile ductility ($\omega' \leq 0.1$) the conversion factor $\lambda_t \approx 1.0$ confirms the use of an average post cracking tensile strength over the whole tensile depth as proposed by [9]. For lower tensile ductility an error is introduced if the full tensile block (λ_{td} =1.0) is used. The above result was also found when using the Rilem tensile model [6] for small variations of the post cracking tensile strengths at service and at ultimate.

3 Flexural Model for SFRC Rectangular Beam

A beam element subjected to flexure may experience three basic stages of stress states shown in Fig. 4. The stages assume that the element will fail in tension. Based on strain compatibility and force equilibrium, equations for the determination of neutral axis, curvature and lever arm for different stages of stress-strain states are derived - see Table 1.

By considering the simplified rectangular stress block (Fig. 4, right), the flexural capacity of the section can be derived. To ensure that moment capacity is derived from strains that do not exceed ultimate limits in either tension or compression, equivalent strains are determined from a strain compatibility equation and a force equilibrium equation. The minimum of the ultimate and equivalent strains dictates the ultimate capacity of the section. It is assumed, based on practical values of strains, that the moment capacity for a SFRC without reinforcement will be limited by the tensile capacity of the section. Therefore, considering that the ultimate tensile strain is reached in either stage two or stage three, the equivalent compressive strains at failure are determined. By replacing ε_{cp} with ε_{cu} and ε_{tp} with ε_{tu} in eqs. (1) and (2) respectively to define the ratio's β_{cu} and β_{tu} the equivalent compressive strains at failure for stages two and three failures can be given by the following expressions:

$$\begin{aligned} \varepsilon_{c2eq} &= \sqrt{2\beta_{tu}\varepsilon_{ty}\varepsilon_{tu}} && \text{Compressive strain state for stage 2 failure} \\ \varepsilon_{c3eq} &= \frac{\beta_{tu}\varepsilon_{ty}\varepsilon_{tu}}{\varepsilon_{cy}} + \frac{\varepsilon_{cy}}{2} && \text{Compressive strain state for stage 3 failure} \end{aligned} \qquad (3)$$

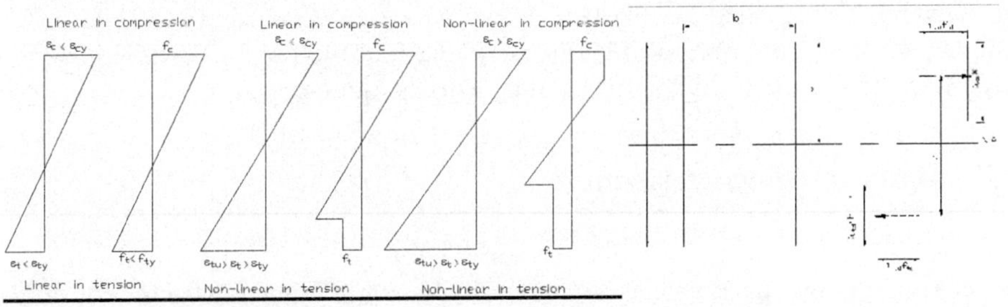

Fig. 4. Left: Three stages of stress-strain states. right: Rectangular stress blocks for design

Table 1. Neutral axis and curvature formulas using bilinear and drop down models

Parameter	Stage 1	Stage 2	Stage 3
$\frac{x}{h}$ (neutral axis ratio)	$\frac{1}{2}$	$\frac{2\beta_t\varepsilon_{ty}}{\varepsilon_c+2\beta_t\varepsilon_{ty}}$	$\frac{\beta_t\varepsilon_{ty}}{\beta_c\varepsilon_{cy}+\beta_t\varepsilon_{ty}}$
$\frac{\varepsilon_c}{x}$ (curvature)	$\frac{2\varepsilon_c}{h}$	$\frac{\varepsilon_c^2}{2h\beta_t\varepsilon_{ty}}+\frac{\varepsilon_c}{h}$	$\frac{\beta_c\varepsilon_{ty}\varepsilon_c}{h\beta_t\varepsilon_{ty}}+\frac{\varepsilon_c}{h}$
$\frac{Z}{h}$ (lever arm ratio)	$\frac{2}{3}$	$1+\frac{x}{h}\left(\frac{\lambda_{td}}{2}-\frac{1}{3}\right)-\frac{\lambda_{td}}{2}$	$1+\frac{x}{2h}(\lambda_{td}-\lambda_{cd})-\frac{\lambda_{td}}{2}$

An appropriate value for equivalent strain calculated from eq. (3) is chosen if it meets the criterion $\varepsilon_{c2eq} \leq \varepsilon_{cy} \leq \varepsilon_{c3eq} \leq \varepsilon_{cu}$.

With the overall depth of the section known, the depth of the neutral axis and the lever arm between compressive and tensile forces are determined using equations in Table 1 and the moment resistance of the section is determined as follows:

$$M = \begin{cases} \beta_c.x.b.Z.f_{cd} & \text{stage 3 failure only} \\ \beta_{tu}.(h-x).b.Z.f_{td} & \text{stage 2 and 3 failure} \end{cases} \quad (4)$$

4 Model Verification

To test the model, a series of experiments was performed. Material characterisation tests involving splitting and compressive test were conducted on 100 mm cubes made of the same SFRC batch as beams used in flexural tests. Three beams 150 x 300 x 750 mm were used while the compressive and splitting tests involved four cubes each. A Contest machine was used for the compression test, while Zwick and Instron Universal Materials Testing Machines were used for the splitting and flexural tests respectively. See Fig. 5 for the testing equipment, specimen dimensions and loading arrangements. All specimens were water cured at 20°C for 28 days before testing. A standard mix of concrete grade C30/37 was used. 1.0%

by volume of Dramix steel fibers with L_f = 30 mm (L_f/d_f=60) was used. Aggregates comprised of local South African natural sand (Malmesbury; 803 kg/m^3), Greywacke stone (6.7 mm (600 kg/m^3) and 13.2 mm (400 kg/m^3)) while binders used were CEM II 32.5 (292 kg/m^3) and fly ash (89 kg/m^3).

4.1 Material Characterisation

The splitting test was used as an indirect tensile test. Tables 2 and 3 summarise data derived from the splitting tests. A conversion factor of 0.9 was used for converting the splitting tensile strength (f_{tsp}) to direct tensile strength while post cracking

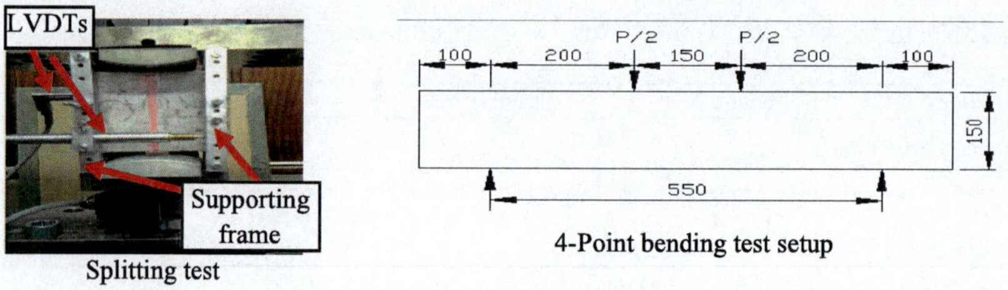

Fig. 5. Testing equipment, specimen dimensions and loading arrangements

stresses were determined by subtracting a uniform stress of $0.1f_{tsp}$ from post cracking splitting stresses. It is acknowledged that the confinement effect of compression force, and the large gauge length on deformation were not compensated for. Equivalent post peak tensile stress, f_{teq} is determined from the expression

$$f_{teq} = \frac{1}{w_{\lim}} \int_0^{w_{\lim}} \sigma(w) dw \tag{5}$$

where w_{lim} is the limiting crack width at an ultimate strain of 0.025. Since post cracking behavior is characterised by the stress-crack width models, a structural characteristic length as suggested by di Prisco *et al* [12] was used to determine post yield strains.

4.2 Experimental Results and Model Verification

4-Point bending tests were performed on three 300x100x750 mm SFRC beams. Table 4 summarises the results. By using parameters derived from average values of material properties, the model over predicts the strength of the beams by 19%. However, when characteristic parameters are used, the model under predicts the flexural strength by about 29%. In this paper the level of reliability has not been studied systematically and, for instance, material factors have not been derived.

From the results shown in Table 4, it is clear that the uncertainty with regard to the material model parameters may lead to significant error in prediction. Another source of inaccuracy has been introduced by the adoption of the bilinear compressive and drop-down tensile stress-strain responses from the onset. A systematic process of quantifying the various sources of uncertainty is a next step towards a flexural design model with acceptable level of reliability.

In addition to comparing the model parameters with experimental data obtained from experiments performed at Stellenbosch University, the model is compared to experimental data obtained from literature and the flexural model developed by Mobasher and Soranakom [3]. Table 5 shows the comparison of the model predictions with those developed by Mobasher and Soranakom, and ex perimental results obtained from literature [13]. Note that compressive yield strain and ultimate strain were taken as 0.133% and 0.35% respectively.

Table 2. Tensile properties of SFRC, obtained from splitting tests

Specimen	f_t=0.9 f_{tsp} (MPa)	Cracking strain, ε_{ty}	f_{teq} (MPa)
C1	4.05	0.00012	3.373
C2	3.78	0.00011	2.163
C3	3.69	0.00011	2.218
C4	3.29	0.00009	2.390
Average	3.70	0.000107	2.536
Stand dev.	0.32	9.163E-06	0.566
Charact. value	3.12	0.00009	1.466

Table 3. Tensile parameters for the model

Description of parameters used	μ	f_t (MPa)	ε_{ty}
Average values	0.684	3.70	0.000107
Characteristic values (5 percentile)	0.469	3.12	0.00009

Table 4. Maximum flexural response for SFRC beams

	Experimental		Model prediction(kN)	
Specimen	Maximum load (kN)	Max moment (kNm)	Avg parameters	Characteristic parameters
C1	69.434	6.944		
C2	67.447	6.746		
C3	62.682	6.269		
Avg	66.53	6.65	79.32	46.96
Stand dev	3.47	0.35		
COV	0.052	0.053		

Table 5. Comparison of current model with experimental results from literature [13] and model by Mobasher and Soranakom [3]

Mix	Fiber con-tent	Model parameters			Bending moment (kNm)		
	kg/m^3	f_{ck}, f_{tk} (MPa)	μ	ε_{ty} (%)	Current model	Model [3]	Exp. [13]
NSC	25	30.2, 3.5	0.31	0.011	4.19	4.25	4.64-5.34
NSC	50	26.6, 4.2	0.48	0.014	7.62	7.44	5.42-7.61
HSC	60	52.9, 6.2	0.5	0.016	11.65	11.71	9.6 -12.68

6 Conclusions

In this paper a model has been presented for the determination of flexural capacity of SFRC beams based on equivalent rectangular stress blocks for both tensile and compressive stresses. The potential use of tensile stress block has been examined through parametric studies of the tensile conversion factors, λ_{td} and η_{td}. The results of the parametric studies show that values of the factor λ_{td} range from 0.95 to 1.05 while conversion factor η_{td} is approximately equal to the normalised post peak tensile strength parameter μ_i. This is, however, valid for relative ductile tensile response. For lower tensile ductility, the assumption of an average tensile stress over the full tension depth of the section may lead to significant errors.

The model has been verified experimentally, by performing compressive and tensile tests to characterise model parameters of compressive strength, tensile strength, post-cracking strength and E-modulus. By subsequent analysis of 4-point bending tests, it was shown that the model over-predicts flexural resistance by 19% if mean values are chosen for the model parameters, but under-predicts flexural resistance by 29 % if characteristic values (5 percentile) values are chosen for the model parameters. Sources of inaccuracy include the simplified tensile model assumed, but also the use of splitting test stress-strain results as opposed to direct tensile test data for characterising tensile model parameters. Future work includes (i) improved assessment of direct tensile behavior, (ii) a systematic reliability study and (iii) inclusion of reinforcing steel bars.

References

[1] Fibre Reinforced Concretes. In: Rossi, P., Chanvillard, G. (eds.) RILEM Proceedings of the 5th RILEM Symposium (BEFIB 2000), PRO15, BEFIB 2000. RILEM Publications S.A.R.L, Bagneux (2000)

[2] Soranakom, C., Mobasher, B.: Closed-form solutions for flexural response of fibre-reinforced concrete beams. J. Eng. Mech. 133(8), 933–941 (2007)

[3] Soranakom, C., Mobasher, B.: Flexural Design of Fiber-Reinforced Concrete. ACI Materials Journal 106(5) (September-October 2009)

[4] Dymond, J.S., Retief, J.V.: Towards a reliability based development program for SHCC design procedures. In: van Zijl, G.P.A.G., Boshoff, W.P. (eds.) Advances in Cement-based Materials. Taylor and Francis Group, London (2010)

[5] Vandewalle, L., Dupont, D.: Test and design methods for steel fibre reinforced concrete: Bending test and interpretation. In: Rilem Proceeding 31, TC 162-TDF Workshop, Bochum, pp. 1–14 (2003)

[6] Vandewalle, L. (ed.): Recommendations of RILEM TC162-TDF: Test and Design Meth-ods for Steel Fibre Reinforced Concrete: bending test (final recommendation). Materials and Structures 35, 579–582 (2002)

[7] ACI Committee 318, Building Code Requirements for Structural Concrete (ACI 318-05) and Commentary (318R-05). American Concrete Institute, Farmington Hills, MI, pp. 430 (2005)

[8] ACI Committee 544, Design considerations for steel Fiber Reinforced Concrete, ACI 544.4R-88. American Concrete Institute, ACI Farmington Hills (1996)
[9] Naaman, A.E.: Strain hardening and deflection hardening fibre reinforced cement composites. In: Naaman, A.E., Reinhardt, H.W. (eds.) High Performance Fibre Reinforced Cement Composites –HPFRCC4,, pp. 95–113. RILEM Publications, PRO 30 (June 2003)
[10] European standard: EN 1992-1-1: Eurocode 2: Design of concrete structures – Part 1: General rules and rules for buildings. Brussels
[11] Lim, T.Y., Paramasivam, P., Lee, S.L.: Bending behaviour of steel-fibre concrete beams. ACI Struct. Journal 84(6), 524–536 (1987b)
[12] Di Prisco, M., Giovanni, A.E., Vandewalle, L.: Fibre reinforced concrete: new design perspectives. Materials and Structures 42, 1261–1281 (2009)
[13] Dupont, D.: Modelling and Experimental Validation of the Constitutive Law (σ-ε) and Cracking Behaviour of Steel Fiber Reinforced Concrete, PhD dissertation, Catholic University of Leuven, Belgium (2003)

Shear Crack Formation and Propagation in Fiber Reinforced Cementitious Composites (FRCC)

I. Paegle and G. Fischer

Technical University of Denmark, Department of Civil Engineering, Kgs. Lyngby, Denmark

Abstract. Knowledge of the mechanisms controlling crack formation, propagation and failure of FRCC under shear loading is currently limited. This paper presents a study that utilized photogrammetry to monitor the shear deformations of two FRCC materials and ordinary concrete (OC). Multiple shear cracks and strain hardening of both FRCC materials was observed under shear loading. The influence of fibers, fiber type, including polyvinyl alcohol (PVA) and polypropylene (PP) fibers, and shear crack angle were investigated. Based upon photogrammetric results, fundamental descriptions of shear crack opening/sliding and subsequent failure are presented.

1 Introduction

There are numerous models available to predict the shear capacity of FRCC [1-3]. However, understanding of the controlling, fundamental mechanisms is lacking, resulting in limited applicability of particular models. Some models may provide accurate predictions for one material and/or load configuration, but minor changes may result in inaccurate predictions. This paper proposes a description of the fundamental mechanisms of shear deformation and failure in FRCC without conventional steel reinforcement in form of stirrups. The proposed mechanisms may be useful in creating a 'universal' model for a wider range of FRCC materials and shear load configurations.

2 Materials and Experimental Methods

The experimental program consists of shear beams, shown in Fig. 1, and corresponding tension and compression specimens of three materials:

- FRCC beams with randomly distributed PVA fibers (PVA-FRCC);
- FRCC beams with randomly distributed PP fibers (PP-FRCC);
- Ordinary concrete (OC).

G.J. Parra-Montesinos, H.W. Reinhardt, and A.E. Naaman (Eds.): HPFRCC 6, pp. 231–238.

Table 1. Properties of PVA and PP fibers

Fiber (Trade name)	Beam description	Fiber type	Ø [μm]	L [mm]	f_t [MPa]	E [GPa]	Strain capacity [%]
REC 15	PVA-FRCC	PVA	40	8	1560	40	6.5
Innegra	PP-FRCC	PP	45	16	550	13.4	8

The properties of PVA and PP fibers used in this study are listed in Table 1. For PVA-FRCC and PP-FRCC beams mortar consisting of fly ash, cement, water, sand (max. grain size 0.18 mm), quartz powder and fibers (2 vol. %) was used. Ordinary concrete consisted of cement, water, fine and coarse aggregates (max. aggregate size 16 mm). The concrete mixture was designed to have a compressive strength similar to that of PVA-FRCC and PP-FRCC. Specimens were demolded after 24 to 48 hours after casting and were air-cured for approximately at least 28 days prior to testing.

Tensile stress-strain responses of PVA-FRCC and PP-FRCC were determined using 'dogbone' specimens with a representative cross section of 25 mm × 50 mm. Deformation controlled tensile tests (0.5 mm/min loading rate) were conducted with linear variable differential transformers (LVDT's) measuring the tensile deformations and a photogrammetric data acquisition system measuring crack deformations. Splitting tensile testing, according to EN 12390-6 [4], was carried out for OC with a loading rate of 1.57 kN/s. The photogrammetry equipment captures images of the representative section of the specimen at a rate of 1 Hz and through post-processing of the captured images of the deformed specimen surface deformations were determined. Additional information on 'dogbone' specimens and the photogrammetric data acquisition system is available elsewhere [5].

Fig. 1 illustrates the specimen geometry and loading configuration, which is similar to the Ohno shear beam [6] used to investigate shear cracking under various crack angles. Longitudinal reinforcement, which is necessary to avoid flexural failure of the beams but may have an effect on the shear cracking behavior (location, angle), was provided by carbon fiber reinforced polymer plates glued on the-top and bottom faces of the specimen (Fig. 1(a)).

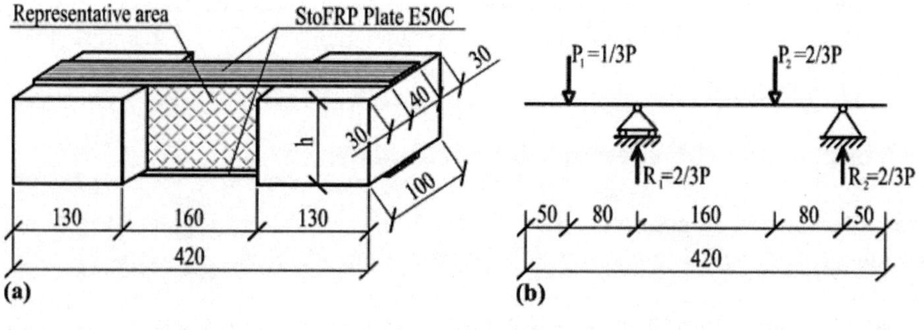

Fig. 1. Shear test setup. Beam's geometry: (a) three-dimensional view; (b) schematic load configuration

Table 2. Specimen height and corresponding targeted and measured shear angles

Beam	Height, h [mm]	Targeted cr. angle [deg]	Measured cr. angle [deg]	Beam	Height, h [mm]	Targeted cr. angle [deg]	Measured cr. angle [deg]
PVA-FRCC60	60	22	30±5	PVA-FRCC150	150	45	45±5
PP-FRCC60	60	22	30±5	PP-FRCC150	150	45	45±5
OC 60	60	22	26	OC 150	150	45	45±2
PVA-FRCC100	100	34	35±5	PVA-FRCC170	170	49	50±5
PP-FRCC100	100	34	35±5	PP-FRCC170	170	49	50±3
OC 100	100	34	40...45	OC 170	170	49	50±3

This configuration restricts formation of flexural cracks, provides minimal addition to shear capacity, and has a reduced influence on the shear capacity compared to embedded rebar. Variations in crack angle were achieved by changing the specimen height while other parameters remained constant. Specimen height and comparison of targeted and measured shear crack angles are shown in Table 2. Beams were loaded in a displacement-controlled procedure with a loading rate of 0.02 mm/s. Shear crack formation and development was monitored using photogrammetry.

3 Results

Typical tensile stress-strain relationships for PVA-FRCC and PP-FRCC are shown in Fig. 2 (a) and (b), respectively. The PVA-FRCC can be considered an Engineered Cementitious Composites (ECC), which is a type of fiber reinforced cementitious material characterized by multiple cracking and strain hardening under uniaxial tension [7-9]. The average first cracking strength taken from six test specimens was 4.1 MPa and 3.2 MPa for PVA-FRCC and PP-FRCC, respectively. Average ultimate tensile strength for the PVA-FRCC, PP-FRCC and OC were 4.5 MPa, 4.1 MPa and 4.1 MPa, respectively, while the compressive strength was 45.9 MPa, 48.8 MPa and 56.0 MPa, respectively.

Fig. 3 shows shear cracks that developed in specimen with a height of 100 mm under shear loading for the different materials. The most apparent benefit of FRCC compared to conventional concrete (OC) is the formation of multiple shear cracking in FRCC resulting in smaller crack widths at the ultimate limit state. Fig. 4 shows typical crack development behavior in terms of crack opening and sliding as measured by photogrammetry for the three tested materials. The schematic in Fig. 4(a) illustrates the two components (opening and sliding) that contribute to the total crack deformation. Opening and sliding were determined by fixing two virtual markers on opposing sides of a single crack in the captured images and subsequently obtaining the relative displacements between them. All graphs show the crack opening and sliding relationships of the 'failure' crack which do not necessarily coincide with the first formed shear crack.

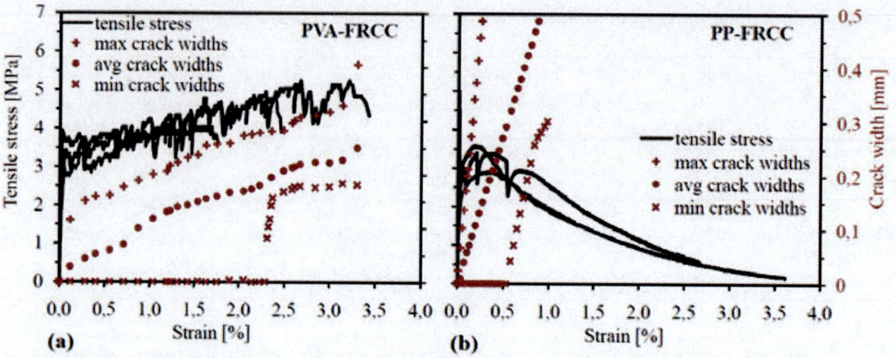

Fig. 2. Tensile properties of: (a) PVA-FRCC and (b) PP-FRCC

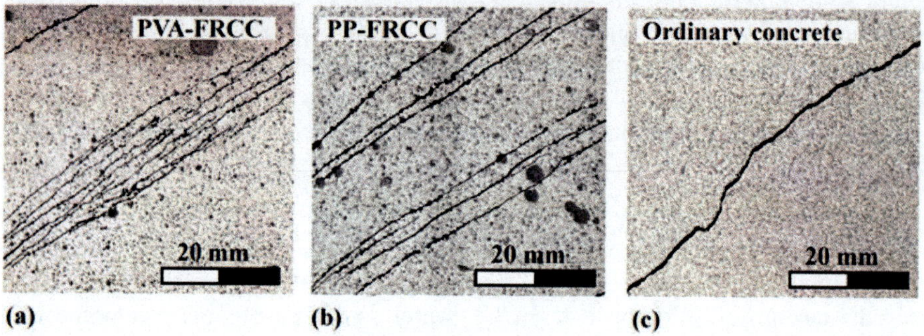

Fig. 3. Shear cracks in PVA (a) and PP (b) FRCCs and in ordinary concrete(c) beams

The graphs in the left column of Fig. 4 show results for beams with shear crack angles between 25-35° (60 and 100 mm beam height), while the right column shows results from beams with 40-50° shear angles (150 and 170 mm beam height).

The effect of fibers is illustrated by comparing results presented in the left column of Fig. 4. Fig. 4(e) shows that the shear stress in OC decreases immediately after crack formation (failure was not completely brittle due to CFRP plates), whereas an increase in shear stress after crack formation due to the addition of fibers resisting shear crack opening and sliding is visible in Fig. 4(a, c). The multiple shear cracking forms in a relatively high range of shear stresses (between 2.5 MPa and failure of the specimen). As previously noted, deformations of 'failure' shear cracks are plotted in graphs; however, this failure crack may not necessarily form as one of the first shear cracks in the beam.

The effect of varying shear angle is illustrated by comparing results in the left column (shear crack angles from 25° to 35°) to results in the right column (shear crack angles from 40° to 50°) in Fig. 4. For OC (Fig. 4(e, f)) with smaller shear crack angles (25-35°), failure occurs upon crack formation, while for larger crack angles (40-50°) immediately after crack formation the shear stress increases.

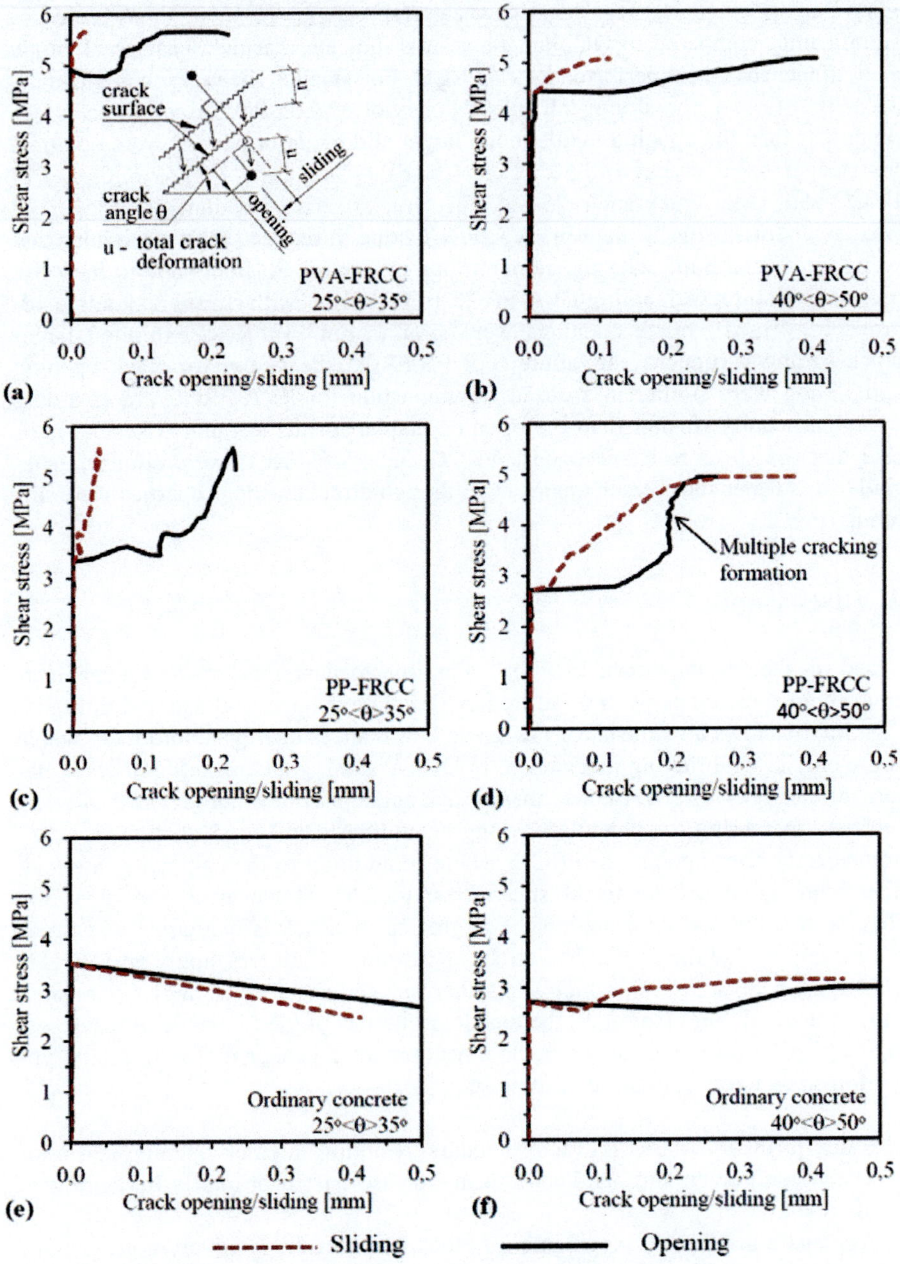

Fig. 4. Characteristic shear stress - crack opening and sliding displacement relationships for: (a) and (b) PVA-FRCC beams for crack angle 25-35° and 40-50°, respectively; (c) and (d) PP-FRCC beams for crack angle 25-35° and 40-50°, respectively; (e) and (f) ordinary concrete beams for crack angle 25-35° and 40-50°, respectively. Schematic in (a) shows the definition of crack opening and sliding displacements

For FRCC (Fig. 4(a-d)) regardless of shear crack angle, the shear stress increases significantly compared to OC, but the results indicate that the shear crack angle also affects structural performance of FRCC. For smaller shear crack angles (25-35°, Fig. 4(a, c)), the sliding of cracks does not exceed 0.05 mm, which is 4-6 times less than the opening, while much larger sliding deformations were obtained for steeper shear cracks (40-50°, Fig. 4(b, d)). Comparing PVA-FRCC to PP-FRCC with shear crack angle 25-35° (Fig. 4(a, c)), the immediate crack opening was larger for PP-FRCC than in PVA-FRCC, due to reduced elastic modulus and softer interfacial bond with the cementitious matrix for PP fibers due to their hydrophobic nature. Comparing PVA-FRCC to PP-FRCC with shear crack angle 40-50° (Fig. 4(b, d)) crack openings were similar, but PP-FRCC exhibited larger crack sliding particularly at failure. For PP-FRCC the values for crack opening and sliding were similar in scale at failure, while for PVA-FRCC crack sliding was significantly smaller than the opening displacements at failure. Results indicate that the shear resistance of PP-FRCC and PVA-FRCC were similar, especially for steeper shear crack angles even though direct tensile properties were different (Fig. 2).

4 Discussion

Based on the results shown in Fig. 4, possible explanations of the mechanisms controlling shear cracking and failure have been developed as illustrated in Fig. 5. Fig. 5(a, b) shows how the shear crack angle influences aggregate interlock, which is the only crack bridging mechanism in OC. Assuming a constant total crack deformation (Fig. 4(a)), the steeper shear crack angle provides more contact of crack surfaces, increasing the potential for aggregate interlock. Fig. 5(c) shows a crack immediately after opening in FRCC, where in addition to the aggregate interlock fiber bridging contributes to the shear stress transfer. However, as shown in Fig. 5(d), with additional deformation the aggregate interlock is overcome and fibers exclusively bridge the crack. The effect of varying shear crack angle on FRCC is illustrated in Figs. 5(d, e), which show that reduced shear crack angles geometrically reduces sliding; therefore, the tensile properties of the fibers are emphasized. Fig. 6(a) illustrates the shear crack development mechanism for ordinary unreinforced concrete consisting of three stages:

1. Crack formation – Brittle fracture occurs, resulting in crack opening and relatively small crack sliding. The deformation is resisted primarily by aggregate interlock.
2. Aggregate interlock – Aggregate interlock can transfer relatively high stresses (higher than at first crack strength) if the crack opening is small.
3. Failure, loss of aggregate interlock – Aggregate interlock failure is dependent on aggregate sizes, crack opening, and shear crack angle.

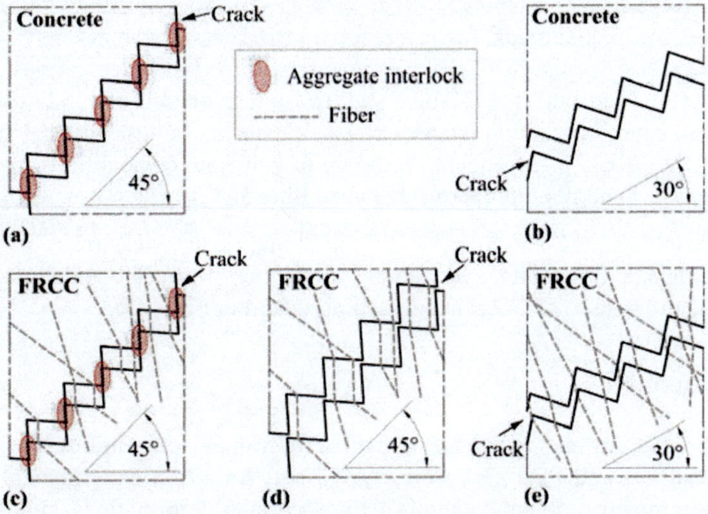

Fig. 5. Illustration of bridging mechanisms for OC with constant crack opening for (a) 45° and (b) 30° shear crack angle, and for FRCC (c) while aggregate interlock is acting and (d,e) while only fiber bridging is acting. In (d) a shear crack angle of 45° is shown, while (e) has a 30° angle

For FRCC, additional load can be transferred across cracks by fiber bridging. The shear crack behavior of FRCC with small aggregate sizes (< 0.18 mm) can be characterized by the following four stages as illustrated in Fig. 6(b):

1. Crack formation – Brittle crack opening and minimal sliding (typically < 0.03 mm) are resisted by fiber bridging and aggregate interlock (Fig. 5(c)).
2. Crack opening and sliding are restrained by fiber bridging and aggregate interlock – Aggregate interlock and fiber bridging allow transfer of additional shear stresses across the crack (Fig. 5(c)). The effect of aggregate interlock is reduced as crack opening exceeds half of the maximum aggregate size.

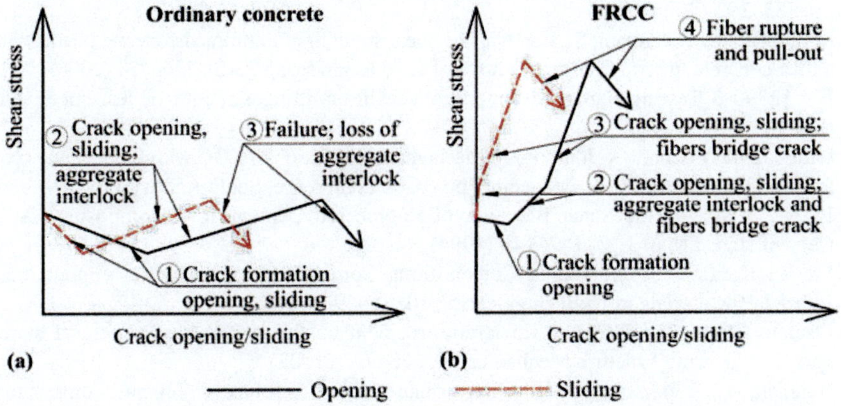

Fig. 6. Shear crack development mechanism for (a) plain concrete (OC) and (b) FRCC

3. Fibers bridge crack and transfer shear stresses – As shown in Fig. 5(d, e), only the fibers bridge the crack, however, the transferred shear stress may continue to increase.
4. Fiber rupture and pull-out – Fibers start to rupture or pull-out. Additional deformation causes reduction in shear stress transfer and ultimately failure. Ductility of the FRCC is dependent on the stiffness of the cementitious matrix, fiber properties and bond properties between fiber and matrix.

Further, during stages 2 and 3 for FRCC, additional shear cracks may form due to their multiple cracking features, temporarily reducing the crack opening rate of the existing shear cracks. This mechanism is highlighted in Fig. 4(d).

5 Conclusions

Shear testing of ordinary concrete and fiber reinforced cementitious composite beams with PVA and PP fibers were carried out. Based on experimental results and photogrammetric documentation of the specimen deformations, fundamental descriptions of shear crack opening, sliding and subsequent failure for plain concrete as well as for FRCC are proposed. The benefits of using FRCC as a structural material for resisting shear loads are improved shear resistance and crack control. Results indicate that the shear resistance of PP-FRCC and PVA-FRCC were similar, especially for steeper shear crack angles, even though direct tensile properties differ between these two types of FRCC. The crack width control provided by FRCC is desirable from a structural and durability viewpoint.

References

[1] Choi, K., Park, H., Wight, J.: Shear strength of steel fiber-reinforced concrete beams without web reinforcement. ACI Structural Journal 104, 12–21 (2007)

[2] Kabele, P.: Fracture Behavior of Shear-Critical Reinforced HFRCC Members. In: Proc. on the 49th Int. RILEM Workshop on HPFRCC in Structural Applications, USA, pp. 383–392 (2006)

[3] Majdzadeh, F., Soleimani, S., Banthia, N.: Shear strength of reinforced concrete beams with a fiber concrete matrix. Canadian Journal of Civil Engineering 33(6), 726–734 (2006)

[4] EN 12390-6 Testing hardened concrete: Tensile splitting strength of test specimens (2009)

[5] Lárusson, L., Fischer, G., Jönsson, J.: Mechanical Interaction of ECC with Fiber Reinforced Polymer (FRP) rebar in tensile loading. In: ACM Conference, South Africa (2009)

[6] Li, V., et al.: On the Shear Behavior of Engineered Cementitious Composites. Adv. Cement Based Mat. 1(3), 142–149 (1994)

[7] Li, V., Kanda, T.: Engineered Cementitious Composites for structural applications. Journal of Materials in Civil Engineering 10, 66–69 (1998)

[8] Fischer, G., Li, V.: Effect of fiber reinforcement on the response of structural members. Engineering Fracture Mechanics 74, 258–272 (2007)

[9] Naaman, A., Reinhardt, H.: High Performance Fiber Reinforced Cement Composites. In: Proceedings PRO 6. RILEM Publications S.A.R.L, France (2003)

Effects of Shear Transfer on the Directions of Principal Strain Field in Cracked Concrete with Hooked Steel Fibers

B. Suryanto, K. Nagai, and K. Maekawa

Department of Civil Engineering, The University of Tokyo, Japan

Abstract. This paper investigates the post-cracking behavior of steel fiber reinforced concrete (SFRC) panels from an analysis perspective based on a smeared, fixed crack approach. The analysis results show that the addition of hooked steel fibers improves the average tensile stress of the concrete and, when added beyond 1% by volume, limits the amount of crack-shear slip in the concrete effectively. The analysis reveals that experimentally observations of smaller angles of inclination of concrete principal strain than those of concrete principal stress at intermediate load levels is due to this limited crack slip. Finally, the analysis identifies that hooked steel fibers tends to be less effective than transverse reinforcement in confining shear cracks, thereby resulting in a lower shear transfer capacity.

1 Introduction

Panel test has provided a comprehensive picture of the post-cracking behavior of reinforced concrete (RC) in shear. Such a test allows for a quantitative assessment of components of cracked concrete in resisting shear such as average tensile stress and strain of cracked concrete, average shear transfer across crack surfaces, and average compressive stresses. To date a wide variety of RC panels have been tested including *normal-strength* and *high-strength* RC panels [1-4]. In this paper, use is made of the results of recent panel tests on Steel Fiber Reinforced Concrete (SFRC) panels by Susetyo [5].

From ten panel tests, Susetyo [5] demonstrated that the addition of hooked steel fibers improved the post-cracking shear performance of the concrete, with the most effective being the fibers with high aspect ratios and short. Susetyo revealed that the addition of hooked steel fibers of 1.0% in volume can achieve comparable shear strength and deformation response to a conventionally reinforced concrete panel with low transverse reinforcement. Interestingly, in contrast to a phenomenon usually observed in RC panels, for example the response of RC Panel PV10 discussed in details by Vecchio et al. [6], the angles of the principal concrete strain after first cracking were smaller than those of the principal concrete stress. Fig. 1, for clarity, compares the angles of inclinations of principal concrete stress and

G.J. Parra-Montesinos, H.W. Reinhardt, and A.E. Naaman (Eds.): HPFRCC 6, pp. 239–246.

strain of RC Panel PV10 and SFRC Panel C2F1V3. One may expect that the angles would differ because of differences of the ability of conventional concrete and SFRC to transmit stresses across cracks.

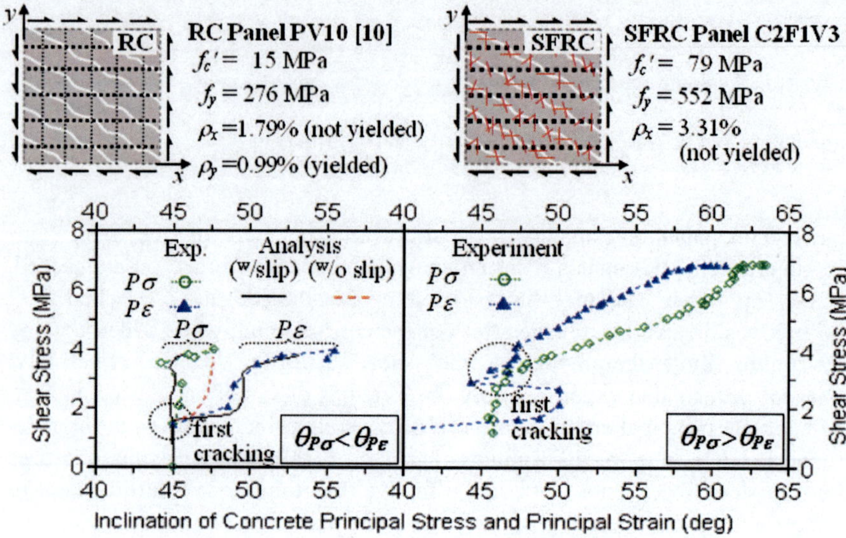

Fig. 1. Comparison of inclination of concrete principal stress and principal strain between RC Panel PV10 [6] and SFRC Panel C2F1V3 [5] subjected to shear

To identify the primary cause of the difference, this paper aims to predict the response of the SFRC panels. The predictions are obtained from a NLFEA for RC that were presented earlier in Suryanto et al. [7]. The RC material models in Maekawa et al. [8] are used, with slight modifications to better modeling the response of the SFRC panels. A parametric analysis is carried out to study the involved mechanisms, focusing primarily on the effects of hooked steel fibers on the ability of the concrete to resist tension and shear across cracks.

2 Overview of the Experimental Program

The experimental program referenced involved the testing of two *high-strength* RC panels and eight SFRC panels [5]. Of the ten panels tested, only the results of five panels are discussed herein, namely Panels C1C, C1F1V1, C1F1V2, C1F1V3, and C2F1V3. The RC panels were orthogonally reinforced, while the SFRC panels were uniaxially reinforced. They were all 890 mm square and 70 mm thick. Details of the test panels are illustrated in Fig. 2. The SFRC used to fabricate the SFRC panels was similar to those developed by Naaman et al. [9]. The property of the material used in the test panels is listed in Table 1. All the panels were tested by applying uniform shear stresses to the shear keys at the perimeter edges of the panels. Test parameters of the five panels include fiber volume fraction V_f (0.5, 1.0, 1.5%) and compressive strength f_c'.

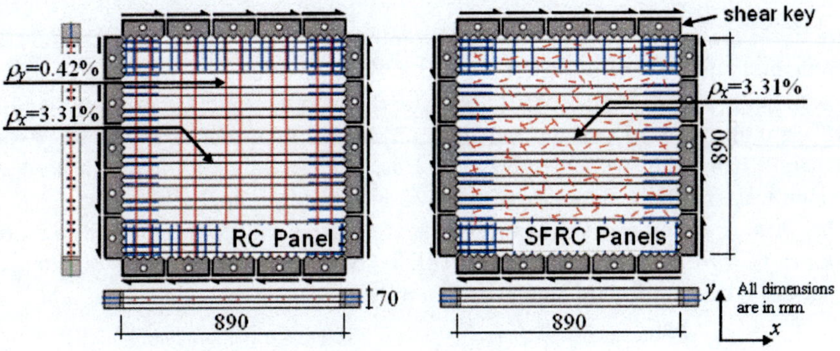

Fig. 2. Typical details of the RC and SFRC panels [5]

Table 1. Overview of the Test SFRC Panels [5]

Panel ID	ρ_x# (%)	$f_{y,x}$& (MPa)	$\rho_y$$ (%)	$f_{y,y}$& (MPa)	f_c'	V_f@ (%)	L_f/d_f L_f
C1C-R	3.31	552	0.42	442	65.7	-	-
C1F1V1	3.31	552	-	-	51.4	0.5	80/50
C1F1V2	3.31	552	-	-	53.4	1.0	80/50
C1F1V3	3.31	552	-	-	49.7	1.5	80/50
C2F1V3	3.31	552	-	-	78.8	1.5	80/50

2-layers of D8 (21mm apart); & taken as the stress at the limit proportionality; $ 1-layer of D4 @ hooked-end type, diameter 0.62 mm, length 50 mm, tensile strength 1,050 N/mm^2.

3 Material Models of SFRC and Modeling of the Panel

The material models used in this paper are the two-dimensional constitutive models of RC, which the full documentation of the models is available in Maekawa et al. [8]. Most aspects of the models are retained. Small changes are made to better modeling the behavior of SFRC and are described below.

The compression model is essentially the same as the elasto-plastic fracture model used for conventional concrete [8]. The only modification is to the compression softening factor ω. For *high-strength* concrete, it is expected that the degree of compression softening is less since the concrete strips between two cracks is not disturbed as a result of smoother crack surfaces. Since the compressive stresses exhibited by the SFRC panels were very low, it is difficult to quantify the appropriate values of ω. Hence ω is assumed to be 1.0.

The tensile model is based on the concrete model with embedded rebar [8]. To include the contribution of steel fibers, the steepness of the post-cracking tensile behavior is reduced by lowering the stiffening factor *c*, as given by:

$$f_t = f_{tu}\left(\frac{\varepsilon_{tu}}{\varepsilon}\right)^c \tag{1}$$

where ε_{tu} is the strain that the tensile softening/stiffening starts, f_{tu} is the tensile strength, and *c* is the tension stiffening factor. A value of *c*=0.4 was found to correlate well with a wide range of RC tests, and a smaller *c* value is proposed for SFRC. It is understood that the model is physically incorrect as the SFRC exhibited strain hardening response. Nevertheless, since the maximum strain hardening value was small, the adoption of Eq. 1 should result in negligible error.

The shear transfer model adopts the shear transfer model proposed by Li and Maekawa for *normal-strength* concrete [8]. To deal with SFRC, two parameters *A* and *B* were introduced to the model as given by:

$$v_{cr} = A\, f_{st}\,\frac{(B\,\beta)^2}{1+(B\,\beta)^2} \tag{2}$$

where: v_{cr} is the shear stress transmitted across cracks, f_{st} is given by $f_{st} = 3.8\, f_c^{'1/3}$ ($f_c^{'}$ in MPa), and β is the ratio of shear strain due to crack slip γ_{cr} to tensile strain due to crack opening ε_t. The parameters *A* and *B* relate to shear transfer capacity and shear stiffness, respectively. Fig. 3(b) shows the example of shear stress and strain responses for *A*=0.5 and various *B* values.

To model the SFRC panels, an 8-node Mindlin plate element was used. The analysis was performed with the NLFE program COM3 [8]. The analysis employed the computational algorithm of the four-way fixed, rotating-mixed crack model since the rebar arrangement in the test panels was highly anisotropic. With this algorithm, new cracks may form at certain angles, if the stress dictates, otherwise slight reorientation of the former crack is allowed. The material properties of the concrete and reinforcement were as reported as previously listed in Table 1, except for the tensile properties of the concrete, which were determined close to the tensile response obtained from the panel tests.

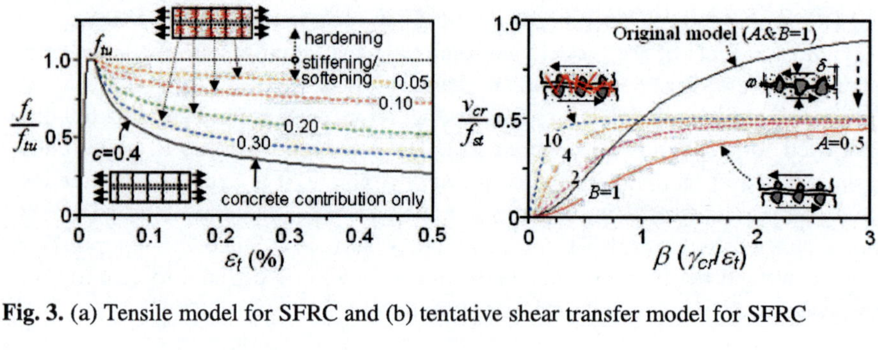

Fig. 3. (a) Tensile model for SFRC and (b) tentative shear transfer model for SFRC

4 Comparison of Experimental and Numerical Results

Fig. 4 shows the comparison of the observed and predicted response of the five SFRC panels in terms of macroscopic response, inclination of principal stress and strain, and principal stress and strain relations. Parameters that result in the best correlation are listed in Table 2. The tension stiffening factor *c* was first determined by comparing the angle of concrete principal stress and the tensile response. Using this as a starting point, it was then possible to identify the parameters of the shear transfer model from the macroscopic response and the angle of concrete principal strain. The summary is given in the following paragraphs.

Panel C1C (RC, f_c'=65.7 MPa, see Fig. 4 (a) to (d))– This panel contained no fibers and is thus suitable to evaluate the shear transfer of *high-strength* concrete. The analysis with *A*=0.75 shows a good agreement to the observed response, in terms of the macroscopic response and the divergence of principal stress and strain directions. When a lower shear transfer reduction factor is considered (*A*=0.35), the response becomes softer and the divergence of the principal stress and strain directions becomes much larger, indicating an overestimation of crack slip.

Panel C1F1V1 (SFRC, V_f=0.5%, f_c' =51.4 MPa, see Fig. 4(e) to (h))– It appears that the use of *A*=0.75 overestimates the panel capacity and wrongly predicts the angles of principal stress and strain fields. The predicted angles of the principal stress and strain remain coincide, indicating of an overestimation of shear transfer. While the panel was reanalyzed with *A*=0.35, a better agreement can be observed. Although the peak load is slightly underestimated, the analysis now correctly predicts the divergence of the principal stress and strain directions. Since the roughness of cracks of Panels C1C and C1F1V1 is comparable, the difference in shear transfer resistance between the two panels suggests that transverse rebar is more effective than hooked steel fibers for confining the cracks in the concrete.

Panel C1F1V2 (SFRC, V_f=1.0%, f_c' =53.4 MPa, see Fig. 4 (i) to (l))– The analysis shows that the addition of 1.0% fibers effectively limits crack slip, although the corresponding shear transfer resistance remains the same (*A*=0.35). Ignoring the contribution of the fibers in limiting slip (*B*=1) results in an underestimation of the plate capacity and an incorrect prediction of the angle of the principal stress and strain. By limiting the crack slip (*B*=4), the analysis replicates the overall response, the divergence of the directions of principal stress and strain fields at intermediate load stages, the gradual coincidence of the two at ultimate stage, and the principal tensile stress and strain responses. The smaller angles of principal strain at intermediate loading suggest that hooked steel fibers, when sufficiently added, are effective in limiting crack slip.

Panel C1F1V3 (SFRC, V_f=1.5%, f_c' =49.7 MPa, see Fig. 4 (m) to (p))– The analysis indicates that by increasing the fiber volume from 1.0% to 1.5%, the resistance of the cracks in the concrete to slip (*B*) slightly increases from 4 to 5. The predicted response with *A*=0.75 and *B*=1, the parameters of Panel C1C, incorrectly predicts both load-deformation and principal stress-strain angles.

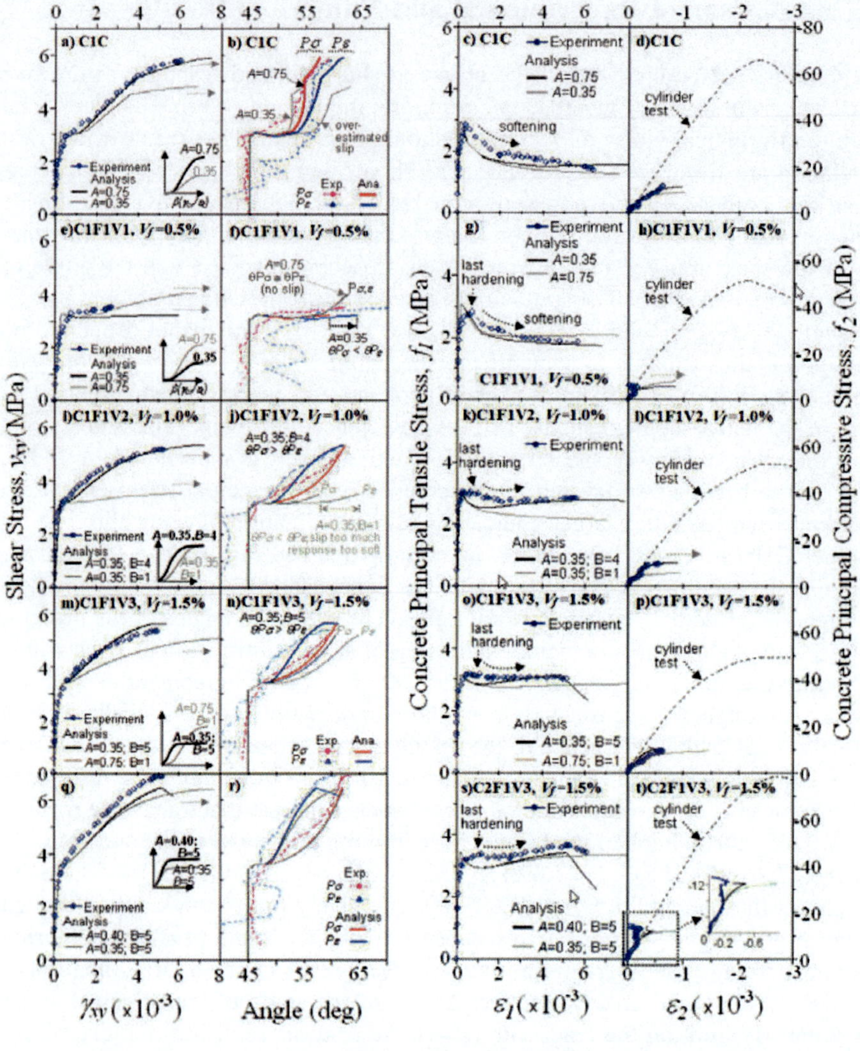

Fig. 4. Comparison of the observed [5] and predicted response of the test SFRC panels

Panel C2F1V3 (SFRC, V_f=1.5%, f_c' =78.8 MPa, see Fig. 4 (q) to (t))– The analysis results show that almost doubling the concrete compressive strength only marginally increases the shear transfer capacity from 0.35 to 0.40 and has no effects to the slip resistance (*B*=5). The slight improvement of shear transfer capacity is likely attributed to the improved fibers bridging as a result of the higher matrix strength since the shear contributed by interlocking of cracks should decrease in high f_c'. That is, the positive and negative effects of higher f_c' tend to cancel each other out.

Table 2. Identified parameters of the tensile and shear transfer model from parametric analysis

Panel ID	*Tensile Model*		*Shear Transfer Model*	
	$f_t^{\$}$(MPa)	c	A	B
C1C	3.00	0.4, 0.2[#]	0.75	1
C1F1V1	3.20	0.25	0.35	1
C1F1V2	3.10	0.15	0.35	4
C1F1V3	3.25	0.10	0.35	5
C2F1V3	3.50	0.125	0.40	5

[$]assumed value; [#]$c_y$, assumed as $\rho_{cr\text{-}y} \cong f_t / f_{y\text{-}y}$

To better illustrate the effects of the tensile and shear stresses, Panel C1F1V3 was reanalyzed by varying the *A, B*, and *c* values and the results are plotted in Fig. 5. Note that parameters *A* and *B* are associated with the shear transfer property, whereas parameter *c* relates to the tensile property. In Analysis Case 1 (*A* varies, *B*=1, c=0.1), only the analysis with *A*=1.5 fairly simulates the macroscopic response, but still incorrectly predicts the principal strain angle. In Analysis Cases 2 (*B* varies, *A*=0.75, c=0.1), it is observed that increasing shear stiffness is more effective to limit the angle of the principal strain field. Perhaps the most interesting result is that the predicted principal stress angles in both analysis cases are always very close to each other, regardless of which shear transfer properties are considered in the analysis. This finding implies that it is possible to identify the mechanisms of crack-shear transfer by observing the angles of principal strain direction.

The effects of concrete tensile property are obvious. As the ability of the concrete to resist tension increases (*c* is decreased), angles of principal stress and strain directions decrease. There is a strong relationship between concrete average tensile stress and principal stress angle, indicating that information on principal stress direction is important determinants of average tensile stress in the concrete. Since none of the result can match the observed response, this analysis case indicates that the improvement of shear transfer is not attributable to the improvement of tensile property alone.

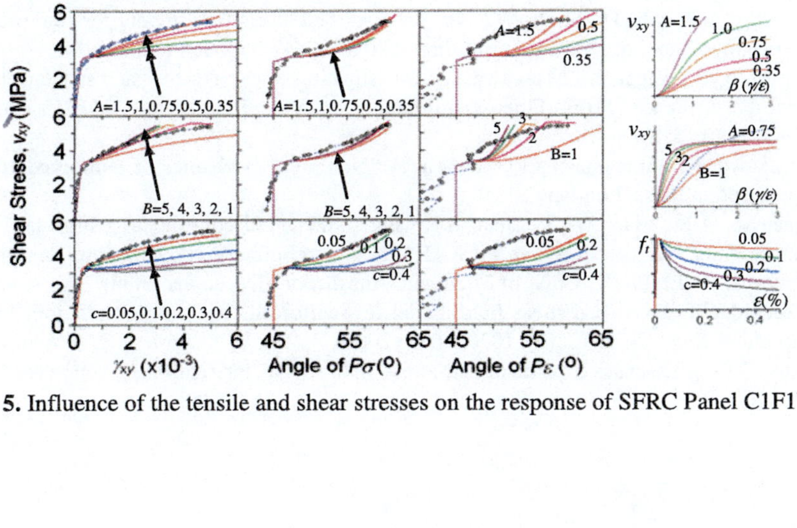

Fig. 5. Influence of the tensile and shear stresses on the response of SFRC Panel C1F1V3

5 Conclusions

1. Crack-shear transfer of the SFRC panels with no transverse reinforcement, when compared to that of *normal-strength* concrete, appears to be higher in stiffness, but lower in capacity. High shear stiffness reflects the contribution of hooked steel fibers in limiting shear slip, while low capacity tends to be caused by the absence of transverse reinforcement.
2. The ability of the concrete to resist tension increases as the fiber volume increases. Increasing fiber volume does not appear to improve the crack-shear resistance, but is effective in limiting crack slip for a fiber volume of larger than 1.0%. However, the use of 1.5% steel fibers results in only a marginal improvement.
3. The inclination of principal stress reflects the average tensile stress of cracked concrete, while that of strain field relates to the average crack-shear transfer. It is found that the smaller angles of principal strain field to that of principal stress field at intermediate load levels observed in the SFRC panels are due to the limited slip of the concrete and are not merely an experimental scatter. This supports the perception by Vecchio [10] and against the argument by Hsu [11].

References

[1] Vecchio, F. J., Collins, M. P.: Response of reinforced concrete to in-plane shear and normal stresses. Report No. 82-03, Dept. of Civil Eng., Univ. of Toronto, Toronto (1982)
[2] Bhide, Collins, M.P.: Influence of axial tension on the shear capacity of reinforced concrete members. ACI Structural J. 86(5), 570–581 (1988)
[3] Vecchio, F.J., Nieto, M.: Shear friction tests on reinforced concrete panels. ACI Structural J. 88(3), 371–379 (1991)
[4] Vecchio, F.J., Collins, M.P., Aspiotis, J.: High-strength concrete elements subjected to shear. ACI Structural J. 91(4), 423–433 (1994)
[5] Susetyo, J.: Fibre reinforcement for shrinkage crack control in prestressed, precast segmental bridges. Ph.D. Thesis, Dept. of Civil Eng., Univ. of Toronto (2009)
[6] Vecchio, F.J., Lai, D., Shim, W., Ng, J.: Disturbed stress field model for reinforced concrete: validation. J. of Structural Eng. ASCE 127(4), 350–358 (2001)
[7] Suryanto, B., Nagai, K., Maekawa, K.: Investigating Shear Transfer across Cracks in High-Performance Steel Fiber-Reinforced Concrete. Proceedings of JCI 32(2), 1285–1290 (2010)
[8] Maekawa, K., Pimanmas, A., Okamura, H.: Nonlinear mechanics of reinforced concrete. Spoon Press, London (2003)
[9] Naaman, A.E., Liao, W.C., Chao, S.H., Park, S.Y.: Self-consolidating high performance fiber reinforced concrete (SCHPFRC) - preliminary investigation. Research Report UMCEE 06-02, Univ. of Michigan, Ann Arbor. (December 2006)
[10] Vecchio, F.J.: Disturbed stress field model for reinforced concrete: formulation. J. of Structural Eng. ASCE 126(9), 1070–1077 (2000)
[11] Hsu, T.T.C.: Discussion of disturbed stress field model for reinforced concrete: formulation by F. J. Vecchio. J. Structural Eng. 128(11), 1487–1488 (2002)

Mechanical Interaction between Concrete and Structural Reinforcement in the Tension Stiffening Process

L. Lárusson, G. Fischer, and J. Jönsson

Department of Civil Engineering, Technical University of Denmark

Abstract. The interaction between structural reinforcement and the surrounding concrete matrix in tension is a governing mechanism in the structural response of reinforced concrete members. The tension stiffening process, defined as the concrete´s contribution to tensile response of the composite, has been investigated using an image-based deformation measurement and analysis system. This allowed for detailed view of surface deformations and the implications on the resulting response of the member in tension. In this study, conventional concrete and a ductile, strain hardening cement composite, known as Engineered Cementitious Composite (ECC), have been combined with steel and glass fiber reinforced polymer (GFRP) reinforcement to contrast the effects of brittle and ductile cement matrices as well as elastic/plastic and elastic reinforcement on the tension stiffening process. Particular focus was on the deformation process and transverse crack formation in the cementitious matrix at increasing tensile strain.

1 Introduction

The research activities presented in this paper are part of a study that focuses on research and design of a continuous expansion joint, also known as link slab [1], for roads and bridge structures. In this context, by using composite materials such as Engineered Cementitious Composites (ECC) and Glass Fiber Reinforced Polymer (GFRP) rebars, the aim of this study is to improve the performance, the production process and applicability as well as durability of current expansion joints.

The presented paper is a continuation of previously presented work on the interactions of ECC with GFRP in direct tension [2]. The test series carried out in this experimental program examine the composite behavior of four different material compositions: steel reinforced ECC, steel reinforced concrete, GFRP reinforced ECC and GFRP reinforced concrete. All test configurations were subject to direct monotonic tensile loading as well as cyclic loading but the current paper will only focus on the results from monotonic loading, i.e. the tension stiffening process,

G.J. Parra-Montesinos, H.W. Reinhardt, and A.E. Naaman (Eds.): HPFRCC 6, pp. 247–254.

crack formation and development, crack widths, and crack spacing of the composite specimens.

The four different materials investigated in this study were tested individually to establish their material properties. Table 1 shows the parameters obtained. Conventional concrete and ECC were tested in compression while ECC, regular steel and GFRP reinforcement were tested in direct tension.

Table 1. Material parameters. f_{cu} and ™$_{cu}$ refer to the ultimate compression strength and strain, f_t is the tensile strength, ™$_{ty}$ and ™$_{y}$ are the tensile strains at yielding, f_y is the tensile yield strength, f_{tu} is the ultimate tensile strength and ™$_{tu}$ is the ultimate tensile strain.

	f_{cu} [MPa]	™$_{cu}$ [%]	f_t [MPa]	™$_{ty}$ [%]	E [GPa]
Concrete	61	0.12	4.1*	-	38
ECC	60	0.22	3.5**	0.03	18
	f_y [MPa]	™$_{y}$ [%]	f_{tu} [MPa]	™$_{tu}$ [%]	E [GPa]
Steel rebars	680	0.35	780	6.9	195
GFRP rebars	-	-	1050	2.7	40

* Value obtained from a split cylinder test (Brazilian test).

** First crack strength, ultimate strength reached 4.0 MPa at 4.0 % strain.

ECC is a subclass of High Performance Fiber Reinforced Cementitious Composite (HPFRCC) [3], which has the ability to exhibit strain hardening with a relatively high strain capacity (~4 %) through the formation of multiple cracking accompanied by limited crack widths (<250 μm). To contrast the highly ductile behavior of ECC, a series of conventional concrete specimens with the same compression strength as ECC were also evaluated. The GFRP reinforcement used in this study (Aslan-100 produced by Hughes Brothers inc.) is a sand coated, low E-modulus, elastic, corrosion free alternative to regular steel reinforcements. For comparison purposes regular elastic-plastic steel reinforcement was also included in a series of composites.

2 Concept and Design

The tension stiffening process of reinforced concrete elements is a previously studied phenomenon [4] governed by the mechanical fracture properties of the bond between rebar and concrete and the material properties of both concrete and reinforcement. In this context tension stiffening is usually defined as the difference between the response of the bare reinforcement and the composite element during tensile loading, either in bending or in direct tension.

By introducing a ductile cementitious composite such as ECC into a reinforced element the supplementary strain and tensile load capacity of the ECC can be utilized. In addition to the load and ductility enhancements the benefits of closely spaced multiple cracking with measurably reduced crack widths, in comparison to conventional brittle concrete, result in increased durability of the structure.

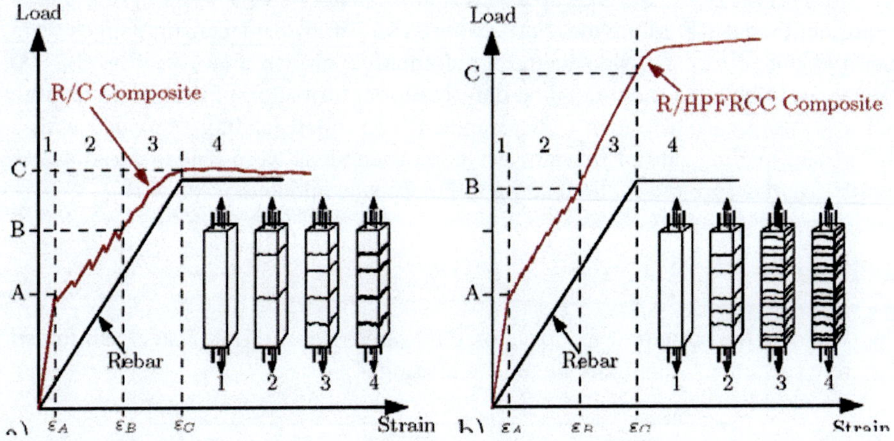

Fig. 1. Schematic illustration of tension stiffening and crack formation, a) shows R/C and b) shows R/HPFRCC. "A" and ™$_A$ are load and corresponding strain of first crack respectively, "B" and ™$_B$ are load and strain after crack saturation respectively, "C" and ™$_C$ are load and strain at yielding of reinforcement respectively.

To emphasize the difference between reinforced concrete and reinforced HPFRCC in direct tensile loading schematic illustrations are given in Fig. 1.

3 Experimental Program

Twelve composite specimens with identical geometry, three for each composition, were prepared and tested. The "dog bone" shaped specimens were 1000 mm long with a 500 mm long middle section with a constant cross section: 100 x 100 mm^2 and four rebars positioned equally in the cross section (see Fig. 2). The rebars extended throughout the entire length of the specimens, protruding from both ends. The diameter of the four rebars were Ø 6 mm resulting in a reinforcement ratio $\rho =$ 1.14 %.

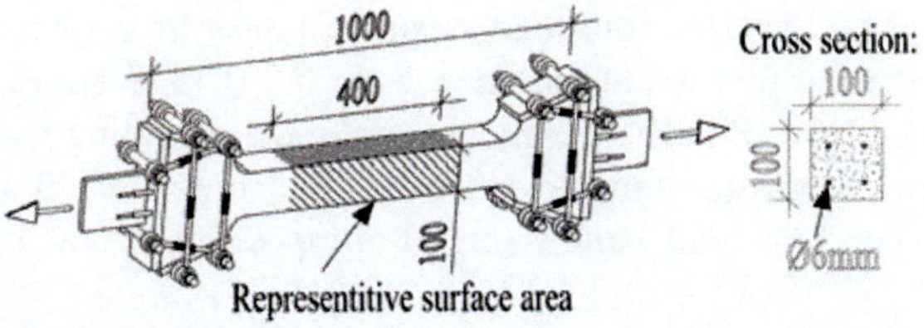

Fig. 2. Test configuration with a clamping system located on both ends to ensure a composite behavior of the representative section. The cross section of the representative section is also shown.

An optical surface deformation measuring system was utilized to deduce surface displacements and quantitative crack opening information of the specimens during testing. The measured surface area was approximately 250 mm along the specimens length and 100 mm wide. Furthermore two linear displacement gauges were also positioned on each side of the representative surface area (see Fig. 2), 400 mm apart, to obtain a more detailed deformation information of the whole section. All tests were carried out in a displacement controlled loading sequence with a monotonic loading rate of 0.5 mm per min. Specimens were loaded up to 1.2 % tensile strain before the cyclic loading sequence was initiated.

4 Results

Fig. 3 shows the structural response of steel reinforced ECC and steel reinforced concrete specimens in monotonic tensile loading.

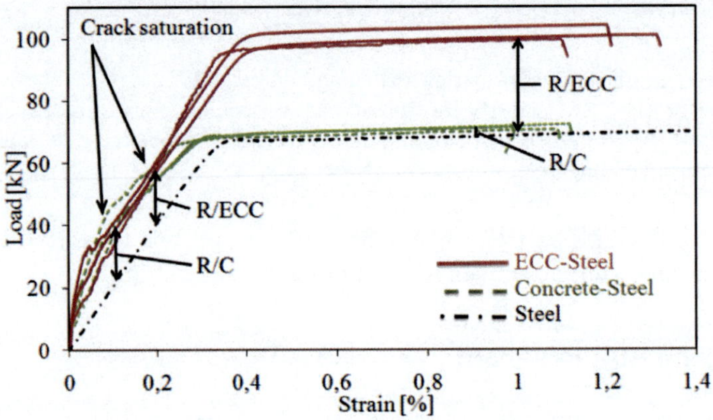

Fig. 3. Structural response of steel reinforced concrete- and steel reinforced ECC specimens during monotonic tensile loading. The tension stiffening effect for steel reinforced concrete (R/C) and steel reinforced ECC (R/ECC) member are shown on graph.

The first cracking in ECC-steel specimens is reached at an approximate strain level of 0.01-0.02 %, corresponding to a load of 10-30 kN. After crack saturation is reached at about 0.09 % strain, the load-strain response of the specimen is linear until yielding of the reinforcement commences at 0.3-0.35 % strain corresponding to 93-98 kN tensile load. Finally, the monotonic tensile loading was discontinued at 1.1-1.3 % strain, corresponding to a tensile load range of 99 kN to 104 kN. Strain and corresponding load values for all composition types are presented in Table 2.

Initially, the response of concrete-steel is similar to that of ECC-steel but as the strain level increases, the stiffness of the composite decreases, which means that the tension stiffening diminishes with increased loading. After yielding of the reinforcement the concrete composite response is only slightly higher than that of the bare rebar as shown in Fig. 3.

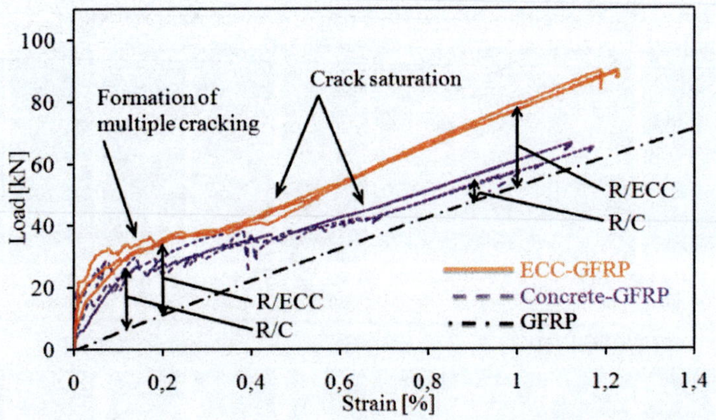

Fig. 4. Structural response of GFRP reinforced concrete- and GFRP reinforced ECC specimens during monotonic tensile loading. The tension stiffening effect for GFRP reinforced concrete (R/C) and GFRP reinforced ECC (R/ECC) member are shown on graph.

The load-strain response of ECC-GFRP and concrete-GFRP is displayed in Fig. 4 together with the response of bare GFRP rebar. ECC-GFRP exhibits crack formation up to about 0.4 % strain after cracking initiated. Beyond crack saturation the stiffness of the composite specimens increases again, showing a slightly stiffer response than the bare GFRP rebar (see Fig. 4).

Table 2. Experimentally obtained values from tensile loading of composite specimens

	First crack		Crack saturation		Rebar yielding		Test stopped	
	Strain [%]	Load [kN]	Strain [%]	Load [kN]	Strain [%]	Load [kN]	Strain [%]	Load [kN]
ECC-steel	0.01-0.02	10-30	0.09	30-40	0.3-0.35	93-98	1.1-1.3	99-104
Concrete-steel	0.01-0.03	14-18	0.15-0.2	50-66	0.3-0.35	68	1.0-1.1	72
ECC-GFRP	0.01-0.03	16-21	0.4-0.45	40-42	-	-	1.2	90
Concrete-GFRP	0.01-0.03	5-20	0.4-0.7	35-42	-	-	1.0-1.2	57-66

Coefficients were assessed from Fig. 3 and Fig. 5 according to Fig. 1.

The concrete-GFRP specimens show a similar behavior to the ECC-GFRP specimens up to about 0.4 % strain but at lower load levels (see Fig. 4). After crack saturation, the composite stiffness is slightly lower than that of the GFRP alone.

Fig. 5 shows deformations of the visible specimen surfaces, that means crack formation and development for all specimen configurations obtained from image analysis. In the figure, a comparison is shown at tensile strain levels from 0.2 % to 1.0 %.

Furthermore, the crack opening development extracted from the analysis is shown in Fig. 6 as a function of strain. For the concrete specimens all visible cracks on the measured surface are depicted while only selected cracks are shown for the ECC specimens.

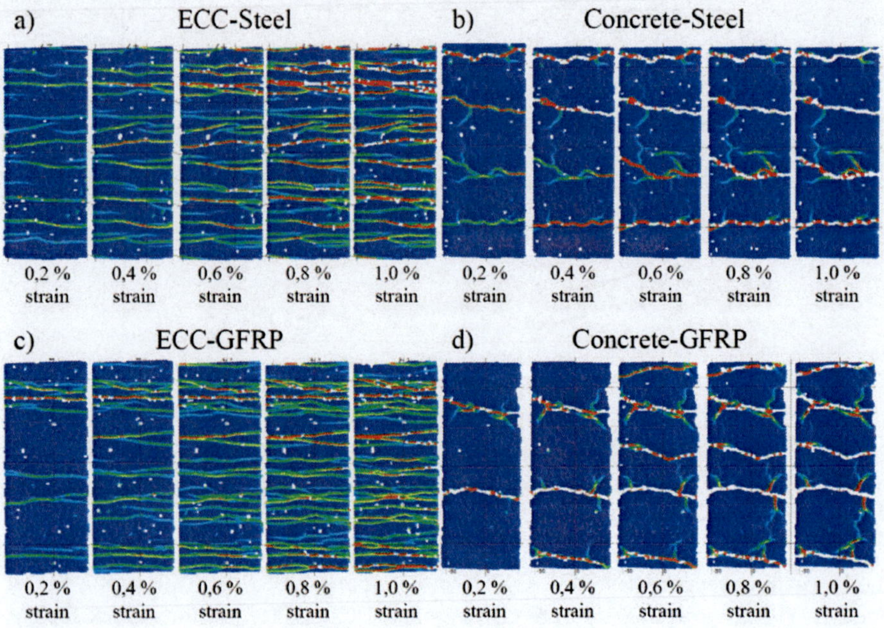

Fig. 5. Surface deformation development on all specimen types during tensile loading at strain levels 0.2-1.0 %. Contour scale used for visualization was put to 0-4 % strain.

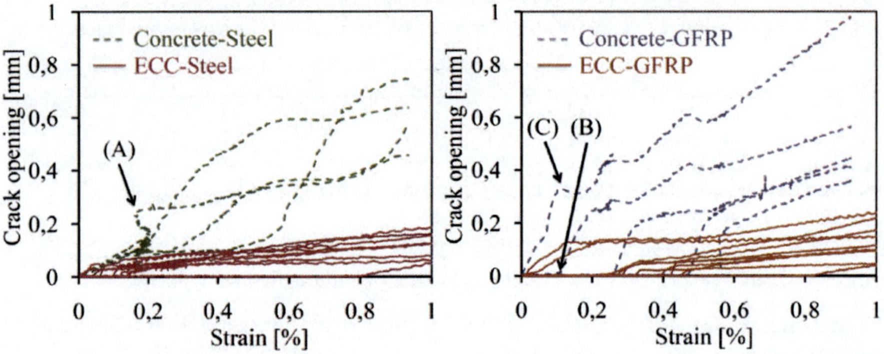

Fig. 6. Development of crack opening vs. strain on a): Steel reinforced concrete in comparison to a steel reinforced ECC, (A) indicates where stain is halted. b): GFRP reinforced concrete in comparison to GFRP reinforced ECC. (B) shows where a new crack is formed while (C) indicates where crack opening is suspended.

To illustrate the cracking behavior of an ECC reinforced member Fig. 7 depicts the range of crack opening and crack spacing as a function of applied composite strain in a representative ECC-GFRP specimen.

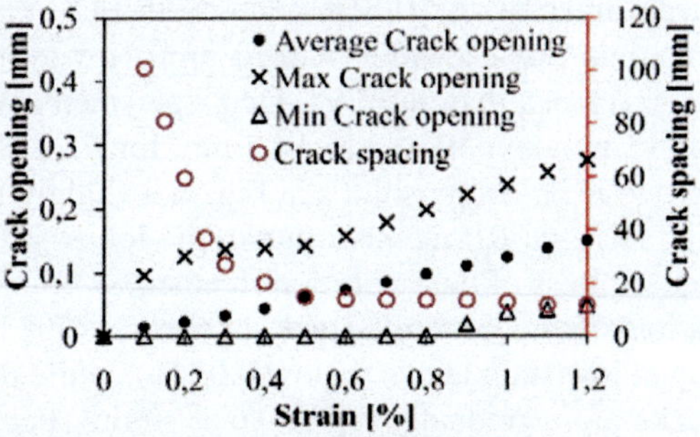

Fig. 7. Average, maximum and minimum observed crack width and crack spacing for a GFRP reinforced ECC element in tensile loading

5 Discussion

The load levels of first crack formation differ somewhat between specimens of all compositions types, due to minor miss-alignment observed in the test configuration causing transverse cracking to occur at lower tensile loads than expected.

Comparing reinforced ECC to reinforced concrete under tensile loading illustrates the additional load bearing capacity exerted by ECC. This is seen in Fig. 3 and Fig. 4 as the difference between load levels of ECC composites and bare reinforcement, particularly in the post-yielding phase, while the load carrying capacity of reinforced concrete decreases towards the capacity of the reinforcement alone. Furthermore, the stiffness of the reinforced ECC members is shown to be as stiff (ECC-Steel) or slightly stiffer (ECC-GFRP) than that of the bare reinforcement after crack saturation is reached. That means that in the tension stiffening process the contribution of ECC is maintaining the heightened load response while tension stiffening in the reinforced concrete members diminishes with increased loading.

Initially, the influence of GFRP on the composite interaction is observed as crack formation occurring over a longer strain interval, that means that crack saturation is reached at higher strain levels than for steel reinforced members, at about 0.7 % and 0.2 % strain respectively (see Fig. 4). Secondly, the tension stiffening effect in GFRP reinforced members is shown to be lower than that of steel reinforced members in the post-yielding region. At 1.0 % strain the difference between ECC-GFRP and GFRP alone in comparison to ECC-Steel and Steel alone is 26.2 kN and 32.8 kN respectively. This difference is due to the softer response of the GFRP as opposed to the stiffer steel reinforcement.

Crack formation and development in reinforced ECC members differ substantially from that of reinforced concrete members as shown in Fig. 5 and Fig. 6. For the ECC-steel specimen, most of the cracks initiate at a low strain level (below 0.15 %) whereas the ECC-GFRP specimen, shown in Fig. 6b, develops cracks over an extended strain interval. At 1.0 % strain, both ECC types show approximately the same maximum crack width of 0.20 mm. However, the ECC-GFRP specimen exhibits more cracks than the ECC-steel specimen resulting in an average crack spacing of 13 mm as opposed to 14.5 mm for ECC-Steel. Furthermore when comparing the two ECC compositions in Fig. 6, a slightly larger initial crack opening is seen in the ECC-GFRP member due to the lower stiffness of GFRP as opposed to the higher stiffness of the steel reinforcement.

The crack formation of the concrete-steel specimen, seen in Fig. 6a, shows most cracks initiating at low stain levels (below 0.02 %), while the concrete-GFRP member initiates cracks more gradually (up to 0.6 % strain). For concrete-steel the average crack spacing was 70 mm while concrete-GFRP exhibited a slightly lower spacing of 60 mm. In Fig. 6a the strain development is momentarily halted, marked as "A" for the concrete-steel specimen, indicating that cracking is forming outside of the area that was measured. In Fig. 6b when a new crack forms in the concrete-GFRP specimen, depicted as "B", the increasing crack opening of the pre-existing crack is suspended, marked as "C", while the new crack opens up. This mechanism is observed throughout the whole crack formation process for the concrete-GFRP specimen.

6 Conclusion

The comparison of reinforced ECC and reinforced concrete has shown that the tension stiffening process can be significantly improved by utilizing ECC. Furthermore ECC was shown to increase the stiffness of the composite and maintaining linear stiffness throughout testing. The cracking process of reinforced ECC consistently showed multiple cracking with considerably smaller crack widths than those in reinforced concrete. Furthermore, the influence of the linear strain deformations of GFRP in the reinforced composites resulted in a larger number of cracks forming when compared to steel reinforced members, i.e. closer crack spacing. Finally, the tensile strain behavior of ductile ECC and low E-modulus GFRP are shown to be compatible resulting in a good composite interaction.

References

[1] Caner, A., Zia, P.: Behavior and Design of Link Slab for Jointless Bridge Decks. Precast Concrete Institute Journal, 68–80 (May-June 1998)

[2] Lárusson, L., Fischer, G., Jönsson, J.: Mechanical interaction of ECC reinforced with FRP rebar in tensile loading. In: Proceedings of ACM, Stellenboch, South Africa (2009)

[3] Naaman, A., Reinhardt, H.: Proposed classification of HPFRC composites based on their tensile response. Materials and Structures 39(5), 547–555 (2006)

[4] Fischer, G., Li, V.: Influence of matrix ductility on tension-stiffening behavior of steel reinforced ECC. ACI Structural Journal, American Concrete Institute 99, 104–111 (2002)

Confinement and Tension Stiffening Effects in High Performance Self-consolidated Hybrid Fiber Reinforced Concrete Composites

W. Trono[1], G. Jen[1], D. Moreno[2], S. Billington[2], and C.P. Ostertag[1]

[1] University of California, Berkeley, USA
[2] Stanford University, Stanford, USA

Abstract. Confinement and tension stiffening effects were studied on self-consolidating, hybrid fiber reinforced concrete (SC-HyFRC) composites and compared to plain self consolidated concrete (SCC) specimens without fibers. The SC-HyFRC composites consist of both steel macrofibers and PVA microfibers with a total fiber volume fraction of 0.015. Cylindrical specimens were confined by continuous steel spirals with transverse reinforcement ratios ranging from 0.32%-1.91% and tested in uniaxial compression. The SC-HyFRC composites despite 2-3 times smaller transverse reinforcing ratios exhibit similar ductility and softening behavior compared to the plain SCC specimens. The effect of SC-HyFRC composites on tension stiffening behavior was investigated using axially reinforced dogbone specimens. The reinforced SC-HyFRC revealed tension stiffening and carried tension to strains far exceeding the yield strain of the reinforcing steel.

1 Introduction

A self-consolidating hybrid fiber reinforced concrete (SC-HyFRC) composite was developed at UC Berkeley for use in high seismic zones where concrete structures are heavily reinforced. The SC-HyFRC mix design is based on HyFRC mixes developed by Blunt and Ostertag [1] that can sustain compatible deformations with reinforcing bars beyond the yield strain of the rebar. The SC-HyFRC incorporates 30 mm long steel macrofibers and 8 mm long polyvinyl alcohol microfibers with aspect ratios of 60 and 200 and volume fractions of 0.013 and 0.002, respectively, and exhibits with slump flow diameters of 500 to 600 mm. The high level of workability in SC-HyFRC improves the flow of concrete between steel reinforcing bars and their enhanced crack resistance and softening behavior reduces damage in members expected to undergo large ductility demands during major earthquakes.

Preliminary results of compression tests on self-consolidating concrete (SCC) and SC-HyFRC cylinders with varying levels of confinement are presented in this

G.J. Parra-Montesinos, H.W. Reinhardt, and A.E. Naaman (Eds.): HPFRCC 6, pp. 255–262.

paper. The behavior of plain concrete confined with steel spirals is well known, but few researchers have investigated the behavior of high performance fiber reinforced composites in a passively confined setting.

This paper also briefly discusses tests conducted in collaboration with researchers at Stanford University on the effect of SC-HyFRC on tension stiffening. Again, only few studies have been conducted on the degree of tension stiffening provided by high performance fiber reinforced composites. Reinforced HyFRC beams tested in flexure by Blunt and Ostertag [2] clearly outperformed conventional reinforced concrete beams of same size and reinforcing ratios in regards to crack control and tension stiffening behavior. In addition, no strain localization was observed in these reinforced HyFRCs beyond reinforcing bar yield and this was attributed to the fact that the softening behavior of the HyFRC matrix was well balanced by the section hardening response of the reinforcing bar. The present study investigates the tension stiffening and tensile properties of reinforced SC-HyFRC up to and beyond rebar yield.

2 Testing Program

For the confinement tests, the reinforcement cages were constructed using 10 gauge (0.135 in. diameter) black annealed steel wire (F_y = 275 MPa) as transverse spiral reinforcement at ratios of ρ_s = 0.32%, 0.48%, 0.95%, and 1.91% corresponding to spiral spacings of 75, 50, 25, and 12.5 mm, respectively. It should be noted that for the specimen geometry of d = 150 mm and h = 300 mm, the ACI code minimum level of transverse reinforcement for the compression member would be ρ_s = 1.5% [3]. Four Grade 40 #3 deformed reinforcing bars (F_y = 375 MPa) were used in each specimen to hold the steel spiral at the correct spacing, giving a longitudinal reinforcement ratio of ρ_l = 1.56%.

Some specimens were instrumented with strain gages affixed to the steel spiral; the gages were placed 90 degrees apart on two consecutive spirals near mid-height of the specimen for a total of 8 gage locations. Axial compression was measured using a compressometer with two LVDT's mounted on opposite sides of the specimen with a 200 mm gage length over the center of the specimens.

Specimens were cast with plain SCC or SC-HyFRC, both of which were proportioned with water to binder ratio (w/b) of 0.45. Pea gravel with maximum aggregate size of 10 mm and Vulcan sand with a measured fineness modulus of 3.2 were used as coarse and fine aggregates, respectively. Unconfined cylinders with no steel reinforcement were also cast and tested for each of the two concrete mixes. The two concrete mixes are given in Table 1.

Table 1. Mix designs for test specimens

Mix	Water	Cement	Fly Ash	Fine	Coarse	SP	VMA	30 mm	8 mm
Plain SCC	238	398	131	1074	430	0.46	2.22	-	-
SC-HyFRC	238	398	131	1046	419	0.46	2.22	1.3	0.2

Water, cement, fly ash, and fine and coarse aggregate are given in kg/m^3. Superplasticizer (SP) and Viscosity Modifying Admixture (VMA) are given as a wt. % of binder materials (cement + fly ash). Fibers are given in volume %.

The effect of SC-HyFRC on tension stiffening was investigated using dogbone shaped SC-HyFRC specimens reinforced with a headed grade 60 #4 reinforcing bar in its center. The steel reinforcement in these specimens was 1.9%. The dogbones were 81 mm in thickness with a total length of 850 mm. The specimen's center region was 213 mm with a width of 81mm. The Grade 60 #4 bars were also tested in direct tension for comparison. Tensile displacements were measured on the dogbones over a 175 mm gage length in the reduced 81 by 81 mm center region. More details on the experimental set-up and testing procedures are given in [4].

3 Results and Discussion

3.1 Confinement

The compressive behavior of SCC and SC-HyFRC, confined and un-confined, is shown in Fig. 1. The average compressive stress carried by the steel and concrete is plotted against axial strain of plain SCC (Fig. 1a) and SC-HyFRC (Fig. 1b), respectively. Stress is calculated as an average value over the 150 mm diameter cross section. A stable softening behavior after peak load was observed even for un-confined SC-HyFRC and it required a transverse reinforcement ratio higher than 0.48% to achieve a similar softening behavior in plain SCC. Increasing the confinement ratio decreased the steepness of the SC-HyFRC softening curves but had little effect on their peak strength. Increasing the transverse reinforcement ratio from 0.32% to 1.91% in plain SCC resulted in a moderate increase in peak strength of 20%, but a decrease in the slope of the softening behavior could only be observed at their highest confinement ratios.

Fig. 2 compares the behavior of SC-HyFRC specimens at ρ_s = 0.95% and plain SCC specimens at ρ_s = 0.95% and 1.91%. The plain SCC specimens showed a much steeper and less stable softening branch than the SC-HyFRC specimens at the ρ_s = 0.95% level. Only one plain SCC specimen at the ρ_s = 1.91% level was able to achieve a similar softening behavior as the SC-HyFRC specimens.

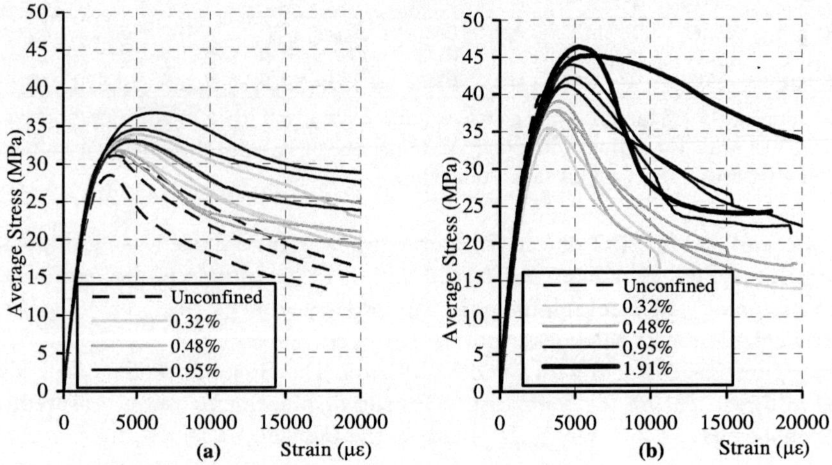

Fig. 1. Average stress vs. strain behavior for (a) SC-HyFRC and (b) plain SCC

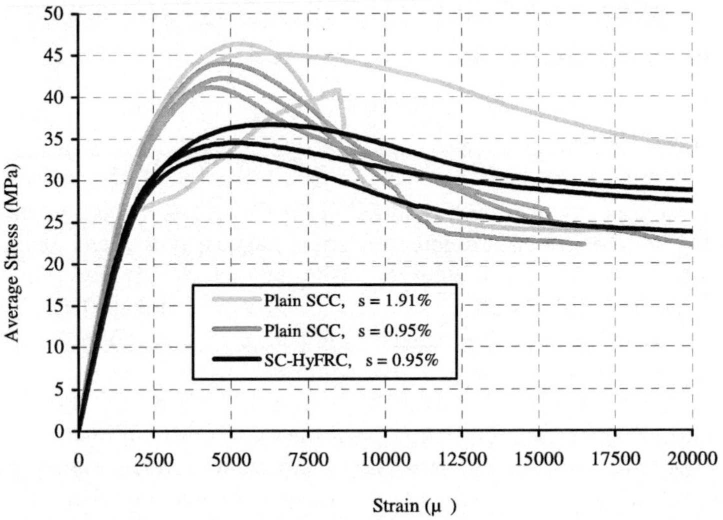

Fig. 2. Average axial stress vs. strain behavior for plain SCC specimens at ρs = 0.95% and 1.91% and SC-HyFRC specimen at ρs = 0.95%

Initiation of tensile yielding in the spiral occurred near the peak load in all specimens; however, the spirals in the SC-HyFRC specimens tended to deform uniformly with few localizations of strain under high axial deformations as shown in Fig. 3. Fig. 3 compares the axial strain vs. tensile strain in the spiral for the plain SCC at ρ_s = 0.95% and 1.91% and for the SC-HyFRC at ρ_s = 0.95%. For the SC-HyFRC specimen, transverse strains at the eight locations developed at a constant, controlled rate with little localization at any one point. For the plain SCC specimen at the same transverse ratio, three gages showed excessive strains immediately after peak load was reached near 0.5% axial strain indicating localized damage and excessive dilation of the confined concrete. This result shows that the internal crack resistance of the fibers is effective in controlling damage localization and spreading plasticity throughout the specimen.

Even at ρ_s = 1.91%, the plain SCC showed higher deformation in the spiral than the SC-HyFRC. The relatively low levels of strain in the SC-HyFRC spiral could be indicative of the fibers' ability to resist internal cracks, leading to less outward dilation of the core material.

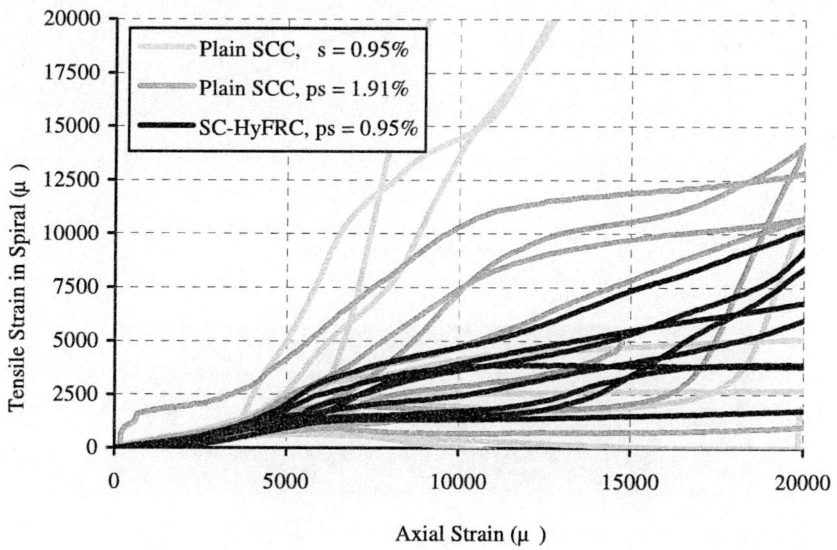

Fig. 3. Tensile spiral strain vs. axial strain for plain SCC specimens at ρs = 0.95% and ρs = 1.91% and SC-HyFRC specimen at ρs = 0.95%

The damage resistance of the SC-HyFRC specimens compared to plain SCC was assessed visually during the tests. While plain concrete specimens spalled their cover and showed excessive signs of crushing and internal cracking, the SC-HyFRC specimens maintained intact due to fiber bridging and crack resistance up to large deformations. Damage progression in plain concrete and SC-HyFRC specimens at ρ_s = 0.48% is shown in Fig. 4 in addition to their stress-displacement behavior at increasing displacement increments measured over the full specimen height. The superior spalling resistance of SC-HyFRC compared to plane concrete was also confirmed by tests on large scale bridge columns conducted at UC Berkeley. Compared to a conventional reinforced concrete column of same dimensions, same longitudinal steel and axial load ratio, SC-HyFRC columns exhibited superior damage resistance with a delay in spalling up to a drift ratio of 4% despite a 50% reduction in transverse reinforcement [5].

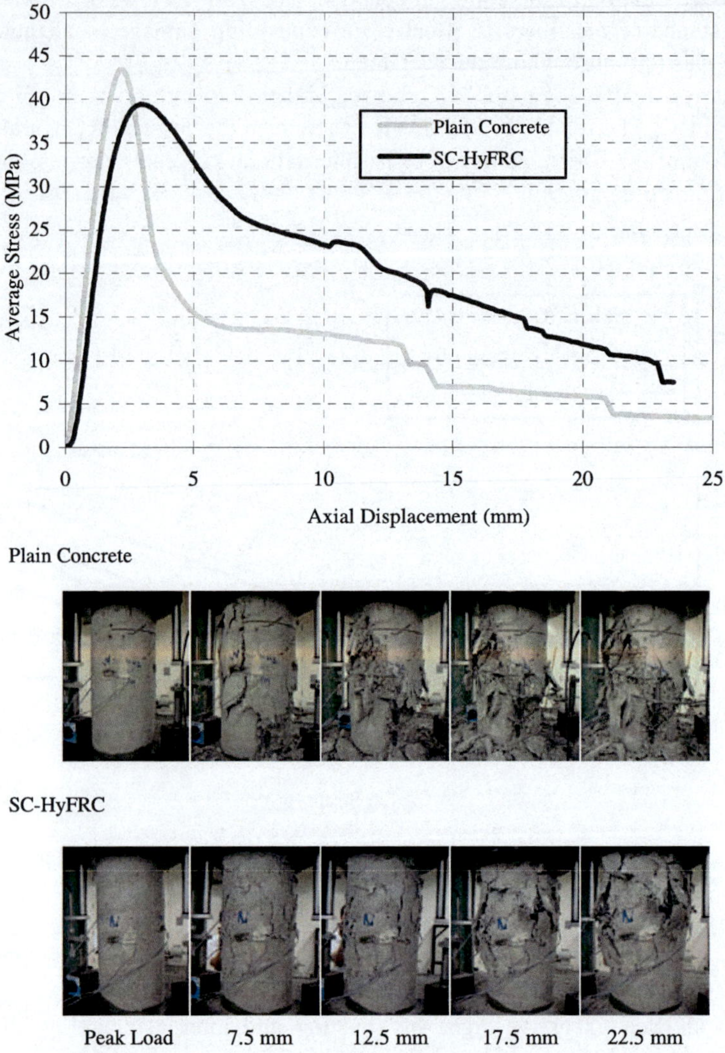

Fig. 4. Damage progression in plain concrete and SC-HyFRC at $\rho s = 0.48\%$

3.2 Tension Stiffening

A photograph of a reinforced SC-HyFRC specimen taken during the tension test reveals multiple cracking prior to crack localization as shown in Fig. 5a. Fig. 5b shows the tension test results of the reinforced SC-HyFRC dogbone, the plain SC-HyFRC dogbone (without rebar), and the plain #4 rebar up to 6% strain. The tension stiffening behavior in the elastic regime is apparent. In addition, reinforced SC-HyFRC with steel reinforcing ratio of 1.9%: i) eliminates strain localization at and beyond rebar yield, ii) increases the tensile load carrying capacity beyond what is expected by superposition of plain SC-HyFRC dogbone and bare steel (see dashed line in Fig. 5b) and iii) leads to a tensile peak stress of 700MPa at 6% strain which coincides with the strain value of the bare steel at its ultimate strength. The plain SC-HyFRC dogbone (see Fig. 5b) reveals pseudo- plastic deformation up to 0.25% strain followed by stable softening behavior with a tensile load carrying capacity of 50% even at 1% strain, which is the strain value where the yield plateau ends and the strain hardening starts in bare rebar.

(a)

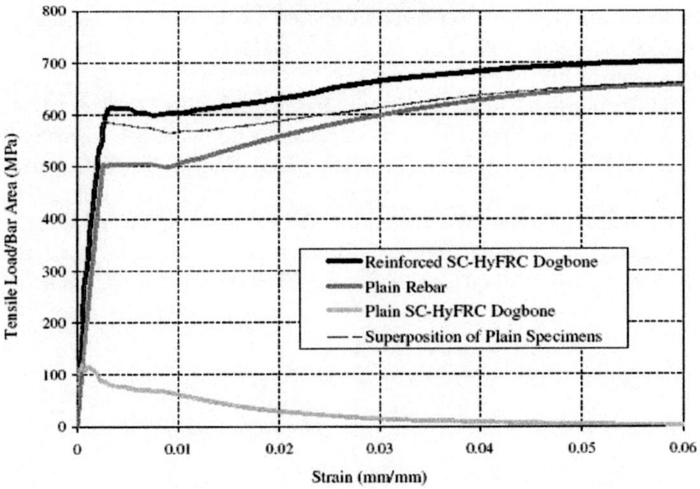

(b)

Fig. 5. a) Multiple cracking of reinforced SC-HyFRC in pure tension; b) tensile behavior of reinforced SC-HyFRC, plain SC-HyFRC (without steel bar), and plain steel bar up to 6% strain. Superposition of plain SC-HyFRC and plain rebar is shown as dashed line

The elimination of strain localization in reinforced SC-HyFRC beyond rebar yield, a necessary requirement for structural applications in seismic regions, may also be achieved in other reinforced HPFRCs if the softening behavior of their matrices (and hence their fiber load carrying capacity) is extended beyond the yield plateau and balanced by the strain hardening of the rebar as discussed in [2].

4 Conclusions

Compression tests on SC-HyFRC specimens with transverse confinement revealed that even at very low reinforcement ratios, the specimens had superior softening behavior beyond the peak load as compared to un-reinforced SCC specimens. Damage due to spalling and crushing was delayed even at much lower confinement ratios compared to the plain SCC samples. SC-HyFRC tension specimens were found to significantly improve the tensile behavior of the reinforcing bar and the specimens were able to carry load beyond the yield load of the plain rebar.

Acknowledgements. The authors would like to acknowledge the financial support received from the Pacific Earthquake Engineering Research (PEER) center.

References

[1] Blunt, J.D., Ostertag, C.P.: Deflection hardening and workability of hybrid fiber composites. ACI Materials Journal 106(3), 265–272 (2009)
[2] Blunt, J., Ostertag, C.P.: A Performance Based Approach for the Design of a Deflection Hardened Hybrid Fiber Reinforced Concrete. ASCE Journal of Engineering Mechanics (135), 978–986 (2009)
[3] America Concrete Institute Committee 318. 318-08: Building Code Requirements for Structural Concrete and Commentary. ACI, Farmington Hills (2008)
[4] Moreno, D.M., Jen, G., Trono, W., Ostertag, C.P., Billington, S.L.: Tension-Stiffening in Reinforced High Performance Fiber-Reinforced Cement-based Composites under Direct Tension. In: Proc. HPFRCC-6 Workshop, Ann Arbor Michigan (2011)
[5] Kumar, P., Jen, G., Trono, W., Lallemant, D., Panagiotou, M., Ostertag, C.P.: Self Compacting Hybrid Fiber Reinforced Concrete Composites for Bridge Columns. PEER Report. Pacific Earthquake Engineering Research Center, UC Berkeley (2010)

Tension-Stiffening in Reinforced High Performance Fiber-Reinforced Cement-Based Composites under Direct Tension

D.M. Moreno[1], W. Trono[2], G. Jen[2], C. Ostertag[2], and S.L. Billington[1]

[1] Dept. of Civil & Environmental Engineering, Stanford University, USA
[2] Dept. of Civil & Environmental Engineering, University of California at Berkeley, USA

Abstract. The unique properties of High Performance Fiber-reinforced Cement-based Composite (HPFRCC) materials provide significant damage reduction and energy dissipation in structural systems and are under investigation by numerous researchers. Unlike traditional fiber reinforced concrete, ductile HPFRCC materials can carry tension to strains greater than the yield strain of reinforcing steel and exhibit distributed compression damage with little to no spalling. Understanding the interaction between the reinforcement and the HPFRCCs through experimental testing is the focus of this research to develop modeling and performance prediction tools. A large-scale dogbone test set-up has been validated and the results for two different HPFRCC materials are presented. The multiple cracking observed in all specimens is hypothesized to have led to both distributed yielding and hardening in the reinforcement, which did not fracture "early" but rather fractured at strains of 14.5-19%, well beyond the typical strains reached at peak stress for grade 60 reinforcement (~10%).

1 Introduction

Research is being conducted to develop robust analytical models and design guidelines to facilitate the implementation into practice of reinforced, high performance fiber-reinforced cement-based composites (HPFRCCs) subjected to cyclic loading. Currently no guidelines exist for modeling and designing with ductile HPFRCC materials when reinforced with mild steel with the exception of a recently adopted option of using steel fiber HPFRCCs in place of minimum shear reinforcement in ACI 318-08 [1]. A tension-stiffening effect provided to the reinforcement by the ductile HPFRCC in tension and flexure has been observed in several experimental studies [e.g. 2-4] as the ductile HPFRCC materials carry tension beyond the yield strain of mild steel reinforcement. This strain compatibility between the HPFRCCs and the mild steel results in better strain distribution in the

G.J. Parra-Montesinos, H.W. Reinhardt, and A.E. Naaman (Eds.): HPFRCC 6, pp. 263–270.

steel, controlled damage under sustained cyclic loading and higher ductility levels at relatively large composite deformations.

Experiments are being conducted to understand the interaction of reinforcement and the HPFRCC in tension and compression and up to large strains. Two HPFRCC materials with varying properties are being considered and a goal is to develop appropriate constitutive models for these performance predictions of HPFRCC used in structural applications, such as those investigated in [5-8]. This paper presents preliminary studies of tension-stiffening experiments.

2 Tension Stiffening Experiments

2.1 Materials

The two HPFRCC materials studied here are a Hybrid Fiber Reinforced Concrete referred to as HyFRC and an Engineered Cementitious Composite referred to as ECC. HyFRC utilizes both steel and polymeric fibers of different sizes and aspect ratios and contains coarse aggregate with 9.5 mm maximum size aggregates [9]. ECC uses 2% volume fraction of polymeric fibers, contains no coarse aggregate and exhibits significantly larger tensile strain capacities than concrete and most other ductile HPFRC materials [10]. The mixture proportions for 1 m^3 of each of the materials are given in Table 1. In addition, the performance of a self-consolidating HyFRC mixture (referred to here as SC-HyFRC) with same fiber volume fraction of 1.5% as the HyFRC but without the 60mm long steel fibers was investigated as well [11].

Table 1. Mixture proportions for 1 m^3

Mix	Binder (kg)		Aggregate (kg)		Water (kg)	Chemical Admixtures (wt. % Binder)		Steel Fibers (% vol.)		PVA fibers (% vol.)	
	C	FA	Fine	Coarse		SP	VMA	60 mm	30 mm	8 mm	12 mm
ECC	547	656	438	–	312	.50	.11	–	–	–	2
HyFRC	484	–	808	743	220	.16	–	.80	.50	.20	–

FA: Class F fly ash, SP: superplasticizer, VMA: viscosity modifying agent

2.2 Specimen Design

Two specimen shapes were initially studied, a dogbone shape (Fig. 1) and a prismatic shape, with the former being presented herein. The large-scale dogbone was designed for a monotonic test set-up based on [12]. Two main issues were considered for the dogbone design: (1) the total length and capacity of the specimens to be fully tested (including strain hardening and fracture of the rebar) in available testing equipment and, (2) the stress concentrations in the transition from the anchoring zone (ends) to the gauged zone. A reduced length of the specimen was achieved by using headed rebar to anchor the steel at the ends of the dogbone specimen (Fig. 1).

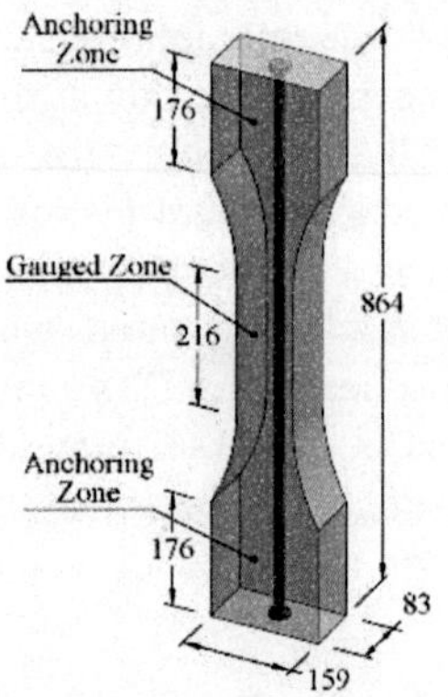

Fig. 1. Dogbone specimen

To reduce stress concentrations, elastic finite element analyses were performed using different values for the radius in the transition zone from the ends of the dogbone to the gauged zone. The effects of the clamping force induced in the specimen as the inserts slide along the curved portion (up and out) when loaded in tension were also considered and varied from 10% to 40% of the applied load in tension [12]. The radius selected was 305 mm. At larger values the reduction in the stress concentration was negligible. From the clamping pressure analyses, it was observed that there was no confinement induced inside of gauged region for confinement pressure up to 40% of the maximum stress in tension applied to the specimen.

A total of twenty large-scale dogbones with and without mild reinforcement have been designed, cast and tested (Table 2). The mix ECC-init referred to in Table 2 is the same as the ECC mix shown in Table 1 with the exception of a different superplasticizer at a slightly higher dosage.

Table 2. Specifications of the specimens tested to date

Specimen Name	Mix	Reinforcement (dia.)	No. of specimens	f'_c (MPa)
ECC3-4	ECC-init	# 4 (12.7mm)	4	53.0
ECC3-4	ECC	# 4 (12.7mm)	4	49.6
SCHy3-4	SC-HyFRC	# 4 (12.7mm)	2	31.3*
Hy3-4	HyFRC	# 4 (12.7mm)	2	40.3*
ECC3	ECC	None	4	36.9
SCHy3	SC-HyFRC	None	2	31.3*
Hy3	HyFRC	None	2	40.3*

* Values based on testing of same mix, different batch

2.3 Test Set-Up and Loading

The test set-up for the dogbone specimens is shown in Fig. 2. The gauged region is within the area of uniform cross section and is 197mm long. The end connections for the dogbone specimens are designed to apply load in the curved portion of the dogbone and are restrained from sliding laterally. The connection regions tie into single rods at the top and the bottom to ensure pure tensile loading and to minimize the ability for restraining moments to be created. The clamping force induced by the transverse threaded rods as the inserts are pulled in tension (and thus slide slightly transverse to the specimen), is assumed not to affect the interaction of the mild reinforcement and the ECC in the gauged region as discussed in Section 2.2.

All specimens were tested in monotonic tension with a 245-kN universal testing machine under displacement control at a rate of 1.5 mm/min. Elongation of the central regions was measured with two linear variable differential transducers (LVDTs) in the first set specimens and three LVDTs with the remaining and ongoing experiments. For the dogbone specimens, the gauging collar and LVDTs are shown in Fig. 2.

Fig. 2. Test set-up

3 Preliminary Dogbone Test Results

The testing results of the 3 types of specimens with reinforcement presented in Table 2 are described here. The average tensile strength of the unreinforced dogbones of ECC was 3.9 MPa at a w/binder ratio of 0.26 and the average tensile strength in the unreinforced dogbones of HyFRC and of SC-HyFRC was 2.8 MPa and 2.6 MPa, respectively, at a w/binder ratio of 0.45. The unreinforced ECC specimens formed localized cracks and began to lose load at approximately 1% strain. The HyFRC and SC-HyFRC specimens developed localized cracks and began to lose load-carrying capacity at strain values between 0.6% and 1%.

3.1 Progression of Cracking in Reinforced Dogbones

The progression of cracking in specimens ECC3-4-7 and SC-Hy3-4-2 are shown in Fig. 3 and 4, respectively as representative specimens. Multiple cracking was observed in the ECC dogbone specimens (Fig. 3). The first visible crack was recorded inside of the gauged zone at a load of 10 kN. For this specimen,

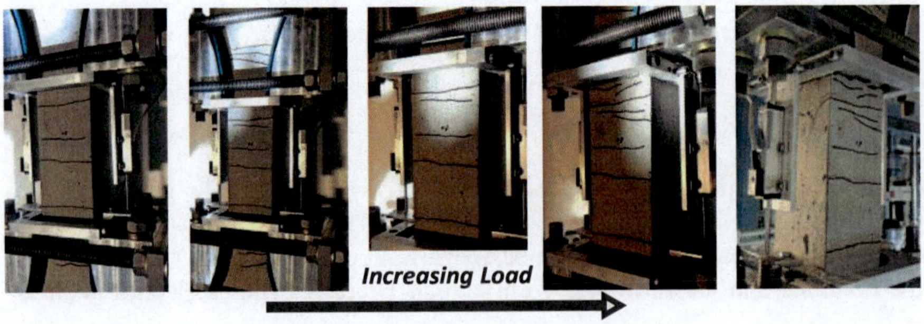

Fig. 3. Progression of cracking in specimen ECC3-4-7

deformations localized in two cracks with one becoming more dominant at a strain of approximately 2% (photo on far right, Fig. 3). At this strain, it was observed that no more fibers were bridging the dominant crack.

The HyFRC and SC-HyFRC specimens also exhibited multiple cracking up to and beyond the yield strain of the rebar but to a lesser extent than in the ECC (Fig. 4). First cracking in these specimens was observed between loads of 13-18 kN. Fiber bridging was observed at strains far beyond the yield plateau of the steel reinforcing bar.

In all specimens, failure occurred when the reinforcement fractured in a dominant crack no longer being bridged by any fibers. For specimens that failed within the gauged region, fracture strains were estimated (using the full gage length) to occur at 14.5% strain for the ECC and 19% strain for the HyFRC. In one SC-HyFRC specimen (SCHy3-4-2) three dominant cracks formed with fracture eventually occurring in one. In two ECC specimens (ECC3-4-2 and ECC3-4-7) two dominant cracks were visible up to fracture of the steel, which again occurred within one of the cracks. Splitting cracks down the side of the dogbone were also observed in two ECC specimens (ECC3-4-7 and ECC3-4-8) at failure. No splitting cracks were observed in the (SC-)HyFRC specimens.

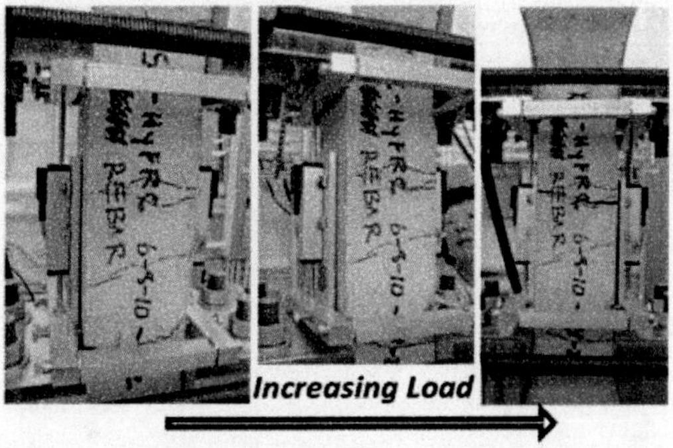

Fig. 4. Progression of cracking in specimen SCHy3-4-2

3.2 Load vs. Strain

The load vs. strain for three of the tested ECC dogbones is shown in Fig. 5. Strain was calculated as the ratio of the average of the displacements measured with each of the two or three LVDT's to the effective gauge length of 197 mm.

The plotted response of the reinforced ECC dogbones accounts for the restrained shrinkage experienced by the dogbones prior to testing [13]. An estimate of the restrained shrinkage strain in the reinforced ECC of 0.00026 is adopted based on a survey of shrinkage strain research on plain ECC [14-15] combined with a strain compatibility analysis of the restrained shrinkage of the reinforced ECC dogbones. The reinforced dogbone shrinkage is expected to be less than the plain ECC shrinkage and also to cause initial tension (and early cracking) in the reinforced ECC, as was observed.

From Fig. 5 it can be observed that the plain ECC dogbone begins to lose tension carrying capacity at approximately 1% strain. In the reinforced dogbones, loads greater than the yield strength of the bar are carried to strains well beyond 1% strain. At approximately 2% strain, the dominant crack that formed was observed to have no more fiber bridging, indicating that all load was being carried by the reinforcement. Therefore the reinforcing bar hardened to be carrying this larger load. While distributed yielding is expected [2], it is hypothesized that the multiple cracking and possibly slight narrowing of the bar during hardening (Poisson effect) facilitated distributed strain hardening as well. As mentioned previously, fracture occurred at a strain of 14.5%, which is well beyond the typical strain at peak stress for a reinforcing bar and essentially the same as the minimum supplier-specified elongation for the bar. Therefore "early" fracture due to hardening was not observed.

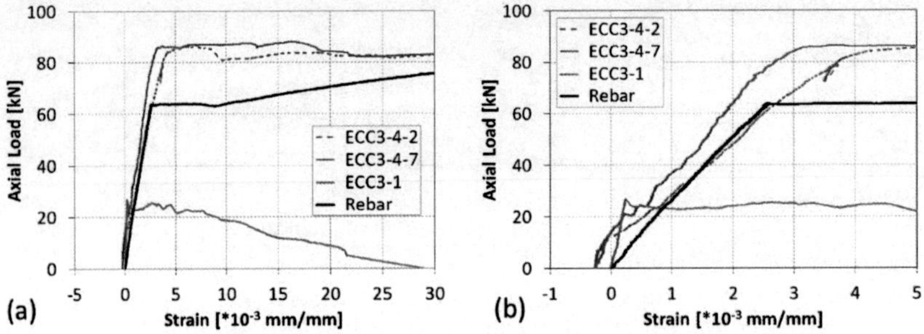

Fig. 5. Load vs. strain response of three ECC dogbone specimens

The reduced stiffness of ECC3-4-2 after initial cracking (at ~8 kN) goes below the stiffness of the reinforcement, which is not reasonable (Fig. 5(b)). This problem is attributed to using only two LVDTs to estimate the deformation of these specimens with saturated multiple cracking that have relatively flexible end supports (single rods connected to the grips). The multiple cracking was observed to open on all sides but not always propagate to the other side. As such, estimating the strain with just an average of two measurements proved to be inadequate. In later testing (i.e. ECC3-4-7), 3 LVDTs were used around the specimen to estimate a plane through the specimen and determine deformations and then strain. The tension stiffening effect of the ECC in specimen ECC3-4-7 is clear.

The stiffness calculated from the data in Fig. 5(b), supports the assumption that the ECC has experienced restrained shrinkage cracking. The average tensile modulus of elasticity of the plain ECC was measured to be 17.9 GPa. Using the rule of mixtures, the reinforced ECC composite stiffness would be 21.4 GPa (assuming a steel modulus of 200 GPa).

However the measured stiffness of specimen ECC3-4-7 gives a composite stiffness of 8 GPa, indicating that shrinkage cracking likely occurred. This assumption is further supported by the first cracking observation in ECC3-4-7, which occurred at a load of 2.7kN (~0.4MPa stress in the ECC).

The load vs. strain for the two forced HyFRC and one of the reinforced SC-HyFRC dogbones is shown in Fig. 6. As mentioned previously, the plain HyFRC and SC-HyFRC specimens were estimated to have lost load-carrying capacity at 0.6 to 1.0% strain. In the reinforced specimens, fiber bridging was still observed beyond 1% strain indicating that the load was shared by both the fibers and the reinforcement. Here again it appears that not only yielding was well distributed along the bar but strain hardening as well. Fracture occurred in Hy3-4-1 (the specimen that failed within the gauged region) at 19% strain.

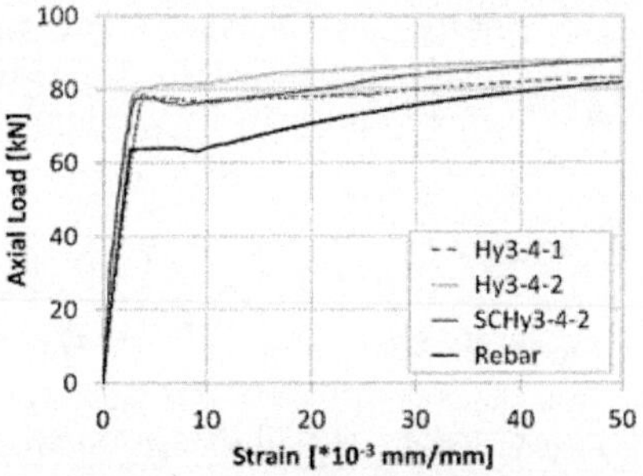

Fig. 6. Load vs. strain response of reinforced HyFRC and SC-HyFRC dogbones

4 Conclusions and Future Work

A large-scale tension-stiffening test set-up has been design and validated on two types of high performance fiber-reinforced cement-based composite (HPFRCC) materials. For this large-scale set-up, three LVDTs are recommended particularly for ECC specimens for more accurate estimations of strain. Shrinkage cracking was observed and should be well characterized for accurate estimations of tensile stiffening provided to the reinforcement by the HPFRCC materials.

Multiple cracking was observed in all reinforced HPFRCC dogbones tested here. It is believed that under the tensile loading, yielding of the bar occurred simultaneously with the multiple cracking and the yielding was spread along the bar, as has been observed in previous research at low strains [2]. The bar strain hardened and it is hypothesized that the hardening was also well-distributed along the bar, which is attributed to the multiple cracking and perhaps slight narrowing of the bar away from the matrix due to the Poisson effect. Once one dominant crack formed, the specimens continued to deform with fracture of the bar not occurring until 14.5% to 19% strain, well beyond the typical strain at ultimate stress for Grade 60 reinforcement (~10%). "Early" fracture of the reinforcement was not observed in these experiments.

Additional testing is underway to understand better the spreading of hardening in the reinforcement. In addition, basic material property characterization on small plates in tension, beams in flexure, and cylinders in compression are being conducted to use in developing appropriate models for steel reinforced HPFRCC materials. Constitutive models and modeling approaches validated on experimental data will be further validated on recent bridge pier experiments using ECC and HyFRC in regions of high moment and shear.

Acknowledgements. This research was sponsored by the Pacific Earthquake Engineering Research (PEER) Center. Funding from both PEER and the Blume Earthquake Engineering Center is gratefully acknowledged. The assistance of undergraduate researcher Sunny Vanderboll at Stanford University is gratefully acknowledged.

References

[1] American Concrete Institute, Committee 318, Building Code Requirements for Structural Concrete (ACI 318-08) and Commentary. Farmington Hills, MI (2008)

[2] Fischer, G., Li, V.C.: Influence of Matrix Ductility on Tension Stiffening Behavior of Steel Reinforced Engineered Cementitious Composites. ACI Str. J. 99(1), 104–111 (2002)

[3] Olsen, E.C., Billington, S.L.: Cyclic behavior of precast, self-compacting ductile concrete infill panels for seismic retrofit of steel frame buildings. ACI Str. J. 108(1), 51–60 (2011)

[4] Blunt, J., Ostertag, C.P.: A Performance Based Approach for the Design of a Deflection Hardened Hybrid Fiber Reinforced Concrete. ASCE J. Eng. Mech. 135, 978–986 (2009)

[5] Rouse, J.M., Billington, S.L.: Behavior of bridge piers with ductile fiber reinforced hinge regions and vertical, unbonded posttensioning. In: Fib Symposium: Concrete Structures in Seismic Regions, Athens, Greece (2003)

[6] Billington, S.L., Yoon, J.K.: Cyclic Response of Unbonded Posttensioned Precast Columns with Ductile Fiber-Reinforced Concrete. ASCE J. Bridge Eng. 9(4), 353–363 (2004)

[7] Saiidi, M., O'Brien, M., Sadrossadat, M.: Cyclic Response of Concrete Bridge Columns using Superelastic Nitinol and Bendable Concrete. ACI Str. J. 106(1), 69–77 (2009)

[8] Kumar, P., Jen, G., Trono, W., Lallemant, D., Panagiotou, M., Ostertag, C.P.: Self Compacting Hybrid Fiber Reinforced Concrete Composites for Bridge Columns. Pacific Earthquake Engineering Research Center (2010)

[9] Blunt, J.D., Ostertag, C.P.: Deflection Hardening and Workability of Hybrid Fiber Composites. ACI Materials Journal 106(3), 265–272 (2009)

[10] Li, V.C.: On Engineered Cementitious Composites: A Review of the Material and its Applications. J. Advanced Concrete Technology 1(3), 215–230 (2003)

[11] Trono, W., Jen, G., Moreno, D., Billington, S.L., Ostertag, C.P.: Confinement and Tension Stiffening Effects in High Performance Self-consolidated Hybrid Fiber Reinforced Concrete Composites. In: Proc. HPFRCC-6 Workshop, Ann Arbor Michigan (June 2011)

[12] Redaelli, D.: Testing of Reinforced High Performance Fibre Concrete Members in Tension. In: 6th International PhD Symposium in Civil Engineering, Zurich, Switzerland (2006)

[13] Abrishami, H.H., Mitchell, D.: Influence of Steel Fibers on Tension Stiffening. ACI Structural J. 94(6), 769–775 (1997)

[14] Rouse, J.M., Billington, S.L.: Creep and Shrinkage of High-Performance Fiber-Reinforced Cementitious Composites. ACI Materials J. 104(2), 129–136 (2007)

[15] Sahmaran, M., Lachemi, M., Hossain, K., Li, V.C.: Creep and Shrinkage of High-Performance Fiber-Reinforced Cementitious Composites. Cem. Conc. Res. 39(10), 893–901 (2009)

Crack Formation in FRC Structural Elements Containing Conventional Reinforcement

J. Deluce, S.-C. Lee, and F.J. Vecchio

Department of Civil Engineering, University of Toronto, Toronto, ON, Canada

Abstract. It is well known that both crack spacings and crack widths in members with conventional reinforcement can be reduced by adding steel fibers to the concrete mix. This behavior is important not only for predictions of the tensile behavior including tensile stresses due to tension stiffening and fibers, but also for evaluations of the serviceability or ultimate capacity of FRC members containing conventional reinforcement (R/FRC). Since test results for the tensile behavior of R/FRC members are scarce in comparison to those for ordinary reinforced concrete members (RC) or FRC members without rebar, an experimental program was undertaken to better understand the effect of fibers in conjunction with conventional reinforcement.In this research, direct tensile tests were conducted on 12 RC and 48 R/FRC specimens of 1.0 m length. The variables included steel fiber type, fiber volumetric ratio, and reinforcement ratio. From the test results it was observed that cracks were more finely distributed in R/FRC members, and that the average crack spacing decreased until the rebar yielded. After yielding of the rebar, the tensile deformation was dominated by one or two cracks. It was also determined that currently available formulae significantly overestimate the average crack spacing in R/FRC members. Based on the test results, a new crack spacing formula is developed in this paper.

1 Introduction

It has been demonstrated over the past several decades that fibers can significantly improve the structural behavior of a member because of its improved tensile behavior and crack control. Despite these advantages, FRC is generally not widely used in structural members; one reason for this is that it is difficult to model or predict the behavior of FRC, particularly in tension. Recently, some models for the tensile behavior of FRC members have been developed [1-3], but the behavior of FRC in conjunction with conventional rebars has not yet been sufficiently demonstrated. In addition, while a number of fiber reinforced concrete (FRC) members subjected to uniaxial tension have been tested by many researchers [4-9], results for FRC members with conventional rebar (R/FRC) are currently very limited [10-12].

G.J. Parra-Montesinos, H.W. Reinhardt, and A.E. Naaman (Eds.): HPFRCC 6, pp. 271–278.

In this paper, a comprehensive test program for R/FRC specimens will be described, and cracking behavior will be investigated. Based on the test results, an improved average crack spacing formula will be presented.

2 Test Program

In order to investigate the tensile behavior of FRC members containing also conventional reinforcement, a total of 60 direct tension specimens were tested. A total of 30 different configurations were used, and each was doubled in order to determine the degree of scatter in the test results. The test parameters included the reinforcement ratio, steel fiber content, fiber length and aspect ratio. To modify the reinforcement ratio, various square cross section dimensions (b mm × b mm) were used in conjunction with a range of steel reinforcing bar sizes. Details are presented in Table 1.

The concrete portion of the typical specimen was 1000 mm in length in order to have an adequate length over which to observe crack development; the steel rebar protruded 250 mm from each end to allow for an adequate gripping length in the test frame as illustrated in Fig. 1. Special care was taken during fabrication to ensure that the reinforcing bar was as centered as possible, in order to minimize flexural stresses caused by bar eccentricities during the uniaxial tension test.

Six different concrete mixes were used in order to vary the mechanical properties of the FRC. These were denoted as PC, and FRC1 to FRC5, depending on the steel fiber content and type. The actual compressive strength at the test date was measured through three 150 mm × 300 mm cylinder tests, and the approximate free shrinkage strain of the specimen was measured with shrinkage prisms. The concrete compressive strength and the fiber contents are presented in Table 2.

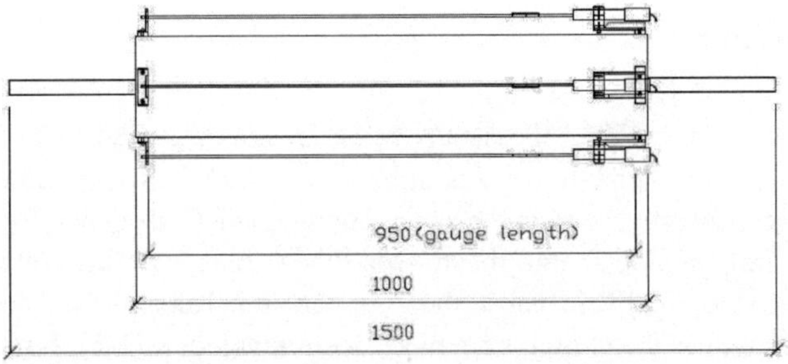

Fig. 1. Instrumentation and dimensions for the direct tension test specimens (units in mm)

Table 1. Geometric conditions of the test specimens and reinforcement materials

Specimen designation	Specimen size (mm×mm)	Reinforcing bar[†]				
		d_b (mm)	ρ_s (%)	f_{sy} (MPa)	f_{su} (MPa)	E_s (GPa)
50/10	50×50	11.3	4.01	442	564	199
80/10	80×80	11.3	1.57	442	564	199
100/20	100×100	19.5	2.99	456	592	194
150/20	150×150	19.5	1.33	456[‡]	592[‡]	194[‡]
150/30	150×150	29.9	3.10	376	558	187
200/30	200×200	29.9	1.74	376	558	187

[†] Material properties were measured through coupon tests. Details will be presented in [13].
[‡] f_{sy} = 525 MPa, f_{su} = 653 MPa, and E_s = 188 GPa for the rebars in FRC4 specimens.

Table 2. Concrete compressive strength and fiber contents

Concrete type	Concrete[†] f_c' (MPa)	Steel fiber				
		Fiber type	l_f (mm)	d_f (mm)	σ_{fu} (MPa)	V_f (%)
PC	91.7-95.6	-	-	-	-	-
FRC1	55.8-91.4	RC-80/30-BP	30	0.38	2300	0.5
FRC2	45.2-63.4	RC-80/30-BP	30	0.38	2300	1.0
FRC3	46.0-63.1	RC-80/30-BP	30	0.38	2300	1.5
FRC4	32.5-52.8	ZP 305	30	0.55	1100	1.5
FRC5	70.3-78.8	RL-45/50-BN	50	1.05	1000	1.5

[†] Details for material properties of concrete will be presented in [13]

Each specimen was instrumented with four LVDTs with 25 mm travel. One LVDT was located on each side of the specimen, so that differences in the readings of the various LVDTs allowed for the determination of unintended flexural effects. To maximize the gauge length of the specimen, while maintaining enough space to affix the LVDT mounts, a gauge length of 950 mm was used except for the 50/10 specimens, in which gauge lengths of 950 mm and 850 mm were used for the two pairs of opposite sides in order to avoid interference between the LVDT mounts at the ends of the specimen.

The specimens were tested vertically in a 1000 kN MTS Universal Test Frame. The loading was displacement-controlled, at a rate of 0.0015 mm/s up to the observance of the first crack, and 0.0025 to 0.004 mm/s prior to yield. At the onset of yielding of the rebar, the loading rate was increased to 0.040 mm/s, where it remained constant until the termination of the test. The tests were paused at periodic load stages to mark crack locations and to measure crack widths. The tests were terminated at an average displacement of 15 mm for the PC specimens and at 25 mm for the FRC1-5 specimens, as measured by the LVDTs.

3 Test Results

3.1 General Behavior

It was observed that new cracks formed up until rebar yielding, and then deformations localized at one or two cracks. Before tension was applied, several cracks were visible in most 50/10 series specimens and in some 100/20 series specimens, since the restrained shrinkage effect was considerable with the high reinforcement ratio. In several specimens with high fiber volumetric ratio and low reinforcement ratio, the protruding rebar ruptured before yielding of the rebar within the concrete section. At the vicinity of the end surfaces in specimens with a relatively high reinforcement ratio, significant splitting cracks occurred at approximately the same time as yielding of the reinforcement.

Fig. 2 shows the final crack patterns of the PC and FRC3 specimens. As shown in the figures, cracks were generally better distributed in the specimens with higher fiber volumetric ratio and higher reinforcement ratio. Unlike PC specimens in which localization of a crack was not clear, one or two cracks were localized in FRC specimens. However, it should be reiterated that the test was terminated at a lower total extension for the PC specimens than for the FRC specimens.

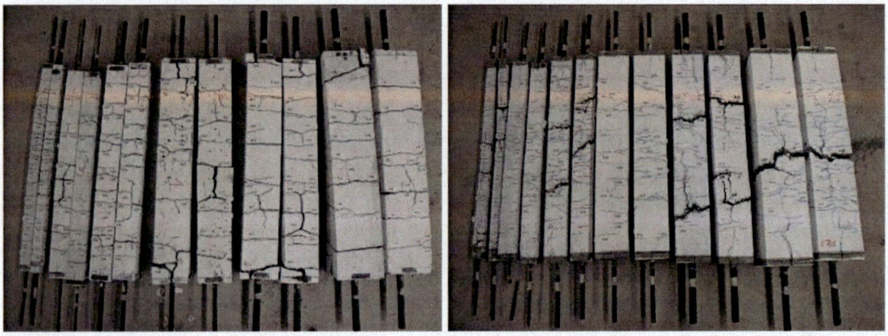

(a) (b)

Fig. 2. Crack patterns after tests; (a) PC specimens, (b) FRC3 specimens

Fig. 3(a) shows the axial tensile force – displacement responses for 150/20 specimens, in which the reinforcement ratio was 1.33%. The displacement was calculated by taking the average of the four LVDT readings. Since there was considerable splitting in the vicinity of the end surfaces in some specimens, the measured data appear scattered near yielding of reinforcement. By subtracting the bare bar response, the average tensile stress of fiber reinforced concrete can be extracted as presented in Fig. 3(b). As shown in the figures, the tensile stress in the concrete in PC specimens significantly decreased after cracking so that the tensile behavior quickly approached the bare bar response. On the other hand, it was observed that considerable tensile stress can be resisted by fiber reinforced concrete even after yielding of the rebar.

It should be noted that restrained shrinkage effects were significant, but these effects were not removed from the displacement reported in the figures. In addition, the uncracked regions near the ends of the specimens were included in the reference length of the LVDTs; thus, the average strains in the middle cracked regions were larger than the displacements shown in this figures would indicate. Therefore, the yield point of the tensile behavior of FRC members in Fig. 3(a) should be closer to that of the bare bar response, since the cracks were well distributed in the FRC members. Further analysis is required for a more accurate evaluation of the actual average tensile strain in the middle region, which can be used for the representative tensile behavior of R/FRC members.

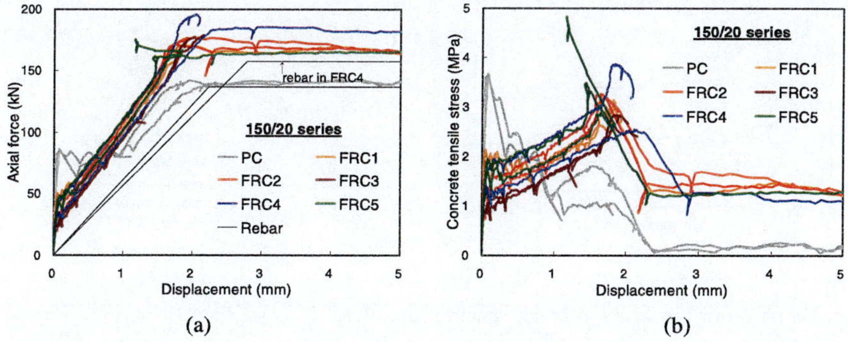

Fig. 3. Load-displacement response; (a) total response, (b) tensile behavior of matrix

3.2 Average Crack Spacing

The average crack spacing was calculated by dividing the length of the cracked region by the number of cracks. The length of the cracked region was evaluated, from photographs, as the distance between the two extreme cracks. As shown in Fig. 4, it can be seen that the average crack spacing decreased until it stabilized at the onset of yielding of reinforcement. It was also observed that the average crack spacing was generally smaller with higher fiber volumetric ratios in the specimens with large reinforcement ratios (see the Fig. 4(a) for the 150/30 series), while it was not as apparent in the specimens with smaller reinforcement ratios (see Fig. 4(a) for the 150/20 series). In addition, through the comparison with FRC3 and FRC4 in Fig. 4(b), it was investigated that larger fiber aspect ratios (l_f/d_f) can reduce the average crack spacing. From Fig. 4(c), it can be seen that a linear relationship between bond parameter and average stabilized crack spacing can be assumed, similar to the CEB-FIP crack spacing formula for ordinary RC members presented in [14].

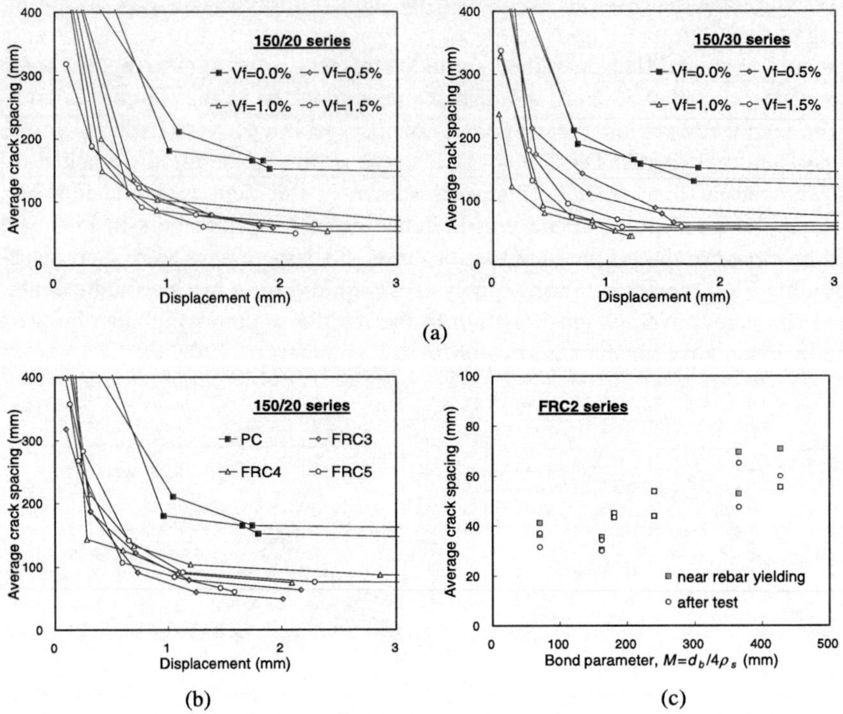

Fig. 4. Variation of the average crack spacing; the effect of (a) V_f, (b) fiber type, and (c) bond parameter

4 Crack Spacing Formulae

In order to consider the effect of fibers on cracking behavior, Dupont and Vandewalle [15] suggested the following average crack spacing model.

$$s_{cr} = \left(50 + 0.25 k_1 k_2 \frac{d_b}{\rho_s}\right)\left(\frac{50}{l_f / d_f}\right) \tag{1}$$

where, $k_1 = 0.8$ for deformed bars, $k_2 = 0.25(\varepsilon_1 + \varepsilon_2)/2\varepsilon_1$, and ε_1 and ε_2 are the largest and the smallest tensile strains, respectively.

From the comparison of the test results with the average crack spacing calculated from the above equation, it was found that Eq. (1) significantly overestimates the average crack spacing, as shown in Fig. 5(a).

To improve the prediction of the cracking behavior in R/FRC members, the following crack spacing model was developed based on the CEB-FIP formulation [14].

$$s_{cr} = 2\left(c + \frac{s_b}{10}\right) + \frac{k_1 k_2}{s_{mi}} \tag{2}$$

where, $s_b = 0.5\sqrt{\frac{\pi d_b}{s_{mi}}}$, $s_{mi} = \frac{\rho_s}{d_b} + k_{eff}\frac{\alpha_f V_f}{d_f}$, $c = 1.5a_{gg}$, $k_1 = 0.4$, $k_2 = 0.25$, a_{gg} is maximum aggregate size, α_f is a fiber orientation factor which can be calculated from DEM [2-3] if considering edge effects or taken as 0.5 for an infinite element, and $k_{eff} = 0.25$.

As shown in Fig. 5(b), it can be seen that Eq. (2) shows good agreement with the test results, especially for the average crack spacing near rebar yielding.

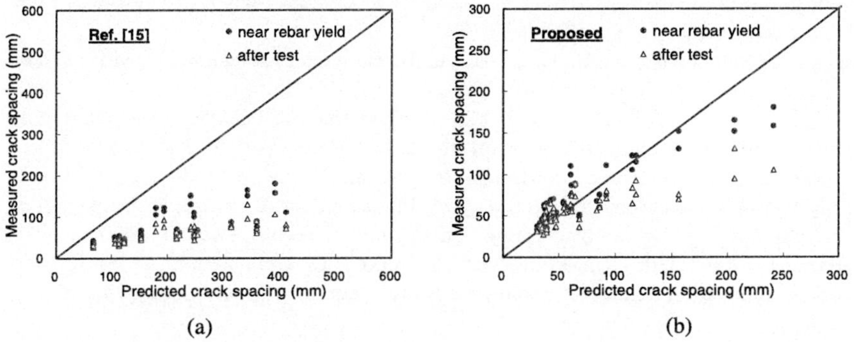

Fig. 5. Comparison of the test results with the formulae on the average crack spacing; (a) Dupont and Vanderwalle [15], (b) the proposed model

5 Conclusions

An experimental program was conducted in which 12 RC and 48 R/FRC specimens were tested in direct tension. Based on the test results, the following conclusions can be drawn:

1. Considerable tensile stress can be resisted by FRC even after yielding of the reinforcement.
2. Steel fibers can effectively control crack widths and reduce crack spacings.
3. Through the comparison of the test results with the currently available crack spacing formula presented by Dupont and Vandewalle [15], it was found that the average crack spacing was significantly overestimated. An improved crack

spacing formula, showing better agreement with the test results, was presented in this paper.

4. The test results can be useful for the evaluation of the tension stiffening effect in R/FRC members. However, further analysis is required in order to evaluate the actual tensile behavior of fiber reinforced concrete in R/FRC members and to separate the tension stiffening effect from fiber stress.

References

[1] Voo, J.Y.L., Foster, S.J.: Variable engagement model for fibre reinforced concrete in tension. Uniciv Report No. R-420, School of Civ. and Envir. Eng, The Univ. of New South Wales, 86 (June 2003)

[2] Lee, S.-C., Cho, J.-Y., Vecchio, F.J.: Diverse embedment model for steel fiber reinforced concrete in tension: model development. ACI Mater. J (2010) (submitted)

[3] Lee, S.-C., Cho, J.-Y., Vecchio, F.J.: Diverse embedment model for steel fiber reinforced concrete in tension: model verification. ACI Mater. J. (2010) (submitted)

[4] Petersson, P.E.: Fracture mechanical calculations and tests for fiber-reinforced cementitious materials. In: Proceedings of Advances in Cement-Matrix Composites, Mater. Res. Soc., Boston, pp. 95–106 (1980)

[5] Lim, T.Y., Paramasivam, P., Lee, S.L.: Analytical model for tensile behavior of steel-fiber concrete. ACI Mater. J. 84(4), 286–298 (1987)

[6] Li, Z., Li, F., Chang, T.Y.P., Mai, Y.W.: Uniaxial tensile behavior of concrete reinforced with randomly distributed short fibers. ACI Mater. J. 95(5), 564–574 (1998)

[7] Groth, P.: Fibre reinforced concrete – fracture mechanics methods applied on self-compacting concrete and energetically modified binders. Doct. thesis, Dept. of Civ. and Mining Eng., Div. of Struct. Eng., Luleå Univ. of Techn., Sweden (2000)

[8] Barragán, B.E., Gettu, R., Martín, M.A., Zerbino, R.L.: Uniaxial tension test for steel fibre reinforced concrete – a parametric study. Cem. and Concr. Comp. 25, 767–777 (2003)

[9] Susetyo, J.: Fibre reinforcement for shrinkage crack control in prestressed, precast segmental bridges. Doct. thesis, the Dept. of Civ. Eng. at the Univ. of Toronto, Toronto, ON, Canada, 307 (2009)

[10] Abrishami, H.H., Mitchell, D.: Influence of steel fibers on tension stiffening. ACI Struct. J. 94(6), 769–776 (1997)

[11] Noghabai, K.: Behavior of tie elements of plain and fibrous concrete and varying cross sections. ACI Struct. J. 97(2), 277–285 (2000)

[12] Bischoff, P.H.: Tension stiffening and cracking of steel fiber-reinforced concrete. J. of Mat. Civ. Eng. ASCE 15(2), 174–182 (2003)

[13] Deluce, J.: Crack development of steel fibre reinforced concrete elements containing a conventional reinforcing bar subjected to direct tension. Master's thesis in preparation, the Dept. of Civ. Eng. at the Univ. of Toronto, Toronto, ON, Canada

[14] CEB-FIP Model code for concrete structures. CEB-FIP International Recommendations, 3rd edn., p. 348. Comite Euro-International du Beton, Paris (1978)

[15] Dupont, D., Vanderwalle, L.: Calculation of crack width the - model. In: Test and Design Methods for Steel Fibre Reinforced Concrete: Background and Experiences – Proceedings of the RILEM TC162-TDF Workshop, RILEM Technical Committee 162-TDF, Bochum, Germany, pp. 191–144 (2003)

Strength and Behavior of SFRSCC and SFRC Wall Panels under One-Way In-Plane Action

N. Ganesan, P.V. Indira, and S. Rajendra Prasad

Department of Civil Engineering, National Institute of Technology, Calicut, India

Abstract. An experimental investigation was carried out to study the strength and behavior of steel fiber reinforced self compacting concrete (SFRSCC) and Steel Fiber Reinforced Concrete (SFRC) wall panels under one-way in-plane action. The volume fraction and aspect ratio of steel fibers used are 0.50% and 60 respectively. Influence of slenderness ratio and aspect ratio of wall panels on the ultimate load were studied. The load-deflection plots indicate that SFRSCC wall panels exhibit softening behavior which means that the wall panels behave in a more ductile manner than that of SFRC wall panels. A method was proposed to predict the ultimate strength of SFRSCC wall panels.

1 Introduction

Reinforced concrete (RC) wall panels are used as load bearing structural elements in a building system. A rectangular wall which is hinged at top and bottom with free vertical edges and subjected to in-plane vertical loads is likely to behave as a panel in one-way action and develops a uniaxial curvature in the direction of loading. In the case of wall panels, the thickness is small in comparison to other dimensions, which introduces the slenderness effect leading to problems of stability. The ratio of height to thickness (h/t) of wall panels is known as slenderness ratio (SR). Also depending on the relative ratio of height to length, the behavior of wall panel under the load would vary from a short wide compression member to a deep narrow slender member [1]. The ratio of height to length (h/L) of wall panels is called as aspect ratio (AR).

Self compacting concrete (SCC) is a recently developed special concrete, which has gained much attention in the concrete industry and used in many applications successfully throughout the world [2]. With the addition of short, discrete steel fibers, the interfacial bond stress between SCC and steel fibers is higher than that of normal concrete due to the increased paste content [3] and SCC mixes possess good fluidity and deformability. Therefore SCC mixes are well suited for fiber addition up to certain volume fraction of steel fibers.

Several studies are available on the strength and behavior of structural elements like Steel Fiber Reinforced Concrete (SFRC) beams, columns,

G.J. Parra-Montesinos, H.W. Reinhardt, and A.E. Naaman (Eds.): HPFRCC 6, pp. 279–286.

beam-column joints etc. A very limited research also has been carried out on the fresh and structural properties of SFRSCC [2-5], but the investigation on the strength and behavior of SFRC and SFRSCC wall panels have not been come across. Considering this gap in the existing literature, an attempt has been made to study the strength and behavior of SFRC and SFRSCC wall panels in one-way in-plane action.

2 Experimental Programme

The experimental programme consisted of casting and testing of 16 SFRSCC and 16 SFRC wall panels under compression. The thickness of the wall panels was kept constant at 40 mm. The heights of walls ranging from 480 mm to 1200 mm were selected to provide a range of SR from 12 to 30. To study the effect of AR on the ultimate load, the height of walls was kept constant at 600 mm and the lengths of walls were varied from 320 mm to 800 mm to provide a range of aspect ratio (AR) from 0.75 to 1.875. The details of overall dimensions, SR and AR of wall panels are presented in Table 1.

Table 1. Details of wall panels and variables

Panel Designation		Panel Size	Variables	
SFRC	SFRSCC	h × L × t (mm)	SR	AR
OWSFN-1	OWSFS-1	480 × 320 × 40	12	
OWSFN-2	OWSFS-2	600 × 400 × 40	15	1.5
OWSFN-3	OWSFS-3	840 × 560 × 40	21	
OWSFN-4	OWSFS-4	1200 × 800 × 40	30	
OWAFN-1	OWAFS-1	600 × 320 × 40		1.875
OWAFN-2	OWAFS-2	600 × 400 × 40	15	1.5
OWAFN-3	OWAFS-3	600 × 560 × 40		1.07
OWAFN-4	OWAFS-4	600 × 800 × 40		0.75

2.1 Materials

Portland Pozzolana Cement (PPC–Fly ash based) conforming to IS 1489 (part 1) - 1991, fine aggregate conforming to grading zone III of IS 383 - 1970 having a specific gravity of 2.67, coarse aggregate having a maximum size of 12.5 mm with specific gravity of 2.78, and potable water were used for casting wall panels. Straight steel fibers having 0.38 mm diameter and 23 mm length were used. The volume fraction and aspect ratio of steel fibers were 0.50% and 60 respectively. Mineral admixtures which consisted of class–C fly ash and silica fume, chemical admixtures comprised of naphthalene based super plasticizer and polysaccharide based viscosity modifying agent (VMA) were used.

The reinforcing mesh consists of 6 mm diameter HYSD bars (Fe 415) placed in a single layer at mid thickness of the wall panels. The spacing of bars (100 mm) in both directions was selected so that it does not exceed three times the panel thickness which is a maximum spacing as per ACI 318-2008 [5] with a clear side cover of 10 mm. The yield strength of reinforcing steel was 445N/mm^2. The percentages of vertical and horizontal reinforcement provided in the panels were 0.88 and 0.74 respectively.

2.2 Mix Proportions

M30 mix was designed for NC wall panels as per IS 10262-1982. The concrete mix for SFRC wall panels was developed by adding straight steel fibers, of 0.38 mm diameter and 23 mm length, to the designed mix of NC wall panels. The addition of steel fibers decreased the workability of concrete. Hence a naphthalene based superplasticiser (HRWRA), at 1% by weight of cement, was used in SFRC to retain the workability.

The mix proportions of steel fiber reinforced self compacting concrete (SFRSCC) were obtained after extensive trials based on the guidelines of EFNARC-2002 [6] for M30 grade concrete. The details of constituents of SFRSCC mix thus obtained are given in Table 2.

Table 2. Mix proportions of SFRSCC

Particulars	Quantity (kg/m^3)	Particulars	Quantity (kg/m^3)
Cement (PPC Fly ash based)	493	Water	246
Fly ash	20	Superplasticiser	5
Silica fume	10	VMA	0.012
Fine aggregate	789	Vol. fraction of steel fibers	0.50%
Coarse aggregate	740	Aspect ratio of steel fibers	60

2.3 Testing of Specimens

All wall panels in the experimental programme were tested under a uniformly distributed in-plane vertical load along their ends. The load was applied at an eccentricity of t/6 to reflect possible eccentric load in actual practice as adopted by other researchers [1, 7-9]. All specimens were subjected to loading in the vertical position in a Compression Testing Machine of 2,940kN (300 tons) capacity. The rate of loading was 24.50kN (2.50 tons) per minute. Fig. 1 shows the details of test set up. Lateral deformations were measured at quarter and mid height points along the central vertical line of the panel at every load increment of 19.60kN (2 ton) using LVDTs. The experimental ultimate loads (P_{ue}) of SFRC and SFRSCC wall panels were recorded and presented in Tables 3 and 4 respectively.

Fig. 1. Test set-up

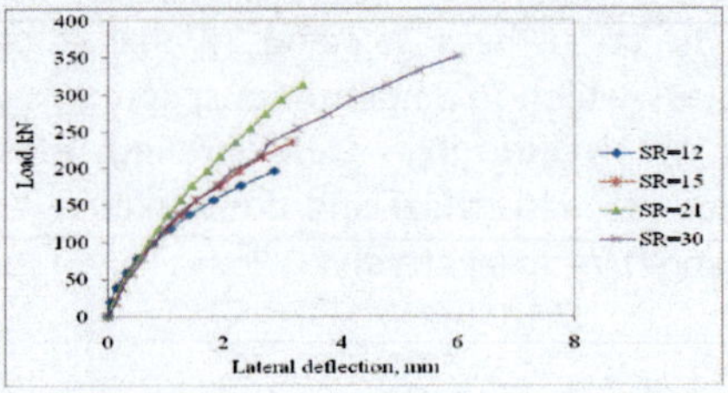

Fig. 2. Effect of slenderness ratio in SFRSCC wall panels

Table 3. Experimental ultimate loads (Pue) of SFRC wall panels

Panel designation	P_{ue} (kN)	$\frac{P_{ue}}{f_c' Lt}$	Panel designation	(P_{ue}) (kN)	$\frac{P_{ue}}{f_c' Lt}$
OWSFN-1	299.20	0.68	OWAFN-1	235.44	0.53
OWSFN -2	343.35	0.62	OWAFN -2	343.35	0.62
OWSFN -3	475.78	0.61	OWAFN -3	485.59	0.63
OWSFN -4	451.26	0.41	OWAFN -4	667.08	0.60

Table 4. Experimental ultimate loads of SFRSCC wall panels

Panel designation	P_{ue} (kN)	$\frac{P_{ue}}{f_c' Lt}$	$P_u / f_c' Lt$ using the equations of			
			Ref. 8	Ref. 9	Ref. 1	Ref. 6
OWSFS-1	309.01	0.69	0.50	0.54	0.56	0.47
OWSFS -2	362.97	0.65	0.45	0.52	0.51	0.43
OWSFS -3	490.50	0.62	0.31	0.47	0.37	0.31
OWSFS -4	475.78	0.42	a	0.36	0.08	0.07
OWAFS-1	250.15	0.56	0.45	0.52	0.49	0.43
OWAFS -2	362.97	0.65	0.45	0.52	0.51	0.43
OWAFS -3	510.12	0.65	0.45	0.52	0.53	0.43
OWAFS -4	711.22	0.63	0.45	0.52	0.54	0.43

Note: 'a' indicates that the equation of Ref.8 yields $P_{ue} = 0$, for walls with SR=30.

3 Results and Discussion

The effect of SR on the ultimate load of SFRSCC wall panels is shown in Fig. 2. It may be noted from the figure that the curves are linear up to the formation of the first crack and beyond which they exhibit non-linearity. In general, as SR increases the load carrying capacity decreases non-linearly and the lateral deflection increases for all the specimens. However a significant increase of lateral deflection can be seen in the case of wall panels with SR equals to 30 when compared to the wall panels with SR less than 30. The continuously increasing values of deflection, as the loading increases, indicate that the wall panels exhibit a smooth ductile type of failure. The load versus lateral deflection plots for different values of AR are presented in Fig. 3. From the figure it may be noted that the wall panel with AR=1.875 suffered maximum deflection than those panels with low values of AR. This type of softening of material may be due to the SCC and presence of steel fibers which absorb energy and arrest cracks. Presence of steel fibers in concrete induces higher degree of compressibility to the concrete specimens which exhibit large softening behavior [10].

In the case of SFRC wall panels, the curves are linear up to the formation of first crack, beyond which they exhibit non-linear characteristic as shown in Fig. 4. However comparing the load–deflection plots of SFRSCC and SFRC wall panels (Figs. 2 and 4), it may be noted that the initial slope of the load–deflection curve is steeper in the case of SFRC panels than those of SFRSCC panels. This indicates that SFRSCC wall panels exhibit softening behavior which means that the SFRSCC wall panels behave in a more ductile manner than the SFRC wall panels.

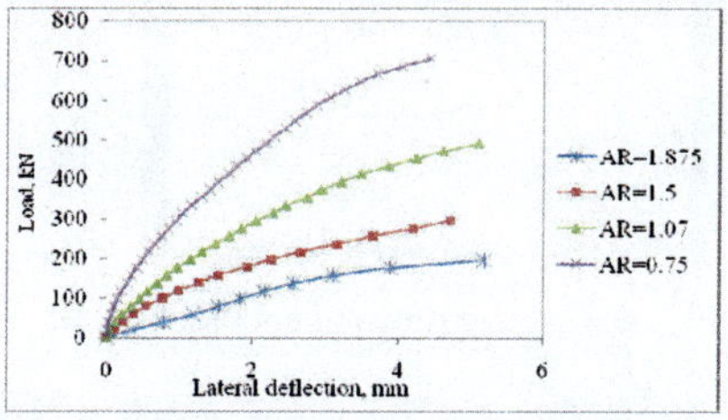

Fig. 3. Effect of aspect ratio in SFRSCC wall panels

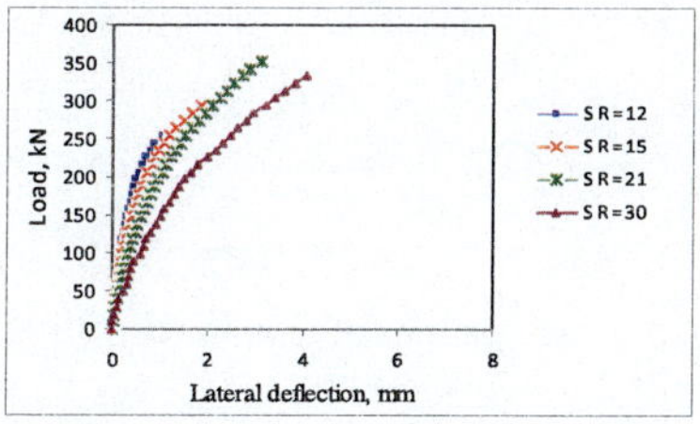

Fig. 4. Effect of slenderness ratio in SFRC wall panels

The experimental ultimate loads (P_{ue}) of the wall panels were normalized as the SR, AR and over all dimensions of wall panels are varying from panel to panel. These normalized values, obtained by dividing the experimental ultimate load (P_{ue}) by the parameter (f_c' Lt), are called as values of experimental axial strength ratio and are included in Tables 3 and 4. Referring to the tables it may be noted that the values of axial strength ratio decrease with increase of SR. In the case of SFRSCC wall panels, the decrease in the values of axial strength ratio is about 39.13% for the increase of SR from 12 to 30 and the reduction in strength is about 11.11% for the increase of AR from 0.75 to 1.875. The values of ultimate load taken by SFRSCC wall panels were higher than SFRC wall panels. Figs. 5 and 6 show the crack pattern of OWSFS-3 and OWSFS-4 respectively.

4 Comparison of Equations with Experimental Results

An attempt is made to compare the equations proposed by (i) Oberlender and Everard [7] (ii) Pillai and Parthasarathy [8] (iii) Saheb and Desayi [1] and (iv) the equation recommended by ACI 318 – 2008 [5] for predicting the ultimate load of RC wall panels, with the experimental results of SFRSCC wall panels of this study. The values of ultimate load (P_u) were computed using these equations and normalized by dividing the same by the parameter (f_c' Lt). These normalized values are called as axial strength ratios and are given in Table 4. This table is reproduced in Fig. 7 for more clarity and quick reference. From the Fig. 7, it may also be noted that the experimental values of SFRSCC wall panels are higher than those values predicted by the methods available in the literature. Hence a model is proposed for predicting the ultimate load of SFRSCC wall panels.

Fig. 5. Crack pattern of specimen with SR=21

Fig. 6. Crack pattern of specimen with SR=30

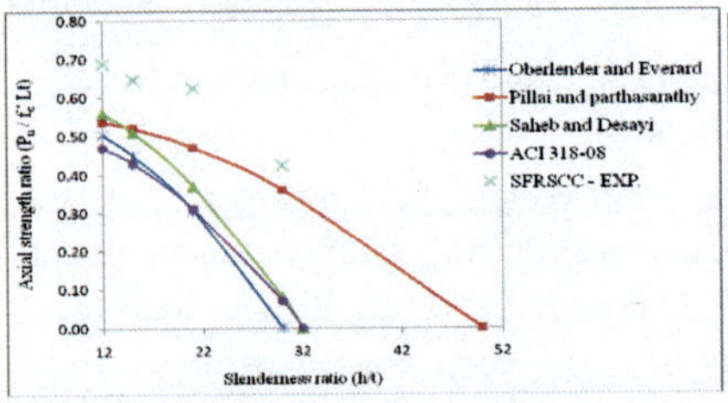

Fig. 7. Comparison of equations available in literature and experimental values

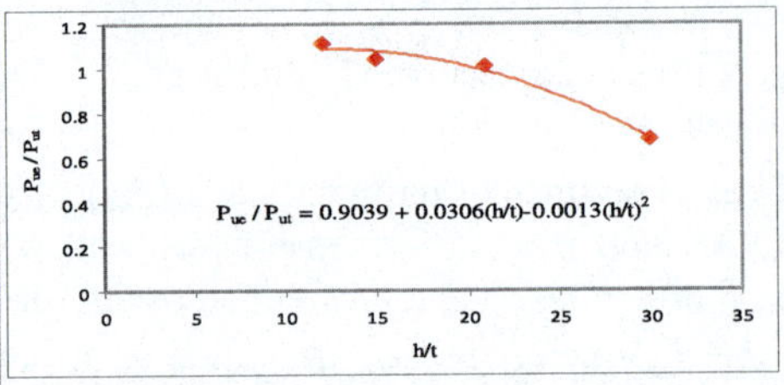

Fig. 8. Influence of SR on ultimate load

5 Prediction of Ultimate Strength of SFRSCC Wall Panels

The proposed method for predicting the ultimate load of SFRSCCC wall panels consists of two steps. The first step involves development of an equation for the ultimate load of wall panel treating it as a short wall. Subsequently this equation is modified to account for the effect of SR and AR on the ultimate load. The following assumptions are made while formulating the equation for predicting the ultimate load of SFRSCC wall panels.

(i) The load on the panel is reasonably concentric: i.e $e/t \leq 1/6$.
(ii) The panel contains at least nominal amounts of steel in vertical and horizontal directions as specified in ACI 318 – 2008 [5].
(iii) The value of SR is limited to 40.
(iv) AR is limited to a maximum value of 2.

(v) The volume fraction and aspect ratio of straight steel fibers are 0.50% and 60 respectively.

The largest eccentricity at which a load can be applied without tension of the wall is t/6. This load case can be approximated by a rectangular stress block extending from the compressed face of the wall for a distance of two-thirds of thickness of the wall [11]. The force per horizontal length of wall, L is

$$P_{ut} = 0.85 f_c' (2/3) Lt \quad (1)$$

$$P_{ut} = 0.56 f_c' Lt \quad (2)$$

The term $[(f_y - f_c') A_{sc}]$ is introduced in Eq. (2) to account for the effect of reinforcement and is modified as follows.

$$P_{ut} = 0.56 [f_c' Lt + (f_y - f_c') A_{sc}] \quad (3)$$

where, P_{ut} – Theoretical ultimate load of a short wall.

In order to introduce the effect of SR on the ultimate load of wall panels, the values of (P_{ue} / P_{ut}) are plotted against (h/t) in Fig. 8. The best fit equation obtained from the figure is

$$(P_{ue} / P_{ut}) = 0.9039 + 0.0306(h/t) - 0.0013(h/t)^2 \quad (4)$$

The above equation is rewritten in terms of (h/kt) as given by the ACI 318 - 2008 and other researchers [1, 8, 9], and is as follows:

$$(P_{ue} / P_{ut}) = [1 + (h / 29t) - (h / 26t)^2] / 1.106 \quad (5)$$

where P_{ue} – experimental ultimate load.

Similarly the effect of AR on the ultimate load of wall panels was obtained by plotting the values of (P_{ue} / P_{ut}) against the values of (h/L) as shown in Fig. 9. The best fit equation obtained from this plot is

$$(P_{ue} / P_{ut}) = 1.1313 - 0.0992(h/L) \quad (6)$$

The effect of AR can be introduced in terms of (h/kL), similar to the procedure followed by Saheb and Desayi [1], and the Eq. (6) is rearranged as,

$$(P_{ue} / P_{ut}) = [1 - (h/11L)] / 0.884 \quad (7)$$

Combining Eq. (3), Eq. (5) and Eq. (7), the following equation is proposed for predicting the ultimate load of SFRSCC wall panels

$$P_{ue} = P_{ut}[1 + (h / 29t) - (h / 26t)^2] [1 - (h/11L)] \quad (8)$$

$$P_{ue} = 0.56 [f_c' Lt + (f_y - f_c') A_{sc}] [1 + (h / 29t) - (h / 26t)^2] [1 - (h/11L)] \quad (9)$$

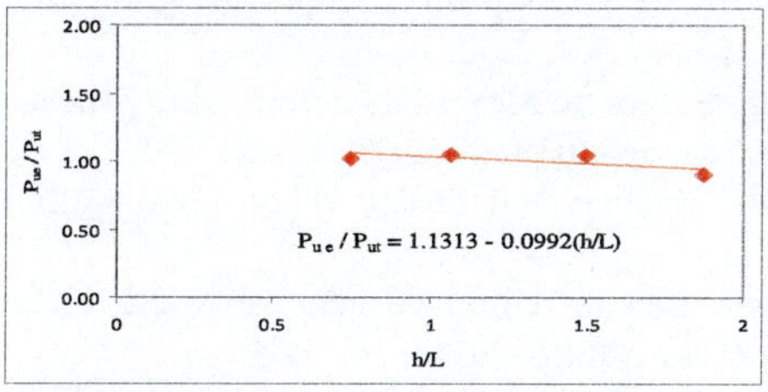

Fig. 9. Influence of AR on ultimate load

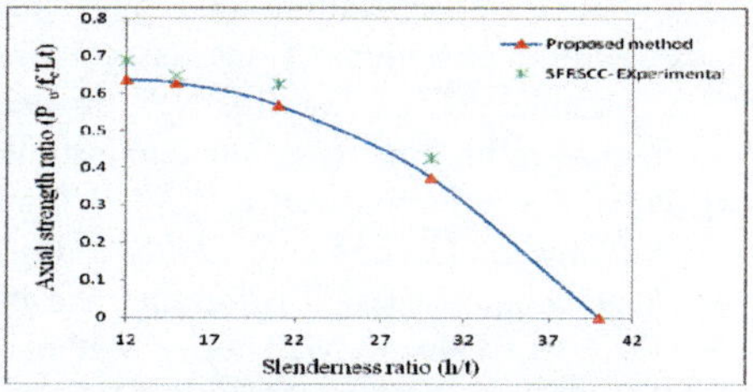

Fig. 10. Proposed method vs experimental

As the value of experimental ultimate load P_{ue} must be equal to the predicted ultimate load P_u, the term P_{ue} in Eq. (9) is replaced by P_u and is given as

$$P_u = 0.56\,[f_c^{'}\,Lt + (f_y - f_c^{'})\,A_{sc}]\,[1 + (h\,/\,29t) - (h\,/\,26t)^2]\,[1 - (h/11L)] \quad (10)$$

where P_u - computed ultimate load, $f_c^{'}$ - cylinder compressive strength of concrete, L - length of the wall panel, t - thickness of the wall panel, f_y - yield strength of steel, A_{sc} - area of compression reinforcement, h- height of the wall panel.

The values of axial strength ratio were calculated using the proposed method (Eq. (10)) and plotted against slenderness ratio (h/t) as shown in Fig. 10. From the figure it may be noted that the values of ultimate load obtained using Eq. (10) were found to compare satisfactorily with the experimental values.

6 Conclusions

1. The incorporation of steel fibers improves the cracking behavior and imparts ductility to the wall panels.
2. SFRSCC wall panels exhibit softening behavior due to the presence of steel fibers and SCC which transforms the material to behave in a ductile manner and induces higher degree of compressibility. Hence SFRSCC wall panels appear to be an ideal structural element in the case of seismic resistant structures.
3. A method is proposed to predict the ultimate load of SFRSCC wall panels and is found to compare satisfactorily with the experimental results.

References

[1] Saheb, S.M., Desayi, P.: Ultimate strength of RC wall panels in one-way in-plane action. Journal of structural Engineering, ASCE 115(10), 2617–2630 (1989)

[2] Domone, P.L.: Self-Compacting Concrete: An analysis of 11 years of case studies. Journal of Cement and Concrete Composites 28, 197–208 (2006)

[3] Ganesan, N., Indira, P.V., Santhoshkumar, P.T.: Ultimate Strength of Steel Fibre Reinforced Self Compacting Concrete Flexural Elements. The Indian Concrete Journal, 8–15 (December 2006)

[4] Grunewald, S., Walraven, J.C.: Parameter-study on the influence of steel fibers and coarse aggregate content on the fresh properties of self-compacting concrete. Cement and Concrete Research 31, 1793–1798 (2001)

[5] ACI–318, Building Code Requirements for Structural Concrete and Commentary (318R-08). American Concrete Institute, Formington Hills MI, ACI 318 (2008)

[6] EFNARC, Specifications and guidelines for self compacting concrete. European Federation of National Trade Associations, Surrey, UK (2002)

[7] Oberlender, G.D., Everard, N.J.: Investigation of reinforced concrete wall panels. Journal of the American Concrete Institute 74, 256–263 (1977)

[8] Pillai, S.U., Parthasarathy, C.V.: Ultimate strength and design of concrete walls. Building and Environment 12, 25–29 (1977)

[9] Fragomeni, S.: Design of normal and high strength reinforced concrete walls, PhD thesis, University of Melbourne, Melbourne, Australia (1995)

[10] Ganesan, N., Ramana Murthy, J.V.: Strength and Behavior of Confined steel Fibre Reinforced Concrete Columns. ACI Materials Journal 87(3), 221–227 (1990)

[11] MacGregor, J.G., Wight, J.K.: Reinforced Concrete–Mechanics and Design, 4th edn., pp. 934–937. Pearson Prentice Hall, London (2005)

Part 5
Impact, Cyclic and Seismic Loading

Drop-Weight Impact Response of Glass-Fiber Reinforced Ceramic Concrete

S.T. Tassew, R. Mutsuddy, V.S. Bindiganavile, and A.S. Lubell

Department of Civil and Environmental Engineering, University of Alberta, Canada

Abstract. This paper reports on the static and impact response of a lightweight concrete fabricated using a chemically-bonded phosphate ceramic binder and expanded clay aggregates. The concrete has a density of 1700 kg/m^3 and exhibits rapid strength gain. Chopped glass fibers having a length of 13 mm were included in the matrix at 0 to 2.0 % mass-fraction of the ceramic concrete. The tests were conducted on notched flexural specimens using configurations of quasi-static four-point bending and drop-weight impact three-point bending. Companion tests were performed to establish reference compressive behavior. The results show that both flexural and compressive strength increased significantly with an increase in the mass-fraction of fibers under quasi-static loading. Under impact, the flexural strength was uniformly higher compared to quasi-static loading, regardless of fiber content. However, as seen from the post-peak flexural toughness, the fiber efficiency was better under the quasi-static condition.

1 Introduction

Chemically bonded ceramics are inorganic solids consolidated by chemical reactions rather than the conventional heat treatment. Use of a magnesium-phosphate cement (MPC) binder in combination with aggregates produces the so-called chemically bonded ceramic concrete [1]. Compared to Portland cement (PC) binders, the MPC binders can be formulated to achieve rapid set times and high early age strength gain, while at the same time having good dimensional stability and long term durability [1]. Further, production of MPC leaves lower environmental impact than PC due to significantly reduced CO_2 emissions and energy usage [2]. The overall environmental impact of ceramic concrete is further lowered by the ability to use higher mass fraction of Fly Ash (FA) in the paste [1].

Prior research on ceramic concrete produced with MPC binder and lightweight expanded clay aggregate (LECA) has demonstrated good compressive strength, but noted that the post-cracking strength in bending may be of concern in some application [3]. Studies have shown that the use of discontinuous, short fibers randomly distributed within concrete imparts an increased flexural strength, post-cracking capacity, impact resistance and energy absorption capacity [4, 5]. The

G.J. Parra-Montesinos, H.W. Reinhardt, and A.E. Naaman (Eds.): HPFRCC 6, pp. 289–296.

selection of fiber material, its geometry and physical properties should be appropriate for the intended application and ensure chemical compatibility with the binder.

Since concrete made with MPC binders have a relatively neutral pH, it favours the use of glass fiber reinforcement [1]. This paper reports research on a ceramic concrete which combined MPC binder with LECA and chopped glass fibers. The primary objective was to investigate the influence of the fiber content on the flexural strength of ceramic concrete and examine the effect of loading rate. In addition, fiber efficiency was evaluated via post-peak energy absorption.

2 Experimental Program

2.1 Materials and Sample Preparation

Wagh [1, 6] developed a magnesium-phosphate cement (MPC) by reacting magnesium oxide (MgO) with monopotassium phosphate (KH_2PO_4) in an aqueous solution according to the acid-base reaction:

$$MgO + KH_2PO_4 + 5H_2O = MgKPO_4.6H_2O \tag{1}$$

In this study, calcined MgO was employed with manufacturer specifications of 97% MgO by weight; specific surface area of 0.3 m^2/g and a minimum of 95% of particles passing the No. 200 mesh size. Fertilizer-grade monopotassium phosphate (MKP) was obtained from a local farm supply warehouse. Class C Fly Ash (FA) was also used as the mechanical properties of this binder system are known to improve by its inclusion [6, 7]. Adapted from the molar ratios in Eqn. 1 and using 50 % FA loading of the total mass, the dry ingredients of the MPC binder were prepared at a mass ratio of MgO:MKP:FA = 1:3.4:4.4. The aggregates consisted of lightweight expanded clay (LECA) in two different gradations (Fig. 1): AR 0-2/880 (FL) and AR 4-10/550 (CL) having a maximum particle size of 2.5 and 10 mm, respectively. The specific gravity and water absorption were evaluated as 0.995 and 24.4% for the CL-LECA and 1.49 and 10.25% for the FL-LECA, respectively, according to ASTM C127 [8]. The LECA was immersed in water for a minimum of 24 hrs and then spread on to paper towels to obtain saturated surface dry conditions prior to use.

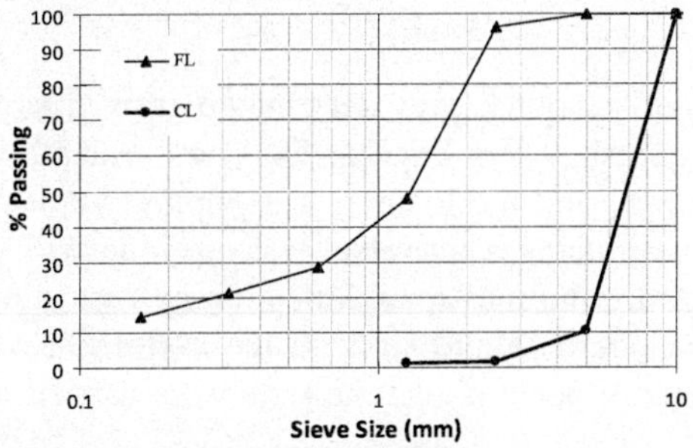

Fig. 1. Particle size distribution of LECA aggregates

The binder:FL:CL mass ratio was 1.0: 0.14: 0.40, wherein the binder refers to the combined mass of MgO+MKP+FA. The water to binder mass (w/b) ratio of all mixes was kept constant at 0.20. A commercially available lignosulphonate admixture was used at 1.5% by mass of binder to control the reaction rate and retard the setting time. Four different fiber contents were used in addition to a plain unreinforced mix so as to study the influence of fiber content on the mechanical properties under quasi-static load and impact loading. The mass fraction of fibers was varied in 0.5% increments between 0-2 % of the total mix. These fibers, made from Glass had a length of 13 mm.

The mixes were prepared by dry-mixing the MgO, MKP and FA for 5 minutes using a 20 *l* capacity portable mixer at 60 RPM. The pre-soaked aggregates were gradually added and mixed for 60 seconds. The water was combined with the lignosulphonate retarder and then added to the dry ingredients, with further mixing for 4 to 5 minutes. At this point, the fibers were introduced and mixed for 60 seconds. The resulting mixes were carefully placed into plastic cube and prism molds and consolidated using a vibrating table. The specimens were removed from the molds after 2 hours and stored in the ambient lab environment (i.e. room temperature of 23±2 °C and relative humidity of 50±5%) until testing.

2.2 *Test Setup*

2.2.1 Compression Tests

The 50×50×50 mm cubes were loaded in compression using a stiff testing machine with a capacity of 3100 kN. The loading rate corresponded to a compressive stress rate of 0.25 MPa/s as per ASTM C39 [9]. The tests were completed at ages of 10 days after casting. Note that it was found that the specimens can gain more than 90 % of their 28 day strength at 7 day.

2.2.2 Flexure Tests

The quasi-static flexural response was determined from 50×50×200 mm prisms under 4-point bending as shown in Fig. 2a. The span-to-depth ratio was 3 according to ASTM C1609 [10] and the loading was applied via a cross-arm displacement rate of 0.1 mm/min using a universal test machine with 30 kN capacity. A yoke was used to measure the mid-span deflection by means of two LVDTs. A 5 mm deep notch was saw-cut at mid-span on the bottom face of each specimen. Three replicates were used per mix and the tests were carried out at the age of 10 days.

The dynamic flexural tests were conducted using a drop-weight impact machine using a three-point bending configuration as shown in Fig. 2b. The system has a 62 kg hammer which can be raised to a height of 2.5 m to generate a maximum energy input at impact of 1 kJ. The specimens were identical to those tested in quasi-static loading and again, three replicates were examined for each data point. Each mix was examined under impact from drop heights of 250 and 500 mm. While the applied load was recorded via a load cell built into the striker, the deflection was measured via a stereoscopic high speed imaging system. Using an infrared signal, the data from the load cell and high speed cameras were continuously captured at the same time stamp.

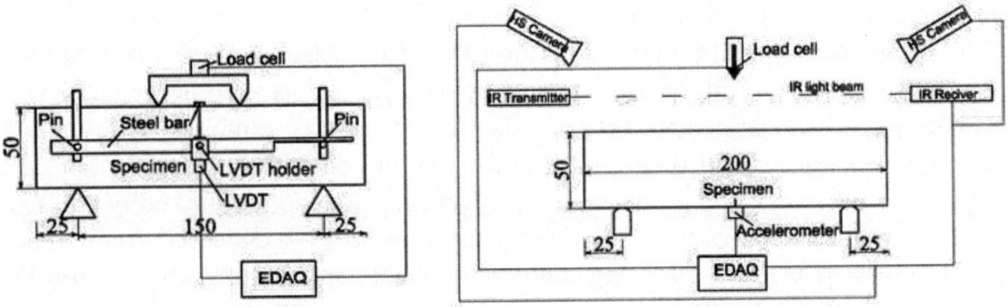

(a) Quasi-static test (b) Impact test

Fig. 2. Schematic sketch of test setup

3 Results and Discussion

3.1 Compression Tests

The relationship between the fiber content and the ceramic concrete density is shown in Fig. 3a. The average density ranged between 1755 kg/m^3 and 1790 kg/m^3 for fiber mass fraction between 0 % and 2 %. As expected, the fiber content had minimal influence on the density. As shown in Fig. 3b, the compressive strength increased with an increase in the fiber content with an improvement by 27 % between fiber mass fraction of 0 to 2 %.

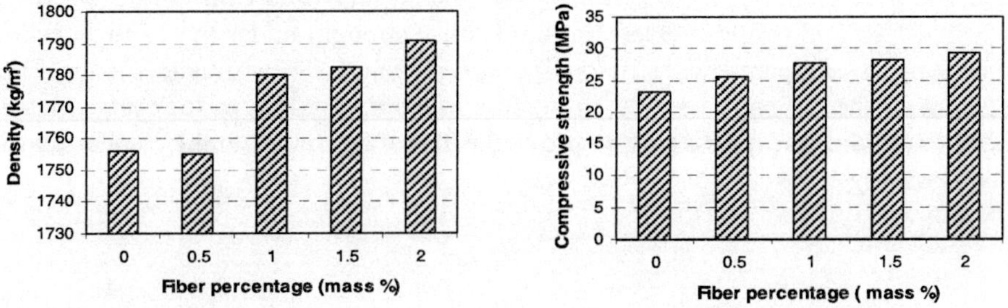

(a) Density (b) Compressive strength

Fig. 3. Influence of fiber content on properties

3.2 *Flexure Tests*

The average load-deflection response under quasi-static four-point bending is shown in Fig. 4a up to a deflection corresponding to span/150 mm. The addition of fibers had negligible influence on the initial stiffness, but resulted in improved maximum load capacity in relation to the fiber dosage. The deflections at peak load were in the range of 0.12 to 0.23 mm, corresponding to span/1250 to span/650. Fig. 4b shows the influence of fiber dosage on the flexural strength. It was observed that there was a linear increase in the flexural strength with the fiber mass-fraction, and increased 175 % when the fiber content increased from 0 to 2 %. The flexural strength increased from 9 % to 13 % of the compression strength when the fiber content was raised from 0 to 2 %.

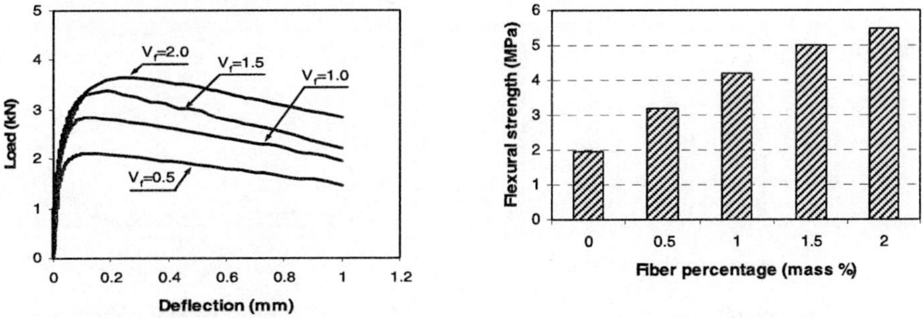

(a) Load-deflection (b) Flexural strength

Fig. 4. Influence of fiber content on the quasi-static flexural response of ceramic concrete

The load-deflection curves under impact loading are presented in Fig. 5 for drop heights of 250 and 500 mm. Note that the bending load of each specimen found adopting a procedure from Banthia et al. [11]. It was observed that the fiber mass-fraction had no discernible effect on the peak load obtained under impact. However, the deflection at peak load increased as the fiber content increased. It increased from 0.02 to 0.19 mm and from 0.01 to 0.06 mm for drop heights of 250 and 500 mm, respectively, when the fiber content increased from 0 to 2 %. There was a 4-fold and 6-fold increase in the flexural strength under the two respective drop-heights compared with the quasi-static response. The influence of the fiber content on the dynamic flexural strength is illustrated in Fig. 6. It appears that fiber dosage did not necessarily improve the flexural strength unlike under quasi-static loading.

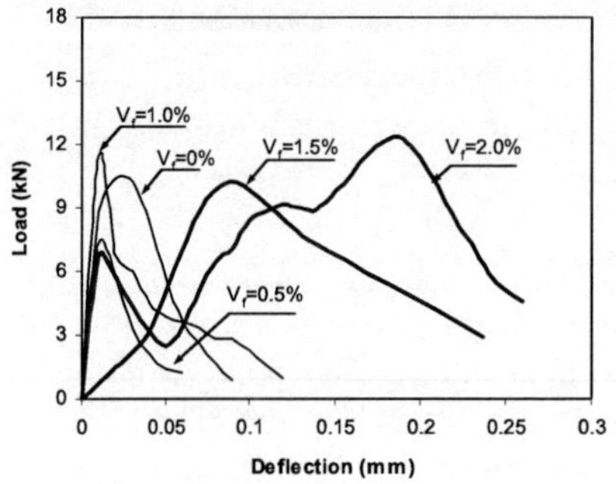

(a) Drop height of 250 mm

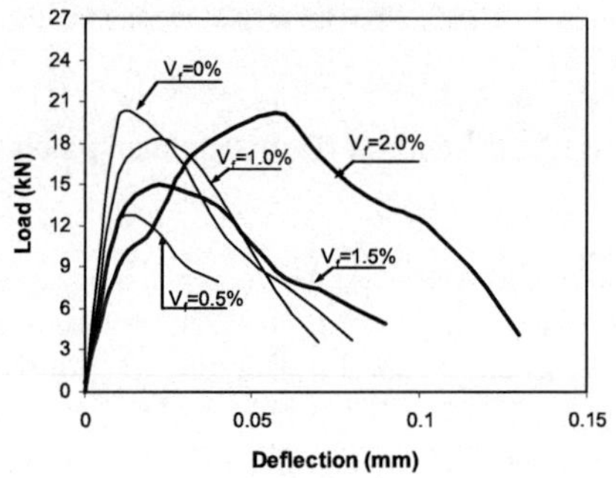

(b) Drop height of 500 mm

Fig. 5. Flexural response of ceramic concrete under impact

On the basis of the initial slope of load-deflection curves the bending stiffness values were obtained and are presented in Fig. 7. It can be seen that the impact loading resulted in an increase in the bending stiffness compared to quasi-static loading, regardless of the fiber content.

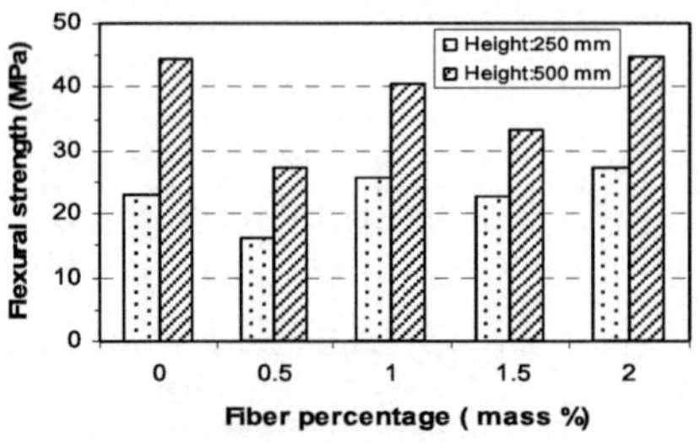

Fig. 6. Effect of fiber content on Modulus of Rupture for impact load

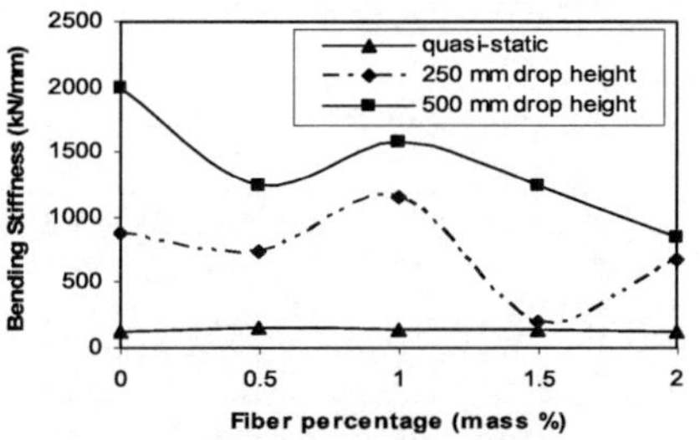

Fig. 7. Stiffness of ceramic concrete

The flexural toughness was calculated using the area under the load-deflection curve with the results summarized in Fig. 8 for different fiber contents. Fig. 8(a) shows that the pre-peak toughness was nearly constant for fiber dosage less than

1% but increased beyond 1 % fibers, irrespective of the loading rate. The fibers had negligible contribution to the pre-peak toughness at low dosage. The maximum pre-peak toughness was obtained for 2 % fibers and a drop height of 250 mm.

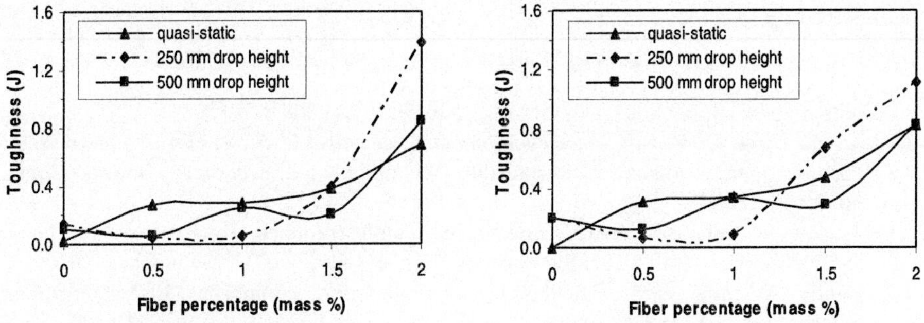

(a) Pre-peak toughness (b) Post-peak response

Fig. 8. Flexural toughness of ceramic concrete

Smaller maximum deflection values were obtained for impact tests (Fig. 5) compared to quasi-static tests (Fig. 4a), indicating embrittlement under higher loading rates. Besides, different failure modes were observed on the two tests samples. For quasi-static loading, the samples failed by gradual fiber pull-out whereas the fiber pullout mechanism was sudden in the case of drop-weight tests. The post-peak toughness was calculated between the peak-load deflection (δ) and 2δ and is summarized in Fig. 8b. The post-peak flexural toughness increased with an increase in the fiber content, irrespective of the loading rate as shown in Fig 9 (b). Again, the fiber efficiency was seen to be better for fiber dosage beyond 1 %. Similar trends between the toughness, fiber content and drop-height were observed between pre-peak and post-peak toughness-response.

4 Concluding Remarks

(1) The compressive strength of the ceramic concrete increased with an increase in the fiber content.

(2) Under quasi-static loading, the flexural strength increased in proportion to the mass fraction of glass fibers. Although it was higher under impact, there was no perceptible influence of fiber dosage. Closer examination showed that the fiber failure mode was predominantly gradual pull-out under quasi-static flexure unlike under impact suggesting an embrittlement at higher loading rates.

(3) As evidenced from the flexural toughness, the fiber efficiency was higher with an increase in the fiber content, especially beyond 1% mass fraction. This was true regardless of the loading rate.

Acknowledgement. The authors greatly acknowledge the financial support from C-Bond Technology and the Natural Sciences and Engineering Research Council of Canada. Donations of material from Nippon Electric Glass America and Argex (Belgium) are also acknowledged.

References

[1] Wagh, A.S.: Chemically bonded phosphate ceramics: Twenty-first century material with diverse applications. Elsevier Ltd., Amsterdam (2004)

[2] Li, Z., Ding, Z., Zhang, Y.: Development of sustainable cementitious materials. In: Proc. of Inter. Workshop on Sustainable Dev. and Conc. Technology, Beijing, China, pp. 55–76 (2004)

[3] Tassew, S., Lubell, A.: Textile reinforced ceramic composite for structural infill slab application. In: 34th IABSE Symposium, Venice, Italy, A-0317 (September 2010)

[4] Bentur, A., Mindess, S.: Fibre reinforced cementitious composites. Elsevier Applied Science, London (1990)

[5] Banthia, N., Sheng, J.: Fracture toughness of micro-fiber reinforced cement composites. Cem. and Concr. Compos. 18(4), 251–269 (1996)

[6] Wagh, A.S., Jeong, S.Y., Singh, D.: High-strength ceramic (cement) using industrial by-product Ashes. In: Proc. of Int. Conf. on High-Strength Concrete, pp. 542–553 (1997)

[7] Ding, Z., Li, Z.: High-early-strength magnesium phosphate cement with fly ash. ACI Material J. 102, 357–381 (2005)

[8] ASTM C127, Standard test method for density, relative density (specific gravity), and absorption of coarse aggregate. ASTM International, West Conshohocken, PA (2007)

[9] ASTM C39/C39M–09a, Standard Test Method for Compressive Strength of Cylindrical Concrete Specimens. ASTM International, W C, PA (2009)

[10] ASTM C1609/C1609M, Standard Test Method for Flexural Performance of Fiber-Reinforced Concrete (Using Beam With Third-Point Loading). ASTM Inter., W C, PA (2007)

[11] Banthia, N., Mindess, S., Bentur, A., Pigeon, M.: Impact testing of concrete using a drop-weight impact machine. Experimental Mechanics 29, 63–69 (1989)

Mechanical Behavior of SHCC under Impact Loading

V. Mechtcherine[1], O. Millon[2], M. Butler[1], and K. Thoma[2]

[1] Institute for Construction Materials, TU Dresden, Germany
[2] Fraunhofer Institute for High-Speed Dynamics, Ernst-Mach-Institut, Freiburg, Germany

Abstract. This paper presents the results of short-time dynamic measurements on a strain-hardening cement-based composite (SHCC) at strain rates 140 to $180s^{-1}$. The dynamic experiments were performed by means of a Hopkinson bar. Uniaxial, quasi-static tensile tests served as reference. The mechanical performances under dynamic and quasi-static loadings are compared and also related to the observed crack patterns and the condition of fracture surfaces.

1 Introduction

In SHCC the generation of numerous cracks and the gradual pullout of crack-bridging fibers allow high energy absorption under both quasi-static and dynamic tensile loads [1, 3]. The mechanical performance of SHCC has been demonstrated numerous times in the case of quasi-static loading (e.g. [1, 2]). Much less information is available on the load-bearing behavior of SHCC at high strain rates. In [3] a pronounced increase in first-crack stress and tensile strength of SHCC made with PVA fibers is reported with increasing strain rates from 10^{-5} to 10^{-1} s^{-1}. At the same time the strain capacity decreased, but multiple cracking of the SHCC could still be observed. It was shown that the strain rate sensitivity was directly influenced by the morphology of the interface between fiber and cementitious matrix. A less ductile SHCC was investigated at strain rates between $2{\cdot}10^{-5}$ to $2{\cdot}10^{-1}$ s^{-1} (cf. [4]). However, multiple cracking was present only at quasi-static loading. A minor influence of interface morphology on the strain rate sensitivity (strain rates ranging from $2{\cdot}10^{-6}$ to $2{\cdot}10^{-1}$ s^{-1}) of SHCC was discussed in [5]. Pullout tests on single PVA fibers and varied the pullout speed between 10^{-2} and 10^{2} mm/min were performed in [6]. An increasing pullout resistance and probability of fiber failure were stated already at these relatively low rates with rising pullout velocity.

G.J. Parra-Montesinos, H.W. Reinhardt, and A.E. Naaman (Eds.): HPFRCC 6, pp. 297–304.

2 Material Composition, Specimen Geometry and Production

The fiber volume content of the investigated SHCC was 2.2% by volume. The diameter of the fiber was 40 µm and the cut length 12 mm. The maximum particle size of the aggregates was 0.2 mm. The binder was composed of cement CEM I 42.5 R and fly ash (cf. Table 1). On the one hand the high portion of fly ash assured good workability of the fresh SHCC. On the other hand the fly ash particles acted as crack initiators in the hardened matrix. The spherical SAP particles (SAP – Superabsorbing Polymers) triggered micro-crack initiation in a similar way.

Table 1. Matrix composition of SHCC [kg/m³]

CEM I 42.5 R-HS	Fly ash	Water	Sand	Stabiliser	SAP	Superplasticizer	PVA fiber
505	614	334	534	3.2	2.0	9.5	29.3

Cylindrical samples (diameter 75 mm, length 250 mm) were used to determine the mechanical performance of SHCC in both the dynamic and quasi-static tests. A part of samples were notched around their circumferences. The width of the notch was 2 mm, its depth 18 mm. The determination of the Young's modulus and the tensile strength was performed using unnotched cylinders, whereas the measurement of the fracture energy was performed on notched samples. The tests were accomplished at a sample age of 28 days. In order to guarantee random fiber orientation in the entire sample volume, cylindrical specimens were extracted from the large SHCC blocks by core drilling.

3 Test Setup of Quasi-static Tensile Test

In the tensile tests, the upper and lower sample holders were fixed as not to rotate. The free length between the sample holders was 170 mm. The deformation of unnotched cylinders was measured between the sample holders by means of two LVDTs. The tests were performed deformation-controlled with a strain rate of 0.001 s^{-1}. At notched cylinders the deformation was measured by spanning the notch with a gauge length of 50 mm. These tests were executed displacement-controlled with a rate of 0.01 mm/s. More details may be found in [7].

From the data recorded in the tensile tests on unnotched cylinders, Young's modulus, the tensile strength, and the total fracture energy were derived. The specific fracture energy (the total fracture energy related to the total surface of all cracks) could not be computed because of the non-quantified crack state inside the specimen's volume. From notched cylinders data the tensile strength and specific fracture energy were evaluated. The specific fracture energy was calculated by relating the total fracture energy to the net area of the notched cross-section of specimen.

4 Test Setup of Hopkinson Bar Experiments

In Hopkinson bar tests, impact loading of material can be investigated. In this research program the spallation configuration of Hopkinson bar was employed where the specimen is fixed on the end of the Hopkinson bar's incident bar (Fig. 1). The load is generated by the shot of the striker bar on the incident bar. Due to the superimposition of the reflected compression wave onto the outgoing decompression wave, high tensile stresses occur in the specimen, leading to tensile failure when the material's tensile strength is reached. Material failure is visible in a development of one or more spallation planes. There, the specimen is fragmented into two or more pieces (spallation) [9].

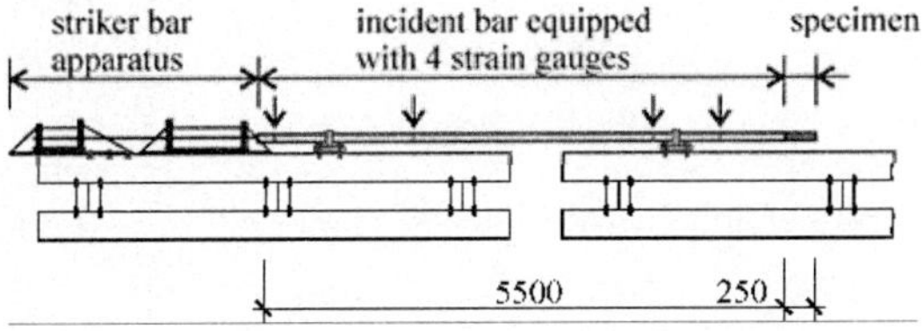

Fig. 1. Sketch of Hopkinson bar in spallation configuration

With the help of spallation experiments, dynamic Young's modulus, dynamic tensile strength, and dynamic fracture energy can be measured. Depending on the properties investigated, notched and unnotched specimens are employed. The specimens are instrumented with strain and acceleration gauges as well as a high-speed extensometer to record the displacement-time histories of the respective fragments. Dynamic Young's modulus and dynamic tensile strength were determined on unnotched and notched specimens based on longitudinal wave propagation velocity and the pullback velocity measurement. Dynamic fracture energy was analyzed only on notched specimens based on extensometer results. A more detailed description of data calculation methods is given in [7, 9].

5 Experimental Results of Quasi-static Test

In addition to tensile tests mechanical properties under compressive loading were evaluated. The compressive strength was 60.7 MPa and Young's modulus was 20.6 GPa. The bulk density of SHCC was 1910 kg/m³. Fig. 2 shows the stress-strain curves obtained from tension tests on unnotched specimens. First cracking occurred at a stress level between 3 and 4 MPa. Pronounced multiple crack formation could be observed on specimen surfaces with increasing strain, the crack pattern is shown in Fig. 4. The tensile strength was 3.8 MPa. The strain capacity scattered between 1 % and 3 %. The total fracture energy yielded large scattering about the mean value of 61 Nm (Table 2).

Fig. 3 shows the stress-deformation curves obtained from tensile tests on notched specimen. The crack initiation takes place only in the notched area at stress levels between 4 and 5 MPa. After crack initiation the tensile stress increases with increasing crack opening. The tensile strength of 5.53 MPa was reached at the deformation of approx. 0.6 mm. The mean value of the total fracture energy was 6.7 Nm, while the specific fracture energy was 5661 N/m (Table 2).

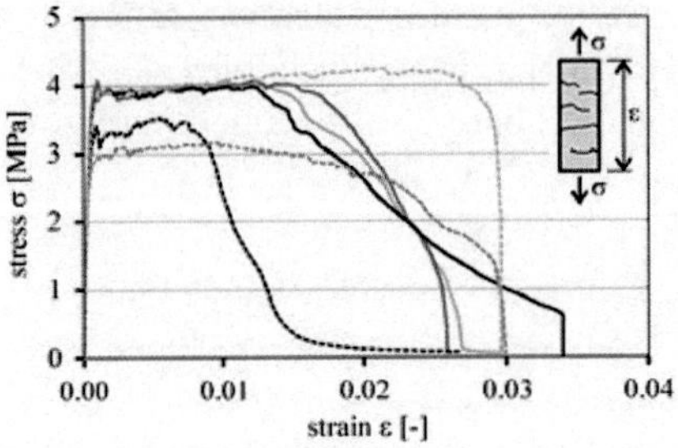

Fig. 2. Stress-strain curves of unnotched cylinders

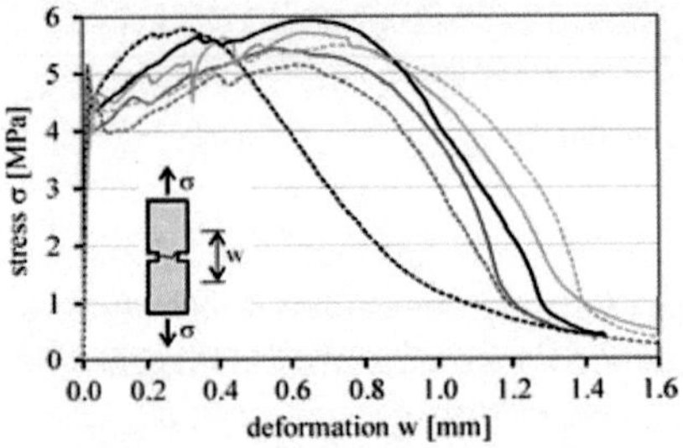

Fig. 3. Stress-deformation curves of notched cylinders

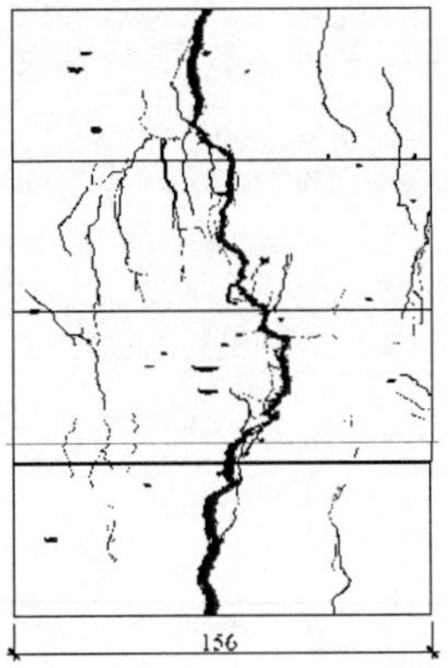

Fig. 4. Wind-off of crack pattern of an unnotched SHCC cylinder after exceeding the failure strain

Table 2. Results of quasi-static tests on SHCC cylinders without and with notch (mean values, standard deviations are given in parentheses)

	Bulk density	Compressive strength	Young's Modulus	Tensile Strength	Total fracture Energy	Specific fracture energy
	[kg/m³]	[MPa]	[GPa]	[MPa]	[Nm]	[N/m]
without notch	1897 (13)	60.7 (1.0)	20.60 (0.36)	3.80 (0.42)	61.0 (18.8)	-
with notch	1897 (13)	-	-	5.53 (0.29)	6.69 (1.14)	5561 (943)

6 Experimental Results of Hopkinson Bar Tests

13 specimens were tested with the Hopkinson bar. For unnotched specimens dynamic Young's modulus of 24.1 GPa and a dynamic tensile strength of 25.7 MPa were determined (Table 3). On notched specimens the dynamic tensile strength

was not only higher (26.6 MPa), but also yields greater scattering. This scattering can be, at least partly attributed to a three-dimensional stress state in the vicinity of notch. The total fracture energy was determined on notched specimens only. Related to the fracture surface, the specific fracture energy reaches a value of 13000 N/m (Table 3). A large scattering was observed for this property as well.

Table 3. Results of Hopkinson bar experiments on SHCC cylinders with and without notch

	Bulk density	Young's modulus	Tensile strength	Crack opening velocity	Specific fracture energy
	[kg/m³]	[GPa]	[MPa]	[m/s]	[N/m]
without notch	1,889 (5)	24.1 (0.8)	25.7 (0.6)	-	-
with notch	1,893 (7)	21.5 (1.6)	26.6 (6.6)	9.03 (0.60)	13,316 (2191)

The unnotched specimens showed multiple cracking. The primary crack appears in the spallation plane mainly orthogonally orientated to the loading direction. Many secondary cracks were detected on the specimens' surfaces, mostly parallel to the loading direction (Fig. 5). The computer-tomographical analysis proofed a particular concentric crack distribution inside the material. The center of the concentric circular cracks was located along the longitudinal axis of the specimen (cf. Fig. 12 in [7]). With increasing distance from the spallation plane, the diameter of the circular cracks and their width became smaller.

In contrast to the unnotched samples the notched cylinders came to complete fragmentation in the experiment. These specimens showed similar multiple cracking, with the primary crack orthogonal to the loading direction and the secondary cracks both orthogonal and parallel to the loading direction as shown in Fig. 6. The main crack leading to the specimen failure appeared in the spallation plane defined by the notch. In comparison to the experiments on unnotched specimens, fewer cracks parallel to the loading direction were observed.

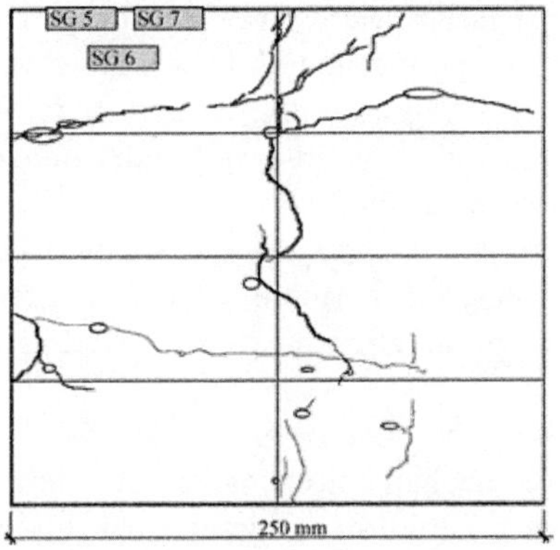

Fig. 5. Crack pattern and wind-off of crack pattern of an unnotched SHCC cylinder after a Hopkinson bar experiment

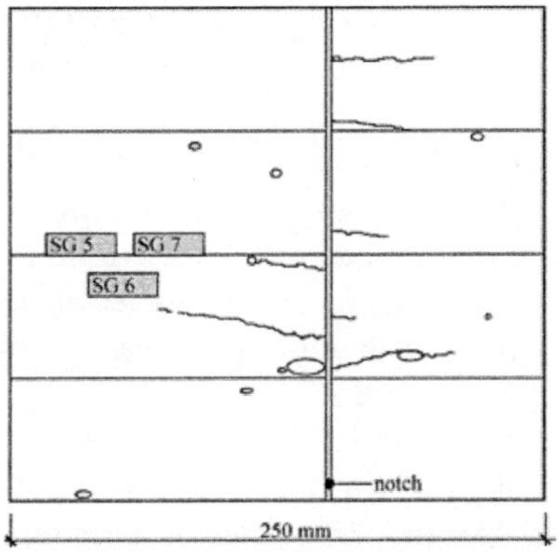

Fig. 6. Wind-off of the crack pattern of a notched SHCC cylinder after a Hopkinson bar experiment

7 Microstructural Analysis and Discussion

In the quasi-static and dynamic experiments, SHCC yielded multiple cracking and pronounced post-cracking load transfer. Multiple cracking was observed under both loading regimes; however, the crack pattern differed considerably. In the case of quasi-static loading a number of more or less parallel cracks developed, all oriented perpendicular to the loading direction. In the dynamic tests a lesser number of visible cracks formed, some of them parallel and some perpendicular to the direction of loading. The latter have been never observed in quasi-static tests.

The specific behavior of SHCC manifests itself not only in the crack pattern but also in the condition of fracture surfaces. After quasi-static experiments, the visible fiber pullout lengths were between 0.2 mm and 0.5 mm (Fig. 7a). The surface of pulled-out fibers was smooth and seemed to be the same as that of the original fiber. Many fibers failed during the crack opening (Fig. 7b).The fiber pullout lengths after dynamic loading reached values of up to 6 mm (Figs. 8a). Fiber failure was seldom observed, most of the fibers showed a clearly defined cut – a result of the production process of the short fibers (Fig. 8b). The rarely visible failed fibers showed a constricted and fibrous fracture surface. In contrast to the smoothly surfaced fibers observed after quasi-static testing, fibers tested under dynamic load were uneven and exhibited, in the axial direction, an irregular, wave-shaped texture indicating plastic deformations. Similar phenomena of fiber pullout were observed in uniaxial high-speed tensile tests on SHCC in [10].

The changes in the fiber surface show that the fibers are particularly subject to plastic deformations at high strain and high crack-opening velocities. It is assumed that in such load cases the bond between fiber and the matrix is damaged, thus favouring fiber pullout. The expected more brittle behavior of the fiber pullout by increasing strain rate, resulting from an increase in the bond strength of the fiber-matrix interaction [3, 6], as proven at low strain rates, seems to be more than compensated by this effect. With increasing strain rates a decrease in the number of cracks and an increase in the probability of fiber failure were observed in previous studies, accompanied by decreasing average fiber pullout length [3, 4]. However, this tendency seems to be valid for a quasi-static regime only. In this study, as well as in [10], a cardinal change in the failure mode was observed under impact loading: very pronounced fiber pullout occurred unexpectedly.

Moreover, it can be concluded that SHCC material design can and should to be adjusted depending on the expected loading rates. Furthermore, particular attention should be given to specimen geometry and preparation methods as well as to testing procedures.

Further investigations are necessary to explain the phenomena of material failure at high strain rates leading to the positive mechanical characteristics during impact loading. In particular questions of crack initiation, of crack growth, and also of the processes related to fiber pullout and failure must be answered.

Answering these questions is of particular importance, especially with regard to the purposeful material design of SHCC for particular applications.

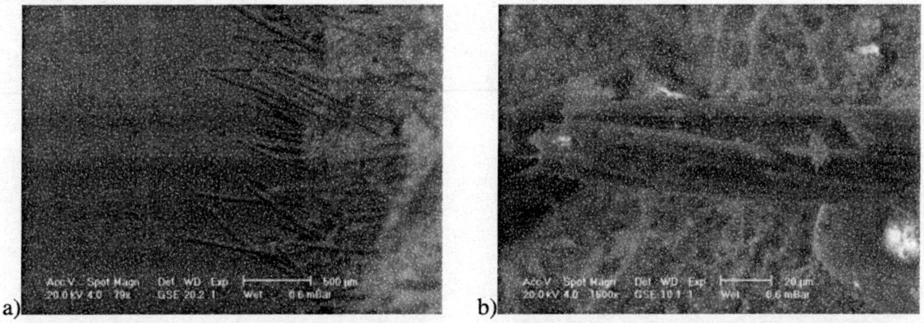

Fig. 7. Pullout lengths of fibers under quasi-static loading (a), typical shape of fiber fracture (b)

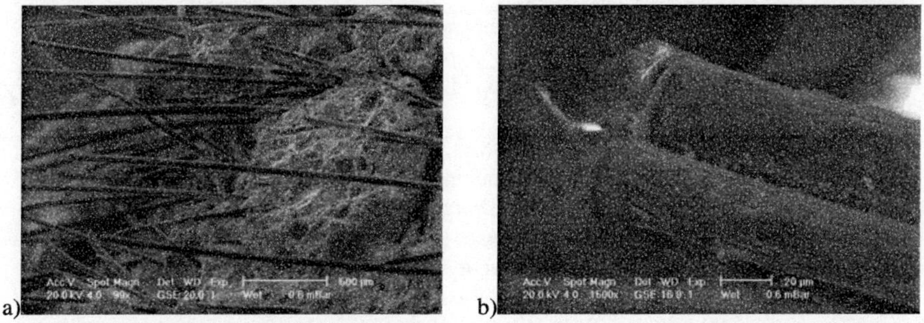

Fig. 8. Pullout lengths of fibers under impact loading (a), typical shape of pulled-out fibers (b)

8 Summary and Outlook

This paper presents the results of static and dynamic experimental investigations on strain-hardening cement-based composite (SHCC). Spallation experiments were performed using a Hopkinson bar device to investigate the dynamic properties of SHCC. Under static loading SHCC shows decidedly ductile behavior due to the formation of multiple cracks and the crack-bridging by PVA short fibers. The energy absorption capacity of SHCC is very high, especially when unnotched specimens are considered. Dynamic experiments with the Hopkinson bar show that the positive material characteristics, like the pronounced quasi-ductile behavior and high energy absorption capacity as observed under quasi-static loading remained and even improved with loading at high strain rates. The reasons for this increase in the performance at high strain rates are assumed to be related to the development of a great number of cracks parallel with and orthogonal to the loading direction and to the extensive plastic deformations of fibers prior to and during fiber pullout. The predominant fiber pullout in the dynamic tensile tests was unexpected since previous investigations showed an increase of fiber failure probability with increasing strain rate. Those results, however, were obtained for a quasi-static loading regime. Obviously the failure mechanisms of SHCC are altered when impact loading is applied. This phenomenon should be investigated in more details.

References

[1] Li, V.C.: On Engineered Cementitious Composites (ECC): A Review of the material and its applications. J. Advanced Concrete Technology 1(3), 215–230 (2003)
[2] Mechtcherine, V., Schulze, J.: Ultra-ductile concrete – material design concept and testing. CPI Concrete Plant International (5), 88–98 (2005)
[3] Yang, E., Li, V.C.: Rate dependence in Engineered Cementitious Composites. In: Int. RILEM Workshop HPFRCC in Structural Applications, pp. 83–92. RILEM Publications S.A.R.L., PRO 49 (2005)
[4] Douglas, K.S., Billington, S.L.: Rate dependence in high-performance fiber reinforced cementbased composites for seismic applications. In: Int. RILEM Workshop on HPFRCC in Structural Applications, pp. 17–26. RILEM Publications S.A.R.L., PRO 49 (2005)
[5] Maalej, M., Zhang, J., Quek, S.T., Lee, S.C.: High-velocity impact resistance of hybridfiber Engineered Cementitious Composites. In: Proc. 5th Int. Conf. on Fracture Mechanics of Concrete and Concrete Structures (FraMCoS-5), pp. 1051–1058 (2004)
[6] Boshoff, W.P., Mechtcherine, V., van Zijl, G.P.A.G.: Characterising the time-dependant behaviour on the single fibre level of SHCC: Part 2: The rate effects on fibre pull-out test. Cement and Concrete Research 39, 787–797 (2009)
[7] Mechtcherine, V., Millon, O., Butler, M., Thoma, K.: Mechanical behaviour of strain hardening cement-based composites under impact loading. Cement and Concrete Composites 33(1), 1–11 (2011), doi:10.1016/j.cemconcomp.2010.09.018
[8] Weerheijm, J.: Concrete under impact tensile loading and lateral compression. Delft University of Technology, Dissertation (1992)
[9] Schuler, H., Mayrhofer, C., Thoma, K.: Spall experiments for the measurement of the tensile strength and fracture energy of concrete at high strain rates. Int. Journal of Impact Engineering 32, 1635–1650 (2006)
[10] Mechtcherine, V., Silva, F.A., Butler, M., Zhu, D., Mobasher, B., Mäder, E.: Mechanical behaviour of strain-hardening cement-based composites (SHCC) subjected to tensile loading at low and high strain rates. Journal of Advanced Concrete Technology 9(1) (2011) (accepted for publication), `http://www.j-act.org/4-25.html`

Shock-Absorbing Blocks Made of HPFRCC for Better Girder-End Structures

K. Rokugo[1], H. Hatano[2], T. Nakashima[1], Y. Sakaguchi[3], M. Yamakami[1], and K. Kobayashi[1]

[1] Gifu University, Gifu, Japan
[2] Nakanihon Engineering Consultants, Nagoya, Aichi, Japan
[3] Maruei Concrete Industry, Hashima, Gifu, Japan

Abstract. The authors intend to propose shock-absorbing blocks to be placed at the girder-ends of bridges with the aim of reducing the joint gaps. These blocks are made to have an effect of alleviating possible colliding forces between main girders and the parapet walls of abutments during an earthquake, as well as an effect of restraining the displacement of superstructures. The post-earthquake restorability of these blocks is also considered. Static loading tests in horizontal and vertical directions were conducted on these blocks to investigate their properties. These shock-absorbing blocks showed excellent deformability and were found to be capable of reducing the girder-end joint gaps.

1 Introduction

Numerous viaducts are currently in service in Japan including those along expressways. Quite a number of such viaducts are located near residential areas. The noise generated by viaducts in many cases provokes demands for being decreased from neighboring communities. A larger joint gap is empirically known to make the expansion joint more prone to be damaged and cause louder impact noise.

Bridges having elastic rubber supports and seismic isolation rubber supports to improve the earthquake resistance have been predominating since the 1995 Hyogoken-Nambu Earthquake. While achieving high seismic resistance, these bridges have become prone to greater horizontal displacements during an earthquake. *Design Specifications of Highway Bridges* [1] require that a joint gap be provided to prevent collision in case of not only level 1 ground motion but also level 2 ground motion. A gap between the superstructure and abutment of a bridge sufficient for preventing collision requires a large expansion joint, such as steel finger joints made of extremely thick steel plate or modular-type joints, posing problems of comfort, noise affecting the environment, and maintenance.

It is therefore required to increase the durability of expansion joints at the ends of bridge girders, rationalize girder-end mechanisms, and enhance ease of maintenance.

G.J. Parra-Montesinos, H.W. Reinhardt, and A.E. Naaman (Eds.): HPFRCC 6, pp. 305–312.

A knock-off mechanism, whereby a part of the abutment slides into the backfill soil when the girder-end collides with the abutment during an earthquake, is effective in absorbing the displacement at the girder-end with only a small joint gap.

The goal of this study is to propose a rational girder-end structures that has a small expansion joint gap to reduce the joint size and, under a level 2 ground motion, absorbs the impact of collision between the superstructure components and between the superstructure and parapet by shock-absorbing blocks made of concrete. This paper reports on the basic structural characteristics and impact force-alleviating effects of shock-absorbing blocks. Note that shock-absorbing blocks dealt with in this study are of a structure whereby damage to the expansion joint does not adversely affect the structural members including the parapet and a secure clearance at girder-ends is ensured. These are characterized by horizontal deformability and vertical load-bearing capability and basically placed under expansion joints as shown in Fig. 1 in consideration of ease of placement and restoration after an earthquake.

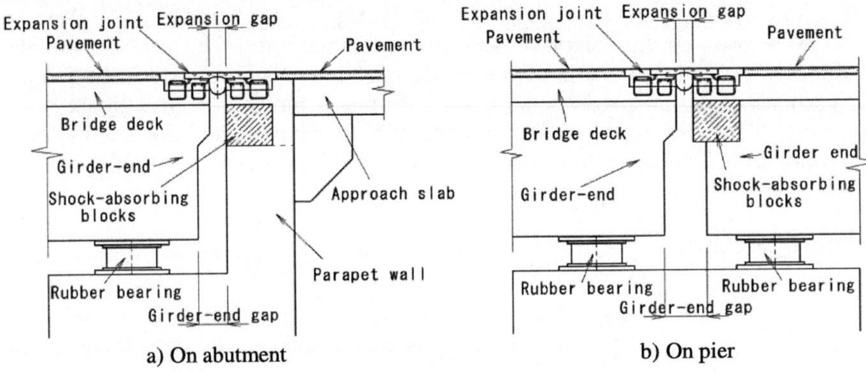

Fig. 1. Placement of shock-absorbing blocks

2 Shock-Absorbing Blocks and Test Procedures

2.1 Characteristics of Shock-Absorbing Blocks Made of HPFRCC

In past studies aiming to alleviate the effect of collision between parapet walls and girders at expansion joint gaps, the colliding force is mostly alleviated by providing buffers at such gaps, but their locations are structurally limited to small spaces, such as the front of parapet walls and shoe seats. The action of impact is generally expressed by forces, deformation, or energy. Shock-absorbing blocks proposed in this study, which are made of a high performance fiber-reinforced cementitious composite characterized by multiple fine cracks (hereafter referred to as a HPFRCC), are expected primarily to absorb displacement of the girders. Because of being made of a kind of concrete, these have wide applications and can be placed at a variety of locations, such as at girder-end gaps and the back of the abutment as shown in Fig. 1. These blocks are also made as precast products to be replaced after failure in case of collision between girder-ends and the parapet wall. The failure of these blocks itself is designed to be ductile, thanks to the bridging effect of fibers, which arrest multiple fine cracks characterizing HPFRCCs, to facilitate subsequent repair, in contrast to normal concrete or mortar (NM), which is prone to brittle failure.

2.2 *Test Specimens*

Fig. 2 shows an image of external forces acting on a shock-absorbing block permanently and during a major earthquake. Fig. 3 shows the shapes, sizes, and material compositions of shock-absorbing blocks used in the experiment. Specimens A and B were used to grasp their basic failure modes, whereas C to E were fabricated using expanded polystyrene (EPS), which serves as both a mold and space filler to impart resilience. Specimens F include one made of NM to increase the practical utility of such blocks. EPS is capable of recovering its original shape even after suffering large displacement. Table 1 gives the mixture proportions of concretes and the specifications of EPS used for the experiment, respectively.

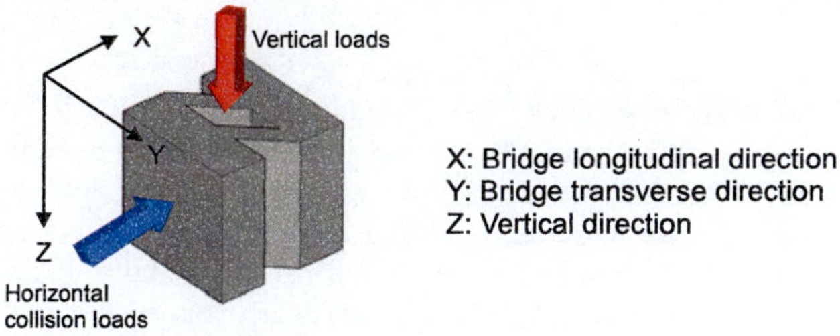

Fig. 2. Direction of external forces

Table 1. Mixture proportions of blocks

Material	Water (kg/m³)	Cement (kg/m³)	Fine aggregate (kg/m³)	Sp (kg/m³)	AD (kg/m³)	Viscosity enhancer Mc (kg/m³)	Fiber content (vol/%)	Fiber type	Com-pressive strength (MPa)	Young's modulus (GPa)	Tensile strength (MPa)	Ultimate tensile strain (%)
NM	273	496		–	1.24	–	–	–	40	50	–	–
HPFRCC - 1.5	342	1264	395	37.9	–	0.9	1.5	PE	88	22	7.9	1.68

Note: Sp=air-entraining and high-range water-reducing agent; AD=air-entraining and water-reducing agent
Fiber : High-strength Polyethylene Fiber (PE), Diameter 12μm, Length 12mm, Tensile strength 2.6GPa, Young's modulus 88GPa
EPS : Foamed polystyrene in a mold, D-30(Unit Weght 0.30±0.02kN/m^3), σ_y=180kN/m^2(Compressive Strain10%)
Reinforcing bar : SD295A

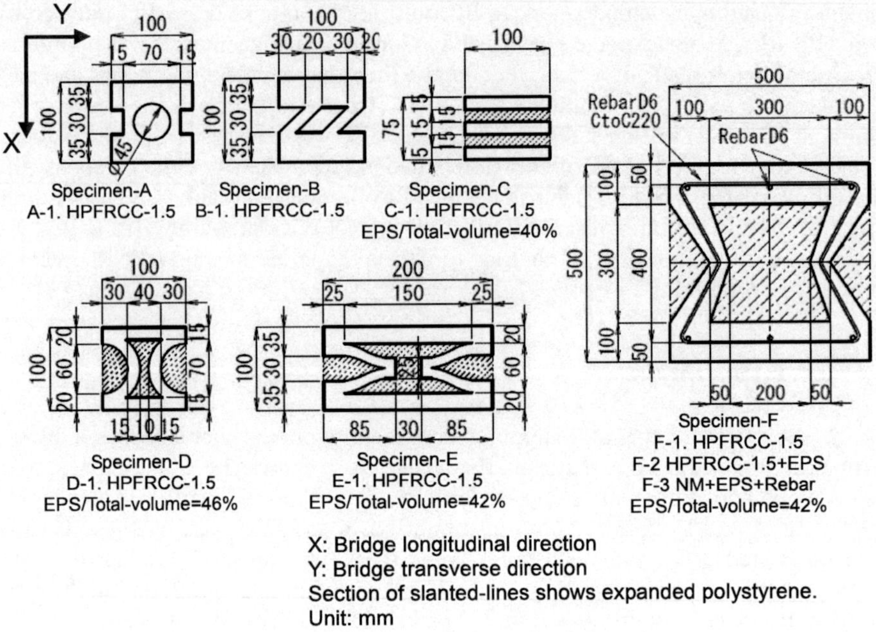

Fig. 3. Shapes, sizes and material compositions of specimens

2.3 *Test Procedures and Investigation Items*

Compression tests were conducted using a testing machine to evaluate the behavior of shock-absorbing blocks when the girder-ends collide with the parapet wall during a major earthquake and when they have to continue to bear the vertical loads after such an earthquake. In the tests, the horizontal load-displacement relationship of the blocks and their vertical load-bearing capacity after failure were to be examined. Specimens were therefore horizontally loaded to block failure, and then only specimens A to D were vertically loaded to measure the load and displacement using a load cell and sensitive displacement transducers, respectively.

3 Test Results and Discussion

The failure mode, conditions after failure, and recovery performance of each specimen were evaluated.

3.1 Failure Mode

In order to find an optimum shape of shock-absorbing blocks for alleviating the colliding forces of main girders, the authors focused on their failure mode and evaluated the post-failure conditions and behavior to failure of each specimen. Their failure mode was either compressive or flexural. The former was brittle, and the latter ductile. Photo 1 shows the state of failure of specimens demonstrating typical modes of failure. Their stress-strain relationships are shown in Fig. 4. Here, the stress is calculated by dividing the applied load by a block's whole cross-sectional area that consists of both mortar and EPS. Therefore, it is an "apparent" stress. In the stress-strain relationship of specimen A, which failed in compression, the post-peak stress is very low compared with the peak stress, and the strain tends to increases while keeping the low stress level. On the other hand, the stress-strain relationship of specimen E, which failed in flexure, is characterized by the post-peak stress that nearly retains the peak level while the strain increases. A shock-absorbing block is therefore required to have a shape that leads to flexural failure from the aspect of the absorption of shock that is forced on structural members.

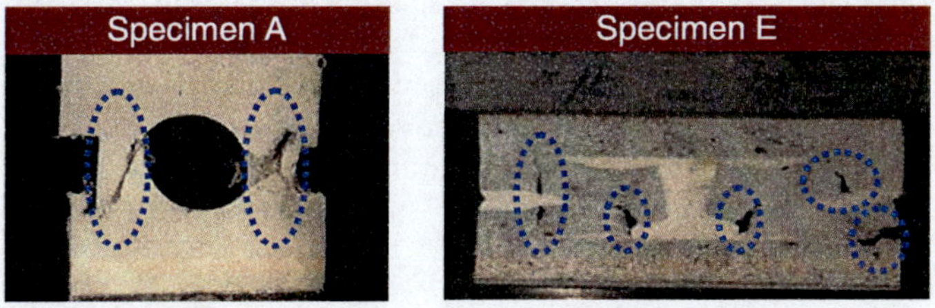

a) Compression failure b) Bending failure

Photo 1. Failure mode of the blocks

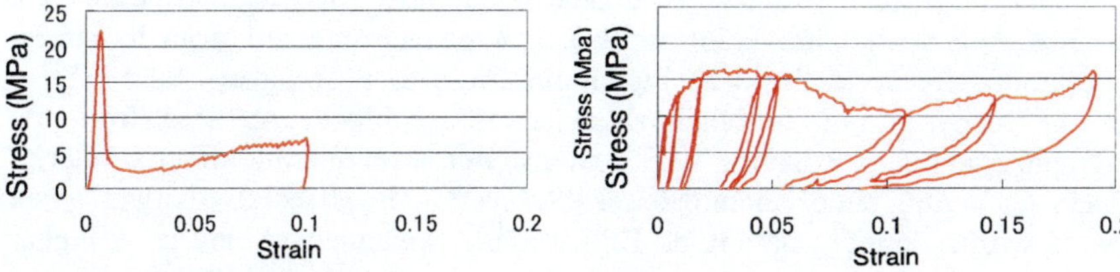

a) Compression failure (Specimen A) b) Bending failure (Specimen E)

Fig. 4. Stress-strain relationships (Specimens A and E)

3.2 Post-Failure Properties

Ease of restoration after an earthquake was assessed focusing on the failure condition of shock-absorbing blocks. Photo 2 shows the state of specimens made of HPFRCC and NM after loading testing.

a) No spalling even after large deformation b) Significant spalling after large deformation

Photo 2. Failure mode (Specimen F)

Whereas HPFRCC specimens led to no spalling even after large deformation thanks to the cross-linking effect of fibers, large deformation of NM specimens was associated with significant spalling, shedding many concrete lumps as seen from the Photo. The use of a HPFRCC for shock-absorbing blocks is evidently effective in ensuring ease of restoration after an earthquake.

3.3 Deformation Recoverability by EPS

The recoverability of specimens by EPS was assessed focusing on the state of shock-absorbing blocks during loading and after being unloaded.

Photo 3 and Fig. 5 show the state of recoverability of the displacement and the stress-strain relationships, respectively, of specimens having EPS. Their recovering performance was also evaluated by defining the residual strain of a shock-absorbing block after being unloaded, which is obtained from its stress-strain relationship, divided by the maximum strain as its recovery ratio. Fig. 6 shows the recovery ratio of each shock-absorbing block. As seen from the photograph, specimens E having EPS demonstrated a recovering effect similar to specimen C having simply laminated EPS layers. The effect is evident when compared with specimens having no EPS, such as specimens A and B. A higher maximum strain tended to lead to a higher recovery ratio of EPS. Accordingly, shock-absorbing blocks can be made to develop high recoverability, when unloaded even after being deformed largely, by using EPS as part of the formwork and leaving it for filling the space in each shock-absorbing block.

a-1) During loading (Specimen C)

a-2) After loading (Specimen C)

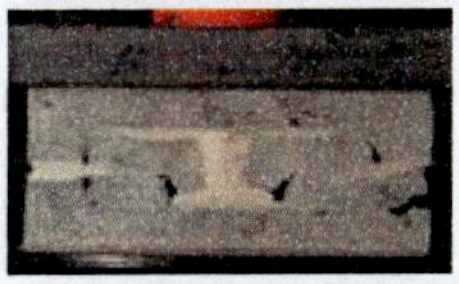

b-1) During loading (Specimen E)

b-2) After loading (Specimen E)

Photo 3. Recoverability by EPS

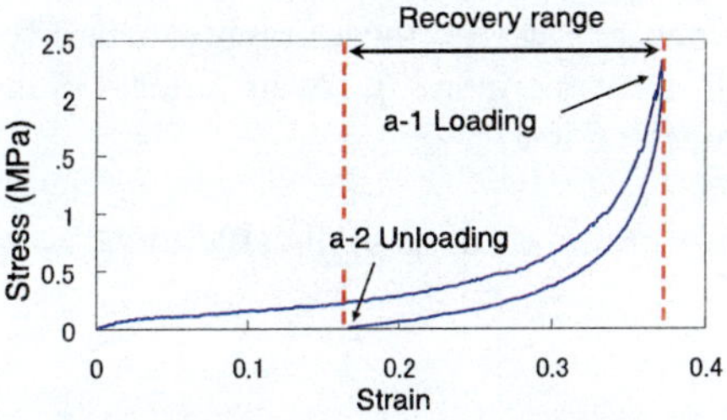

a) Recoverability by EPS (Specimen C)

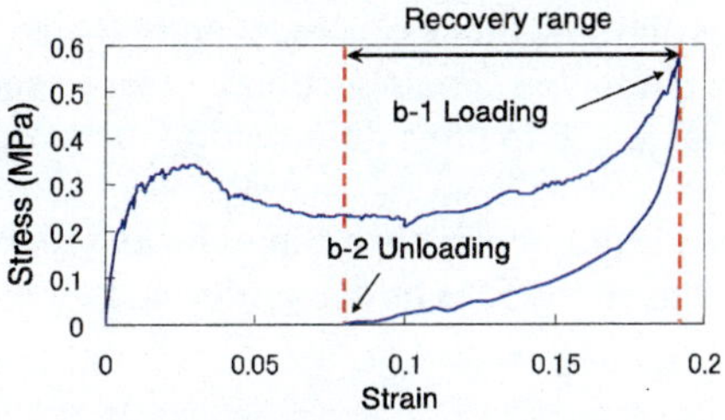

b) Recoverability by EPS (Specimen E)

Fig. 5. Stress-strain relationships (Specimens C and E)

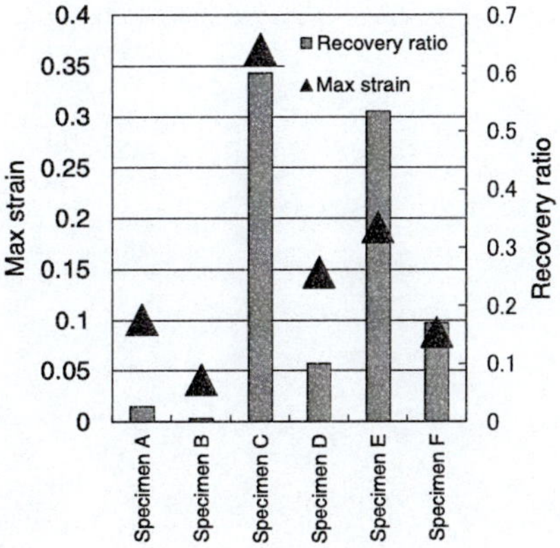

Fig. 6. Recoverability and max strains of blocks

4 Conclusions

With the aim of proposing shock-absorbing blocks made of a fiber-reinforced cementitious composite, their failure mode and load-displacement relationship were examined, using the materials, shape, and dimensions as test parameters. The following were found in this study:

1. The failure modes of the shock-absorbing blocks are classified into two types: compressive and flexural. In regard to specimens leading to failure in compression, the displacement in the load-displacement relationship increases while the post-peak load keeps an extremely low value compared with the peak load. In regard to specimens leading to failure in flexure, the displacement increases while the load keeps a high level after reaching the peak.
2. By using a fiber-reinforced material, shock-absorbing blocks can prevent spalling even after suffering large deformation, demonstrating fine emergency serviceability and restorability after an earthquake.
3. When expanded polystyrene is used as formwork and a space filler for shock-absorbing blocks, a recoverability is around 50% for those having a highly deformable cross-sectional shape such as specimen E, even after being subjected to large deformation in a compression test.

The idea of shock-absorbing blocks proposed in this study is applicable to various uses in addition to bridge girder-ends.

References

[1] Japan Road Association: Design Specifications of Highway Bridges, Part V, Seismic Design (2002)

[2] Japan Road Association: The Documents on Seismic Design of Highway Bridges, pp.1–139 (1997) (in Japanese)

Post-Peak Cyclic Behavior of Steel Fiber Reinforced Concrete under Bending

F. Germano and G.A. Plizzari

Department of Civil, Architectural, Environmental and Land Planning Engineering, University of Brescia, Italy

Abstract. The paper deals with an experimental study on the fatigue behavior of plain and Steel Fiber Reinforced Concrete (SFRC). The research was carried out by studying the fatigue behavior of cracked concrete by means of three point bending tests. The effects of two different fiber contents and three different load levels were investigated and the fatigue performances of plain and SFRCs are compared. Despite the variable information regarding the fatigue behavior of concrete reported in the literature, this research study highlights that the inclusion of fibers can improve the fatigue performance of SFRC for high tensile stresses, guaranteeing a longer structural life.

1 Introduction

The utilization of steel fibers in civil engineering construction has considerably increased over the years due to the recognized ability of Steel Fiber Reinforced Concrete (SFRC) to provide toughness, guaranteeing to resist tensile stresses after cracking [1, 2]. Many concrete structures, such as highway and railroad bridges, airport pavements, marine structures, etc., are subjected to large number of cycles at lower stress level (high-cycle fatigue), whereas low-cycle loading involves the application of a few load cycles at high stress level typical of earthquakes.

Experimental results from fatigue tests show that the process of unloading and reloading in compression causes significant additional cracking or crack widening [3] which, in the case of high-cycle fatigue, mainly concerns the matrix-aggregate interface. For low-cycle fatigue, however, the additional cracking or crack widening occurs in the matrix itself [4, 5]. As far as concrete in tension is concerned, it has been demonstrated that damage mainly occurs in the microcracked zone present at the crack tip, known as Fracture Process Zone (FPZ) [6]. As a consequence, the behavior of concrete structural elements subjected to fatigue in tension or bending can be correctly assessed only if the presence of FPZ is taken into account [7].

For this reason, Plizzari and co-workers [8, 9] performed fatigue tests on cracked specimens and found that the envelope curves obtained from cyclic test

G.J. Parra-Montesinos, H.W. Reinhardt, and A.E. Naaman (Eds.): HPFRCC 6, pp. 313–320.

match well with the curves obtained from static ones; furthermore, the effectiveness of steel fibers in improving fatigue life turned out to be greater in HSC than in NSC matrix.

Moreover, an extended overview of the main results concerning the fatigue performance of SFRC is given by Lee and Barr [10], who collected the results of several researchers and observed that the presence of fibers does not seem to enhance the fatigue life of concrete under compressive fatigue loading. On the other hand, fiber addition enhances the fatigue performance under flexural fatigue loading; the extent of these advantages can be expected to depend upon the SFRC toughness which is related to the fiber volume fraction (V_f), fiber type and geometry, as well as to the mechanical properties of the concrete matrix.

The aim of the present work is to investigate the fatigue behavior of SFRC by performing three point bending tests (3PBT) on notched beams [11]. In particular, the influence of the volume fraction and the load levels were analyzed and tests were performed on pre-cracked specimens to simulate concrete behavior in the FPZ. The results will be discussed by means of cyclic creep curves and S-N diagrams, by devoting special attention to the two main parameters that seem to have the major influence on the fatigue life, namely the crack opening rate and the material toughness (post-cracking strength).

2 Experimental Program

The experimental program consisted of both monotonic quasi-static tests to decide the fracture properties of the material and of cyclic tests to study the cracked concrete behavior under fatigue loading.

Five sets of experiments were performed, by using two concrete batches: for each batch, cubes, cylinders and beams of plain and SFRCs (V_f=0.5% and 1%) were produced. In Table 1, the main characteristics of the concrete mix, the material properties and the fatigue load levels are reported. About 400 kg/m^3 of cement type II/AL 42.5R and a total amount of about 1750 kg/m^3 of aggregates were employed. The expected concrete class was C35/45. Hooked-end steel fibers (aspect ratio L_f/D_f = 35/0.55 = 64) were used with two contents: 39.25 kg/m^3 and 78.5 kg/m^3, which correspond to a volume fraction of 0.5% and 1.0%, respectively. The fracture properties of SFRC were determined according to the European Standard EN14651 [11] which requires bending test (3PBT) on small beam specimens (150 x 150 x 550 mm^3). Fracture tests were carried out with a closed-loop hydraulic testing machine by using the Crack Mouth Opening Displacement (CMOD) as control parameter; the latter was measured by means of a clip gauge positioned astride a notch (having a depth of 25 mm) at mid-span. Additional Linear Variable Differential Transformers (LVDTs) were used to measure the Crack Tip Opening Displacement (CTOD) as well as the vertical displacement at mid-span under the load point.

The load history for cyclic tests was planned by means of a special software (Fig. 1) [8]. After the monotonic branch (OA), the reference signal is inverted so the load decreases up to a preset lower limit. Then the CMOD is cycled at 3Hz (BC) between two preset load limits. In the third stage, the cyclic loading continues on the envelope curve (CD) until the maximum load detected became lower than a fixed value and the CMOD increases monotonically (DE).

Table 1. Main strength properties and load levels of fatigue load applied

SERIES	Material	w/c	D_{max}	$f_{cm,cube}$	E_{cm}	Fatigue load P_{upp}/P_{max}	Fatigue load P_{low}/P_{max}
		[-]	[mm]	[MPa]	[MPa]	[-]	[-]
1st series	PL	0.47	15	57.2	36570	0.75	0.25
	SFRC-0.5%			58.7	34687		
	SFRC-1.0%			31.4	30226		
2nd series	PL			49.8	32530		
	SFRC-0.5%			55.2	34611		
	SFRC-1.0%			46.5	34233		
3rd series	PL			44.9/51	30125/33881		
	SFRC-0.5%			49.2	29855		
	SFRC-1.0%			55.6	33835		
4th series	PL		10	44.8	27678	0.85	0.35
	SFRC-0.5%			47	28707		
	SFRC-1.0%			44.3	29516		
5th series	PL			54.24	31802	0.65	0.15
	SFRC-0.5%			51.22	30712		
	SFRC-1.0%			42.25	31569		

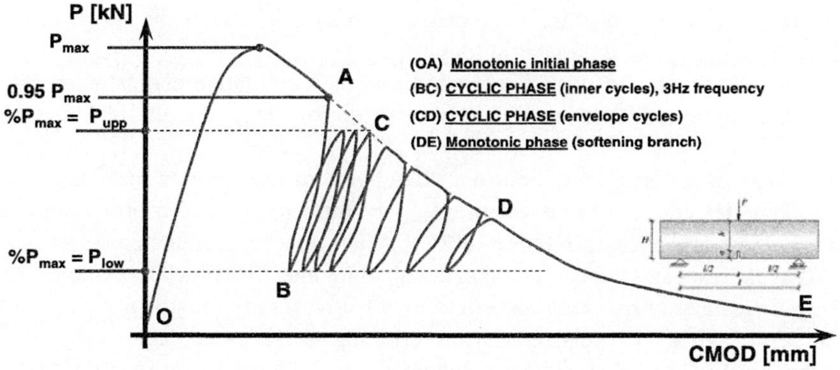

Fig. 1. Loading history for the cyclic tests

3 Test results and Discussion

The main results from the cyclic tests are presented in this section. Prior to the fatigue test, some quasi-static tests were performed under monotonically increasing deformation; eventually, it was demonstrated that the envelope curves from cyclic tests match the static curves quite well (Fig. 2), as reported by other authors [8, 9, 12].

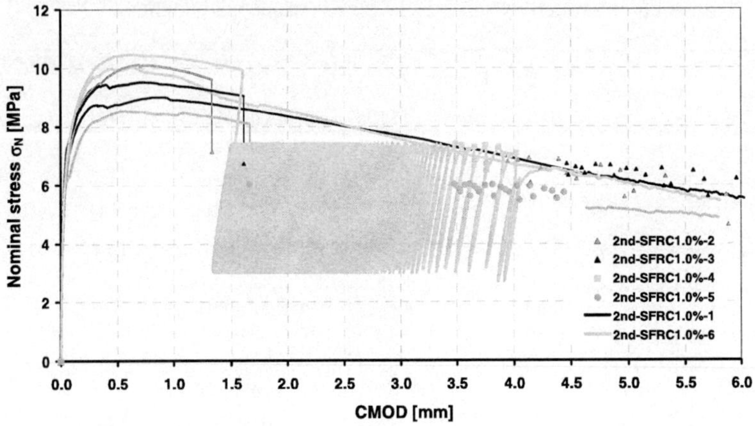

Fig. 2. Diagrams of nominal stress σ_N vs. CMOD as obtained from cyclic tests and quasi static tests (2nd-SFRC-1.0% phase)

The fatigue life of a specimen, that can be identified as the number of cycles at failure (N_{max}), mainly depends on the following parameters (Figs. 3 and 4):

- the crack increment per cycle (dw/dn), that represents the slope of the linear branch of the well known cyclic creep curves;
- the maximum crack opening that can be covered in the cyclic stage (Δw), which is a measure of the material toughness.

In addition, the secant stiffness degradation and the cumulative energy were also investigated herein (Fig. 4); the complete set of experimental results is discussed in [13].

From Table 3 it clearly turns out that the choice to use a lower maximum aggregate diameter (D_{max}=10 mm versus D_{max}=15 mm) seems to be promising; in fact, in the first two phases, only the SFRC-0.5% is able to guarantee a better fatigue life with respect to the plain concrete, while the SFRC-1.0% exhibits the worst performances. Analyzing the effect of different fatigue loads, it can be noticed that, at high load levels (85/35, representing to the percentage of the peak load), the number of cycles decreases whereas, for the lowest load levels (65/15), the plain concrete specimens exhibit the longest life.

Results show that Δ_{CMOD} increases with the fiber content and the lower load levels (Fig. 5). As a matter of fact, in the literature it is well known that fibers enhance material toughness, with a larger softening behavior after the peak; since failure occurs when the cyclic curve reaches the envelope curve, the higher is the upper load level the shorter is the crack opening that can be covered before reaching the envelope curve.

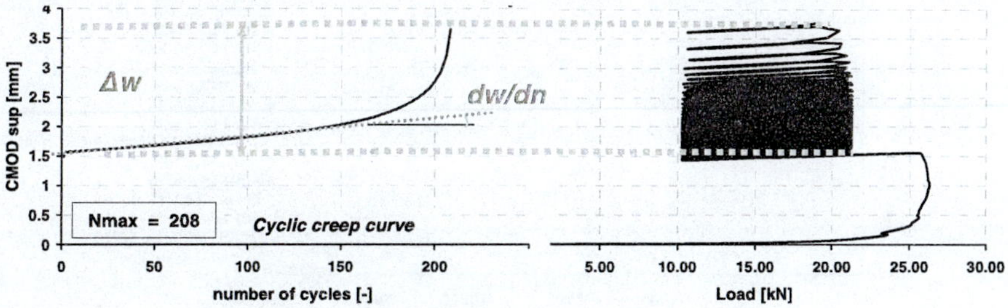

Fig. 3. Cyclic creep curve for the SFRC 1.0% - 1 85/35

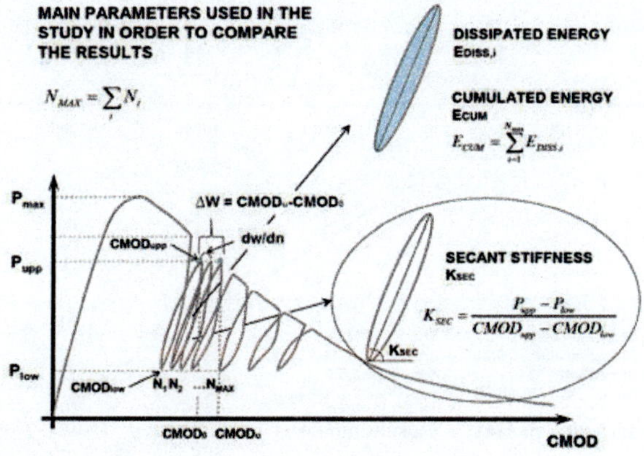

Fig. 4. Parameter studied during the inner cycles stage

Table 3. Mean values for N_{max}, Δ_{CMOD} and d_{CMOD}/dn

PHASES		N_{max} [-]	Δ_{CMOD} [μm]	d_{CMOD}/dn [μm/cycle]
	PLAIN	5790	66.3	0.00416
1st-phase	SFRC - 0.5%	7903	582.3	0.02971
	SFRC - 1.0%	3298	1296.7	0.15427
	PLAIN	1777	72.9	0.01186
2nd-phase	SFRC - 0.5%	7905	2554.3	0.09965
	SFRC - 1.0%	1299	4524	0.7074
	PLAIN	2207	70.3	0.00947
3rd-phase	SFRC - 0.5%	34115	2341	0.01642
	SFRC - 1.0%	9376	2417.7	0.072
	PLAIN	70	48.1	0.24069
4th-phase	SFRC - 0.5%	1122	1662.2	0.4357
	SFRC - 1.0%	405	2500	2.70025
	PLAIN	145562	131.6	0.00038
5th-phase	SFRC - 0.5%	71974	4106.4	0.00968
	SFRC - 1.0%	47119	4629.6	0.00978

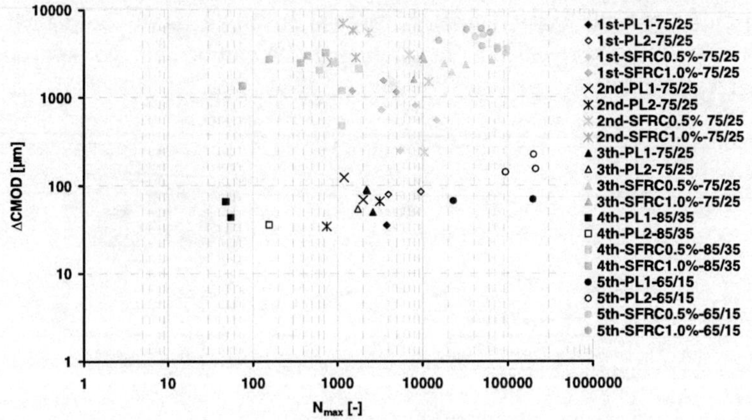

Fig. 5. Crack opening that can be covered (Δ_{CMOD}) vs. N_{max}

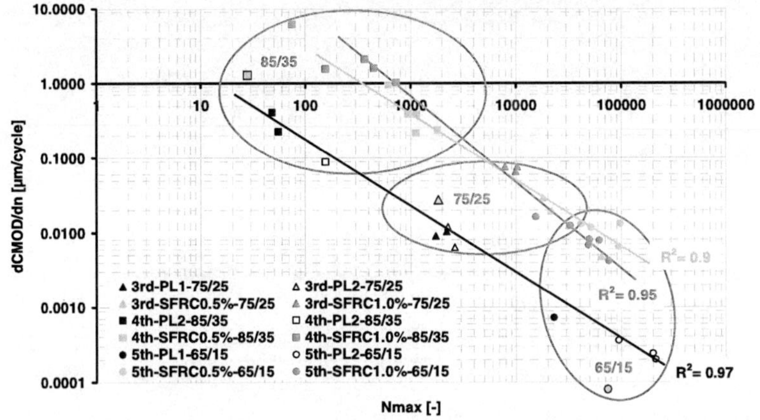

Fig. 6. Crack opening rate and fatigue life (3rd, 4th, and 5th phases)

When considering the crack opening rate as a function of the number of cycles, it was found out that there is a clear correlation between the fatigue life and the crack opening rate (d_{CMOD}/dn; Fig. 6): when the latter increases, the fatigue life decreases with a correlation factor R^2 of about 0.9 (higher than plain concrete.

As expected, by increasing the load level there is a significant increase of the crack opening rate and, hence, a lower fatigue life; in fact, due to the increased damage, the maximum number of cycles decreases. For a chosen load level, plain concrete has the lowest crack opening rate and only the SFRC-0.5% shows about the same values at 85/35 and 75/25 ranges. When applying a load level 0.65/0.15, SFRC-0.5% and SFRC-1.0%, the specimens have the higher fatigue life. This tendency turned out with a decrease of the cycles number at failure for SFRCs with a load level of 65/15 while, due to crack opening rate values (about the same or little higher for SFRCs), together with a remarkable higher Δ_{CMOD}, SFRC guarantees the longest fatigue lives at 85/35 and 75/25 load levels.

Finally, the results are presented by using the S-N or Wöhler diagrams. Fig. 7 shows the lines corresponding to a linear regression of the logarithms of each number of cycles at failure for each tests, at the preset load level. Three lines are presented, one for each mix; it can be observed that none of them provides a perfect fit, but a certain trend for a straight line on a semi-logarithmic scale can be seen. Furthermore, the correlation is much better for the plain concrete mix whereas, contrarily to what expected, SFRC-0.5% exhibits the worse fitting. For high load levels, SFRC-0.5% and SFRC-1.0% are able to guarantee a longer fatigue life than plain concrete; when the load level reduces, the benefits due to the fiber progressively diminishes. At 75/25 load level, SFRCs (especially with V_f=0.5%) guarantees better performances than plain concrete whereas, at 65/15 load level, the plain concrete mix exhibits the highest number of cycles (only SFRC -0.5% shows similar performances).

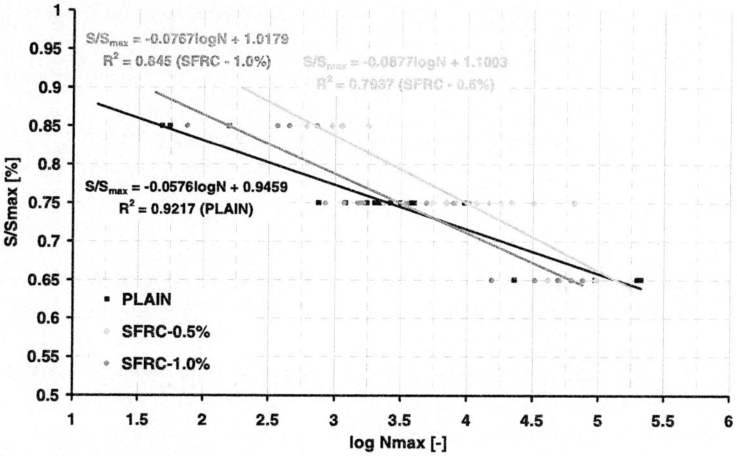

Fig. 7. S-N lines for Plain, SFRC0.5% and SFRC1.0% (all the phases)

4 Concluding Remarks

Static and fatigue tests have been performed on three different mixes (plain concrete, and SFRC with a volume fraction of fibers of 0.5% and 1%).

Cyclic tensile were performed after cracking of the concrete matrix, by applying different load levels with a constant amplitude.

Even though SFRC-1.0% shows the better properties from static tests, such an improvement is not visible from fatigue tests with lower load levels. Among the three mixtures tested, only the SFRC-0.5% exhibited a better fatigue performance than plain concrete.

The crack opening available for the fatigue damage (Δ_{CMOD}) as well as the crack opening rate (d_{CMOD}/dn) have been demonstrated to have great influence on the fatigue life; SFRCs show the highest crack opening rates but, due to the larger crack opening available for the fatigue crack development (due to the enhanced toughness of SFRC), a better performance of SFRC was observed for higher load levels.

Acknowledgements. The Authors would like to thank Andrea Delbarba and Domenico Caravaggi of the Laboratory for Testing Materials of the University of Brescia, and the Bekaert Corp., for providing the steel fibers. A special acknowledgement goes to all the students who collaborated in carrying out the tests and in the data reduction.

References

[1] di Prisco, M., Felicetti, R., Plizzari, G.A. (eds.): Proceedings of the 6th RILEM Symposium on Fiber Reinforced Concretes (FRC), Varenna (Italy), Bagneaux, France, vol. 39. (September 20-22, 2004)

[2] Gettu, R.: In: Proceedings of the Seventh Intnl. RILEM Symp. on Fibre Reinforced Concrete: Design and Applications, BEFIB-2008, September 17-19, p. 1154 (2008)

[3] Reinhardt, H.W.: Fracture Mechanics of an Elastic Softening Material like Concrete. Heron 29(2), 43 (1984)

[4] RILEM Committee 36-RDL, Long Term Random Dynamic Loading of Concrete Structures. Materials and Structures, Research and Testing (RILEM, Paris) 17(97), 1–28 (1984)

[5] Yin, W., Hsu, T.T.C.: Fatigue Behavior of Steel Fiber Reinforced Concrete in Uniaxial and Biaxial Compression. ACI Materials Journal 92(1), 71–81 (1995)

[6] Hillerborg, A., Modeer, M., Petersson, P.E.: Analysis of crack formation and crack growth in concrete by means of fracture mechanics and finite elements. Cement and Concrete Research 6, 773–782 (1976)

[7] Slowik, V., Plizzari, G.A., Saouma, V.: Fracture of Concrete under Variable Amplitude Fatigue Loading. ACI Materials Journal 93(3), 272–283 (1996)

[8] Plizzari, G.A., Cangiano, S., Alleruzzo, S.: The fatigue behaviour of Cracked Concrete. Fatigue & Fracture of Engineering Materials & Structures 20(8), 1195–1206 (1997)

[9] Plizzari, G.A., Cangiano, S., Cere, N.: Post-peak Behavior of Fiber-Reinforced Concrete under Cyclic Tensile Loads. ACI Material Journal 97(2), 182–192 (2000)

[10] Lee, M.K., Barr, B.I.G.: An overview of the fatigue behaviour of plain and fiber reinforced concrete. Cement and Concrete Composites 26, 299–305 (2004)

[11] UNI EN 14651, Test method for metallic fiber concrete - Measuring the flexural tensile strength (limit of proportionally (LOP), residual). European Committee for Standardization 18 (2005)

[12] Hordijk, D.A.: Local approach to Fatigue of Concrete. Ph.D Thesis, Delft University of Technology, p. 210 (1991)

[13] Germano, F.: Cyclic Behavior of Steel Fiber Reinforced Concrete: from material to seismic columns. Ph.D Thesis, University of Brescia, Department of Civil, Architectural, Environmental and Land Planning Engineering, p. 433 (in press, 2011)

Seismic Strengthening of Piers by Using High Ductility Cement

K. Kosa, H. Shimizu, M. Kusano, and H. Goda

Department of Civil Engineering, Kyushu Institute of Technology, Kitakyushu, Japan

Abstract. To find an effective use of high ductility cement for seismic strengthening of bridge piers, a loading test was conducted using four specimens. Among the two specimens for type 3 with high ductility cement applied to the cover concrete range only, Specimen No. 3-3 that has spiral ties around the main reinforcement, showed a better deformation capacity. As to the other two specimens with high ductility cement applied up to beyond the main reinforcement, even Specimen No. 4-2 having high ductility cement in the range from the bottom to the height of 350 mm, showed an excellent deformation capacity.

1 Introduction

In the Japanese Specifications for Highway Bridges: Seismic Design, the cover concrete is ignored when calculating the ultimate deformation of ordinary RC piers, assuming that it is unable to carry the stress in the ultimate stage by spalling off from the pier. But, we considered that if the cover concrete is constructed of high ductility cement, it will suffer little damage and can carry the stress even in the ultimate stage. Based on this concept, our study aimed to confirm that the seismic resistance of RC piers can be improved by the partial application of high ductility cement. Four specimens were prepared. Two specimens were constructed with high ductility cement applied to the cover concrete area only. The other two specimens were constructed with high ductility cement applied up to beyond the main reinforcement. Using these four specimens, improvement of deformation capacity was evaluated.

G.J. Parra-Montesinos, H.W. Reinhardt, and A.E. Naaman (Eds.): HPFRCC 6, pp. 321–328.

2 Evaluation by Calculation

2.1 Attributes of Specimens

Fig. 1 and Table 1 show the configuration and attributes of specimens, respectively. The specimens were constructed to 1/8 scale of an ordinary RC single column pier. The main reinforcement ratio and the hoop tie ratio were identical to those of ordinary piers. The specimens were designed to fail by bending.

Fig. 2 shows the cross sections studied. Specimen No. 1 is the control type with its entire cross section constructed of ordinary concrete. In Specimens Nos. 3 and 4, high ductility cement was applied to the cover concrete area only, but their application thickness varied. Because at least 50 mm width of cover concrete is required for casting concrete, the width of specimen is varied from 400 mm to 440 mm. These specimens were intended to evaluate the effect of the cover concrete width on the deformation capacity of the specimen. In specimen No. 3-1, high ductility cement was applied to the cover concrete range only. This specimen showed debonding at the boundary of high ductility cement and ordinary concrete. Therefore, to improve the bonding effect, specimen No. 3-3 was produced by placing the spiral ties in the cover concrete. Specimens No. 4 series was produced by applying high ductility cement up to beyond the main reinforcement. Compared with Specimen No. 4-1, application range of high ductility cement along the column was changed from 700 mm (2 d, d: effective depth) to 350 mm (1 d) in the Specimen No. 4-2 to evaluate if the volume of high ductility cement, was reduced. Properly anchored transverse reinforcement, which changed the bending angle, is also provided in specimen No. 4-2 to increase the confinement effect.

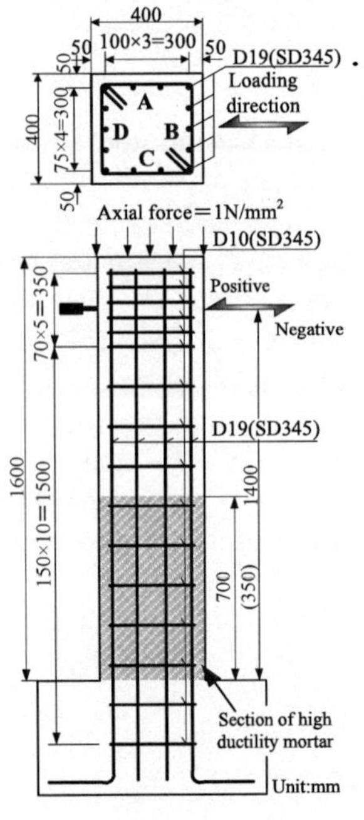

Fig. 1. Configuration of specimen

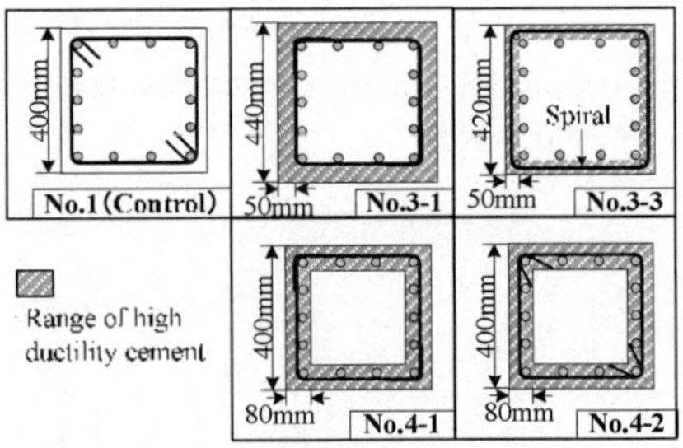

Fig. 2. Cross section studied

Table 1. Attributes of specimens

No.		No.1	No.3-1	No.4-1	No.3-3	No.4-2	Value for calculation
Cross section [mm]		400×400	440×440	400×400	420×420	400×400	—
Thickness [mm]		30	50	30	50	30	—
Shear span ratio [mm]		1400					
Ordinary concrete strength [N/mm^2]		30.3	28.1	24.8	24.9		27.7
High ductility cement [N/mm^2]	Compressive stress	—	53.4	81.4	46.0		67.4
	Tensile strength	—	4.3	—	—		4.3
Axial stress [N/mm^2]		1.0					

No.		No.1 No.4s	No.3-1	No.3-3
Main reinforcement	Grade	SD345		
	Diameter	D19		
	Tensile rein.ratio[%]	1.43	1.12	1.33
Hoop tie	Grade	SD345		
	Diameter	D10		
	Spacing [mm]	150		
	Volume ratio of fiber [%]	0.63	0.59	0.53

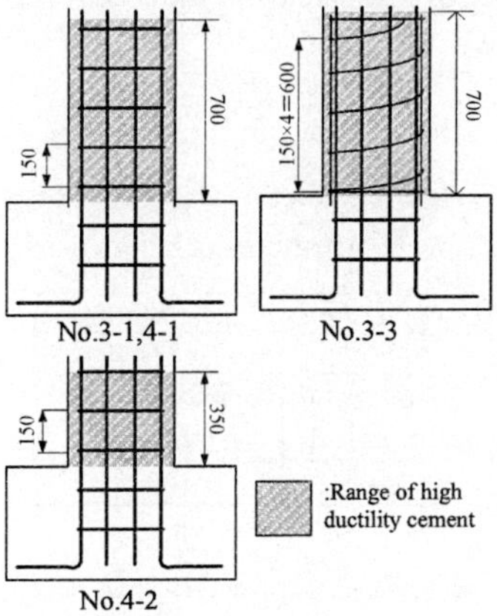

Fig. 3. Added specimen

2.2 Calculation method

For comparison with experimental results, the ultimate strength and deformation of columns was calculated in accordance with the Japanese Specifications for Highway Bridges: Seismic Design [1]. Although the cover concrete is ignored in this specification, it was taken into account in the current calculation, assuming that the cover concrete constructed of high ductility cement could carry the stress even in the ultimate stage [2]. As the thickness of cover concrete (50 mm) is relatively large against the fiber length, the resistance in the tensile zone is small [2]. So, the stress-strain relationship on the tensile side was not taken into account. The ultimate strain is defined as the strain when the stress decreases to 30% of the maximum compressive stress. This is based on the analysis results that a value very close to an actual ductility factor was obtained when the stress decreased to 30% of the maximum stress [2]. The stress-strain curve used for the calculation was extracted from reference [3].

2.3 Calculation Results

Fig. 4 shows the load-displacement relationship obtained from calculation. Table 2 shows the yield load and maximum load obtained from calculation. Compared with an ordinary specimen, Specimens No. 3 and 4 increased the maximum load about 5-16%, and deformation capacity about 130-150%.

As the high ductility cement can carry the stress in the compression zone, even though cracking occurs, it leads to an increase of the ultimate load and ultimate strain of the specimen. Using a frame model, the difference of deformation capacity between Specimen No. 4-1 and Specimen No. 4-2 was evaluated. As the results were almost the same, the same deformation capacity can be expected even though the range of height of ductility cement is reduced from 700 mm to 350 mm.

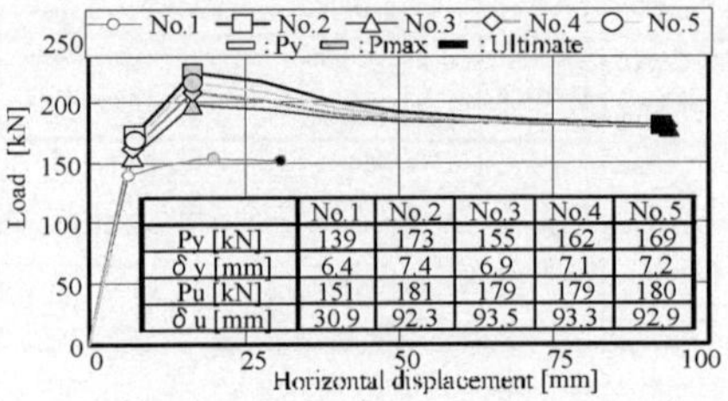

a) When the cover concrete thickness is taken into account

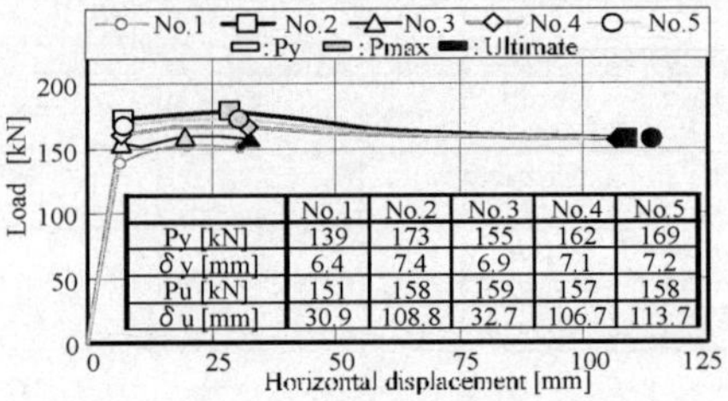

b) When the cover concrete thickness is ignored

Fig. 4. P-δ relationship by calculation

Table 2. Yield load and maximum load (calculation)

	No.1	No.3s	No.4s
Yield load (kN)	161	170	158
Yield displacement(mm)	7.1	7.0	7.5
Maximum load (kN)	195	227	208
Ultimate	64.5	85.0	103.2

Table 3. Attributes of fibers used

Fiber	Diameter [μm]	Length [mm]
	40.0	8.0
Vinylon	Volume ratio of fiber [%]	Breaking strength [N/mm^2]
	2.0	1600

3 Experimental Program

Table 3 shows the material properties of fibers. To produce high ductility cement, Vinylon fibers, 8 mm long, were added at a volumetric ratio of 2%. The application range of this cement along the column height is basically the plastic hinge section only. But, the application range of Specimens No. 3-1, 3-3, 4-1 was made to 700 mm, which is the plastic hinge section plus the transition range. The application range of Specimen No. 4-2 was shortened to 350 mm, which is the plastic hinge section only.

Loading was applied by the reverse manner, with the load control method up to the yielding load obtained by calculation and then with the displacement control method at each integral multiple of the yield displacement (δ_y). Each loading step was repeated just once and loading was terminated when the load decreased to 0.5 P_y. In consideration of the dead load of an actual structure, a uniform axial load equivalent to 1.0 N/mm^2 was applied to the top of the column. The displacement meters were installed at the bottom of column sides B and D of each specimen to find the effect of main reinforcement pullout from the footing.

4 Experimental Results

4.1 Comparison between Specimens No. 3-1 and No. 3-3

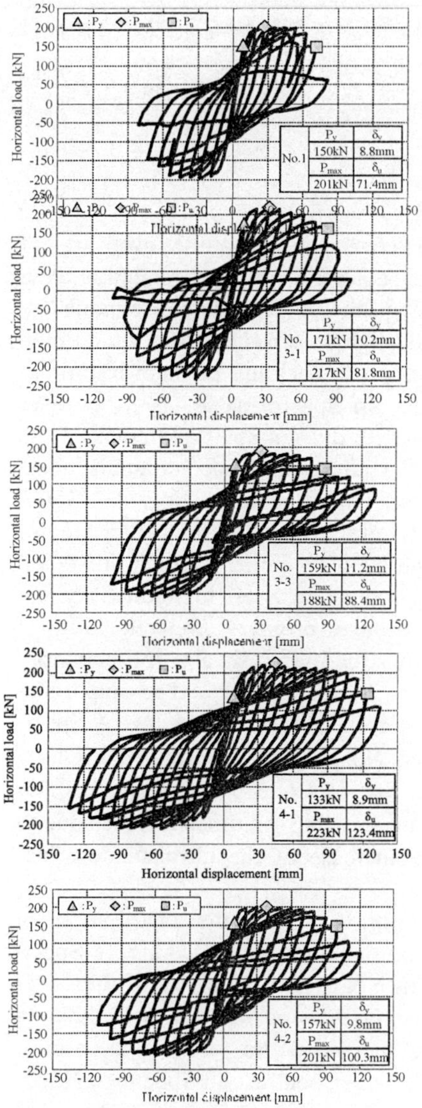

Fig. 5. Load-displacement curve (experiment)

Fig. 5 shows the load-displacement curve of each specimen. Specimen No. 3-1 reached yield strain at 170 kN, and maximum load (217 kN) at 3 δ_y. Afterwards, the load reduced gradually up to 6 δ_y, and the buckling of main reinforcement and diagonal cracks occurred at the lower section of column sides A and C. Finally, with the spalling of the cover concrete, the load reduced rapidly to the P_y value at 8 δ_y.

Specimen No. 3-3 reached the yielding strain at 159 kN and the maximum load (188 kN) at 3 δ_y. Afterwards, the load was kept constant value up to 7 δ_y (78.4 mm), and reduced to P_y at 8 δ_y (88.4 mm) by the buckling of the main reinforcement at the lower section of the column. As the load exceeding 0.5 P_y was still maintained, even after 10 δ_y displacement, a negative stroke reached an application limit. Therefore, only a monotonic positive stroke was applied after 10 δ_y displacement.

Compared with Specimen No. 3-1, Specimen No. 3-3 showed a larger deformation capacity up to ultimate deformation, and gradual reduction of load after ultimate deformation. This improvement of deformation capacity may be caused by the effect of spiral ties placed around the main reinforcement.

Compared with Specimen No. 3-1 and calculated value, Specimen No. 3-3 showed small maximum load. It is because the actual compressive strength of high ductility cement used for the specimen was smaller compared with those of

No. 3-1 specimen and calculated value. Fig. 6 shows damage to the specimens at the end of ±9 δ_y loading. In Specimen No. 3-1, loading was terminated at this load.

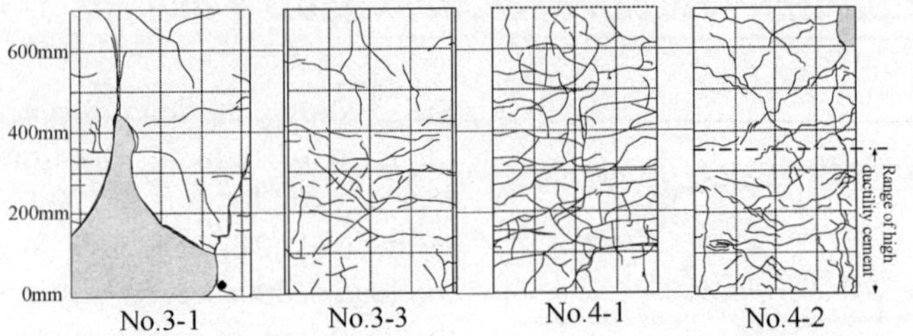

Fig. 6. Damage at the end of ± 9δy loading

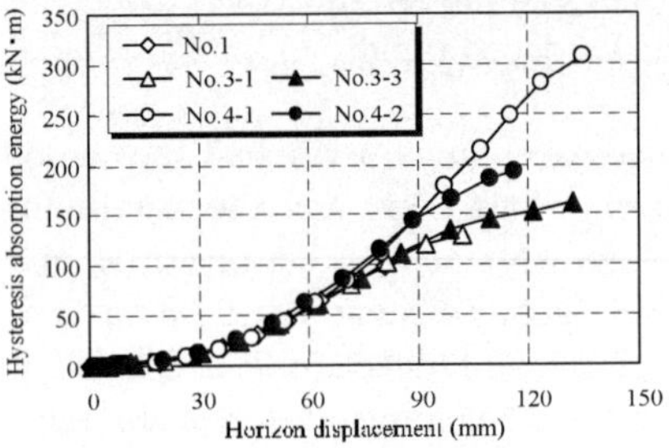

Fig. 7. Hysteresis absorption energy at each loading step

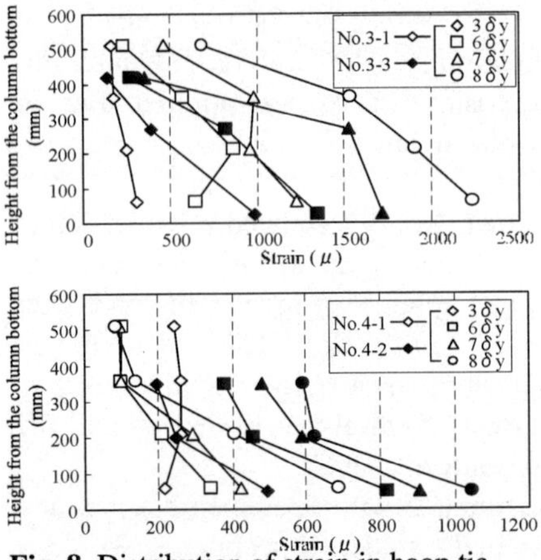

Fig. 8. Distribution of strain in hoop tie

The cover concrete at the lower area of the column mostly spalled. Even though the cover concrete remained, it was slightly detached from the column due to the buckling of main reinforcement. In Specimen No. 3-3, the cover concrete did not spall and cracks were dispersed around the column, compared with Specimen No. 3-1. Spiral ties were arranged at the center area of high ductility cement, which prevented debonding between high ductility cement and ordinary concrete. Fig. 7 shows the hysteresis absorption energy at each loading step. The maximum hysteresis absorption energy of Specimen No. 3-1 was about 130 kN · m, but that of No. 3-3 was about 160 kN · m, showing 1.2 times increase due to the increase of the ultimate deformation.

Fig. 8 shows the distribution of hoop tie strain on the D side. In Specimen No. 3-1, the spalling of the cover concrete was occurred at 6 δ_y, however, the hoop tie strain was almost 1000 μ. In contrast, in Specimen No. 3-3, the spalling of the cover concrete occurred at 8 δ_y, when the strain of the spiral tie reached over 2000 μ. This means that spiral ties are more effective than hoop ties for confining the column, which leads to an increased deformation capacity.

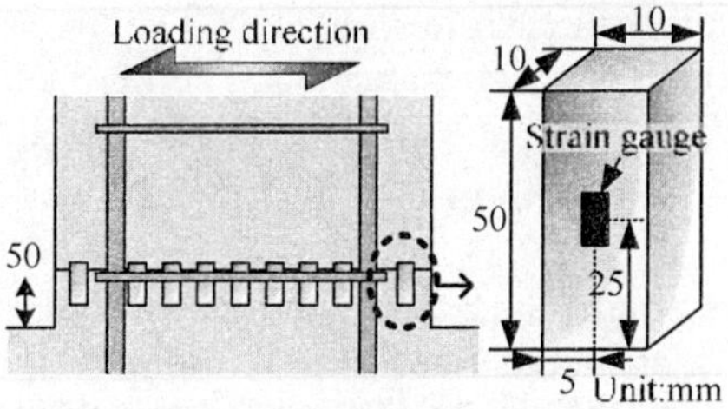

Fig. 9. Position of acrylic plates

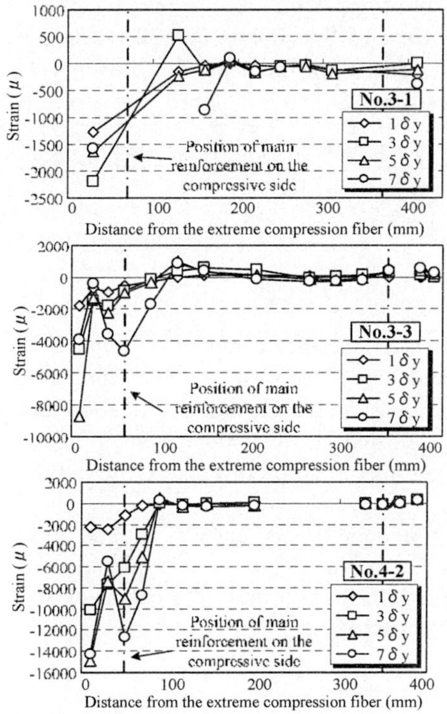

Fig. 10. Strain distribution in section

Fig. 9 shows the arrangement of acrylic plate to which strain gauges are attached. By measuring the strain at each acrylic plate, strain distribution in the section were plotted. The results are showed in Fig. 10.

In Specimen No. 3-1, compressive strain increased with an increase of deformation, and the maximum compressive strain reached almost 2200 μ at 7 δ_y. In contrast, the maximum compressive strain of acrylic plate of Specimen No. 3-3 reached over 9000 μat 7 δ_y. This specimen carried a large load because its damage was relatively small.

4.2 Comparison between No. 4-1 and No. 4-2

Fig. 5 shows the load-displacement curve of each specimen. Specimen No. 4-1 reached yield strain at 136 kN, and maximum load (210 kN) at 5 δ_y. Afterwards, the load was carried constantly up to 13 δ_y. The load reduced to below P_y at 14 δ_y (123.4 mm) due to the buckling of main reinforcement of the column.

Specimen No. 4-2 reached yield strain at 157 kN, and maximum load (210 kN) at 4 δ_y. Afterwards, the load was carried constantly up to 9 δ_y. The load reduced to below P_y at 10 δ_y (100.3 mm) due to the buckling of main reinforcement of the column. Compared with Specimen No. 4-1, Specimen No. 4-2 reduced both the maximum load and a deformation capacity. This is probably because the compressive strength of high ductility cement is 81.4 N/mm^2 for Specimen No. 4-1 and 46.0 N/mm^2 for Specimen No. 4-2 as shown in Table 1.

In Specimen No. 4-1, many fine cracks appeared widely, and uniformly as shown in Fig. 6. In specimen No. 4-2, many fine cracks also appeared widely in the range of high ductility cement. In the range of ordinary concrete, only a few large cracks occurred and the spalling of the cover concrete was observed at 9 δ_y.

Fig. 7 shows the hysteresis absorption energy at each loading step. The maximum hysteresis absorption energy of Specimen No. 4-1 was about 310 kN・m, but that of Specimen No. 4-2 was about 180 kN・m, showing a relatively small value.This may be caused by the difference of strength of each high ductility cement, which was already described before.

Fig. 8 shows the distribution of hoop tie strain. Specimen No. 4-2 showed a large strain distribution, due to the properly anchored transverse reinforcement. However, the maximum strain of hoop ties of Specimen No. 4-2 was almost 1000 μ, which is relatively small compared with Specimen No. 3-1. As the Specimen No. 4-2 has a relatively large thickness of high ductility cement (80 mm) , it has a larger effect in preventing buckling of main reinforcement.

As shown in Fig. 10, the compressive strain of Specimen No. 4-2 reached a large value such as 14000 μat 7 δ_y. This is because, Specimen No. 4-2 showed relatively small damage of cracks, and it can sustain the large compressive force in the compressive zone.

5 Conclusions

The following conclusions were drawn from this experiment.

(1) In Specimens No. 3-1 and No. 3-3, the high ductility cement was applied to the cover concrete range only. Compared with Specimen No. 3-1, Specimen No. 3-3 showed a larger deformation capacity up to ultimate deformation and gradual reduction of load after ultimate deformation. This improvement of deformation capacity may be caused by the effect of spiral ties placed around the main reinforcement.

(2) In Specimens No. 4-1 and No. 4-2, the high ductility cement was applied up to beyond the main reinforcement. In Specimen No. 4-2, even if application of high ductility cement is limited in the range from bottom to the height of 350 mm, it showed an excellent deformation capacity.

References

[1] Road Association of Japan, Specifications for Highway Bridges, Seismic Design (2002)

[2] Taguchi, J.: Study on the Deformation Capacity of RC Piers, Thesis for MS, Kyushu Institute of Technology (2003) (in Japanese)

[3] Suwada, H., et al.: Basic Research on the Restoring Force Characteristics of Response Control Elements Constructed with High Ductility Cement-based Composite Material. Proc. of JCI 25(2), 1375–1380 (2003) (in Japanese)

Drift Limits of Concrete Frame Members Reinforced with High-performance Steel Bars and Fibers

H. Tavallali[1], A. Lepage[1], J. Rautenberg[2], and S. Pujol[2]

[1] Department of Architectural Engineering, Penn State University, University Park, PA
[2] School of Civil Engineering, Purdue University, West Lafayette, IN

Abstract. Concrete test specimens reinforced with ultrahigh strength steel, $f_y > 80$ ksi (550 MPa), were subjected to large displacement reversals. Each specimen consisted of two cantilever beams (or columns) connected to a central loading stub. The applied displacement history included cycles of increasing deflection reversals followed by a final push to failure. Experimental data are presented for concrete specimens conventionally reinforced and for concrete specimens reinforced with ultrahigh strength steel bars with and without fibers. The study shows that the use of ultrahigh strength longitudinal reinforcement in frame members is a viable option for earthquake-resistant construction.

1 Introduction

For many years, the seismic design of reinforced concrete structures in the U.S. has been dominated by the use of steel reinforcement with specified yield strength, f_y, of 60 ksi (410 MPa). Although higher values of f_y are allowed for non-seismic applications, it has been limited to 80 ksi (550 MPa) since the 1971 version of ACI 318 [1]. ACI 318-08 permits designs with f_y of 100 ksi (690 MPa) but only if used for confinement reinforcement.

The term ultrahigh strength steel (UHSS) is used here to designate steel bars having yield strength in excess of 80 ksi (550 MPa) and a fracture elongation, ε_{su}, of 6% or more in 8 inches (200 mm). Fig. 1 shows representative stress-strain curves of both conventional Grade 60 (410) and UHSS steels. After the introduction of ASTM A1035 [2], there has been growing interest in UHSS bars. However, there is paucity of test data describing the behavior of concrete members reinforced with UHSS bars as longitudinal reinforcement. A series of experiments was designed to determine the deformation capacity of UHSS-reinforced concrete frame members subjected to displacement reversals. Specimens with high-performance fiber reinforced concrete (HPFRC) were included. HPFRC is defined here as a class of fiber reinforced concrete that shows strain-hardening behavior after first cracking [3].

G.J. Parra-Montesinos, H.W. Reinhardt, and A.E. Naaman (Eds.): HPFRCC 6, pp. 329–337.

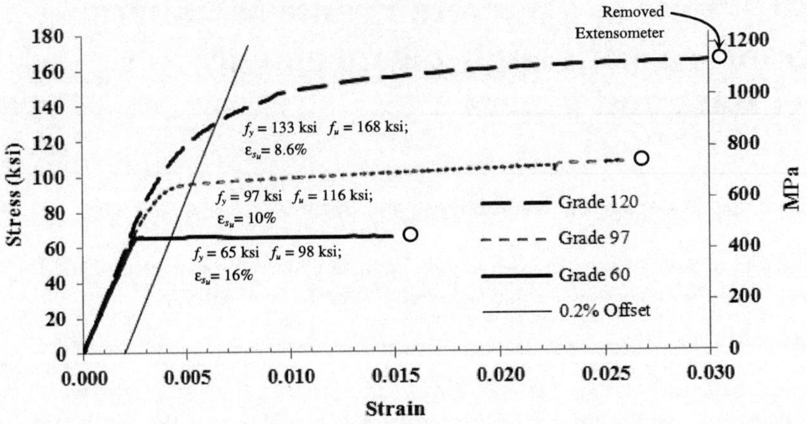

Fig. 1. Measured tensile properties of reinforcing steel, 1 ksi = 6.9 MPa

2 Background

There is no evidence in the U.S. of structures built or rehabilitated using UHSS as longitudinal and transverse reinforcement and designed to take full advantage of f_y in excess of 80 ksi (550 MPa) for stresses induced by combined shear, flexure, and axial forces. This under-utilization is predominantly due to the shortage of experimental data and the lack of coverage in existing building codes.

Limits on the specified yield strength of reinforcing bars in the ACI 318 Code were originated in the 1963 edition. Their introduction was primarily related to the prescribed limit on compressive strain of 0.003 for concrete and to control crack width at service load.

A strategy for the effective use of UHSS bars as primary reinforcement is to control the crack width and to enhance the usable compressive strain of concrete by adding fibers to the concrete mix. High tensile and compressive strain capacities are attainable by concrete reinforced with dispersed fibers. Recent developments in high-performance fiber reinforced cementitious composites include formulations of fibers and concrete matrices to achieve strain hardening behavior with fiber contents of 1.5% by volume [4].

Studies on the compressive properties of cementitious composites have shown that the introduction of fibers into the matrix delays spalling of the cover and increases the load capacity and the ductility of columns over that of comparable non-fiber-reinforced specimens [5]. Several research projects have explored the application of fiber-reinforced composites in earthquake-resistant construction [3]. Results from these studies revealed HPFRC to be effective in increasing shear strength, deformation capacity, and damage tolerance in members subjected to cycles of large inelastic deformations, but none of these studies incorporated UHSS reinforcement.

3 Description of Experiments

To evaluate the mechanical behavior of concrete beams and columns reinforced with UHSS bars, a series of experiments was designed at Penn State University and Purdue University. The purpose of this series of tests was to evaluate the deformation capacity of such members when subjected to a cyclic loading protocol. Table 1 shows a summary of the specimens considered. The test variables include:

- Nominal yield strength of longitudinal reinforcement, f_y = 60, 97, or 120 ksi (410, 670, or 830 MPa)
- Applied axial force, $P = 0$ or 20% $f_c' A_g$, where f_c' is the concrete compressive strength, and A_g is the gross cross-sectional area of the specimen
- Volume fraction of fibers, V_f = 0 or 1.5%

Table 1. Description of test specimens (1 ksi = 6.9 MPa; 1 inch = 25.4 mm)

Specimen		Axial Load $A_g f'_c$(%)	Longitudinal Reinforcement[a] Top	Bottom	f_y, ksi	Transverse Reinforcement, (2)#3 Spacing, in	f_y, ksi
Beams Tested at Penn State (b = 16", h = 10", d = 8")							
#1	CC4-X	0	4#7	4#7	60	2	60
#2	UC4-X	0	4#6[b]	4#6[b]	97	2	60
#3	UC2-F	0	4#6[b]	4#6[b]	97	4	60
Columns Tested at Purdue (b = 9", h = 12", d = 10")							
#4	CC-3.3-20	20	3#7	3#7	60	2.5	60
#5	UC-1.6-20	20	2#6	2#6	120	2.5	60
#6	UC-1.6-20F	20	2#6	2#6	120	4.5	60

[a] Provided by SAS Stressteel (f_y = 97 ksi) and by MMFX Technologies (f_y = 120 ksi)
[b] Actual bar diameter is 0.70 in. with an area of 0.39 in^2

HPFRC Specimens #3 and #6, with ½" (13 mm) coarse aggregate had a 1.5% volume fraction of Dramix RC-80/30-BP steel hooked fibers, with length-to-diameter ratio of 80 and a length of 1.2 inches (30 mm). These fibers are commercially available and manufactured by Bekaert Corporation. The tensile strength of the fibers is 330 ksi (2300 MPa).

The target concrete compressive strength was 6 ksi (41 MPa). The geometry of typical beam and column specimens are shown in Fig. 2. The specimens were loaded through the central stub so that they were in single-curvature bending. The shear span-to-effective depth ratio (a/d) for all specimens was 3. In the column specimens, the axial force was kept constant throughout the loading protocol.

The amount of longitudinal reinforcement in each specimen was chosen so that the nominal flexural strength of all specimens was nearly identical. Thus, Specimen #4 with 60-ksi (410 MPa) steel used about twice the amount of longitudinal reinforcement as used in Specimens #5 and #6 with 120-ksi (830 MPa) steel. A similar relationship exists between Specimens #1, #2, and #3.

The nominal shear demand corresponding to the probable moment, M_{pr}, calculated according to ACI 318-08, was $5\sqrt{f_c', \text{psi}}\,bd$ $(0.42\sqrt{f_c', \text{MPa}}\,bd)$ for the beams and $7\sqrt{f_c', \text{psi}}\,bd$ $(0.58\sqrt{f_c', \text{MPa}}\,bd)$ for columns. Specimens #1, #2, #4, and #5, without fibers, used Grade-60 transverse reinforcement spaced at *d*/4, in compliance with Chapter 21 of ACI 318-08. Specimens #3 and #4, cast using HPFRC, used Grade-60 transverse reinforcement spaced at approximately *d*/2. The spacing was increased to account for the enhanced shear strength and confinement provided by fibers.

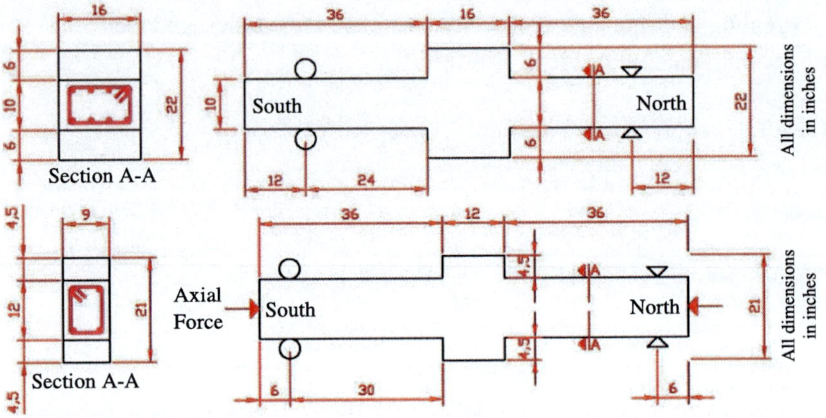

Fig. 2. Geometry of test specimens, beams (top) and columns (bottom), 1 inch = 25.4 mm

4 Loading Protocol

The drift ratio was defined as the lateral displacement of the specimen divided by the shear span, corrected for the rotation of the loading stub. The drift ratio history applied to each specimen follows the protocol of FEMA 461 [6] in Table 2. Two cycles at each drift target were applied to the stub at increasing amplitudes. After step 12, the displacement was increased monotonically until failure (defined here as a reduction in the lateral-load capacity of more than 20% from the peak value).

Table 2. Loading protocol

Step[a]	1	2	3	4	5	6	7	8	9	10	11	12
Drift(%)	0.15	0.2	0.3	0.4	0.6	0.8	1.0	1.5	2.0	3.0	4.0	5.0

[a] Two cycles of loading in each step

5 Experimental Results

The measured shear vs. drift data of the controlling end of the specimens (north or south) are presented for the 12 steps of the loading protocol. The data for beam specimens are included in Fig. 3 and for column specimens in Fig. 4.

Beams:
Specimen #1, CC4-X, was designed with 60-ksi (410-MPa) rebar according to the requirements in ACI 318-08 for special moment frame beams. Fig. 3(a) shows the measured shear-drift response. In the final push, not shown in Fig. 3(a), the north beam exceeded 10% drift without failure.

Specimen #2, UC4-X, had similar properties to Specimen #1 with the exception that it was reinforced longitudinally with UHSS. Fig. 3(b) shows the measured shear-drift response without the final push. Compared with Specimen #1, Specimen #2 showed reduced post-cracking stiffness and increased yield deformation. The south beam controlled the loading protocol and was capable of resisting the final push through a drift ratio exceeding 10%.

Specimen #3, UC2-F, had similar properties to Specimen #2 except that it was cast using HPFRC. Spacing of the transverse reinforcement was relaxed to $d/2$ to investigate the effects of fibers. Fig. 3(c) shows the measured shear-drift response without the final push. The south beam deviated from the loading protocol at the first cycle of step 9 because the north beam was controlling through step 8. The specimen completed the loading protocol without failure and was capable of resisting the final push through a drift ratio exceeding 10%. At each cycle, the secant slope of the shear-drift curve, measured from peak drift to zero shear, was about 20% higher in Specimen #3 than in Specimen #2.

Both concrete specimens without fibers had similar behavior. Specimen #1 showed a small increase in flexural strength after yielding while Specimen #2 had a nearly flat post-yield shear-drift curve resembling the stress-strain curve of the rebar. The conventional Grade-60 (410) bars are characterized by a tensile-strength-to-yield-strength ratio (f_u/f_y) of 1.5, while the UHSS Grade-97 (670) bars have a ratio of 1.2. The hysteretic energy dissipated in Specimen #2 during a given post-yield hysteretic loop was smaller than the energy dissipated in the same cycle for Specimen #1. With the addition of fibers to Specimen #3, the amount of energy dissipated increased. The reduction of transverse reinforcement by 50% in Specimen #3 had no negative impact on its behavior. HPFRC was effective in enhancing the shear capacity and confinement. However, at drift ratios exceeding 3% crack openings concentrated in a single flexural crack near the face of the stub. The presence of this wide flexural crack resulted in the accumulation of residual strains in the longitudinal bars crossing the crack. Figs. 5 and 6 show Specimens #2 and #3 at the end of the second cycle at 5% drift. Figs. 7 and 8 show the strain measurements in the longitudinal bars of Specimens #2 and #3 (at face of the central stub) during steps 10 and 11. The strain values measured at the same location for the same drift ratios are approximately 50% larger in Specimen #3, an indication of higher strain concentration in the plastic hinge of the HPFRC specimen. As a result of strain accumulation, the longitudinal bars fractured at 11% drift in Specimen #3, while similar bars in Specimen #2 fractured at a drift ratio of 15%. In any case, the beam specimens all reached drift ratios well in excess of what would be expected for a modern building structure subjected to strong ground motion.

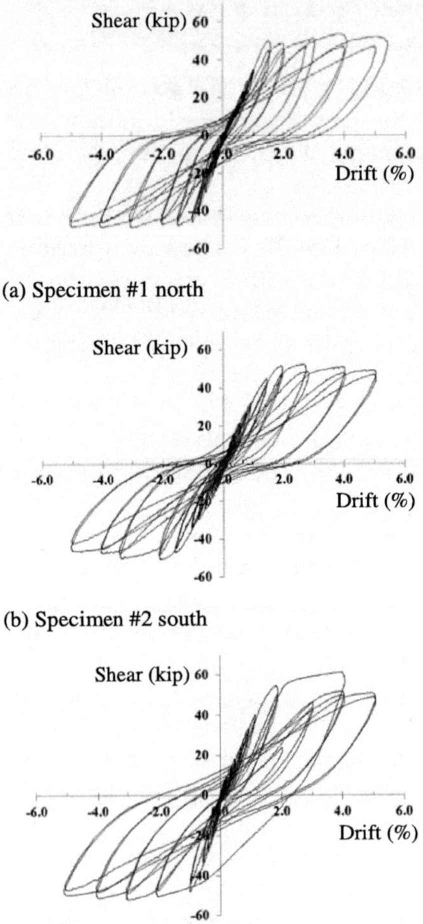

(a) Specimen #1 north

(b) Specimen #2 south

(c) Specimen #3 south

Fig. 3. Shear vs. drift, beams (1 kip = 4.45 kN)

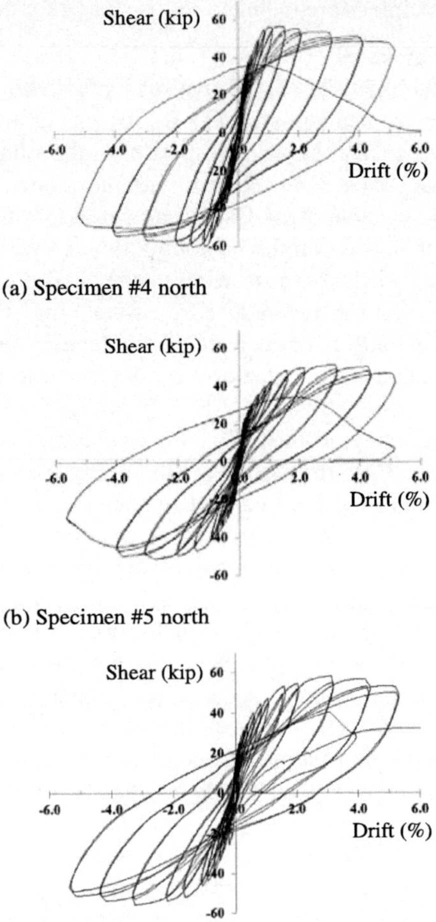

(a) Specimen #4 north

(b) Specimen #5 north

(c) Specimen #6 north

Fig. 4. Shear vs. drift, columns (1 kip = 4.45 kN)

Columns:

Specimen #4, CC-3.3-20, was designed according to ACI 318-08 requirements for special moment frame columns. Fig. 4(a) shows the measured shear-drift response. The north column completed the first cycle of step 12 of the loading protocol (Table 2). During the second cycle to 5% drift, the longitudinal bars buckled at a drift ratio of about 1%. A plausible explanation for the specimen failing at a drift ratio near zero is that the cracks on both sides of the specimen were open.

The axial load was carried predominantly by the longitudinal reinforcement, which led to buckling.

Specimen #5, UC-1.6-20, had properties that were similar to Specimen #4, with the exception that it was reinforced longitudinally with UHSS using about half as much longitudinal reinforcement. Fig. 4(b) shows the measured shear-drift response. Compared with Specimen #4, Specimen #5 showed reduced post-cracking stiffness and increased yield deformation. The north column completed the first half-cycle to 5% drift but the longitudinal bars buckled during the second half-cycle at that drift ratio. Testing was continued and the remaining longitudinal bars buckled at a small drift ratio (approximately 2%) during the second cycle to 5% drift. Again, a plausible explanation is that the longitudinal bars need to carry a larger fraction of the axial load when the cracks in the concrete were open.

Specimen #6, UC-1.6-20F, had properties that were similar to Specimen #5, with the exception that the concrete matrix consisted of HPFRC. The spacing of the transverse reinforcement was nearly doubled to evaluate the influence of HPFRC in shear strength, confinement, and bar buckling. Fig. 4(c) shows the measured shear-drift response. This specimen successfully completed the 12-step loading protocol. The fibers were effective in reducing the amount of spalling of the concrete cover and providing lateral support to the longitudinal reinforcement. Therefore, unlike Specimen # 4 and #5 the reinforcement did not buckle. However, in the final push, one of the longitudinal bars fractured in tension at a drift of approximately 3%. Thus adding fibers changed the mode of failure from buckling of the compression bars to fracture of the tension bars.

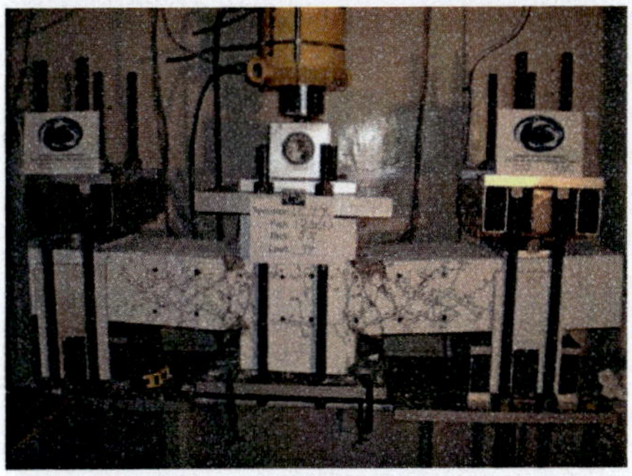

Fig. 5. Specimen #2 at 5% drift

Fig. 6. Specimen #3 at 5% drift

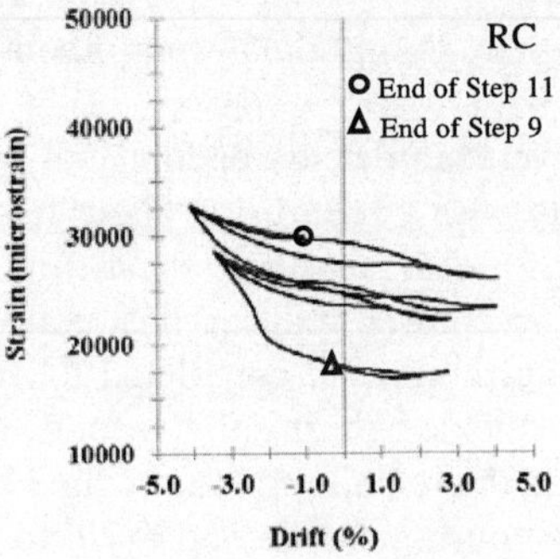

Fig. 7. Measured strain at top longitudinal bar at face of stub, Specimen #2 south

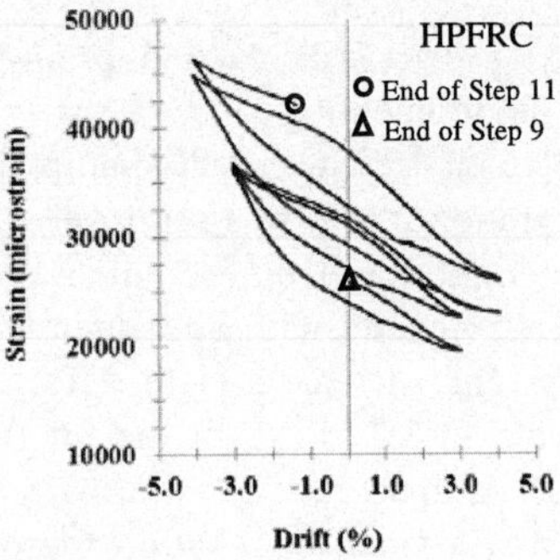

Fig. 8. Measured strain at top longitudinal bar at face of stub, Specimen #3 south

6 Conclusions

Based on the experimental data presented and within the scope of variables tested the following conclusions are made:

(1) For reinforced concrete beams and columns without fibers:

a) Replacing conventional longitudinal reinforcement with reduced amounts of ultrahigh strength reinforcement maintained flexural strength and did not decrease the deformation capacity. The tested beams tolerated drift ratios in excess of 10% without failure while the column specimens tolerated drift ratios of 5%. Column failures were due to bar buckling.

b) Reducing the amount of longitudinal reinforcement and increasing the yield strength of the reinforcement reduced post-cracking stiffness and increased yield deformation, leading to a reduction of the area inside the load-deformation hysteresis loops.

(2) For high-performance fiber-reinforced concrete beams and columns:

a) Reducing the amount of transverse reinforcement to about half the amount used in the specimens without fibers did not reduce the deformation capacity. HPFRC beams and columns with ultrahigh strength longitudinal reinforcement tolerated drift ratios in excess of 10% and 5%, respectively. Specimen failures were due to bar fracture.

b) The stiffness reduction associated with increased cyclic deformation was less pronounced in the HPFRC specimens than in those without fibers. For cycles above 1% drift, the ratio of the unloading stiffness to first-yield stiffness (shear-drift curve) is about 20% higher in HPFRC specimens.

References

[1] ACI 318-08, Building Code Requirements for Structural Concrete and Commentary, Reported by ACI Committee 318, American Concrete Institute, Farmington Hills, MI, p. 465 (2008)
[2] ASTM A1035/A1035M-05. Standard Specification for Deformed and Plain, Low-Carbon, Chromium, Steel Bars for Concrete Reinforcement. ASTM International, West Conshohocken, PA, p. 5 (2005)
[3] Parra-Montesinos, G.J.: High-Performance Fiber-Reinforced Cement Composites: An Alternative for Seismic Design of Structures. ACI Structural Journal 102(5), 668–675 (2005)
[4] Liao, W.-C., Chao, S.-H., Park, S.-Y., Naaman, A.E.: Self-Consolidating High-Performance Fiber Reinforced Concrete (SCHPFRC) – Preliminary Investigation. Report UMCEE 06-02, University of Michigan, Ann Arbor, p. 68 (2006)
[5] Foster, S.J.: On Behavior of High-Strength Concrete Columns: Cover Spalling, Steel Fibers, and Ductility. ACI Structural Journal 98(4), 583–589 (2001)
[6] ATC . FEMA 461 / Interim Testing Protocols for Determining the Seismic Performance Characteristics of Structural and Nonstructural Components. Applied Technology Council, Redwood City, CA, p. 113 (2007)

Dynamic Behavior of HPFRCC at High Strain Rate: The Fiber Role

A. Caverzan[1], E.Cadoni[2], and M. di Prisco[1]

[1] Department of Structural Engineering, Politecnico di Milano, Milan, Italy
[2] Department of Environmental Constructions and Design, SUPSI, Lugano, Switzerland

Abstract. High Performance Fiber Reinforced Cementitious Composites are characterized by high ductility and tension hardening behavior in statics. The high energy absorbed during the fracture process is associated to multiple cracking and pull-out phenomena. The good response highlighted in the static field by many researchers suggests to use these materials also for dynamic loading applications. However, very few investigations have been carried out in order to guarantee the improvement of the dynamic response. Current understanding of the impact resistance of cementitious composites, and especially of high-strength concrete, is very limited. An experimental research aimed at contributing to the understanding of the behavior of advanced fiber-reinforced cementitious composites subjected to low and high strain rates was carried out. The material investigated is a steel fiber reinforced mortar. Straight high carbon steel micro-fibers were used. The material behavior was investigated at three strain rates (0.1, 1 and 150 s^{-1}) and the tests results were compared with their static behavior. Tests at intermediate strain rates (0.1–1 s^{-1}) were carried out by means of a hydro-pneumatic machine (HPM), whilehigh strain rates (150 s^{-1}) were investigated by exploiting a modified Hopkinson bar (MHB). A comparison between static and dynamic tests highlighted several relevant aspects regarding the influence of fibers on the peak strength and post-peak behavior at high strain rates.

1 Introduction

The mechanical behavior of fiber-reinforced cementitious composites when subjected to impact or blast still has many aspects requiring further investigation [1], with specific reference to large and socially-sensitive structures, as sheltering structures, high-rise buildings, bridges, off-shore platforms, pipelines, gasification reactors, secondary containment shells for nuclear power plants, and tunnels. The mechanical response of these structures exposed to blast and impact loading could only be predicted - and controlled - by formulating proper material models for cementitious composites, including strain rate effects. The results available in the

G.J. Parra-Montesinos, H.W. Reinhardt, and A.E. Naaman (Eds.): HPFRCC 6, pp. 339–346.

literature are also often contradictory. For example, while some researchers [2-4] reported a progressively reduced sensitivity to strain rate with the growing of the static compressive strength of concrete, other studies contrast this opinion [5].

High performance fiber cementitious composites are characterized by high toughness and at least a hardening behavior in bending. Fibers enhance the ductility of brittle materials like concrete, and this improvement is strictly related to the process by which load is transferred frommatrix to fibers and the bridging effect of fibers across the cracks. Hence fiber pull-out is the principal mechanism contributing to the high toughness of the material, thus preventing fiber failure mechanism. Many researchers studied the pull-out sensitivity to loading rates for different types of fiber (polypropylene, hooked, twisted and smooth steel fibers) [6, 7]. They concluded that polypropylene and "deformed" steel fibers were sensitive to the loading velocity, while smooth steel fibers were insensitive to it. Since, the behavior of the fiber, the cement matrix and the bond interface (or the pull-out mechanism) are likely dependent on the strain rate, it is expected that the strain rate sensitivity moves to the composite level. Moreover, due to the interaction for HPFRCC, at low strain rate, crack growth stability and pull-out mechanism, the only investigation of the pull-out rate dependency could be not sufficient to predict the rate sensitivity of the material response at high strain rate.

Very few studies concern with the effect of strain rate on the tensile behavior of HPFRCC [8, 9]. Kim et al. have investigated the tensile response of HPFRCC reinforced with twisted and hooked steel fibers [8]. They reported that, in general, HPFRCC specimens with twisted fibers are generally sensitive to strain rate, whereas their counterparts with hooked fibers are generally not. While Kim et al. focused their attention to a range of strain rates proper of the seismic problems (0.1 s^{-1} and lower strain rate), Cadoni et al. studied the behavior of two HPFRCC (reinforced with two different types of fiber: polyvinyl-alcohol and steel) at high strain rates (50 s^{-1} and 120 s^{-1}) by means of a modified Hopkinson bar (MHB) [9]. They observed, after the dynamic failure, that steel fibers were pulled-out from the matrix, while the majority of polyvinyl-alcohol fibers were subjected to tensile failure: according to the authors opinion this could originate the difference in terms of fracture energy. On the other hand, the tensile strength of the steel fiber composite grows with the strain rate while the tensile strength of PVA composites seems to be less sensitive to strain-rates even though it remarkably increasesin relation to the static values.

However, the research quoted was not focused on the influence of fiber reinforcement in the strength increase at high strain rates, if any. Furthermore, it is not known at what extent such increase could vary with strain rate. In this paper the fiber role is investigated at different strain rates comparing the results obtained for a plain and fiber reinforced HPCC.

2 Mix Design and Manufacturing Process

The constituents of the HPCC material, sand (0-2 mm), slag, water, superplasticizer and cement were mixed in the following mass ratios: 1.64/ 0.83/ 0.33/ 0.06/ 1. The steel fibers considered for the reinforced material were high carbon straight fibers, 13 mm long with a 0.16 mm diameter (aspect ratio l_f / d_f equal to 80); their proportion in the mix was equal to 100 kg/m^3.

Two 1.6 m x 0.60 m in plane slabs, 30 mm thick, were manufactured. The slabs were cast by applying a unidirectional flow. In order to guarantee a certain fiber orientation, when the reinforced material is considered, the properties of the self compacting material were used, taking advantage of the flow direction. Three prismatic beam samples, 40 mm wide and 600 mm long, were sawed from the slab and tested in bending to perform a proper mechanical characterization of the material according to Italian Guidelines.

The high fiber content and the favorable orientation imposed by the casting flow control allow us to guarantee a hardening behavior in uniaxial tension characterized by multi-localization (micro-cracking) and a small dispersion of the response before and even after the final single macro-crack localization [10, 11]. From the bent specimens, several small cylinders 20 mm long were cored in the direction of tensile stresses. These specimens are the object of the present work. Their diameter was nominally equal to 20 mm (Fig. 1). Each cylinder cored from each prismatic specimen was notched (notch depth = 1.5 mm), to be tested in uniaxial tension at different loading rates.

3 Static Tests

Uniaxial tensile tests on notched cylindrical specimens (S0 series) were performed with an electro-mechanical testing machine INSTRON 5867 at the laboratory of the Politecnico di Milano - Polo Regionale di Lecco. The press has a maximum loading capacity of 30 kN. The samples were glued to the press platens by means of an epoxy resin. Two aluminum cylinders, connected to the press by a knuckled joint (Fig.1), were used as press platens. In both cylinders, a 5 mm deep cylindrical cavity with a 22 mm diameter was made in order to increase the glued sample surface. Stroke was considered as feedback parameter during the tests. The displacement rate imposed during the tests was equal to 5×10^{-5} mm/s up to 1.5 mm and then it was progressively increased to 10^{-3} mm/s.

Test results are shown in Fig. 2 in terms of nominal stress (σ_N) versus crack opening displacement (COD) for the HPFRCC material, while in Table 1 are reported the peak strength values, the effective cross section area and the number of fibers counted in the failed sections for the HPFRCC and the HPCC as well. Due to a defect identified in the sample 3 only two specimens were tested for the

HPCC. Considering the cavity depth of the platens (5 mm for each side, where samples are glued) as a rigid zone, a clear span equal to 10 mm can be used to compute the strain, starting from the COD values.

Fig. 1. Press platens and sample

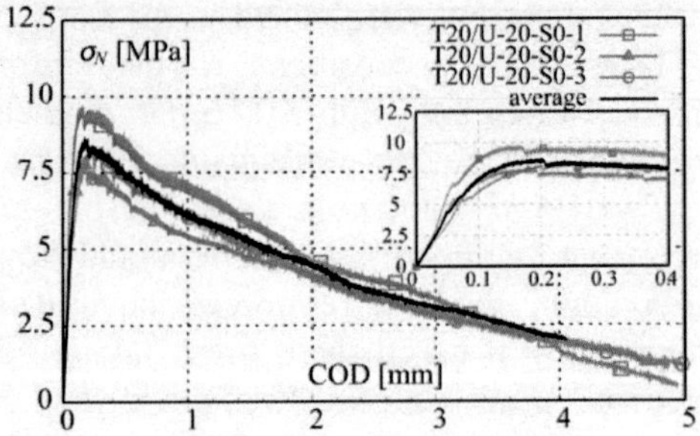

Fig. 2. Tests results for the HPFRCC

Table 1. Static peak strength

Material type	Sample n°	Fiber n°	Area [mm^2]	f_{peak} [MPa]	$f_{peak,ave}$[MPa] (std)
	1	75	181.70	9.71	
HPFRCC	2	73	184.82	7.83	8.62 (0.97)
	3	77	173.66	8.33	
	1	-	154.71	6.81	
HPCC	2	-	166.96	5.50	6.15 (--)
	3	-	150.11	--	

On the basis of this assumption, it is interesting to observe that steel fibers are able to favor stable propagation up to a strain of about 2%. The pre-peak behavior is well described by a parabola rectangular model. This would mean that, even if a notch depth ratio equal to 0.15 was used, the material is hardening and it distributes over the whole clear length the cracking process. Of course, the geometrical defect introduced by the notch and the only stroke measure prevents a correct evaluation of the effective ductility measured in terms of strain of the peak plateau. New tests without any notch are in progress. The low values of the elastic stiffness obtained in these tests could be caused by a visco-elastic behavior of the glue. In order to account for the contribution of this behavior, some preliminary tests aimed at evaluating the elastic modulus of the material were carried out. These tests showed an upper limit for the elastic modulus equal to 40 GPa. By comparing at the elastic limit the effective stiffness of the glued samples with the one computed on the basis of 40 GPa E modulus it was possible to estimate the maximum value of the error affecting the strain measures. The difference between the two strain values discussed at the peak is equal to about 1%. The peak strains could be thereafter overestimated as maximum of 1%. A peak strength increase of about 40% can be pointed out comparing the results obtained for the HPCC. The fibers favor a more stable crack propagation leading to the increase of peak strength in the static field.

4 Dynamic Tests

In order to identify the influence of fibers on the material response under dynamic conditions an experimental program was carried out. Two different mechanical testing machines were used to investigate the dynamic field: a hydro-pneumatic machine (HPM) was employed to investigate the strain rate equal to $1s^{-1}$ while a modified Hopkinson bar (MHB) [12] was exploited to carry out high strain rate tests (150 and $300s^{-1}$).

The HPM functioning is widely described in [13], here only a briefly description of the device is reported. At the beginning of the test, a sealed piston divides the cylindrical tank into two chambers, one being filled with gas at high pressure (viz. 150 bars), and the other with water. An equal pressure is initially established in the water and gas chambers so that the forces acting on the two faces of the piston are in equilibrium. The test starts when the second chamber discharges the water through a calibrated orifice that is activated by a fast electro-valve. The piston starts then to move, expelling the water. The specimen (S1 and S2 series) is connected on one side to the piston shaft and, on the other side, to the end of an elastic bar which is rigidly fixed to a supporting structure. The elastic bar is instrumented with a strain-gauge that provides, through the elastic properties of the bar, the force acting on the specimen during the test. Two targets are attached at both ends of the specimen and their movement is measured by two contact-less displacement transducers.

The MHB consists of two circular aluminum bars, called input and output bars (with a diameter of 20 mm and having length of 3 and 6m, respectively) between which the specimen (S3 series) is glued using a bi-component epoxy resin. The input bar is connected to a high strength steel pretension bar (having 6 m length and 12 mm diameter), used as pulse generator. A test with the MHB starts pulling the high strength steel bar by means of a hydraulic actuator (of maximum loading capacity of 600 kN); the pretension stored in this bar is assured by the blocking device. In the second operation the fragile bolt in the blocking device is the rupture, giving rise to a tensile mechanical pulse of duration 2.4 ms and with a linear loading rate during the rise time. The pulse then propagates along the input and output bars, leading the specimen to failure.

In Figs. 3, 4 and 5 the stress versus COD curves of the HPCC specimens tested at 0.1, 1 and 150 s^{-1} respectively are reported. In Fig. 6 only the average curves for the HPFRCC specimens are presented [14]. The peak strength values, the effective cross section area and the number of fibers counted in the failed cross sections are listed in Table 2. Analyzing the test results, several interesting considerations can be highlighted. First of all the scatter in the stiffness for all and only the HPCC samples tested at intermediate strain rates carried out by means of HPM device is quite high. A possible reason for that could be related to the occurrence of wrong displacement measurement of the specimen ends because only one gauge was used for each end. It is worth observing the increasing of the initial stiffness with the growth strain rate. This increase is much more evident when the strain rate is equal to 150 s^{-1}.

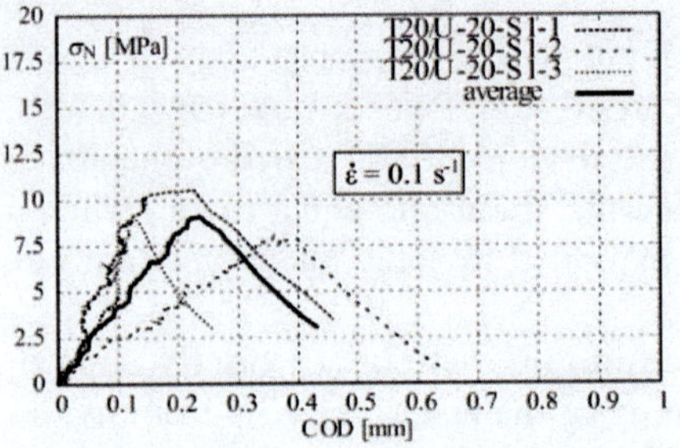

Fig. 3. Test results for the HPCC: $\dot{\varepsilon}$ =0.1s^{-1}

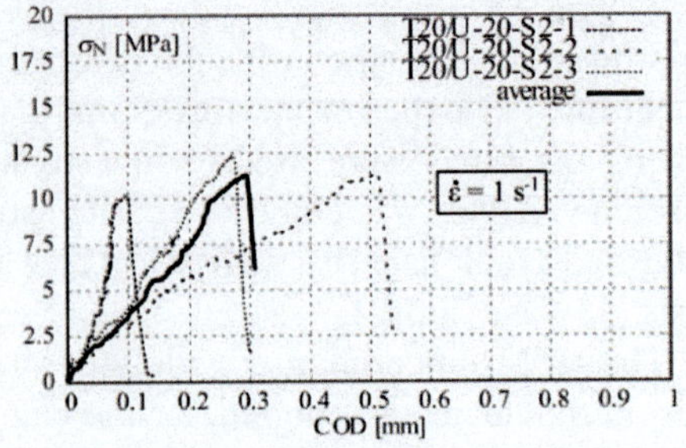

Fig. 4. Test results for the HPCC: $\dot{\varepsilon}$ =1s^{-1}

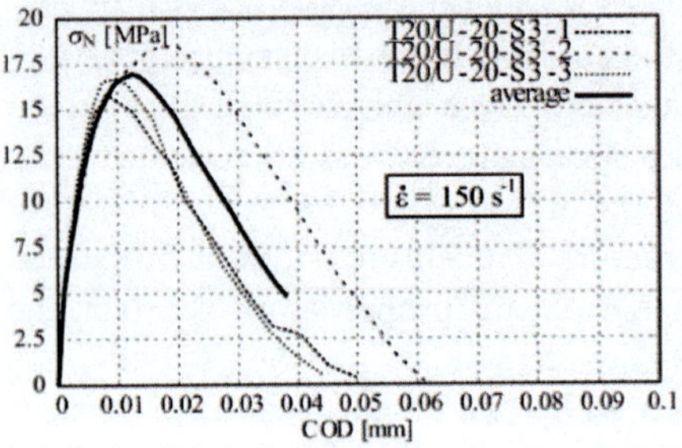

Fig. 5. Test results for the HPCC: $\dot{\varepsilon}$ =150s^{-1}

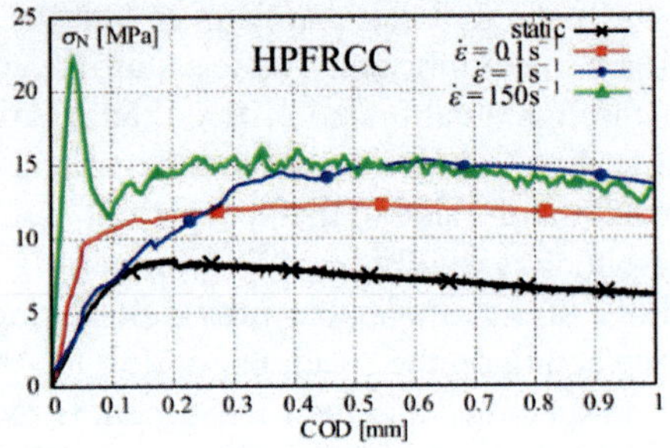

Fig. 6. Average curves for the HPFRCC

When the strain rate grows, the initial stiffness increases together with the peak. In order to evaluate the peak strength trend in the dynamic field the dynamic increase factor (DIF) can be considered. The dynamic increase factor is computed as the ratio between the dynamic strength of the material ($f_{p,d}$) and the static strength of the material ($f_{p,s}$). In Fig. 7 DIF versus strain rate is plotted for the HPFRCC and the HPCC as well.

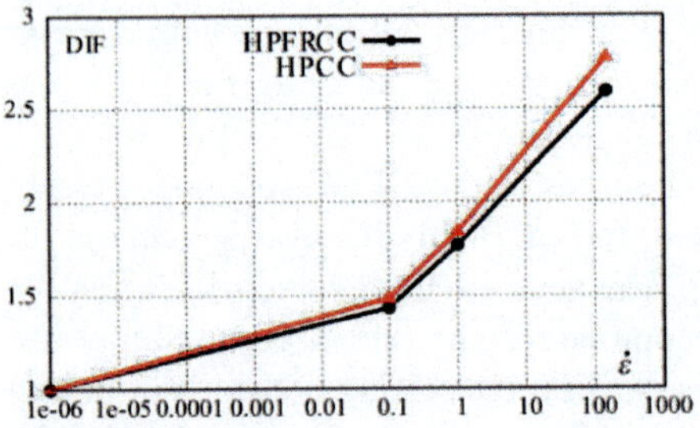

Fig. 7. DIF versus strain rate

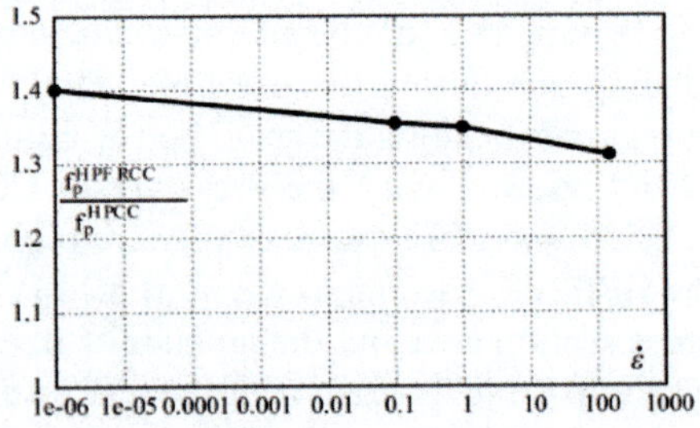

Fig. 8. Peak strength ratio versus strain rate

The DIF parameter can be used to quantify the rate sensitivity of these materials which appears comparable in the strain rate range investigated. However the HPCC material denoted a higher values of DIF for all the strain rates investigated; by growing the strain rate, the difference in terms of DIF between HPCC and HPFRCC increases (passing from 0.1 to 150 s-1 the difference/DIF_{HPCC}- DIF_{HPFRCC}/ increases from 0.05 to 0.18).

Table 2. Dynamic peak strengths

Material type	Strain rate $[s^{-1}]$	Sample n°	Fibers n°	Area $[mm^2]$	f_{peak} [MPa]	$f_{peak,ave}$[MPa] (std)
HPFRCC	0.1	1	71	173.90	13.07	12.36 (1.81)
		2	76	167.64	10.30	
		3	72	166.96	13.71	
HPFRCC	1	1	76	181.46	14.29	15.25 (1.65)
		2	71	188.69	14.09	
		3	73	175.77	17.67	
		4	77	163.09	14.96	
HPFRCC	150	1	79	183.85	20.46	22.36 (1.89)
		2	90	194.58	22.28	
		3	79	191.13	24.96	
		4	81	165.13	21.74	
HPCC	0.1	1	-	161.28	10.54	9.12 (1.34)
		2	-	159.48	7.87	
		3	-	147.41	8.95	
HPCC	1	1	-	158.37	10.22	11.31 (1.07)
		2	-	159.04	11.33	
		3	-	154.38	12.36	
HPCC	150	1	-	155.92	15.89	17.05 (1.32)
		2	-	159.48	18.48	
		3	-	162.41	16.77	

To clarify fiber contribution to peak strength increase, the ratio between HPFRCC and the HPCC peak strengths (f_p^{HPFRCC}/f_p^{HPCC}) for all the strain rates investigated is plotted in Fig. 8. Observing the Fig. 8 several interesting considerations can be highlighted. In the static field fibers strongly influence the stable crack propagation leading to an increase of the peak strength of about 40%. When the strain rate increases a reduction of the fibers influence can be observed. Passing from 0.1 to 150 s^{-1} peak strength grows of about 35-30%.

4 Conclusions

The behavior of HPFRCC and HPCC materials when subjected to different strain rates was investigated. From the experimental results presented in this research work some important conclusions can be drawn:

- The comparison between static and variable strain rate tests, carried out by means of three different mechanical devices, exhibits high values of the dynamic increase factor for both the HPCC and the HPFRCC.

- The rate sensitivity of HPCC seems to be higher than what observed for HPFRCC. In fact, HPCC material has denoted a weakly higher values of DIF for all the strain rates investigated.
- Fibers increase the peak strength in static of about 40%, favoring the stable crack propagation. When the strain rate is increased, fiber influence on the stable crack propagation phenomena reduces: at high strain rates the peak strength ratio between HPFRCC and HPCC decreases to 30%.

References

[1] Bindiganavile, V., Banthia, N.: Impact and blast resistance of fiber reinforced concrete. Technical Report. ACI 544 Technical Commission (2007)

[2] Banthia, N., Mindess, S., Bentur, A., Pigeon, M.: Impact testing of concrete using a drop-weight impact machine. Experimental Mechanics 29, 63–69 (1989)

[3] Bentur, A., Mindess, S., Banthia, N.: Behavior of reinforced concrete under impact: The effect of concrete strength. Society for Experimental Mechanics, 449–458 (1987)

[4] Ross, C.: Review of strain rate effects in materials. In: Structures under Extreme Loading Conditions, ASME Pressure Vessels and Piping Conference, pp. 255–262 (1997)

[5] Bischoff, P.H., Perry, S.H.: Impact behavior of plain concrete loaded in uniaxial compression. J. Engrg. Mech. 121, 685–693 (1995)

[6] Banthia, N., Trottier, J.F.: Deformed steel fiber-cementitious matrix bond under impact. Cement and Concrete Research 21, 158–168 (1991)

[7] Kim, D., El-Tawil, S., Naaman, A.: Loading rate effect on pullout behavior of deformed steel fibers. ACI Materials Journal 105, 576–584 (2008)

[8] Kim, D.J., El-Tawil, S., Naaman, A.: Rate-dependent tensile behavior of high performance fiber reinforced cementitious composites. Materials and Structures 42, 399–414 (2009)

[9] Cadoni, E., Meda, A., Plizzari, G.: Tensile behaviour of frc under high strain-rate. Materials and Structures 42, 1283–1294 (2009)

[10] Ferrara, L., Ozyurt, N., di Prisco, M.: High mechanical performance of fibre reinforced cementitious composites: the role of casting-flow induced fibre orientation. Materials and Structures, 1–20 (2010)

[11] Caverzan, A., Colombo, M., di Prisco, M.: High performance steel fibre reinforced concrete: Residual behaviour at high temperature. In: Proceedings of the 2nd workshop on Performance, Protection and Strengthening of Structures under Extreme Loading (PROTECT2009), Hayama, Japan (2009)

[12] Cadoni, E., Albertini, C., Solomos, G.: Analysis of the concrete behaviour in tension at high strain-rate by a modified hopkinson bar in support of impact resistant structural design. J. Phys. IV France 134, 647–652 (2006)

[13] Asprone, D., Cadoni, E., Prota, A., Manfredi, G.: Dynamic behavior of a Mediterranean natural stone under tensile loading. Int. J. Rock Mech. Min. 46, 514–520 (2009)

[14] Caverzan, A., Cadoni, E., di Prisco, M.: Tensile behaviour of High Performance Fibre Reinforced Cementitious Composites at High Strain Rates. submitted to Int. J. Impact Eng.

Beam-Column Connections for Precast Concrete Frames Using High Performance Fiber Reinforced Cement Composites

L.F. Maya and L. Albajar

Department of Continuum Mechanics and Structures,
Universidad Politécnica de Madrid, Spain

Abstract. Precast construction system interest has been growing given the emphasis on improving work zone safety, reducing construction time and environmental impact, while maintaining the quality. The connections are the more important singularity of precast construction systems, being the general behavior of the precast structures related to their design, construction and performance. A rigid beam-column connection typology for precast construction is proposed. This typology takes advantage of the outstanding properties of the High Performance Fiber Reinforced Cement Composites (HPFRCC) and in particular its better bond conditions. The principal mechanism of connection is the splice of longitudinal beam reinforcement, which takes place at the end of the beam elements, outside the joint region. The configuration proposed avoids the interference between longitudinal and transverse reinforcement, reduces the *in situ* work and makes possible to define an efficient and safe construction process.

1 Introduction

Precast construction is regarded as an appealing alternative to be considered in a wide range of construction projects. It is due among other factors to the advantages related to the reduction in construction times, work force and in situ labors, as well as a more favorable cost-benefit relation, less environmental impacts, and greater control and final quality of the elements. However, the use of precast construction is sometimes limited by an inappropriate assessment of several typical singularities of these construction systems. Being the connections a critical factor and one of the most important singularities, they can determine the general structure behavior. Besides, the connections represent the construction process stage developed *in situ*, during which common problems at the structure assemblage process have to be faced.

Several studies have been conducted over the last 25 years to evaluate the efficacy of using Fiber Reinforced Concrete (FRC) to improve the behavior of beam-column connections. More recently, different proposals to incorporate High Performance Fiber Reinforced Cement Composites (HPFRCC) in beam-column

G.J. Parra-Montesinos, H.W. Reinhardt, and A.E. Naaman (Eds.): HPFRCC 6, pp. 347–354.

connections have been successfully tested [1, 2]. This research deepens on the same line by considering the incorporation of HPFRCC to develop continuity connections among precast elements. In particular, the use of these materials in reinforcement splices using short splice lengths is studied. A two stage experimental study was carried out in collaboration with the Technical Department of PRAINSA. During the initial stage, four beam flexion tests were performed to experimentally assess the behavior of short splices in simple flexural elements. In the second stage, a new interior beam-column connection for moment resisting frames was proposed and tested. This paper is devoted to describe the experimental results of the second stage of the study. A detailed description and theoretical analysis for first experimental stage is presented in [3].

2 Test Program

Four beam-column connections of precast elements were fabricated and tested. The elements represent a typical interior beam-column joint of a multi-storey frame bounded by the point of contra-flexure in the members. The tested elements consisted of a single column element and two beam elements framing at opposite sides of the joint, which were connected through the splice of the longitudinal reinforcement in the beam elements. The design process was performed according to the *strong column and weak beam* approach, as suggested in most of the modern design codes [4, 5]. A design detail of the test elements is shown in Fig. 1. The design of the elements was defined by the construction process proposed. The beam elements had channels at the top and the bottom, where the splices of the reinforcement took place. The channels were connected by two vertical ducts to allow the pouring of the HPFRCC *in situ*. The beam ends at each side of the column were different for practical erection-building purposes. The columns had two windows at the level of the beam channels to allow the splice bars to pass through the joint. Additionally, two other windows were left below the joint section to fix provisional supports during the construction process. Once the column had been erected, the beam with the longer channel was rested on a provisional support and the splice bars were slid through the windows in the column until the bar ends did not protrude from the column outer face. Then, the second beam was rested on the support at the opposite side of the column and the splice bars were slid to their final position, so that they spliced the longitudinal reinforcement of both beams.

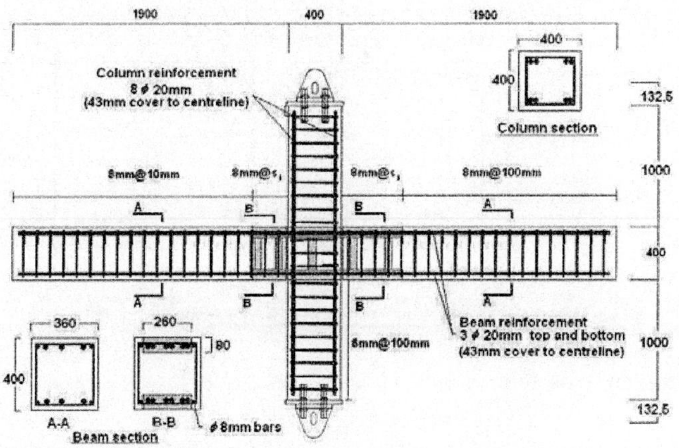

Fig. 1. Test element configuration and reinforcement details (unit:mm)

The configuration and construction process proposed enable to overcome some common drawbacks of beam-column connections for precast structures. Firstly, connection takes place outside the joint region, which is usually subjected to high stress demands. Furthermore, it is possible to provide closed stirrups all along the beam, including the beam ends close to the column faces, and into the joint core, where the confinement and shear strength are important to assure the proper element behavior and to fulfill the ductility requirements. Likewise, the HPFRCC poured *in situ* enables to establish some continuity among the precast elements, not being the sections at the column faces completely discontinuous.

The test elements were designed such that the shear demand was high but it did not limit the *strong column and weak beam* approach. Two main variables were considered; the splice length and the transverse reinforcement spacing in the splice region. The splice lengths tested were 200 mm ($10d_b$) and 300 mm ($15d_b$), while the spacing of the transverse reinforcement in the splice region was set as 100 mm and 150 mm, according to the beam test carried out previously [3]. Table 1 summarizes the test units fabricated. The reaction frame used in this study is an adaptation from the test setup described in [2], Fig. 2. The beams and the column were pin supported at their ends. The vertical displacements at the beam ends were restricted by double hinged supports, which also acted as pseudo-load cells. The top of the column was restrained in the horizontal plane but allowed vertical displacements, while the loads were applied through the actuators at the bottom of the column. A rail was set on the floor to prevent the test elements from moving transversally. The test setup did not allow axial loads to be applied to the column. The cyclic loading history used is illustrated in Fig. 3. Two cycles of each load step were applied, starting at 0.5% and increasing by steps of 0.5% until a 2.0% drift was achieved. Then, the step increment was set to 1%.

All the precast elements were cast with self-compacting concrete, which average compressive strength tested on 150 mm x 300 mm cylinders and the age of testing are given in Table 1.

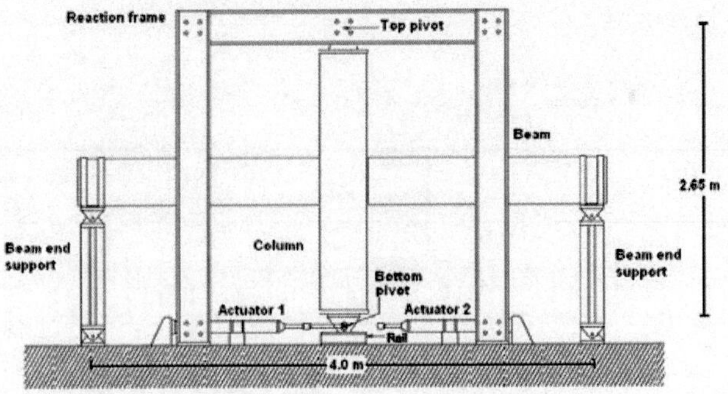

Fig. 2. Experimental test setup

The reinforcement consisted of bars of quality B500SD. According to the Spanish Code [5], the characteristic yield stress and ultimate strength of the reinforcement are 500 and 575 MPa, respectively. Compact Reinforced Composite (CRC), a HPFRCC marketed by *Hi-Con*, was poured *in situ* to configure the splice regions. The average compressive strength tested on 100 mm side cubic specimens and the age of testing are also given in Table 1.

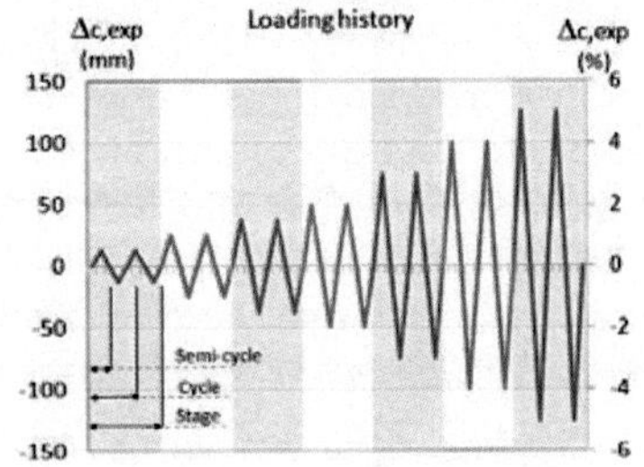

Fig. 3. Loading history

The lateral drift at the bottom of the column was measured by a linear variable displacement transducer (LVDT). Four LVDTs were located at the beam ends close to the column faces to measure the width of the foreseen cracks at the beam-column junctions. Additionally, other four LVDTs were located at the beam ends to measure the average strain on a base length of 400 mm in the CVP-1A element and 300 mm in the rest of the elements (see Fig. 5). A pseudo displacement control was implemented for the load steps, according to the actuator and testing facilities available. The load was first applied in one direction by means of an actuator. Then, the unit was unloaded and loaded in the opposite direction by means of another actuator to complete the cycle. Unfortunately, due to the characteristics of the actuators used, it was possible to register the load steps but not the unloading steps. Although the load history was registered by the load cell in the actuator, the vertical supports at the beam ends were also instrumented to act as pseudo-load cells, since the friction on the rail was an uncertainty.

Table 1. Element and material properties

Element	Splice configuration		Material properties			
	Splice length (mm)	Transverse reinforcement spacing s (mm)	Precast Elements		HPFRCC	
			f_c (MPa)	Age of testing (days)	$f_{c,HPFRCC}$ (MPa)	Age of testing (days)
CPV-1A	300	100	53.6	203	147.1	190
CPV-2A	300	150	52.6	216	145.0	205
CPV-1B	200	100	59.7	226	127.0	215
CPV-2B	200	150	59.5	231	133.4	219

3 Test Results

The general behavior of the four tested elements corresponded with the expected behavior for rigid frame elements subjected to lateral forces. From the first loading steps, the tested elements were cracked at the beam-column junction. Additional flexural cracks were observed along the beam elements outside the HPFRCC poured region, increasing their width and depth during the test. Horizontal cracks were observed in the column elements with approximately the same spacing of the transverse reinforcement, as well as a pair of horizontal cracks that defined completely the joint region. Furthermore, diagonal shear cracks were observed in the joint but they did not compromise the integrity of this detailed region. The load-drift curves of all the elements are shown in Fig. 4.

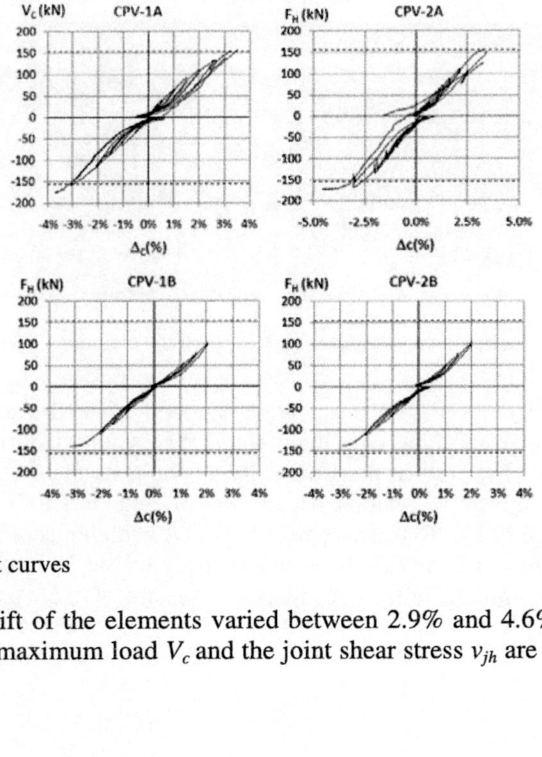

Fig. 4. Load-drift curves

The story drift of the elements varied between 2.9% and 4.6%. The measured and calculated maximum load V_c and the joint shear stress v_{jh} are listed in Table 2.

The calculated strength is based on the flexural strength of the reinforced beams, assuming monolithic behavior of the frame element. The actual concrete properties listed in Table 1 and reinforcement bars of quality B500SD were considered. Furthermore, for the reinforcing steel an overstrength factor λ of 1.25 was considered.

Table 2. Experimental results

Element	$V_{c,max}$ (kN)	$\frac{V_{c,max}}{V_{c,n}}$	$v_{jh,max}$ (MPa)	$\frac{v_{jh,max}}{\sqrt{f_c}}$
CPV-1A	174.7	1.10	7.6	1.04
CPV-2A	172.2	1.09	7.5	1.03
CPV-1B	138.1	0.87	6.0	0.78
CPV-2B	135.7	0.86	5.9	0.77

Elements CPV-1A and CPV-2A attained approximately 9% higher lateral load than the calculated value, while elements CPV-1B and CPV-2B attained approximately 85% of the strength calculated. The failure of elements CPV-1B and CVP-2B was determined by the short splice length, 200 mm ($10d_b$). The crack width at the beam-column junction increased due to the progressive loss of bond capacity in the splices as the load and drift were also increased with the cycles. Element CVP-2B, which had a lower transverse reinforcement ratio, exhibited a more fragile failure than element CVP-1B, being observed an extensive cracking in the splice region. A single but wider crack was observed at the beam-column junction in the splice region of element CPV-1B, as is show in Fig. 5.

Fig. 5. Splice regions under tension. Element CVP-2B (left), element CVP-1A (right)

Element CVP-1A and CVP-2A tests were stopped once the loading process became unsafe, given the type of actuators and the pseudo-displacement control process employed. However, at the final stage of the tests there was no evidence of bond failure in the splice regions. The aspect of element CPV-1A at the 3.7% drift cycle is shown in Fig. 6. Both element CPV-1A and element CVP-2A did not exhibit fragile failure, which indicates the adequacy of the 300 mm ($15d_b$) splice length for the proposed precast connection system. Other precast yielding connections but with different typologies than the element tested in this research have been satisfactorily assessed for moment-resisting frames [2, 6, 7, 8]. Unlike the cast in place systems, the complete formation of plastic hinges in the precast beams was not observed, in agreement with the expected behavior of yielding connections. Moreover, the continuity established among the precast elements by the HPFRCC poured *in situ* and the better bond condition enable a more uniform distribution of the steel strains in the critical region of the beam elements. The tension chord behavior is defined by the composite action between bar reinforcement and the HPFRCC, being the post-cracking behavior positively influenced by the action of the HPFRCC.

Finally, beam-column joint behavior involves an interaction between strut and truss mechanisms, where large shear and bond strength can be required [9]. The better bond condition of the reinforcement in the splice regions and into the joint core enabled to develop adequately the joint resistance mechanisms. The strut mechanism of the forces flow through the joint was observed in the element CVP-2A, as is shown in Fig. 6. However, the joint behavior did not limit the load capacity of the whole frame, despite the beam flexural overstrength.

Fig. 6. Element CVP-1A. Δc=3.7% (left), element CVP-2A, Δc=4.6% (right)

4 Concluding Remarks

A beam-column connection typology for precast construction is proposed and the following conclusions were obtained from the experimental results:

- The behavior of the precast connection proposed was adequate and corresponds with the expected behavior of yielding connections for moment-resisting frames.
- The use of HPFRCC in the splice region enables to reduce the splice length required. As a result, it is possible to propose a simple and efficient construction process.
- The construction process proposed aims to reduce *in situ* labors and construction time, while overcoming some common drawbacks of beam-column connections for precast structures.
- Closed stirrups can be provided all along the beam and into the joint core, where confinement and shear strength are important to assure the adequate element behavior.

- Joint damage is not extensive, given the localization of the connection outside the joint core. However, the strut and truss mechanisms were observed at the joint region, indicating adequate bond and shear strength.
- A splice length of $15d_b$ (300 mm) was enough to provide the required connection performance, preventing the occurrence of fragile bond failures. Other bonding materials will influence the ratio between splice length and bar diameter. A theoretical model has been developed to assess this ratio.

Additional analysis of the complete experimental results is being performed. Alternative proposals of bond connections for precast structures will be the subject of subsequent studies.

Acknowledgements. The authors gratefully acknowledge the financial support of this research from PRAINSA, Eduardo Torroja Institute for Construction Science, *Hi Con*, ACHE, and ANDECE. The technical support and advice of the Technical Department of PRAINSA and *Hi Con* were fundamental in carrying out this study. The authors would also like to extend special thanks to the technicians of the laboratory of the Eduardo Torroja Institute for Construction Science for their invaluable assistance.

References

[1] Parra-Montesinos, G., Peterfreund, S., Chao, S.: Highly damage-tolerant beam-column joints through use of high-performance fiber reinforced cement composites. ACI Structural Journal 102(3), 487–495 (2005)

[2] Brooke, N., Ingham, J.: Advanced fiber reinforced precast concrete beam-column joints. In: Proceedings of the New Zealand Concrete Industry Conference, Christchurch, New Zealand (2006)

[3] Maya, L.F., Albajar, L., Moran, F., Portabella, J.: Ultra High Performance Fiber Reinforced Concrete (UHPFRC) for connections between precast elements. Informes de la Construcción 62(510), 27–41 (2010) (in Spanish)

[4] American Concrete Institute, Building code requirements for structural concrete and commentary. ACI 318-08. ACI Committee 318, Farmington Hills, MI, USA (2008)

[5] Comisión Permanente del Hormigón, Instrucción de Hormigón Estructural EHE 2008 (2008)

[6] Castro, J., Yamaguchi, T., Imai, H.: Seismic performance of precast concrete beam-column joints. Journal of Structural Construction Engineering AIJ 455, 113–126 (1994)

[7] Alcocer, S., Carranza, R., Perez-Navarrete, D., Martinez, R.: Seismic test of beam-to-column connections in a precast concrete frame. PCI Journal 47(3), 70–89 (2002)

[8] Khaloo, A., Parastesh, H.: Cyclic loading response of simple moment-resisting precast concrete beam-column connection. ACI Structural Journal 100(4), 440–445 (2003)

[9] Paulay, T., Park, R., Priestley, M.: Reinforced concrete beam column joints under seismic action. ACI Structural Journal 75(11), 585–593 (1978)

A Summary of Ten Years of Research on HPFRC Coupling Beams

G.J. Parra-Montesinos, J.K. Wight, R.D. Lequesne, and M. Setkit

Department of Civil and Environmental Engineering,
University of Michigan, Ann Arbor, USA

Abstract. The design of coupling or "link" beams connecting structural walls in earthquake-resistant construction requires the use of intricate diagonal and transverse reinforcement detailing to ensure adequate strength, stiffness and energy dissipation during an earthquake event. The tensile strain-hardening behavior exhibited by high-performance fiber reinforced concretes (HPFRCs), along with their compression behavior that resembles that of well-confined concrete, led the senior writers to consider their use as a means to simplify the required reinforcement detailing in coupling beams, while leading to comparable or even enhanced seismic performance. Short coupling beams with a span-to-depth ratio (ℓ_n/h) of 1.0 were first investigated. Test results showed that HPFRC provides confinement to the diagonal reinforcement and increases coupling beam shear strength and drift capacity. This allows for a substantial reduction in both diagonal and confinement reinforcement without compromising deformation capacity. A follow-up study on coupling beams with $\ell_n/h = 1.75$ showed that a ductile flexural mechanism with high damage tolerance can be achieved through the use of HPFRC. A precast scheme with a short embedment length was shown to effectively anchor the beam into the wall without interfering with the wall reinforcement. Also, an HPFRC mixture with high-strength (2300 MPa) hooked steel fibers in a 1.5% volume fraction was found to be the most promising in terms of structural performance, economy and ease of construction. In order to cover the range of ℓ_n/h ratios common in practice, additional studies were conducted on more slender coupling beams, with $\ell_n/h = 2.75$ and 3.3. It was shown that slender precast HPFRC coupling beams can develop a high drift capacity and damage tolerance, even when diagonal reinforcement is eliminated. The results from this work thus provide structural engineers and contractors a viable design alternative for use in earthquake-resistant coupled wall construction.

1 Introduction

Structural walls are commonly used for lateral strength and stiffness in earthquake-resistant construction. Due to the need for window and/or door openings,

G.J. Parra-Montesinos, H.W. Reinhardt, and A.E. Naaman (Eds.): HPFRCC 6, pp. 355–362.

these walls are typically "pierced", which leads an otherwise solid wall to be "split" in two or more walls connected by relatively short beams referred to as "coupling" or "link" beams. Because of their low span-to-depth ratio, typically between 1 and 4, these beams require special detailing requirements to ensure adequate deformation capacity during earthquakes.

Current design practice for coupling beams is based on findings from research conducted in the early 1970s [1] and consists of the use of diagonal bars designed to resist the entire shear demand, combined with heavy amounts of confinement reinforcement (Fig. 1). While this reinforcement detailing has been shown to lead to adequate stiffness, strength, and deformation capacity under displacement reversals [2], it is labor intensive and often controls the construction schedule. There has thus been a need for simpler yet structurally efficient coupling beam designs that can be constructed more quickly and with less material and labor.

Fig. 1. Reinforcement detailing in diagonally reinforced coupling beam

Because of the strain-hardening behavior of HPFRC materials when subjected to direct tension, as well as their compression stress-strain response that resembles that of well-confined concrete, it was hypothesized that the use of HPFRC materials in coupling beams would allow a substantial reduction in both diagonal and transverse reinforcement without compromising strength, stiffness and deformation capacity. This hypothesis was confirmed through research conducted over the past ten years at the University of Michigan. A summary of the main findings from this research is presented herein.

2 HPFRC Materials Investigated

Three different HPFRC mixtures were used during the various experimental phases to evaluate their ability to increase shear strength and ductility, as well as to serve as partial replacement for diagonal and transverse reinforcement. The properties of the three types of fibers used are summarized in Table 1 and the HPFRC mixture proportions are summarized in Table 2. All three mixtures exhibited the desired strain-hardening response in tension and led to adequate structural performance in terms of shear strength, stiffness and deformation capacity. However, both ultra-high molecular weight (Spectra) and high-strength twisted steel (Torex) fibers had drawbacks that led to the adoption of a high-strength hooked steel fiber mixture for the majority of the coupling beam tests. Specifically, the mixture with Spectra fibers was expensive and difficult to mix, and the high-strength Torex fibers, although easier to mix, are not commercially available. As a result, the highly workable hooked steel fiber mixture design developed by Liao et al. [4] was adopted. Additional data regarding the mechanical properties of these mixtures can be found in References [3, 4].

Table 1. Fiber properties (specified by manufacturer)

Fiber Type	Length (in./mm)		Diameter (in./mm)		L/d	Tensile Strength (ksi/MPa)		Elastic Modulus (ksi/GPa)	
Spectra (Mixture 1)	0.5	13	0.0015	0.038	340	375	2570	17000	117
Torex (twisted steel) (Mixture 2)	1.2	30	0.012	0.3	100	360	2470	29000	200
Hooked Steel (Mixture 3)	1.2	30	0.015	0.38	80	333	2300	29000	200

Table 2. HPFRC mixture proportions by weight of cement

	Cement	Fly Ash	Sand	Agg.	Water	SP[1]	VMA[2]
Mixture 1 Spectra ($V_f = 2\%$)	1	0.15	1	0	0.4	0.02	0
Mixture 2 Torex ($V_f = 1.5\%$)	1	0.15	1	0	0.4	0.02	0
Mixture 3 Hooked ($V_f = 1.5\%$)	1	0.875	2.2	1.2*	0.8	0.005	0.038

[1] Super-plasticizer (Glenium 3200HES); [2] Viscosity Modifying Agent (Rheomac VMA 362)

* 13 mm (1/2 in.) maximum size

3 Coupling Beam Detailing

The first four coupling beam specimens tested had an aspect (span-to-depth) ratio of 1.0. The reinforcement of these specimens is shown in Fig. 2. Specimen 1 was constructed with conventional reinforced concrete and reinforced according to the requirements of the ACI Building Code [5]. Specimens 2 and 3 were constructed with HPFRC Mixture 1 (Table 2), and modified reinforcement detailing.

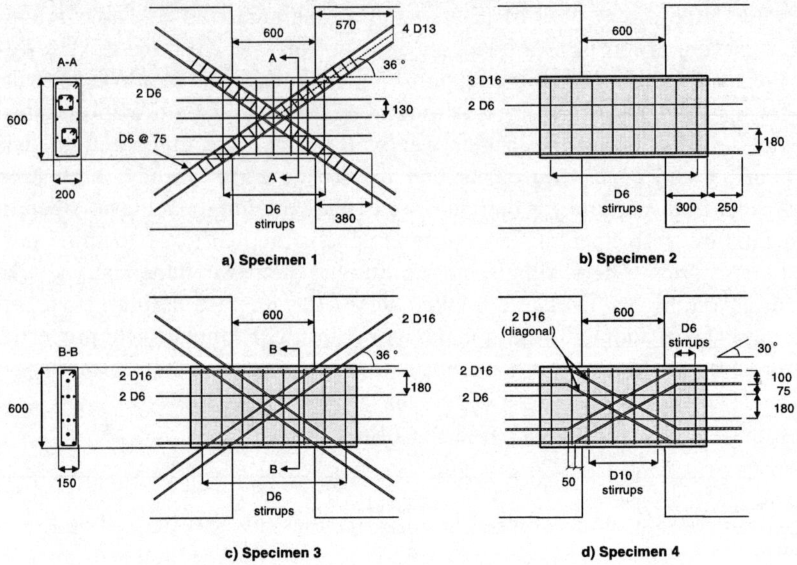

Fig. 2. Reinforcement of specimens with $\ell_n/h = 1.0$ [3]

Results of these tests [3] demonstrated that HPFRC can replace the confinement reinforcement around the diagonal bars, and showed that the HPFRC coupling beam can be precast and embedded into the adjacent walls. The test results of Specimen 4, which was constructed with HPFRC Mixture 2 (Table 2), showed that the diagonal reinforcement can be bent within the precast HPFRC section and exit the beam horizontally. This detail significantly simplifies placement of the precast section on the jobsite. Comparing the shear stress versus drift response of Specimens 1 and 4 (see Fig. 3), a significant improvement in shear strength is evident with the use of HPFRC, even though the area of diagonal reinforcement used in Specimen 4 was approximately 80% of that used in Specimen 1. For two vertical walls rotating the same amount when subjected to lateral displacements, drift is defined as the angle between a tangent passing through the beam end and the horizontal (beam chord).

Based on the results from the tests of coupling beams with an aspect ratio of 1.0, eight more coupling beam specimens with aspect ratios of 1.75, 2.75 and 3.3 were tested [6]. The reinforcement layouts proposed on the basis of these tests for coupling beams with aspect ratios between 1.5 and 2.5, and greater than 2.5, are shown in Figs. 4a and 4b, respectively. Special column-type transverse reinforcement is only provided at the beam ends because fiber reinforcement is sufficient to confine the remaining coupling beam span. The shear stress versus drift response of these specimens is shown in Fig. 5. The specimens were shown to develop a stable flexural response with energy dissipation and stiffness retention capacities comparable to those of well detailed diagonally-reinforced concrete coupling beams. This performance was achieved with a 70 and 100% reduction in diagonal

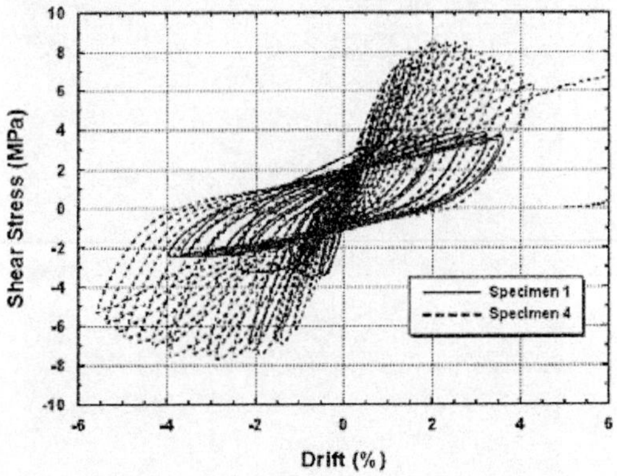

Fig. 3. Behavior of specs. 1 (RC) and 4 (HPFRC) [3].

reinforcement for moderate and slender coupling beams, respectively, when compared to a conventional reinforced concrete design. The shear force not resisted by diagonal reinforcement is assumed to be resisted by transverse reinforcement and the HPFRC material. A shear stress of $5\sqrt{f_c'}$, [psi] ($0.42\sqrt{f_c'}$, [MPa] was found to be a conservative limit for the contribution of the HPFRC material to shear strength.

A major advantage of the proposed HPFRC coupling beams is the ability to precast the HPFRC section (shaded grey in Fig. 4) and embed it into the walls without interfering with the wall reinforcement.

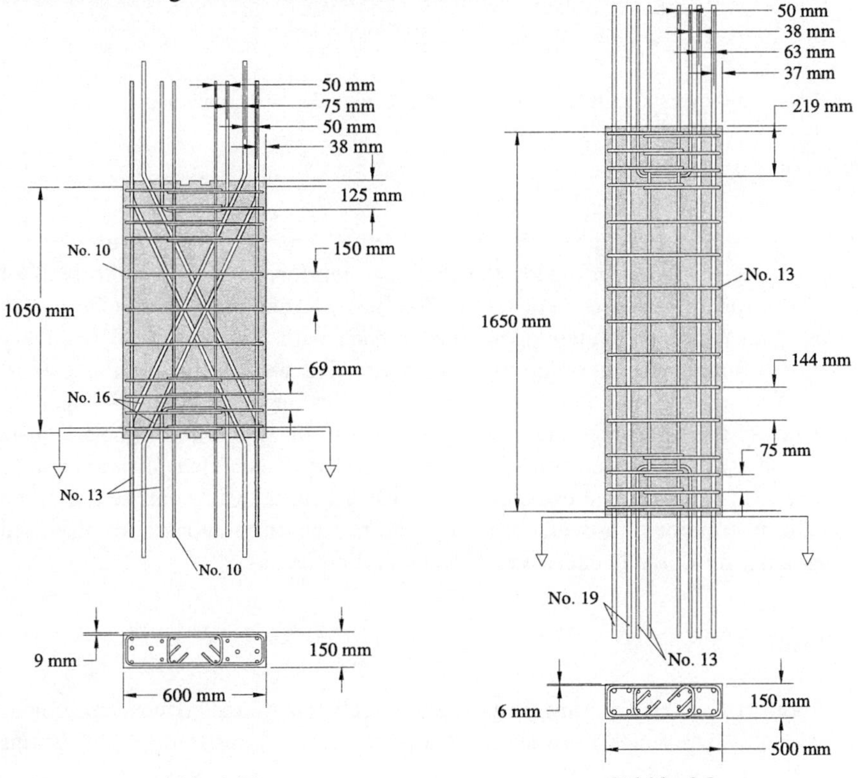

Fig. 4. Reinforcement detailing for "short" and "slender" HPFRC coupling beam specimens

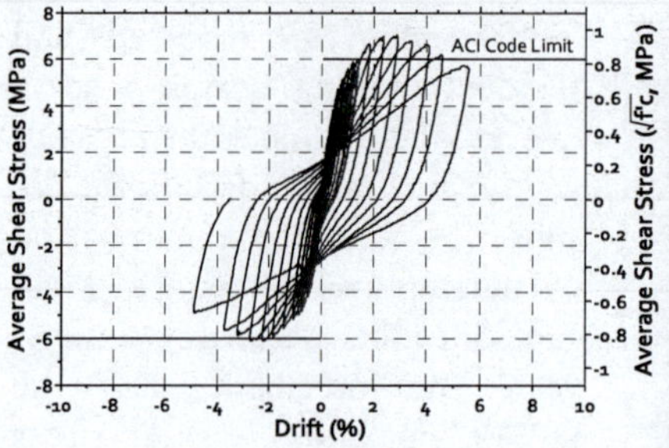

a) $l_n/d = 1.75$

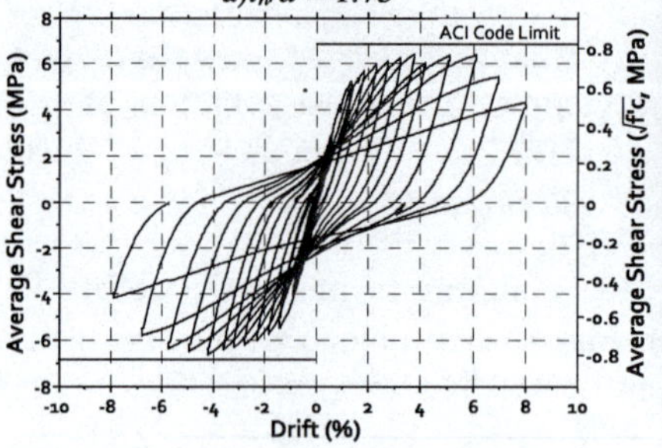

b) $l_n/d = 3.3$

Fig. 5. Shear stress versus drift response of "short" and "slender" coupling beam specimens

Fig. 6. Coupled wall specimen

This is achieved by embedding the precast HPFRC section only as deeply as the wall cover. Also, the bending of the diagonal reinforcement allows for all of the coupling beam reinforcement to exit the beam horizontally for anchorage in the wall. This limits the potential for interference with the dense wall boundary element reinforcement. In order to prevent damage localization at the precast beam-wall interface, and thus a premature sliding shear failure, U-shaped or straight dowel bar reinforcement crossing the cold joint was found to be adequate to force most of the beam inelastic deformations to occur away from the cold joint. The ease with which these precast HPFRC coupling beams can be placed on the jobsite is a major improvement over the construction methods currently used for diagonally reinforced concrete and steel coupling beams.

4 Coupled Walls

In addition to tests of coupling beam components, two coupled wall specimens (approximately 1/3-scale) were tested. These specimens consisted of four beams

($\ell_n/h = 1.7$) linking two T-shaped structural walls (Fig. 6). In each specimen, three of the coupling beams were precast with HPFRC and reinforced similarly to the specimen shown in Fig. 4(a). The coupling beam at the second story was precast with regular concrete. Slabs were built at the second and fourth levels to facilitate application of lateral displacements. These slabs also allowed the evaluation of the precast beam-slab-wall interaction during earthquake-induced displacements.

The reinforcement of the first two stories of the wall specimens differed. The walls in the first specimen were designed and detailed to satisfy the requirements of the ACI Building Code [5], whereas HPFRC was used in the first two stories of the second specimen. In the HPFRC walls, the boundary element confinement reinforcement was reduced and a shear stress of $4\sqrt{f_c'}$, [psi] ($0.33\sqrt{f_c'}$, [MPa] was assumed to be resisted by the HPFRC, which is twice the value assumed for conventional reinforced concrete walls. Additional details on these tests are available in Reference [7].

Both specimens exhibited the high strength and stiffness characteristic of coupled walls, with excellent strength retention and energy dissipation up to a system drift of approximately 3.0% (Fig. 7). The HPFRC portions of the specimens exhibited narrower crack spacing and significantly improved damage tolerance (Fig. 8), despite simplified reinforcement detailing. These tests demonstrated that placement of the precast coupling beams, with the beam reinforcement threading through the adjacent wall reinforcement, was straight-forward. This proposed method is believed to be a viable alternative method for construction of coupled wall systems.

5 Summary and Conclusions

Experimental evidence indicates that the use of steel HPFRC materials in coupling beams is a viable alternative to simplify reinforcement detailing without compromising seismic performance. In coupling beams with span-to-depth ratios less than approximately 2.5, a nearly 70% reduction in diagonal reinforcement is possible, and elimination of diagonal reinforcement is possible in more slender coupling beams. In all cases, special column-type confinement reinforcement is only needed at the beam ends due to the confinement provided by the fiber reinforcement. For further construction simplification, the proposed HPFRC coupling beams can be precast, eliminating the need for cast-in-place HPFRC.

Results from two four-story coupled wall specimens subjected to lateral displacement reversals indicate that coupled wall systems with HPFRC coupling beams exhibit a stable seismic behavior with drift capacities on the order of 3%.

The construction of the coupled walls with precast HPFRC coupling beams was shown to have potential for substantial reductions in construction labor and time.

Acknowledgements. Research on HPFRC coupling beams at the University of Michigan has been funded by the US National Science Foundation through grants CMS 0001617 and CMMI 0530383. The latter grant was part of the Network for Earthquake Engineering Simulation (NEES) Program. The support of Bekaert Corporation and Erico is also greatly appreciated. The ideas and conclusions expressed are those of the writers, and do not necessarily represent the views of the sponsors.

References

[1] Paulay, T., Binney, J.R.: Diagonally Reinforced Coupling Beams, pp. 579–598. Special Publication SP-42, American Concrete Institute, Detroit, Michigan (1974)

[2] Naish, D., Wallace, J.W., Fry, J.A., Klemencic, R.: Reinforced Concrete Link Beams: Alternative Details for Improved Construction. UCLA-SGEL Report 2009-06, Structural & Geotechnical Engineering Laboratory, University of California at Los Angeles, p. 103 (2009)

[3] Canbolat, B.A., Parra-Montesinos, G.J., Wight, J.K.: Experimental Study on the Seismic Behavior of High-Performance Fiber Reinforced Cement Composite Coupling Beams. ACI Structural Journal 102(1), 159–166 (2005)

[4] Liao, W.C., Chao, S.H., Park, S.Y., Naaman, A.E.: Self-Consolidating High Performance Fiber Reinforced Concrete (SCHPFRC) – Preliminary Investigation, Report No. UMCEE 06-02, University of Michigan, Ann Arbor, MI, p. 68 (2006)

[5] Committee, A.C.I.: 318, Building Code Requirements for Structural Concrete (ACI 318-08) and Commentary, American Concrete Institute, Farmington Hills, MI, p. 465 (2008)

[6] Lequesne, R.D., Setkit, M., Parra-Montesinos, G.J., Wight, J.K.: Seismic Detailing and Behavior of Coupling Beams With High-Performance Fiber Reinforced Concrete. In: Naaman, A.E. (ed.) Symposium – Four decades of progress in prestressed concrete, fiber reinforced concrete, and thin laminate composites, SP-272, pp. 205–222. American Concrete Institute, Farmington Hills, MI (2010)

[7] Lequesne, R.D., Wight, J.K., Parra-Montesinos, G.J.: Seismic Detailing and Behavior of Coupled-Wall Systems with High-Performance Fiber-Reinforced Concrete. In: 9th National and 10th Canadian Conference on Earthquake Engineering Proceedings, Toronto, Canada, July 25-29, p. 10 (2010)

Part 6
Ultra High Performance Fiber Reinforced Concrete

Size and Shape Effect of UHPFRC Prisms Tested under Axial Tension and Bending

B. Frettlöhr[1], K.-H. Reineck[1], and H.-W. Reinhardt[2]

[1] Institute for Lightweight Structures and Conceptual Design, University of Stuttgart, Germany
[2] Department of Construction Materials, University of Stuttgart, Germany

Abstract. Tests on prism out of ultra-high-performance fiber reinforced concrete (UHPFRC) were performed in order to further clear up the size effect, i.e. the strength decrease with increasing depth for axial tension and flexure as well as for combined bending and axial forces. Further aim was to investigate the influence of the prism shape. i.e. the width to height ratio b/h. The tests were carried out with Ductal® from Lafarge with an average compressive strength of fcyl,m = 211 MPa as well as with Duracrete Plus from Schwenk Zement with fcyl,m = 169 MPa.

For axial tension the elastic limits as well as the maximum tensile stresses exhibited a clear decrease with increasing depth from 25 to 100 mm. These values also decreased with increasing ratio b/h from 1 to 3 and 5 of the prisms. For pure bending of prisms from h = 25 to 150 mm the elastic limits as well as the maximum bending stresses (the bending tensile strengths) decreased with increasing depths. The width to height ratio b/h had no influence on the elastic limit of the flexural tensile stress but the bending tensile strengths decreased with increasing ratio b/h from 1 to 5.

1 Material Characteristics of UHPFRC

Tests were performed by Frettlöhr and Reineck [1] on two Ultra High Performance Fiber Reinforced Concrete UHPFRC with a content of 2 % steel fibers by volume.

One material was Ductal® G2 FM delivered by Lafarge, Paris with steel fibers of the type Redaelli Tecna with Ø 0.175 mm and l_f = 13 mm length. The test specimens were cast at the Otto-Graf-Institut, MPA Universität Stuttgart and were heat treated for 48 h at about 90°C in a water basin. The 59 compressive tests of cylinders (Ø 100, h = 200 mm) yielded an average strength of f_{cm} = 211 MPa with a coefficient of variation of only v = 4.4 %. The modulus of elasticity in compression was on average E_{cm} = 53.070 MPa, and that under tension was E_{ctm} = 48.575 MPa and thus 8.5 % lower. Die 99 flexural 4-point bending tests on control prisms 40 x 40 mm yielded the following average values for the tensile stress

G.J. Parra-Montesinos, H.W. Reinhardt, and A.E. Naaman (Eds.): HPFRCC 6, pp. 365–372.

$\sigma_{ct,fl}$ = M/W: $\sigma_{ctfl,el,m}$ = 21.6 MPa for the elastic limit with v = 12.5 %, and $f_{ct,fl,m}$ = 42.3 MPa for the bending tensile strength with v = 8.9 %.

The second material was material Duracrete Plus® delivered by Schwenk Zement AG, Ulm. The test specimens were cast by the firm Sebastian Wochner GmbH & Co.KG, Dormettingen. The steel fibers were Stratec FM13/0.2 with Ø 0.2 mm and l_f = 13 mm from the firm Stratec GmbH, Heme. The test specimens were heat treated as described above. The average compressive strength of the 12 cylinders (∅ 100, h = 200 mm) tested was f_{cm} = 168.6 MPa with a coefficient of variation of only v = 4.4 %. The modulus of elasticity in compression was tested on 6 cylinders (∅ 100, h = 200 mm) and yielded an average value of E_{cm} = 42.065 MPa with v = 8.7 %, whereas that under tension was E_{ctm} = 38.245 MPa and thus 9 % lower. The flexural tests on the 21 control specimens with 40 x 40 mm yielded the following values: $\sigma_{ctfl,el}$ = 21.7 MPa with v = 9.4 % for the elastic limit, and f_{ctfl} = 33.2 MPa for the bending tensile strength with v = 14.8 %.

2 Structural behavior of UHPFRC in Tension

2.1 *Stress-Strain Curves*

The typical behavior for tests in axial tension is demonstrated in Fig. 1, where for both materials the axial tensile stress σ_{ct} = N/A is plotted versus the strain measured over a gauge length of 150 mm. For all evaluations special attention was paid to assess the axial tensile stress $\sigma_{ct,el}$ at the elastic limit (first cracking stress) and of course the maximum value of the axial tensile strength f_{ct}. UHPFRC with 169 MPa attains slightly lower levels than the UHPFRC with 211 MPa.

The reason for this could be that even though the fiber contents and fiber lengths were the same the diameter of the fibers were different Ø 0.175 mm for Ductal® and Ø 0.2 mm for Duracrete Plus®. This means that a member out of Ductal® contains 31% more fibers and also that the surface of the fibers is 15% higher than for an identical member out of Duracrete Plus® with the same fiber content. However, the tensile strengths of the prisms out Ductal® is about 25 to 45 % higher than that for Duracrete Plus®, so that most probably the bond strength is higher for the fibers in Ductal® than for the fibers in Duracrete Plus®.

All tests performed by Frettlöhr and Reineck [1], like those shown in Fig. 1, exhibited a distinct stress increase after the elastic limit $\sigma_{ct,el}$ and in the range of micro-cracking before the axial tensile strength f_{ct} is attained, where localisation occurs. After that the stress decreases while fibers are pulled out in this descending branch. The strain measured over a certain gauge length is then governed by the crack width of the normally only one crack.

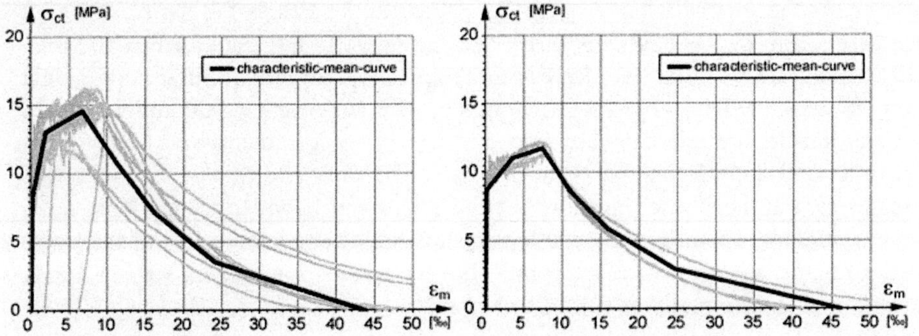

a) UHPFRC with 211 MPa

b) UHPFRC with 169 MPa

Fig. 1. Stress-strain curves in axial tension (gauge length 150 mm), h = 25 mm, b = 3· h = 75 mm

2.2 *Size and Shape Effect in Tension*

In the following the size and shape effect is investigated by plotting for both materials the axial tensile stress $\sigma_{ct,el}$ at the elastic limit (first cracking stress) (Fig. 2a) and the tensile strength f_{ct} (Fig. 2b) versus the height h for different ratios b/h of the prisms.

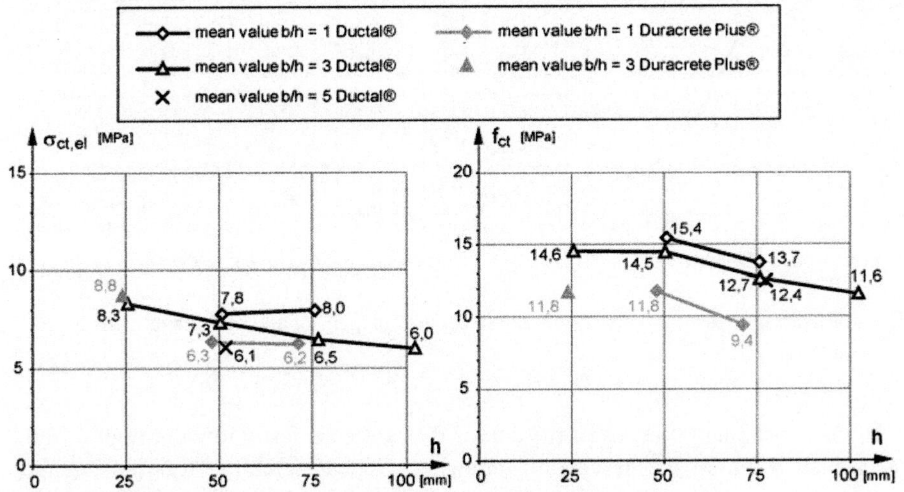

a) Axial tensile stress $\sigma_{ct,el}$ at the elastic limit

b) Axial tensile strength f_{ct}

Fig. 2. Size and shape effect in axial tension

For both cases the mean values for the UHPFRC with 169 MPa are lower than that for the UHPFRC with 211 MPa. The size effect is clearly visible in both figures for the prisms with b/h = 3. For Ductal® the values for the elastic limit decreases from $\sigma_{ct,el}$ = 8.3 MPa for h = 25 mm down to 6.0 MPa for h = 100 mm, i.e. by about 28 %. Likewise decreases the tensile strength with increasing height from f_{ct} = 14.5 MPa for h = 25 and 50 mm to 11.6 MPa for h = 100 mm, i.e. by 20 %. The tensile strength decrease also occurs for the specimens with b/h = 1 between h = 50 and 75 mm for both materials. The only exceptions are the tensile stresses $\sigma_{ct,el}$ at the elastic limit for the prisms with b = h, which were almost equal for h = 50 and 75 mm. Fig. 2a and b also show that there is an effect of the shape of the prisms, i.e. with increasing ratio b/h the values for $\sigma_{ct,el}$ as well as for f_{ct} decrease when b/h increases from 1 to 3 and 5.

3 Bending Tests on UHPFRC Prisms

3.1 Structural Behavior in Bending

The flexural behavior is assessed by plotting the flexural stress $\sigma_{ct,fl}$ = M/W versus the deflection w_m at midspan, and Fig. 3a and b gives examples for the two materials tested for prisms with h = 50 mm and the ratio b/h = 3. Some more examples for other prism sizes are presented by Reineck and Frettlöhr [2].

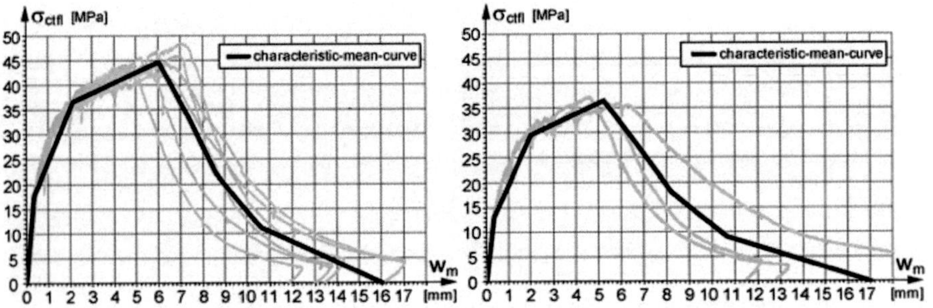

a) prisms out of Ductal®   b) prisms out of Duracrete Plus®

Fig. 3. Flexural stress σ_{ctfl} versus deflection w_m at midspan for the specimens with h = 50 mm and b/h = 3

In the ascending branches of the curves the scatter is fairly low but high differences may occur in the descending branches. Special attention was paid to the flexural tensile stress $\sigma_{ctfl,el}$ at the elastic limit, apart from of course the maximum value, the bending tensile strength $f_{ct,fl}$. This characteristic curve applied to both materials and only the stress levels are lower for Duracrete Plus® (see Fig. 3a and b).

3.2 *Size and Shape Effect in Bending*

3.2.1 Elastic Limit of the Flexural Tensile Stress

The flexural tensile stress $\sigma_{ctfl,el}$ at the elastic limit is plotted in Fig. 4a versus the depth h of the prisms for both materials tested, whereby the values for b/h = 1, 3 and 5 are marked differently.

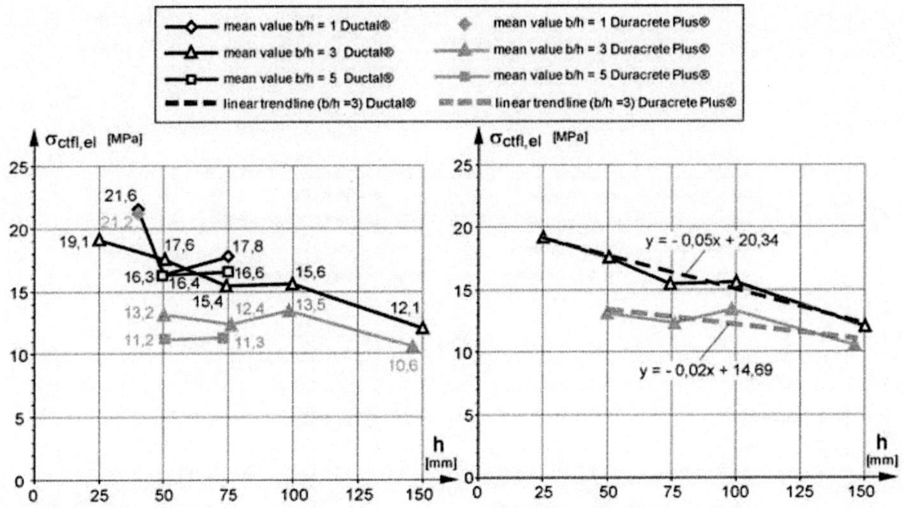

a) $\sigma_{ctfl,el}$ versus h for b/h = 1,3 and 5

b) Trend lines for $\sigma_{ctfl,el}$ with b/h = 3 versus h

Fig. 4. Flexural tensile stress $\sigma_{ctfl,el}$ at the elastic limit versus height h for different ratios b/h for both materials

Altogether there is a clear size effect that $\sigma_{ctfl,el}$ decreases with increasing depth h, apart from the prisms with b/h = 5. This size effect is especially pronounced for the series with b/h = 3, where for example for Ductal® $\sigma_{ctfl,el}$ drops by 36% from 19.1 MPa for h = 25 mm down to 12.1 MPa for h = 150 mm. The following linear trend lines for the ratio b/h = 3 plotted in Fig. 4b fit well to these test data:

- Ductal®:	$\sigma_{ctfl,el} = 20 - h/20$	with h [mm] $\leq$ 150 mm
- Duracrete Plus®:	$\sigma_{ctfl,el} = 14.7 - h/50$	with h [mm] $\leq$ 150 mm

The flexural tensile stress $\sigma_{ctfl,el}$ at the elastic limit is about 26 % higher for Ductal® than for Duracrete Plus®. For the prisms with the ratios b/h = 1 and h = 40 mm the difference is only 2 %, whereas the maximum difference is about 45 % for the series with b/h = 5 and h = 50 respectively 75 mm.

3.2.2 Bending Tensile Strength

The bending tensile strength is the maximum value for the stress curves as shown in Figs. 5a and 5b. For the test series with different values for the ratio b/h the bending tensile strength $f_{ct,fl}$ is plotted versus the depth h in Fig. 5a for both materials. There is a clear size effect and this decrease with increasing height h is well predicted by the following straight lines for the tests with the ratio b/h = 3 as shown in Fig. 5b:

- Ductal®: $f_{ct,fl} = 48 - h/7$ with h [mm] ≤ 150 mm
- Duracrete Plus®: $f_{ct,fl} = 41 - h/7$ with h [mm] ≤ 150 mm

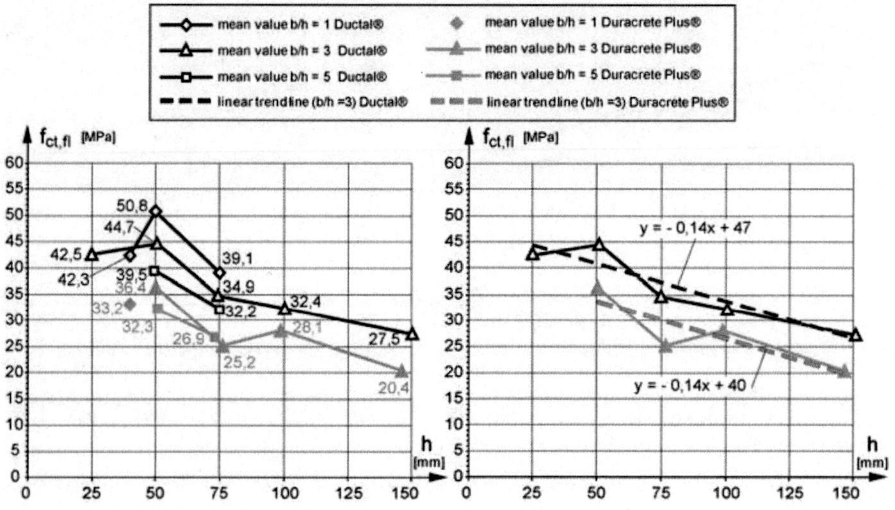

a) $f_{ct,fl}$ versus h for b/h = 1,3 and 5 b) Trend lines for $f_{ct,fl}$ with b/h = 3 versus h

Fig. 5. Flexural tensile strength $f_{ct,fl}$ versus height h for different ratios b/h for both materials

Also the shape of the prisms expressed in terms of the ratio b/h plays a role as can be seen in Fig. 6. Clearly the bending tensile strength $f_{ct,fl}$ decreases with increasing height h, as shown by the two marked lines for the tests with Ductal® with h = 50 mm (upper line) and h = 75 mm as well as for Duracrete Plus® for h = 50 mm. The only exception are the tests with h = 75 mm on prisms out of Duracrete Plus®.

The conclusion is therefore, that both the size effect as well as the influence of the shape needs to be considered in design.

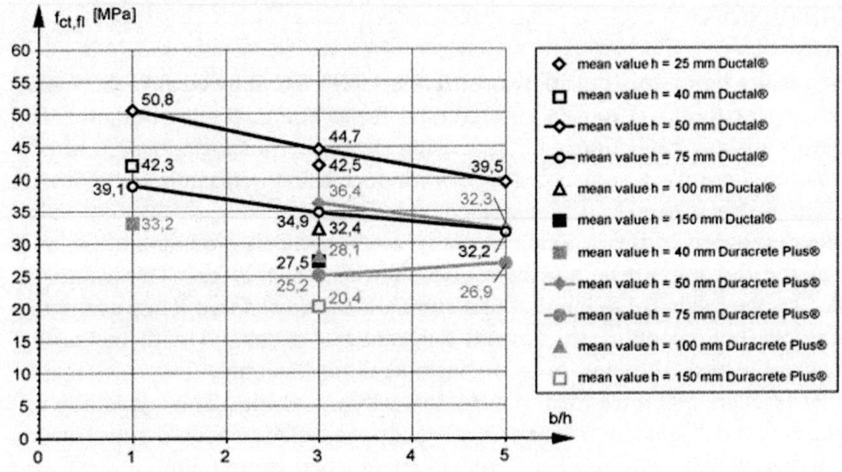

Fig. 6. Flexural tensile strength $f_{ct,fl}$ versus the ratio b/h for prisms with different heights for both materials

4 Comparisons of Elastic Limits with Proposed Relationships

The dependency of the elastic limit of the flexural stress was compared with the theory by Weibull and an empirical relationship of AFGC/SETRA [3] by Frettlöhr and Reineck [1] as well as by Reineck and Frettlöhr [2, 4], so that only the results and conclusions are presented here.

The theory of Weibull may explain the decrease of the elastic limit of the flexural stress due to the increase of the volume. However, the Weibull modulus m was found not to be constant but decreased with increasing volume from about m = 26 to 12. In addition, completely different values of about m = 2 had to be set in order to predict the ratio between the flexural stress to the axial tensile stress of the relevant tests in axial tension. Therefore, contradictory results were found when applying the theory of Weibull so that it appears not to apply.

The test results were also compared with the relationship between the axial tensile stress and the elastic limit of the flexural tensile stress proposed in AFGC/SETRA [3]. This formula is based on a relationship given in the CEB-FIP MC 90 which was empirically derived for the maximum value of the bending tensile strength. For the test series with b/h = 3 the ratio between the elastic limit of the flexural tensile stress (see Fig. 4a) and the elastic limit of the axial tensile stress of the tension tests (see Fig. 2a) varied only slightly between $\sigma_{ctfl,el} / \sigma_{ct,el}$ = 2.2 and 2.6. However, the relationship of AFGC/SETRA [3] predicts a decrease from 2.3 for h = 25 mm to 1.38 for h = 150 mm, so that this clearly contradicts the test results.

5 Conclusions

The test results on prisms out of two different UHPFRC showed that the maximum values for the axial tensile strength and the bending tensile strength exhibited both a distinct reduction with increasing depth, i.e. displayed a size effect. Furthermore, these values also reduced with increasing ratio b/h of width to height. These effects need to be considered in design.

Another emphasis of the evaluations was placed on the elastic limits of the axial and of the flexural stress, and both also showed a size effect. This could be explained by the theory of Weibull, but a comparison showed that this was not the case. The Weibull modulus m was not constant but decreased with increasing volume, and this was different for axial tension and for bending.

Further tests are recommended in order to verify the results. This especially refers to tests in axial tension, because some specimen failed outside the measuring region so that there may be some uncertainty about a premature failure.

References

[1] Frettlöhr, B., Reineck, K.-H.: Versuche zum Maßstabseinfluss bei kombinierter Beanspruchung aus Biegung und Längskraft von dünnen Bauteilen aus ultrahochfestem Faserfeinkornbeton. (Tests on the size effect of thin members out of Ultra-High-Performance fibre reinforced concrete subjected to combined bending and axial forces). Abschlussbericht zum Forschungsvorhaben RE 813/6-1 der Deutschen Forschungsgemeinschaft (DFG). Universität Stuttgart, Institut für Leichtbau Entwerfen und Konstruieren (ILEK) (September 2009)

[2] Reineck, K.-H., Frettlöhr, B.: Tests on scale effect of UHPFRC under bending and axial forces. In: 3rd Fib International Congress, Washington DC, aPaper 54, p. 14 (2010)

[3] AFGC/SETRA: Bétons fibrés à ultra-hautes performances. Recommandations provisoires (Ultra high performance fibre-reinforced concretes) Documents scientifiques et techniques. In: Association Française de Génie Civil, SETRA, Bagneux Cedex, pp. 1–152 (January 2002)

[4] Reineck, K.-H., Frettlöhr, B.: Versuche zum Maßstabseinfluss bei kombinierter Beanspruchung aus Biegung und Längskraft von UHFFB mit einer Druckfestigkeit von 211 MPa (tests on the scale effect under combined bending and axial forces of UHPFRC with 211 MPa). Bauingenieur 86(H.1), 42–52 (2011)

Characterization of Bending and Tensile Behavior of Ultra-High Performance Concrete Containing Glass Fibers

S. Rigaud, G. Chanvillard, and J. Chen

Lafarge Centre de Recherche, Saint Quentin Fallavier, France

Abstract. Ultra High Performance Concrete (UHPC) is used to design structural elements with unique combinations of aesthetic, lightness, self-placing and ductility properties. The difficulty to design such a ductile composite is obtaining an optimum compromise between reinforcements, cementitious matrix strengths, rheology and costs.

An ultra high performance reinforced concrete containing glass fibers (GF-UHPRC) is described for thin plate geometries. Including complete mechanical characterization, based on four point bending tests, and reverse analyses to extract the post-cracking behavior in tension (e.g. the tensile strength versus strain relationship of the glass fibers). The results of accelerated aging tests in 50°C water showed that its durability and ductility were maintained.

1 Introduction

Ultra-high performance fiber reinforced concrete allows the design of complex and very thin structural elements by combining aesthetic, lightness, self-placing and ductility properties. The difficulty to design such a ductile composite is obtaining an optimum compromise between reinforcements, matrix strengths, rheology and costs.

The purpose of this paper is to present an ultra high performance reinforced concrete containing glass fibers (called here GF-UHPRC) for thin-plate applications. The mechanical performances of such material were evaluated in bending tests. A reverse analysis method was used, based on the mechanical equilibrium of the cracked cross section to obtain the direct tensile behavior. An important point in all glass fiber reinforced cementitious composites is a loss of ductility due to wet aging. This phenomenon is attributed to the growth of portlandite at the glass/fiber mortar interface [2, 7]. Results showed that the ductility was maintained after accelerated aging tests.

G.J. Parra-Montesinos, H.W. Reinhardt, and A.E. Naaman (Eds.): HPFRCC 6, pp. 373–380.

2 Glass Fibers and Ultra-High Performance Concrete

From a rheological perspective, the aim is to design a GF-UHPRC with self-placing capability. The glass fiber used is a bundle of numerous microfilaments held together by a sizing. An appropriate choice of sizing is needed to ensure good fluidity and adequate bonding between the fiber and the matrix. The mechanical results shown in this paper are based on commercially available hydrophilic fibers. The physical characteristics of the glass fibers are shown below.

- Young´s modulus: 72 GPa
- Tensile strength: 1200 MPa–1700 MPa
- Length: 6 mm
- Diameter: 0.15 mm–0.2 mm
- Slenderness: 30-40
- Density: 2.6 g/cm^3

Two other aspects must be taken into account to design a ductile GF-UHPRC. The cementitious matrix strength must be adapted to the tensile strength and aspect ratio of the glass fibers to avoid an early failure of the composite. The composites in this study have a W/C ratio from 0.25 to 0.30, a compressive strength of 120-140 MPa after 28-day hydration at ambient temperature, and up to 170 MPa after accelerated aging at 50°C. The volume content of glass fibers must ensure an adequate mean spacing of the fibers compatible with the maximum sand size of the matrix to ensure good rheological properties and efficient crack bridging. Moreover, each fiber must have sufficient mobility, without forming clusters. The volume content of glass fibers in GF-UHPRC is thus limited to 2%-2.5% to ensure a self-placing composite (Fig. 1).

3 Characterization Tests

The tensile behavior was characterised by reverse analysis carried out on four point bending tests. The specimens were 450 x 145 x 20 mm^3 plates cut from a larger plate (Fig. 1) after demolding at 24 hours. The samples were stored in a room at 20°C and 100% HR for 28 days. A device fixed to the plate measured the real deflection (Fig. 2). The latter was controlled during the test by a LVDT sensor at a rate of 0.1 mm/min. Fig. 4 shows two series of tests performed on composites with varying glass fiber contents and matrix strengths.

Fig. 1. Self-placing properties. The plates are sawn in the direction of casting after demolding

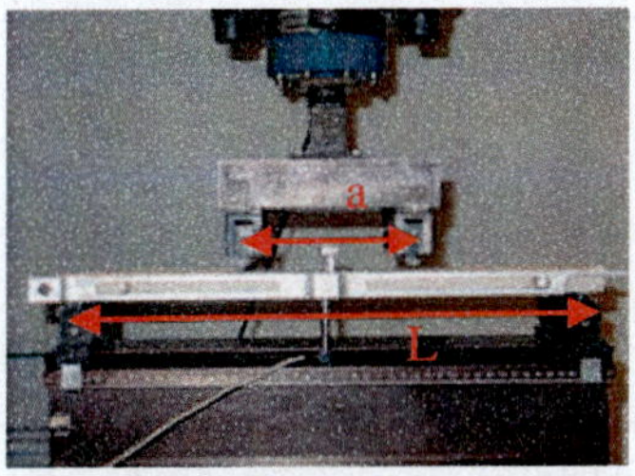

Fig. 2. Deflection measuring device

Fig. 3 illustrates very low dispersion for the elastic behavior or the first cracking stress, and the ductile nature of the composite. The limit of proportionality was reached at a deflection of 0.4 mm and 0.5 mm in both systems; the maximum load (modulus of rupture) corresponded to a deflection of approximately of 3.5 mm and 2.3 mm for the composites with 2.0% and 2.5% volume fibers, respectively. Such ductile behavior is due to fine multiple cracking (Fig. 4) in the area submitted to the highest flexural moment.

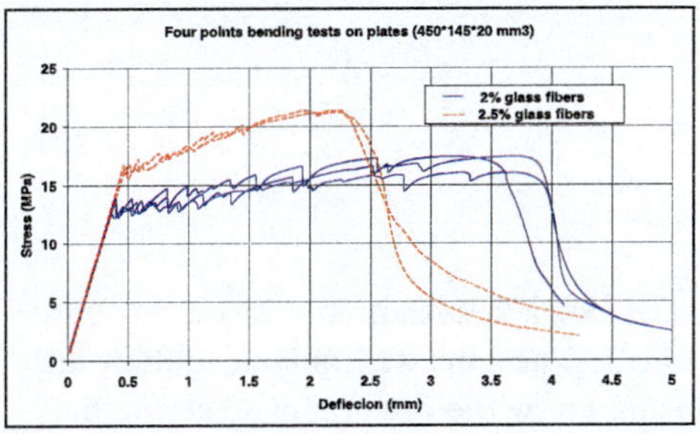

Fig. 3. Flexural tests on two types of composites

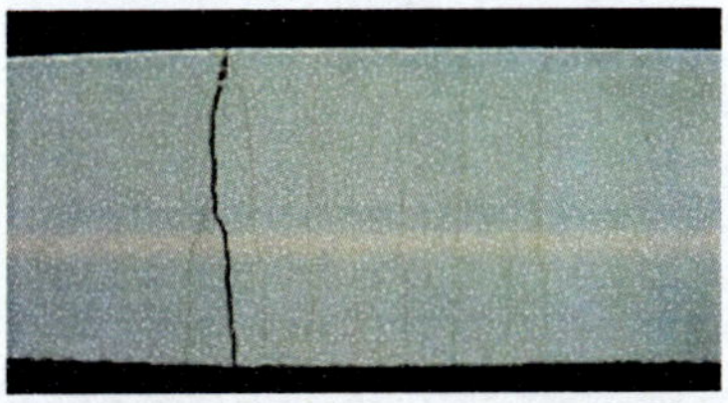

Fig. 4. Multiple cracking on the tensile face after the flexural test

The location of the main crack was established after the peak; the opening mechanism depended directly on the anchoring of the fibers in the matrix and how well the fibers were located regarding the main crack. Fig. 3 shows a systematic failure of the fibers characterised on the curves by a steep slope just after the peak. It is important to highlight that the ultimate deflection obtained before the failure of the fibers is not a material but a structural property. As soon as a uniform crack distribution was established in the central area of the specimen, the ultimate deflection was reached. Consequently, a longer plate would directly increase this deflection and the multiple cracking. Table 1 summarizes the obtained mechanical properties. Young´s modulus (E) could be estimated with Eq. 1 from the elastic behavior and the real deflection measurement. It took into account the effect of shear stress occurring outside the central area of the plate. F represents the load, h the thickness of the plate, δ the deflection, I the inertia moment, ν the Poisson coefficient and for L and a see Fig. 2.

$$E = \frac{Fa}{48\delta I}\left[\left(3L^2 - 4a^2\right) + 4.8h^2(1+\nu)\right] \tag{1}$$

Table 1. Flexural results

	E (GPa)	Elastic limit σ_{fe} (MPa)	Max stress(MPa)
Composite 1 - 2% glass fibers	45	13.5	17
Composite 2 - 2.5% glass fibers	47	16	20

The lower side of the sample is in tension during the flexural test. The limit of proportionality (σ_{fe}) in bending is known to not match the elastic limit in pure tension. This is called the scale effect phenomena and arises from micro-cracking occurring above the crack front, leading to a higher apparent limit of proportionality compared to the real tensile strength. Models based on the cohesive crack concept [5] describe this phenomena very well; the CEB Model Code [3] is a simplified model describing the scale effect on beam-geometry given in Eq. (2). The coefficient β depends on the type of matrix and varies between 1 and 2. Coefficient β increases with the brittleness of the matrix; and h (in mm) represents the height of the sample.

$$\sigma_t = \sigma_{fe} \cdot \frac{1 + \beta . \left(\frac{h}{100}\right)^{0.7}}{\beta . \left(\frac{h}{100}\right)^{0.7}} \tag{2}$$

Three 70 x 70 x 280 mm^3 prisms were made for each system and tested in four-point bending test to illustrate the scale effect. The aim was only to obtain the elastic limit to apply the CEB Model Code and know the direct tensile strength of the two GF-UHPRC studied in this paper. The limit of proportionalities (σ_{fe}) were 13 MPa and 15 MPa for composites 1 and 2, respectively. It seems that β=2 is a good compromise for the GF-UHPRC systems. The direct tensile strengths of the two composites were 8 MPa and 9 MPa, respectively, when applying the relation (2). Additional mechanical tests would be necessary to evaluate these results more accurately.

4 Reverse Analysis

Calculations were done for a rectangular cracked section (height h, width b), loaded in flexure. This section is divided into two parts (Fig. 5). First, the composite has an elastic behavior in the compressive zone and at the beginning of the tensile zone. Second, the material is in tension and damaged by an unknown behavior law. The objective is to determine this tensile post-cracking behavior.

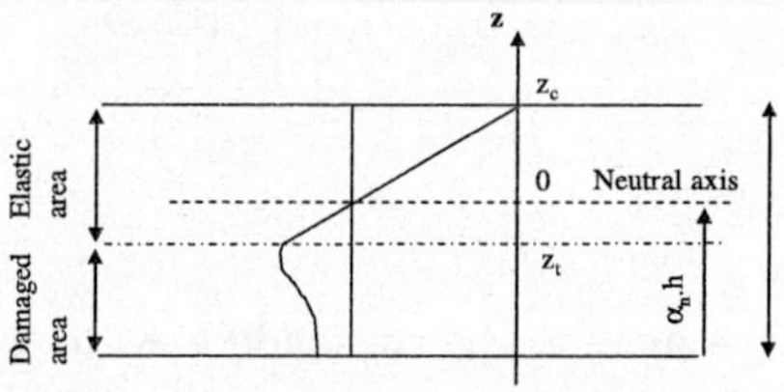

Fig. 5. Stress distribution in the section (z_c: height of the compressive area; z_t: height where the tensile strength is reached, before damage z_t=$h/2$; α_n: relative height of neutral axis, before damage α_n=0.5.)

This approach is similar to the French recommendations for UHPFRC [1], without explicitly introducing the cracking concept. From the crack distribution in the constant moment zone, we propose a mechanical model based on the assumption of a constant curvature in this area. Consequently, the experimental deflection can be converted into a curvature using Eq. (3). This equation is considered valid until crack localization. The curvature (χ) was chosen to describe the yield strains in the section. All the calculations were done from the neutral axis; the sign convention was negative for tensile stresses and positive for compressive stresses.

$$\chi_i = \frac{216\delta_i}{23L^2} \tag{3}$$

The mechanical equilibrium of the section led to the following Eq. (4). The sum of axial loads in the section is equal to zero; the sum of the moments was equal to the applied external bending moment (M_{ext}).

$$N = N_e + N_d = 0 \qquad M_{ext} = M_e + M_d \tag{4}$$

Thus, in the elastic area, Eq. (6) was obtained by relation (5).

$$\varepsilon = \chi.z \tag{5}$$

$$N_e = \int_{z_t}^{z_h} \sigma.b.dz = \frac{E.b.\chi}{2}\left(z_c^{\,2} - z_t^{\,2}\right)$$
$$M_e = \int_{z_t}^{z_h} \sigma.b.z.dz = \frac{E.b.\chi}{3}\left(z_c^{\,3} - z_t^{\,3}\right) \tag{6}$$

From geometry consideration, the heights z_c and z_t are given as:

$$z_c = h(1-\alpha_n) \qquad z_t = -\frac{6M_0}{Ebh^2\chi} \tag{7}$$

M_0 corresponds to the bending moment required to reach the elastic limit on the bottom side, ε_0 corresponds to the strain.

The same calculations were reproduced in the damaged zone. The obvious relation $d\varepsilon = \chi.dz$ (from Eq. (5)) led to:

$$N_d = \int_{-\alpha_n h}^{z_t} \sigma.b.dz = \int_{\varepsilon_m}^{\varepsilon_0} \frac{\sigma.b}{\chi} d\varepsilon \quad M_d = \int_{-\alpha_n h}^{z_t} \sigma.b.z.dz = \int_{\varepsilon_m}^{\varepsilon_0} \frac{\sigma.b.\varepsilon}{\chi^2} d\varepsilon \tag{8}$$

ε_m is the strain of the bottom side, $\varepsilon_m = -\chi.\alpha_n h$.

An incremental approach was used to calculate the integrals in Eq. (8) considering two successive loading stages. Stage *i* of loading is characterized by ε_m^i, χ_i, α_n^i, and introducing σ_i, the tensile stress associated to ε_m^i. This gives:

$$N_d^i = \int_{\varepsilon_m^i}^{\varepsilon_0} \frac{\sigma.b}{\chi_i} d\varepsilon \qquad M_d^i = \int_{\varepsilon_m^i}^{\varepsilon_0} \frac{\sigma.b.\varepsilon}{\chi_i^2} d\varepsilon \qquad \varepsilon_m^i = -\chi_i.\alpha_n^i h \tag{9}$$

Equations (9) can be expressed between two successive steps as follows (10).

$$\begin{aligned} N_d^{i+1} &= \left(\frac{\chi_i}{\chi_{i+1}}\right).N_d^i + \frac{b}{\chi_{i+1}} \frac{(\sigma_i + \sigma_{i+1})}{2}.(\varepsilon_m^i - \varepsilon_m^{i+1}) \\ M_d^{i+1} &= \left(\frac{\chi_i}{\chi_{i+1}}\right).M_d^i + \frac{b}{\chi_{i+1}^2} \frac{(\sigma_i.\varepsilon_m^i + \sigma_{i+1}.\varepsilon_m^{i+1})}{2}.(\varepsilon_m^i - \varepsilon_m^{i+1}) \\ \varepsilon_m^i - \varepsilon_m^{i+1} &= h.(\chi_i \alpha_i - \chi_{i+1} \alpha_{i+1}) \end{aligned} \tag{10}$$

α_n^{i+1} and σ_{i+1} are unknown in each step; the reverse analysis consists of determining these two parameters to ensure the mechanical equilibrium of the section. The variables of step 0 were initialized to start the calculation:

$$N_e^0 = 0 \quad M_e^0 = -\frac{b.h^2.\sigma_0}{6} \qquad \sigma_0 = \sigma_{fe} \qquad N_d^0 = 0 \quad M_d^0 = 0 \quad \alpha_n^0 = 0.5$$

Thus, step-by-step and after a reasonable smoothing process of the experimental data to avoid numerical instabilities, the tensile behavior versus strain could be described (Fig. 7). An interesting plastic behavior before crack localization was observed, leading to the failure of the fibers.

Fig. 6 compares the bending moment, for the two composites from the first cracking, obtained experimentally and by direct modeling using the tensile behavior calculated by reverse analyses.

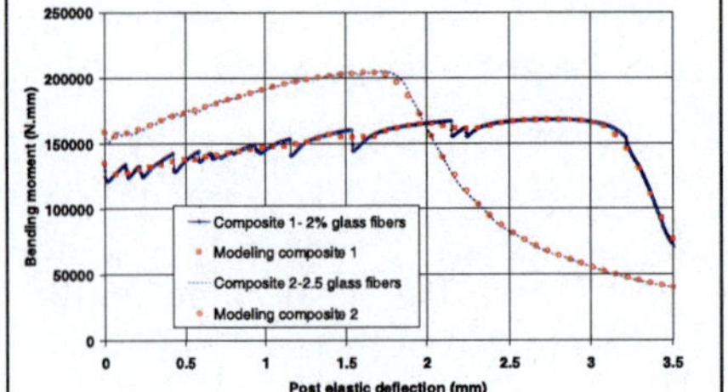

Fig. 6. Comparison between experimental data and the modeling of the flexural test

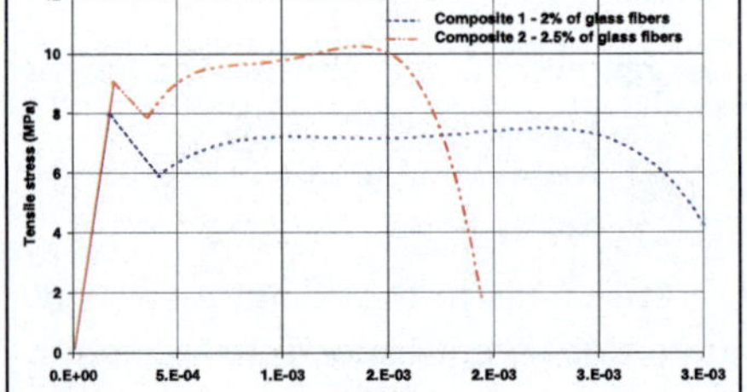

Fig. 7. Tensile behavior obtained by reverse analysis

The elastic limit in bending was indeed submitted to a scale effect (§3) and does not equate with the direct tensile strength. Therefore, the first point σ_{fe} of the reverse analysis was corrected by the real tensile strength of the composite.

At this stage, it was interesting to compare the reverse analyses results with a simplified approach of the potential tensile strength [4], σ_p, of a fiber-reinforced concrete (11).

$$\sigma_p = v_f.\sigma_f.k.\omega \tag{11}$$

v_f is the volume content of the fibers (2.0% or 2.5%). σ_f is the direct tensile of the fibers (~1500 MPa). k is a coefficient for the orientation of the fibers in the matrix. The orientation coefficient is 0.5, 2/π or 1 for a 3D, 2D or 1D distribution, respectively. We assumed k=0.6 for an intermediate solution. ω is a coefficient representing the effectiveness of the fiber/matrix combination. All the fibers are therefore perfectly centered relative to the crack for a value equal to one (but in our case, ω=0.5).

Equation (10) gives an estimated tensile strength equal to 9 MPa for the composite with 2.0% fibers and 11 MPa for the composite with 2.5% fibers. These results agree with the reverse analyses and show that the fibers' tensile strength is fully exploited.

5 Durability

An accelerated aging method was used to assess the long-term performance of GF-UHPRC, using the common immersion in hot water method for various periods. The following tests were carried out. The samples were stored in a humid chamber at 20°C-100%HR for one month, then placed in hot water at 50°C for three months after demolding at 24h.

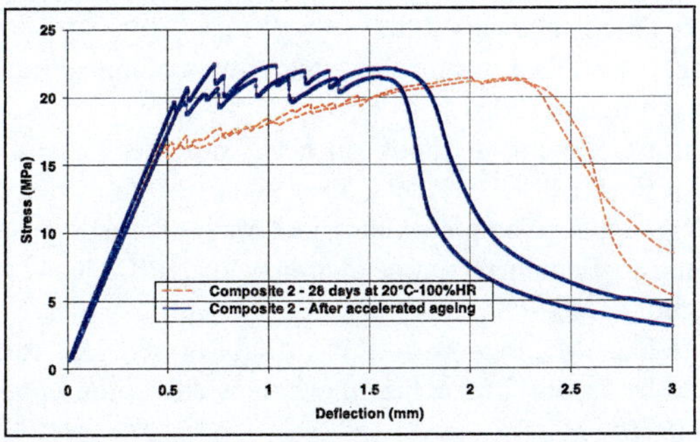

Fig. 8. Results of four-point bending test after accelerated ageing, composite 2

Fig. 8 illustrates the maintenance of ductility of this composite after wet aging. This aging resistance is due to the optimum use of a pozzolanic filler in the matrix, limiting the formation of portlandite at the fiber interface. Multiple cracking was observed and the maximum bending strength was reached for a deflection of ~1.5 mm. According to previous remarks on structural ductility, this impact on the ultimate deflection should not be considered as a material property. Indeed, structural ductility is largely governed by the length of the specimen. The limit of proportionality increased and was equal to 20 MPa; the elastic modulus remained unchanged. This was due to cement hydration and pozzolanic reactions, both of which increased during the accelerated aging. The compressive strength of the composite with 2.5% fibers was 170 MPa after the aging treatment.

6 Conclusions

A self-placing, ductile, ultra high performance reinforced concrete was designed containing glass fibers. Ductility was maintained on thin structural elements after accelerated aging tests by an optimum combination of types of fibers and matrix compositions. Reverse analysis of the flexural results was proposed to extract the mechanical behavior of the composite in direct tension after taking into account the scale effect. The curvature was used as a description parameter; the cracking concept is therefore not explicit. According to the presented model, the tensile behavior is unique and could be used in codes to design thin structural elements. Considering the very high mechanical performance of this new GF-UHPRC, in terms of strength, ductility and durability, we firmly believe that innovative structural elements could be developed with these composites, thereby facilitating the current drive in construction industries for more sustainable and energy-efficient practices.

References

[1] AFGC, Ultra High Performance Fibre Reinforced Concretes, Recommendations, AFGC Publication, France (January 2002)
[2] Bentur, A., Mindess, S.: Fibre reinforced cementitious composites, 2nd edn., vol. 15. Taylor & Francis, Abingdon (2007)
[3] CEB-FIP, Structural Concrete, textbook on behaviour, design and performance, updated knowledge of the CEB-FIP Model Code 1990 FIB Publication (1999)
[4] Chanvillard, G., Rigaud, S.: Complete characterization of tensile properties of Ductal® UHPFRC according to French recommendations. In: HPFRCC-4, June 16-18, p. 14. Michigan, Ann Arbor (2003)
[5] Hillerborg, A., Modéer, M., Petersson, P.E.: Analysis of crack formation and crack growth in concrete by means of fracture mechanics and finite element. Cem. Concr. Res. 6, 773–782 (1976)
[6] Leonard, S., Bentur, A.: Improvement of the durability of glass fiber reinforced cement using blended cement matrix. Cem. Concr. Res. 14, 717–728 (1984)
[7] Yilmaz, V.T., Glasser, F.P.: Reaction of alkali-resistant glass fibres with cement: part 1, review, assessment and microscopy. Glass Technol. 32, 91–98 (1991)

Strain Rate Dependent Tensile Behavior of Ultra-High Performance Fiber Reinforced Concrete

K. Wille[1], S. El-Tawil[2], and A.E. Naaman[2]

[1] Civil and Environmental Engineering, University of Connecticut, USA
[2] Civil and Environmental Engineering, University of Michigan, USA

Abstract. Ultra High Performance Fiber Reinforced Concretes (UHP-FRC) can be designed to resist increasing tensile loading after first matrix cracking, which results from strain hardening tensile characteristics accompanied by multiple cracking. Previous investigations carried out under static loading conditions have clearly shown that matrix composition, fiber material and geometry as well as fiber volume fraction and fiber orientation influence the strain hardening tensile behavior. This paper describes research that was conducted to study the direct tensile behavior of UHP-FRC loaded at various speeds. A hydraulic test machine was used to apply load up to 10^3 times faster than static loading, i.e. up to a strain rate of $\dot{\varepsilon} = 0.1\,s^{-1}$. The test setup was designed to permit reliable measurement of direct tension test results at the different loading speeds considered, taking into consideration a reasonable gage length for multiple crack development while minimizing the inertial effects associated with the specimen and attached measurement equipment. The strain rate dependent tensile behavior is analyzed in terms of peak strength, strain at peak strength, hardening modulus and energy absorption capacity prior to softening. The results show the strain rate sensitivity of each of these parameters at fiber volume fractions of 2, 2.5 and 3%.

1 Introduction

Ultra-high performance concretes (UHPC) represent a class of cement composites characterized by a low water/binder ratio, a high particle packing density, and a compressive strength in excess of 150 MPa (22 ksi). The term UHP-FRC (ultra-high performance fiber reinforced concrete or cement composite) is used for an UHPC containing fibers. Here straight very high strength steel fibers are used. The addition of fibers leads to significant improvement in the material ductility under direct tensile loading, which also affects the ductility under compression, shear or torsion. Prior research by the authors has shown that the ductility of UHP-FRC can be significantly increased by utilizing a ternary optimization (matrix, fiber, interface properties), leading to an energy absorption capacity of about g =130 kJ/m^3

G.J. Parra-Montesinos, H.W. Reinhardt, and A.E. Naaman (Eds.): HPFRCC 6, pp. 381–387.

prior to softening; which exceeds by at least 5 times comparable energy values reported by other researchers [1]. High energy absorption capacity is a result of increased post-cracking strength (in excess of 20 MPa) and increased strain capacity at stress-peak (in excess of 0.5%), accompanied by multiple cracking with an average crack spacing as low as 2 mm. These results suggest that such UHP-FRCs have the potential to be particularly useful for structures that could be subjected to extreme events such as blast, impact or seismic loading. Therefore the material's tensile behavior under higher strain rate loading is of particular interest.

Previous research on high performance fiber cement composites (HPFRCC) up to a compressive strength of 84 MPa has clearly shown that the tensile strength [2] as well as the single fiber pull-out resistance [3] is strain rate sensitive. Habel & Gauvreau [4] performed two uniaxial tension tests on UHP-FRC up to a strain rate of $\dot{\varepsilon} = 0.02\,\mathrm{s}^{-1}$, which refers to seismic loading or vehicle collision on bridge piers [5], and obtained a 25% increase in tensile strength up to 14 MPa in comparison to the tensile strength of 11 MPa under static loading. While limited information exists about the effect of strain rate on HPFRCC, very little information is available on the rate-dependent behavior of UHP-FRC, which is the motivation for the work reported herein.

2 Specimen Preparation and Testing

The uniaxial tensile behavior of three different UHP-FRCs composites was investigated under four different strain rates. These composites only varied in Their composition is summarized in Table 1, where it can be seen that the composites were developed from the base UHPC by volumetric replacement of some sand by steel fibers. The volume of fiber content varied in $V_f = 2.0$, 2.5, and 3.0%. The UHPC composition in Table 1 is a result of particle packing and strength optimization performed in prior research [1, 6]. In order to cover the rate effects associated with seismic loading, normally considered to be correspond to $\dot{\varepsilon} = 0.001 - 0.1\,\mathrm{s}^{-1}$ [7], four different strain rates of $\dot{\varepsilon} = 0.0001 - 0.1\,\mathrm{s}^{-1}$ were applied to the specimens using a hydraulic servo-controlled testing machine (MTS-810). This follows the same procedure reported in a prior study on HPFRCC [2]. Three to six specimens were tested for each series.

All tests were performed after 28 days of age and no special heat or pressure curing was applied. The workability of the UHP-FRC mixtures was designed to exhibit self-consolidating properties without the risk of fiber segregation. No vibration was applied during or after casting. The workability of the mixes and the casting method preferably aligned the fibers in the direction of loading. The test set-up is shown in (Fig. 1) and simulates a similar set-up described in [8]. Mechanical anchorage and rotational boundary conditions allowed for quick specimen installation and alignment. An infrared-based motion measurement system was used (Optotrak) with accuracy of up to 0.1 mm, resolution of 0.01mm and a maximum marker frequency of up to 4600 Hz. Prior comparison using LVDTs had proven the accuracy of this system. These sensors are particularly suitable for high strain rate tests due to their small weight, which introduces negligible inertial effects, and their high frequency. Additionally, the markers could be quickly deployed, removed and reused.

Table 1. Mixture proportions by weight

Type	UHPC	UHP-FRC-2%	UHP-FRC-2.5%	UHP-FRC-3%
Cement	1.00	1.00	1.00	1.00
Silica Fume	0.25	0.25	0.25	0.25
Glass Powder	0.25	0.25	0.25	0.25
Water	0.19	0.19	0.19	0.19
Superplasticizer[1]	0.011	0.011	0.011	0.011
Sand A[2]	0.31	0.29	0.28	0.28
Sand B[3]	0.72	0.67	0.66	0.65
Fiber[4]	0.00	0.18	0.22	0.27
Fiber by Vol.%	0	2.0	2.5	3.0
$f'_c[cube,28d]$ (MPa)	230	248	246	250
$f_t[tension]$ (MPa)	7.4 - 8.5*	15	16.5	18

[1] solid content; [2] max. grain size 0.2 mm (1/128 in.), [3] max. grain size 0.8 mm (1/32 in.), [4] straight steel fiber, length/diameter = 13 mm/0.20 mm, tensile strength ≈ 2600 MPa (377 ksi), * at first cracking, followed by immediate failure

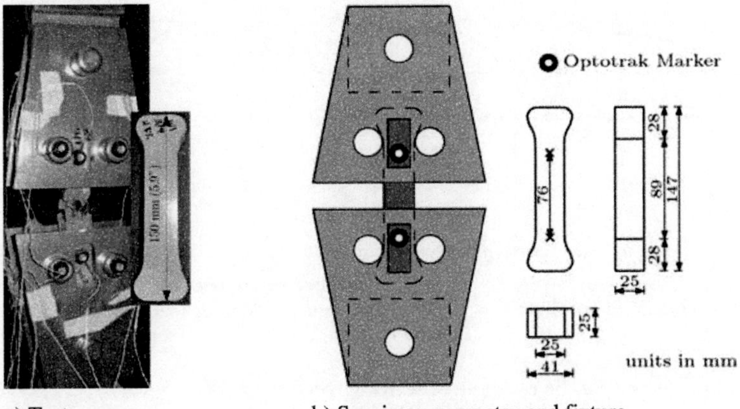

a) Test b) Specimen geometry and fixture

Fig. 1. Direct tensile test setup used in this research with Optotrak markers on front and back face of the specimen

3 Test Data Analysis and Results

Fig. 2a shows the tensile stress versus strain curves of the three UHP-FRC-2.5% specimens tested at a strain rate of $\dot{\varepsilon} = 0.0001\,\mathrm{s}^{-1}$ that is considered quasi-static loading. Each curve is then idealized into a best fit polygon made up of 3 segments: the first corresponds to the elastic response, the second for the strain-hardening behavior and the third for the plateau that occurs just prior to softening (Fig. 2b). The definitions of the tensile parameters are given in Fig. 2b. The idealized curves and their parameters are then averaged (Fig. 2c). Fig. 2d compares the averaged idealized curves of UHP-FRC-2.5% at the four different strain rates up to $\dot{\varepsilon} = 0.1\,\mathrm{s}^{-1}$.

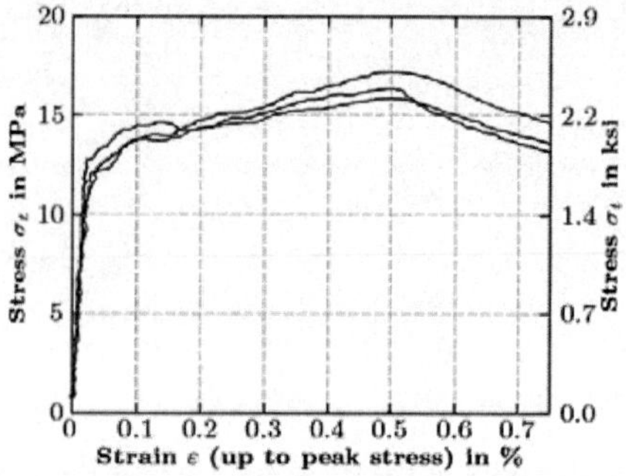

a) Specimens 1-3 experimental data

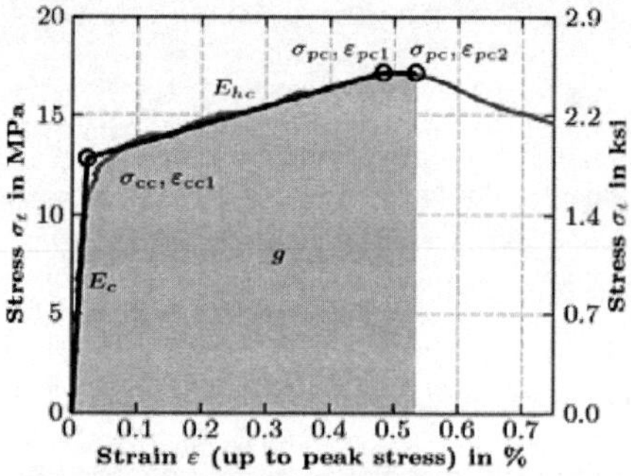

b) Specimen 2 and idealized tensile behavior

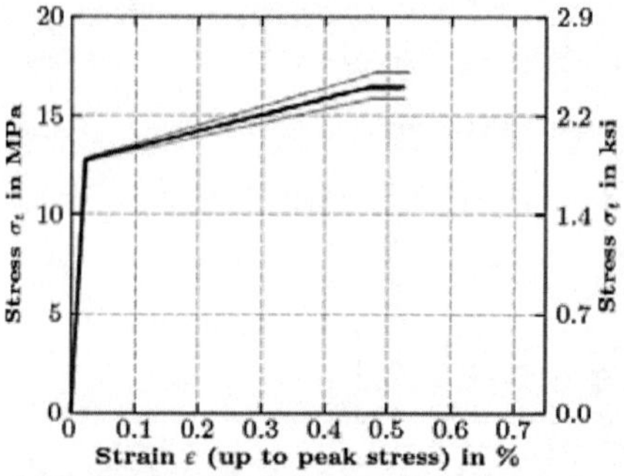

c) Idealized tensile behavior of specimens 1-3 and average curve

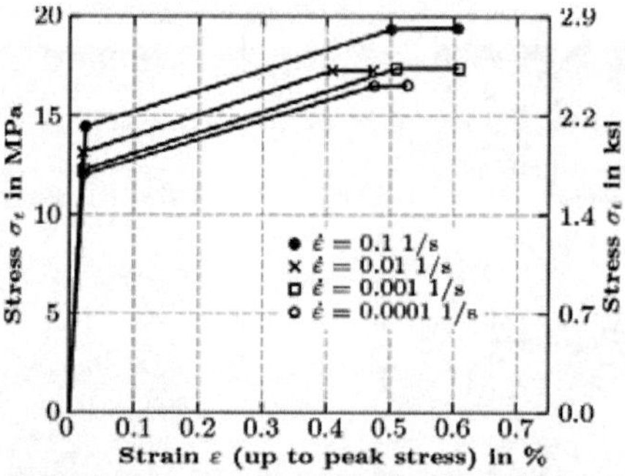

d) Strain rate dependent idealized tensile behavior

Fig. 2. Analytical procedure for the example of series UHP-FRC-2.5%

Figs. 3a-f show the values of the various tensile parameters for the idealized average curves at different strain rates for UHP-FRCs with straight steel fibers of 2, 2.5 and 3% fibers by volume. In general it can be observed that the strain rate effect is not influenced by fiber volume fraction.

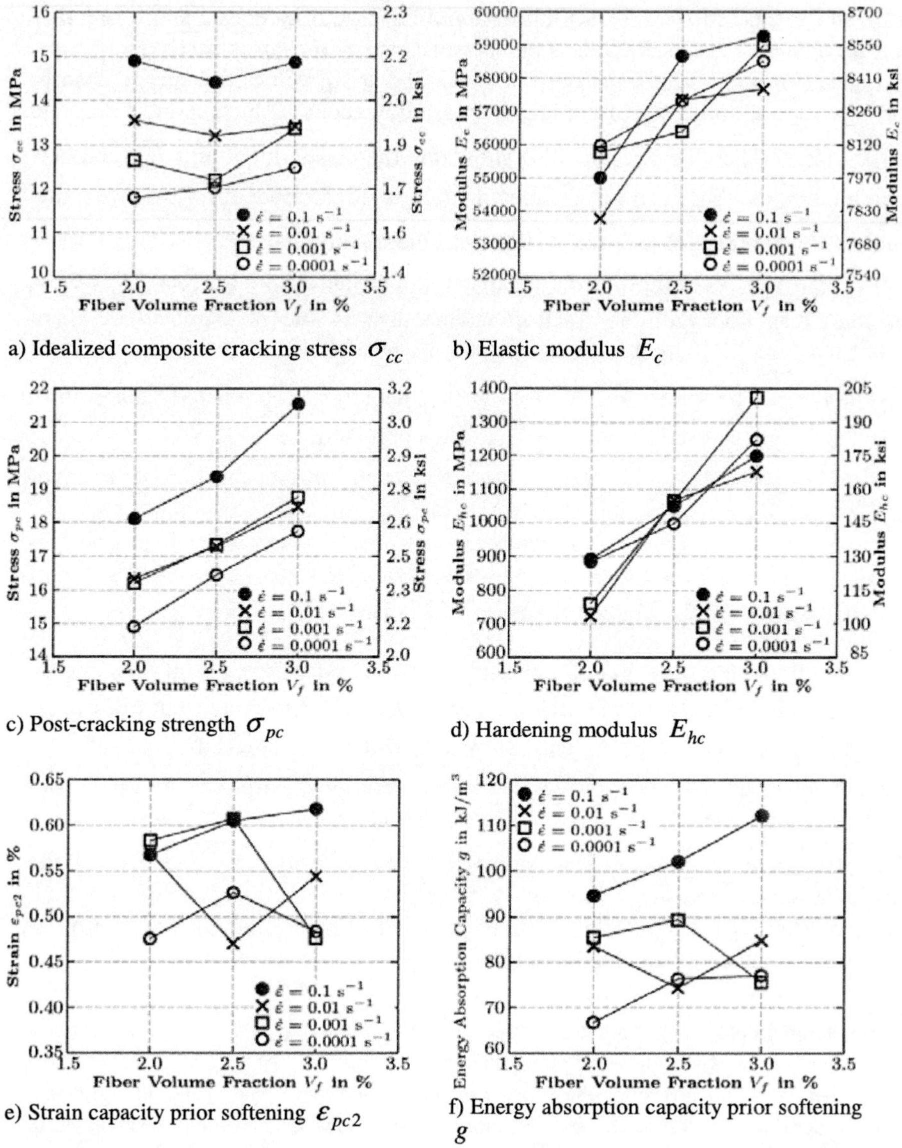

a) Idealized composite cracking stress σ_{cc}

b) Elastic modulus E_c

c) Post-cracking strength σ_{pc}

d) Hardening modulus E_{hc}

e) Strain capacity prior softening ε_{pc2}

f) Energy absorption capacity prior softening g

Fig. 3. Strain rate dependent tensile parameters of UHP-FRC with straight steel fibers of 2, 2.5 and 3% fiber by volume

A strong strain rate dependency is observed for the stress based parameters σ_{cc} and σ_{pc}. The increase in strain rate by 10^3 up to $\dot{\varepsilon} = 0.1\,\mathrm{s}^{-1}$ increases the post-cracking strength by about 20% for all volume fractions. This is in agreement with the preliminary test results reported in [4]. Strain based parameters like ε_{pc2}

show an increase in their values up to about 25% with increased strain rate, but their trend at various strain rates is not as consistent as the stress based parameters. As a result of strain rate sensitivity of stress and strain values the energy absorption capacity, g, prior to softening is increased by about 40% when the strain rate is 10^3 times higher. Test results also show that the elastic modulus, E_c, and hardening modulus, E_{hc}, can be considered strain rate independent for the strain rate range of $\dot{\varepsilon} = 0.0001 - 0.1\,\mathrm{s}^{-1}$ for all fiber volume fractions.

It is also worth noting that the idealized composite cracking stress σ_{cc} can be considered as fiber volume fraction independent in the investigated range of $V_\mathrm{f} = 2.0 - 3.0\%$. A slight increase in strain capacity ε_{pc2} with an increase in V_f is noticed. The values of E_c follow the theoretically predicted values from elastic mechanics of the composites, that is a small increase of E_c with an increase in V_f. However, σ_{pc}, E_{hc} and g are significantly influenced by the fiber volume fraction V_f.

4 Summary and Conclusions

In this study the tensile behavior of strain-hardening UHP-FRCs with three different fiber volume fractions was investigated under four different strain rates ranging from $\dot{\varepsilon} = 0.0001 - 0.1\,\mathrm{s}^{-1}$. The test results show an increase in post-cracking strength up to about 20% and an increase in energy absorption capacity up to about 40% with an increase in strain rate from $\dot{\varepsilon} = 0.0001\,\mathrm{s}^{-1}$ to $\dot{\varepsilon} = 0.1\,\mathrm{s}^{-1}$. For a given strain rate, the post-cracking strength, the strain-hardening modulus, and the energy absorption capacity increase with an increase in fiber volume fraction. These results suggest that UHP-FRC is particularly promising for application in structures subjected to seismic loading.

Acknowledgements. This work was supported by a fellowship within the Postdoctoral-Programme of the German Academic Exchange Service (DAAD). The senior authors would also like to acknowledge the support of the University of Michigan and that of the National Science Foundation under grant No. CMMI 0754505. The opinions expressed in this paper are those of the writers and do not necessarily reflect the views of the sponsors.

References

[1] Wille, K., Naaman, A.E., El-Tawil, S.: Ultra High Performance Fiber Reinforced Concrete (UHP-FRC) Record Performance under Tensile Loading. ACI Concrete International Journal (2010) (submitted)

[2] Kim, D.J., El-Tawil, S., Naaman, A.E.: Rate-dependent tensile behavior of high perform-ance fiber reinforced cementitious composites. Materials and Structures 42(3), 399–414 (2009)

[3] Kim, D.J., El-Tawil, S., Naaman, A.E.: Loading Rate Effect on Pullout Behavior of De-formed Steel Fibers. ACI Materials Journal, 1–9 (2008)

[4] Habel, K., Gauvreau, P.: Response of ultra-high performance fiber reinforced concrete (UHPFRC) to impact and static loading. Cement & Concrete Composites 30, 938–946 (2008)

[5] CEB-FIP, Concrete Structures under impact and impulse loading, Synthesis Report, Bulle-tin d'information 187, Lausanne, Switzerland (1988)

[6] Wille, K., Naaman, A.E., Parra-Montesinos, G.: Ultra High Performance Concrete with Com-pressive Strength Exceeding 150 MPa (22ksi): A Simpler Way. ACI Materials Journal 108(1), 46–54 (2011)

[7] Tai, Y.-S.: The behavior of reactive power concrete at high strain rates. Magazine of Concrete Research 62(11), 763–772 (2010)

[8] Sujiravorakul, C.: Development of High Performance Fiber Reinforced Cement Compos-ites Using Twisted Polygonal Steel Fibers. Ph.D. dissertation, University of Michigan, Ann Arbor, p. 230 (2002)

Dynamic Properties and Damage Model of Ultra-High Performance Fiber Reinforced Cement Composites Subjected to Repeated Impacts

J. Lai[1], W. Sun[2], S. Xu[1], and C. Yang[1]

[1] College of Materials Science and Engineering, Nanjing University of Science and Technology, Nanjing, P.R. China
[2] College of Materials Science and Engineering, Southeast University, Nanjing, P.R. China

Abstract. Ultra-high performance fiber reinforced cement composites (UHPFRCC) is a new type of concrete with very high strength, ductility and durability and is expected to be used for innovative construction structures. The dynamic properties of UHPFRCC subjected to repeated impacts were investigated using the split Hopkinson pressure bar. The dynamic damage to the UHPFRCC was measured by ultrasonic wave velocity methods after each impact. The influence of different fibers and the number of impacts was researched on the stress-strain curves and dynamic damage of UHPFRCC. A damage model was proposed to simulate the stress-strain curves of UHPFRCC subjected to repeated impacts. Results showed that the simulation values were very close to those of the experiments.

1 Introduction

There are many engineering structures that experience dynamic loads such as impact, penetration and blast. The material behavior under a dynamic load is different from that under a static load so it is important to investigate the dynamic strength, dynamic damage and dynamic stress-strain relationship of engineering materials. The split Hopkinson pressure bar (SHPB) is effective in obtaining the dynamic stress-strain behavior of materials [1-3]. Defensive structures are required to have the ability to resist repeated impacts of accurately guided missiles in a high-tech war. Therefore, it is necessary to investigate the dynamic behavior of and damage process to defensive materials subjected to repeated impacts. Ultra-high performance fiber reinforced cement composites (UHPFRCC) is a type of new concrete with a very high strength and durability and is expected to be used in the containment structures of nuclear power plants and defensive facilities. The dynamic compression behavior of ultra-high performance concrete was studied using SHPB [4-6]. In this study, UHPFRCC was prepared and the high velocity repeated impact experiments were conducted to assess the resistance of UHPFRCC. The dynamic properties of UHPFRCC were investigated when subjected to

G.J. Parra-Montesinos, H.W. Reinhardt, and A.E. Naaman (Eds.): HPFRCC 6, pp. 389–396.

repeated impacts in three different impact modes using the SHPB, and the dynamic damage to the UHPFRCC was measured using the ultrasonic wave velocity method. The effects of the number of impacts and impact modes on the stress-strain curves, and the dynamic damage to the UHPFRCC, were investigated. The dynamic stress-strain behavior of UHPFRCC was simulated using a visco-elastic damage model.

2 Materials Preparation

Four cementitious materials were used in the preparation of UHPFRCC including Portland cement, silica fume, fly ash and blast-furnace slag. The strength grade of the cement is P·II 52.5 according to the relevant Chinese standard. Ground fine quartz sand used for normal UHPFRCC was totally replaced with natural sand. The maximum particle size of natural sand is 2.5 mm with a fineness modulus of 2.6. The polycarboxylate based superplasticizer was produced by the Grace Company in Shanghai China with a water-reducing ratio of more than 40%. Three types of fibers were used, including steel fibers, PVA fibers and basalt fibers. The cementitious materials and sand were put in a forced mortar mixer at the same time and mixed for 3 minutes. The water and superplasticizer were mixed together and then the solution was added into the mixer and mixed for another 6 minutes. Finally, fibers were added to the mortar and mixed for 3 minutes to ensure that fibers were well distributed in the mortar. After mixing, the UHPFRCC was cast into steel moulds and was compacted on a standard vibrating table. The specimens were stored in the standard conditions (20°C ± 2°C, R.H. > 90 %) and removed from their moulds after 24 hours. The specimens were cured in the standard conditions for 60 days before testing. The mix proportion and static uniaxial compressive strength of the UHPFRCC with different fibers are listed in Table 1.

Table 1. Mix proportions of UHPFRCC

Materials	Binder (wt.%)				Fiber volume fraction (%)			f_c (MPa)
	Cement	Silica fume	Fly ash	Slag	Steel fiber	PVA fiber	Basalt fiber	
$UPCV_0$	40	10	25	25	0	0	0	143
$UPCV_3$	40	10	25	25	3	0	0	186
$UPCV_4$	40	10	25	25	4	0	0	204
$UPCV_3P_1$	40	10	25	25	3	1	0	227
$UPCV_3B_1$	40	10	25	25	3	0	1	211

Note: the superplasticizer to binder ratio, sand to binder ratio and water to binder ratio are 0.02, 1.2 and 0.2 respectively. f_c is the static compressive strength of UHPFRCC.

3 Experimental Method with Repeated Impacts

A split Hopkinson pressure bar was used to obtain the dynamic stress-strain relationships of the UHPFRCC at high strain rates. The cylinder specimens are 70 mm in diameter and 35 mm in length. Fig. 1 shows an illustration of the SHPB. The SHPB test is based on the theory of one-dimensional wave propagation in an elastic bar. The specimen is placed between the input and output bars. The projectile impacts on the input bar and a compressive incident wave is generated in the input bar. The incident wave then impacts on the specimen between the two bars. Part of the incident wave is reflected back into the input bar and part of it is transmitted through the specimen into the output bar. Strain gages are placed on the bars and the impact signals are recorded by a computer. The average stress, strain and strain rate of specimens can be calculated using the following equation:

$$\sigma = E_0 \varepsilon_t(t) \frac{A_0}{A_s}, \ \dot{\varepsilon} = \frac{2C_0}{L_s}[\varepsilon_i(t) - \varepsilon_t(t)], \ \varepsilon = \int_0^t \dot{\varepsilon}(t)dt \tag{1}$$

where $\varepsilon_i(t)$ and $\varepsilon_t(t)$ are incident wave and transmitted wave; E_0 and C_0 are the Young's modulus and the elastic wave speed of the bar; A_0 and A_s are the cross section areas of the bar and the specimen; and L_s is the length of the specimen.

Materials and structures in regular use usually suffer more than a single impact, so it is important to investigate the dynamic strength, stress-strain relationship and dynamic damage of materials subjected to repeated impacts. The strain rates of concretes are usually within $10^2 s^{-1}$ on impact by the SHPB. Repeated impacts using three types of loading mode for different projectile speeds were as follows:

Mode A: the projectile speed on the first impact was 22 ms^{-1}, the others were 22 ms^{-1} on the following impact;
Mode B: the projectile speed on the first impact was 26 ms^{-1}, the others were 22 ms^{-1} on the following impact;
Mode C: the projectile speed on the first impact was 29 ms^{-1}, the others were 22 ms^{-1} on the following impact.

The strain rates of the UHPFRCC were 10 s^{-1} to $10^2 s^{-1}$ on impact for the three types of mode. The projectile speed on the first impact increased from Mode A to Mode C to ensure the specimens had different levels of initial damage. The projectile speed was the same for the following impact through which the change of dynamic properties and damage to the UHPFRCC was investigated.

The wave velocity of UHPFRCC was measured by the ultrasonic wave velocity method after each impact [7]. The material damage after repeated impacts is defined as:

$$D = 1 - \frac{E_n}{E_0} = 1 - \frac{V_n^2}{V_0^2} \tag{2}$$

where E_0 and V_0 are the initial elastic modulus and ultrasonic wave velocity of the specimen; and E_n and V_n are the elastic modulus and ultrasonic wave velocity of the specimen after the *n*th impact.

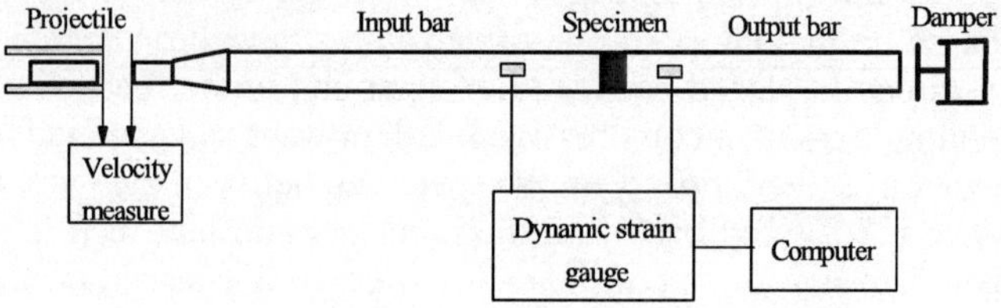

Fig. 1. The SHPB test set-up

4 Results and Analysis

4.1 Dynamic Properties of UHPFRCC Subjected to Repeated Impact in Different Modes

Fig. 2 shows the strain-stress relationships of UHPFRCC subjected to repeated impacts in the different modes. Results show that the strain rate, elastic modulus, peak stress and peak strain increase gradually on the first impact from Mode A to Mode C because the impact speed increases with the changes in impact mode. Moreover, the damage to the UHPFRCC on first impact increases from Mode A to Mode C. The stress-strain relationship on the second impact changes in a manner contrary to that on the first impact, namely, the elastic modulus and peak stress on second impact decrease from Mode A to Mode C.

Results show that the elastic modulus and peak stress decrease while the strain rate and peak strain increase, with the number of impacts increasing in the same impact mode. In Mode A and Mode B, the elastic modulus and peak stress of UHPFRCC subjected to repeated impacts decrease less when the fiber volume fraction increases. In Mode C, the strain rates of different UHPFRCC are over 90 s^{-1} and the strains are over 0.02 on the first impact. The elastic modulus and peak stress decrease significantly on the second impact in Mode C because the bond between fibers and mortar is broken on the first impact and the fiber bridging effect becomes weak. It can be concluded that the serious damage on the first high speed impact leads to the similar weak dynamic properties of a different UHPFRCC on the second impact. In other words, the effects of different fibers on the resistance to repeated impacts are insignificant if the initial damage to the UHPFRCC is very serious.

In high speed impact loading, the initiation and propagation of cracks in UHPFRCC are restrained by fibers. These fibers, different in dimensions and elastic modulus, cooperate to restrain cracks in the UHPFRCC on impact. Compared with steel fibers, the number of smaller PVA fibers or basalt fibers per cubic meter is much larger given the same fiber volume fraction, while the average fiber distance of smaller fibers is much shorter than that of steel fibers. The micro-cracks are mainly controlled by smaller fibers (PVA fiber or basalt fiber) and when the cracks expand, the larger fibers (steel fibers) start to have an effect. Therefore, the cracks in hybrid fiber reinforced UHPFRCC are much smaller than those in single steel fiber reinforced UHPFRCC.

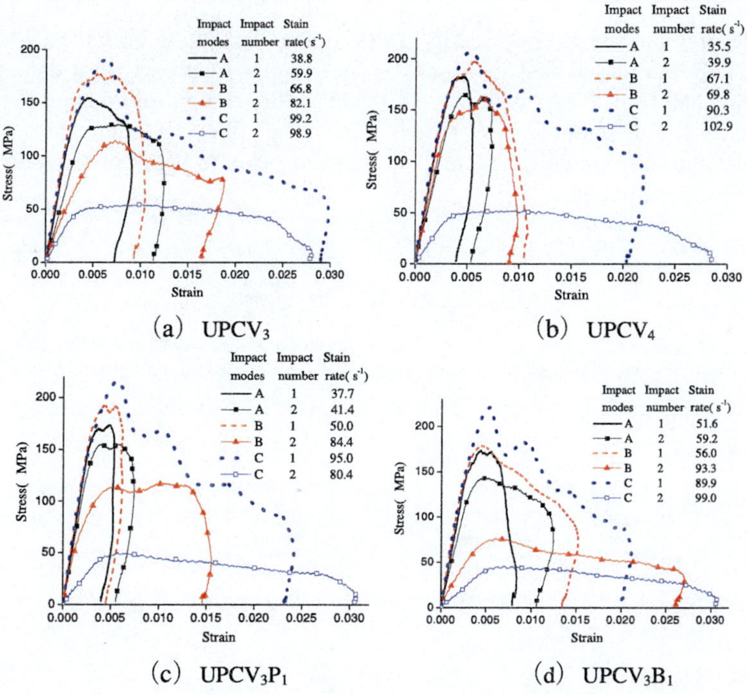

Fig. 2. The stress-strain curves of UHPFRCC subjected to repeated impacts in the different modes

4.2 Dynamic Damage to UHPFRCC Subjected to Repeated Impacts in Different Modes

Fig. 3 shows the influence of the number of impacts and impact modes on the dynamic damage to the UHPFRCC. The damage to unreinforced UHPFRCC was significant on the first impact and the specimen broke into pieces because of its weak ability to resist impacts. Therefore, a second impact could not be performed. The damage to the UHPFRCC on first impact was reduced using the reinforcement of fibers. The damage to the reinforced UHPFRCC increased gradually with increases in the number of impacts until it lost impact resistance. In Mode A, $UPCV_4$ showed the best ability to resist repeated impacts. The first impact damage to the $UPCV_4$ was within 0.05 and the second was within 0.11. The damage to the different materials obviously increased with the increase of the initial impact speed in Modes B and C. The damage was over 0.2 and 0.5 on the first and second impacts in Modes B and C. The specimens were completely broken and the damage was over 0.9 after the third impact in Mode C. Results showed that the damage on the first impact had obvious effects on the resistance to repeated impacts. The repeated impact damage was accelerated with the cracks increasing after the first impact. The damage to hybrid fiber reinforced UHPFRCC is shown in Figs. 3(c) and (d). The damage to the $UPCV_3P_1$ was within 0.11 and 0.25 on the first and second impacts in Mode A. The damage to the $UPCV_3B_1$ was within 0.13 and 0.3 on the first and second impacts in Mode A. The damage to both the $UPCV_3P_1$ and $UPCV_3B_1$ increased fast in Modes B and C, but there was less damage to $UPCV_3P_1$. Therefore, PVA fibers have a better ability to control damage than basalt fibers in the different impact modes. The basalt fibers are more brittle than PVA fibers and they are half the length so they are easier to break or separate from the UHPFRCC on impact.

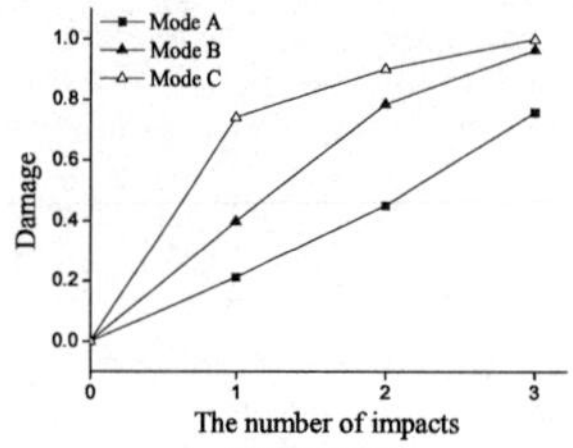

(a) $UPCV_3$

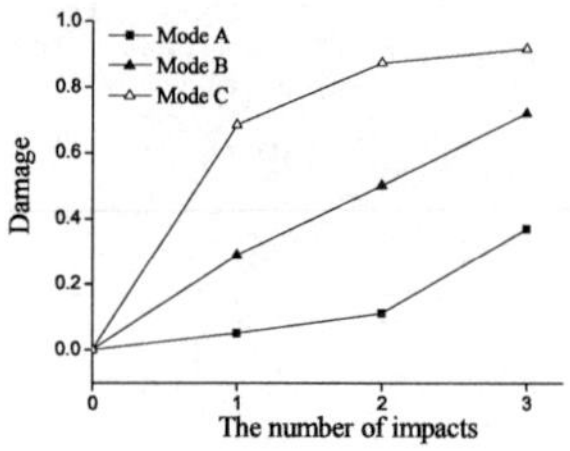

(b) $UPCV_4$

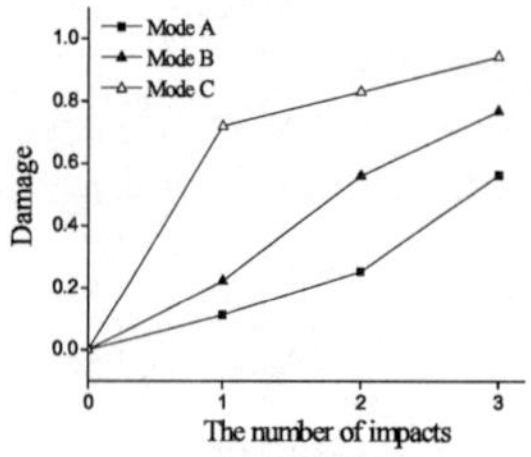

(c) $UPCV_3P_1$

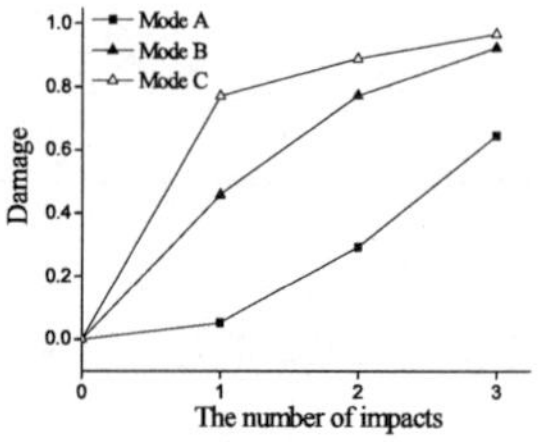

(d) $UPCV_3B_1$

Fig. 3. The influence of the number of impacts and impact modes on the dynamic damage to the UHPFRCC

4.3 *Visco-elastic Damage Model of UHPFRCC Subjected to Repeated Impacts*

Researching the dynamic behavior of engineering plastic, Zhu, Wang and Tang proposed a nonlinear visco-elastic model as follows - the ZWT model [8]:

$$\sigma = E_0\varepsilon + \alpha\varepsilon^2 + \beta\varepsilon^3 + E_1\int_0^t \dot{\varepsilon}\exp\left(-\frac{t-\tau}{\theta_1}\right)d\tau + E_2\int_0^t \dot{\varepsilon}\exp\left(-\frac{t-\tau}{\theta_2}\right)d\tau \tag{3}$$

where, E_0, α and β are elastic constants; τ is the time variable and t is the loading time; and E_1 and θ_1 are the elastic constant and relaxation time of the low frequency Maxwell element while E_2 and θ_2 are those of the high frequency one.

Based on the principle of strain-equivalence in damage mechanics and the ZWT model, Lai and Sun proposed a visco-elastic damage model as follows to simulate the dynamic behavior of concretes [7]:

$$\sigma = (1 - D)\left[E\varepsilon + E_2\theta_2\dot{\varepsilon}\left(1 - \exp\left(-\frac{\varepsilon}{\dot{\varepsilon}\theta_2}\right)\right)\right] \tag{4}$$

where E, E_2 and θ_2 are fitting parameters; E and E_2 show the elastic properties of concretes at low and high strain rates; θ_2 is the relaxation time and it shows the viscous properties of concretes at high strain rates; and D is the damage to the concretes subjected to impacts. In this study, the D of UHPFRCC subjected to repeated impacts was measured by ultrasonic wave velocity methods and as previously discussed.

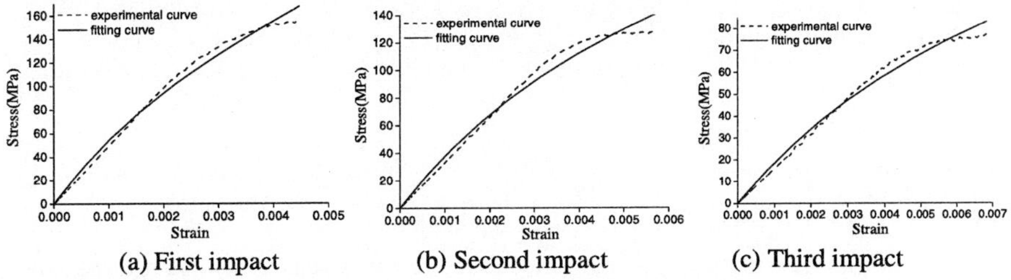

Fig. 4. Comparison between the fitting and experimental stress-strain curves of $UPCV_3$ in Mode A

Table 2. The fitting parameters of the visco-elastic damage model for UPCV3

Impact mode	Impact number	Damage	Strain rate(s^{-1})	E(GPa)	E_2(GPa)	θ_2(μs)
	1	0.212	38.8	30.7	51.7	40.0
A	2	0.451	59.9	18.3	59.7	50.0
	3	0.759	87.6	15.6	68.7	50.0

The fitting and experimental stress-strain curves for different numbers of impacts are shown in Fig. 4 by nonlinear regression analysis. Results show that the theoretical and experimental data are similar. Therefore, the given model is suitable for simulating the dynamic stress-strain relationships of UHPFRCC subjected to repeated impacts. Table 2 gives the values of the fitting parameters. Results indicate that the damage and strain rate increase with the number of impacts. E_2 shows the dynamic elastic property of the UHPFRCC and it increases with strain rate.

5 Conclusions

The dynamic stress-strain relationships and damage to the UHPFRCC with different hybrid fibers and fiber volume fractions were studied in three types of repeated impact modes using the SHPB and the ultrasonic wave velocity method. The stress-strain curves of the UHPFRCC subjected to repeated impacts were simulated using a visco-elastic damage model and the theoretical and experimental data were similar. Results show that the peak stress and elastic modulus decrease and the strain rate and peak strain increase gradually with increasing numbers of impacts. The damage to the UHPFRCC increased with the number of impacts but it decreased with hybrid fiber reinforcement. The ability of the UHPFRCC to resist repeated impacts is improved by increasing the fiber volume fraction. The repeated impact modes have significant effects on the dynamic stress-strain curves and damage to the UHPFRCC. The dynamic strength of the UHPFRCC decreases and the dynamic damage increases much faster with increases of the projectile speed on the first impact. The serious damage on the first impact leads to similar dynamic properties for different UHPFRCCs on the second and third impact.

Acknowledgements. This work is supported by the National Natural Science Foundation of China (No.50808101), the China Postdoctoral Science Foundation (No.20080431100), Jiangsu Provincial Program for Basic Research (Natural Science Foundation) (No.BK2008417), the NUST Excellence Plan "Zijin Star" and NUST Research Funding (No.2010ZYTS087).

References

[1] Han, Z.: Study on testing techniques for concrete-like materials under compressive impact loading. Cem. Concr. Compos. 20(4), 293–299 (1998)

[2] Grote, D.L., Park, S.W., Zhou, M.: Dynamic behavior of concrete at high strain rates and pres-sures: I.experimental characterization. Int J. Impact Engng. 25(9), 869–886 (2001)

[3] Lok, T.S., Asce, M., Zhao, P.J.: Impact response of steel fiber-reinforced concrete using a split Hopkinson pressure bar. J. Mater. Civ. Eng. 16(1), 54–59 (2004)

[4] Tai, Y.S.: Uniaxial compression tests at various loading rates for reactive powder concrete. Theor. Appl. Fract. Mech. 52(1), 14–21 (2009)

[5] Rong, Z.D., Sun, W., Zhang, Y.S.: Dynamic compression behavior of ultra-high performance ce-ment based composites. Int. J. Impact. Engng. 37(5), 515–520 (2010)

[6] Zhang, Y.S., Sun, W., Liu, S.F., Jiao, C.J., Lai, J.Z.: Preparation of C200 green reactive powder con-crete and its static-dynamic behaviors. Cem. Concr. Compos. 30(9), 831–838 (2008)

[7] Lai, J.Z., Sun, W.: Dynamic behavior and visco-elastic damage model of ultra-high perform-ance cementitious composite. Cem. Concr. Res. 39(11), 1044–1051 (2009)

[8] Chu, C.H., Wang, L.L., Xu, D.: A nonlinear thermo-viscoelastic constitutive equation for ther-moset plastics at high strain rates, in: Chien WZ, editor. In: Chien, W.Z. (ed.) Proceedings of the International Conference on Nonlinear Mechanics, Shanghai (China), pp. 92–97 (1985)

CARDIFRC – From Concept to Industrial Application

B.L. Karihaloo

Laing O'Rourke Chair of Hybrid Concrete Innovation,
Cardiff University, School of Engineering, UK

Abstract. This paper gives an overview of the development of CARDIFRC, which is a high-performance fiber-reinforced cement-based material, from the stage of conception to its eventual application in industry. It will highlight the important changes that had to be made to the original CARDIFRC mixes in order for them to be acceptable to the industry. The changes were necessitated not only by cost considerations, but also by health and safety concerns.

1 Introduction

CARDIFRC is a class of ultra high performance fiber-reinforced cement-based composite (UHPFRCC), in the same category as DENSIT [1], DUCTAL, RPC [2], characterised by high compressive strength (in excess of 200 MPa), splitting/flexural strength (up to 30 MPa) and high energy absorption capacity (in excess of 17,000 J/m^2). This has been made possible by the use of large amounts (up to 6% by volume) of brass-coated short steel fibers of two lengths (6 and 13 mm long, 0.16 mm diameter) in a cement matrix densified by the use of silica fume. Extensive investigations were undertaken to examine the role of key parameters: (i) grading of quartz sands, (ii) fineness of micro-silica, (iii) amounts of short and long steel fibers, and of water and superplasticiser. As a result, the matrix of CARDIFRC contains only very fine graded quartz sand, instead of ordinary river sand and coarse aggregates. By optimizing the grading of these fine quartz sands, the water demand was considerably reduced without affecting the workability of the mix. This was achieved using novel mixing procedures described in the patent GB2391010 [3], as follows: (i) mix the coarsest quartz sand and the finest constituent (micro-silica), then next finest constituent (cement), next coarsest quartz sand and the least coarse quartz sand (for mix Type II only). Before each addition mix thoroughly for more than 2 min; (ii) place 25% of 13 mm fibers in 12 mm sieve and agitate it over opening in the mixing pan. Fibers are separated and fall in a 'shower' into the pan. Add next 25% of fibers and mix in the same way. Repeat until all 13 mm fibers added. Add 6 mm fibers in same incremental manner (using a 5 mm aperture size sieve); (iii) mix water with

G.J. Parra-Montesinos, H.W. Reinhardt, and A.E. Naaman (Eds.): HPFRCC 6, pp. 397–404.

two-thirds of SP. Add one half of this mixture to the mixed dry constituents and mix for ca. 2 min. Add half of the remaining mixture and mix for ca. 2 min prior to adding half of the remaining, and so on. Add the remaining one-third of SP and mix until the constituents are thoroughly mixed and wetted.

The method of its production and its mechanical and fracture properties were reported in a series of papers [4, 5, 6]. The mixes can be cured at ambient temperature for 28 days or at the elevated temperature of 90°C for just 7 days with no noticeable difference in the mechanical properties. Extensive tests have shown that it has remarkable durability and resistance to thermal cycling and cyclic loading [7, 8]. Computerized tomography imaging and sectioning of specimens have confirmed that the production procedures ensure a remarkably homogeneous mix with a uniform distribution of fibers in thin sections (< 50 mm thick) [6].

Moreover, it improves with age and has self-healing properties because of un-hydrated cement and silica fume. Tests have shown that its toughness increased in 2 years to more than 32,000 J/m^2. In view of its cost (primarily as a result of the use of brass-coated thin steel fibers and of various grades of quartz sand), it was initially intended only for small volume applications, e.g. retrofitting of concrete structures or for jointing of pre-cast concrete elements [9, 10].

However, recent cost analysis by the industrial partner Laing O'Rourke plc, has shown that CARDIFRC can be adapted to make it highly competitive even in a variety of hitherto-unforeseen very large volume applications. Jointly we have created an industrially competitive self-compacting variant of CARDIFRC for manufacturing pre-cast pre-stressed concrete elements without shear reinforcement for use in structural applications. The advantage of this variant over the competing steel is lower self weight (which forms a large part of design load), corrosion resistance, and smaller carbon footprint. We have managed to reduce the cost by replacing expensive brass-coated small steel fibers with less expensive and safer, and smaller amounts of, longer steel fibers and to create a self-compacting mix not requiring vibratory compaction. All these improvements have been achieved at a tolerable sacrifice (around 20%) to the above-mentioned strengths but with practically no reduction in the toughness. A further improvement has been the inclusion of 0.2% (by volume) polypropylene sacrificial fibers for enhancing the fire resistance of this extremely dense mix.

The optimum design of UHPFRCC mixes tries to balance the intrinsic competition between the tensile strength f_t and toughness G_F – toughness decreases with increasing strength. The simultaneous maximisation of f_t and G_F requires the solution of a dual-objective optimization problem in which each objective (f_t and G_F) is given a weighting depending upon its importance. As both objectives are equally important in the development of optimum UHPFRCC mixes, the dual-objective problem is replaced by a single design objective involving the maximisation of a ductility measure – the so-called characteristic length of the mix [11]

$$l_{ch} = \frac{EG_F}{f_t^2} \tag{1}$$

where E is the Young modulus of the mix. The mathematical optimisation strategies require a knowledge of the constitutive equations relating the mechanical response parameters (E, f_t, G_F and compressive strength f_c) to the mix and fiber parameters (water to binder ratio, w/b, maximum size of fine aggregate, g, volume fraction of aggregate, V_a, surfactant to water ratio, sp/w, fiber length, L, fiber diameter, d, and fiber volume fraction, V_f), collectively called the micro-structural parameters. These constitutive equations have been obtained by Lange-Kornbak and Karihaloo [12], and Wang and Karihaloo [13] using micromechanical principles.

$$E = \frac{E_m[n+\Theta+V\Theta(n-1)]}{[n+\Theta-V(n-1)]} \tag{2}$$

$$f_t = \frac{K_{Ic,m}}{\sqrt{\pi a_o}}(1-V_f) + \frac{h}{4}LV_f\alpha\, K_{Ic,m}\sqrt{\frac{E_f}{dE_b}} \tag{3}$$

$$G_F = \frac{h}{24}L^2V_f\alpha\, K_{Ic,m}\sqrt{\frac{E_f}{dE_b}} \tag{4}$$

In equation (2), V is the volume fraction of the discrete phase, Θ is a factor depending on the geometry (i.e. aspect ratio) and stiffness of the discrete phase, and n is the ratio of Young modulus of the discrete phase to that of a continuous phase E_m (i.e. of the matrix). In equations (3) and (4), $K_{Ic,m}$ is the fracture toughness of the brittle UHPFRCC matrix without fibers, E_f is the Young modulus of steel fibers, and E_b the Young modulus of the cement-silica binder. h is a dimensionless factor that takes into account the bending of steel fibers during their pull-out from the matrix, α is a constant (with the dimension of inverse length) that takes into account crack trapping and deflection induced by the presence of fibers, and a_o is proportional to the maximum size (in mm) of fine aggregate used in the mix. The mathematical optimization problem that is solved by standard nonlinear mathematical programming techniques is to maximise l_{ch} $(\underline{X})$ by choosing the vector of mix and fiber variables $\{\underline{X}\} = \{V_a, g, L, d, V_f, w/b, sp/w\}^T$ (i.e. micro-structural variables) in such a way as to satisfy the constitutive relations and any prescribed bounds (inequality constraints) on these variables, e.g. $0.15 \le w/b \le 0.25$, $0.01 \le V_f \le 0.08$. Moreover, these variables may be restricted to have certain fixed values (equality constraints) as, for instance, L = 6 or 13mm, d = 0.16 mm. It should be mentioned that we have used the fracture mechanics based concepts to derive equations (4) and (5) which introduces the matrix fracture toughness $K_{Ic,m}$. It is however possible to take a strength-based approach when instead of the matrix fracture toughness, the matrix tensile strength and the fiber-matrix bond strength would appear in these equations [14].

CARDIFRC mixes can be broadly grouped into two classes depending upon the maximum size of the quartz sand and the relative amounts of short (6 mm) and long (13 mm) fibers (the total volume fraction is 6%), designated mixes I and II in Table 1. It is worth mentioning that the short fibers primarily enhance the strength, whereas the long fibers enhance primarily the toughness of CARDIFRC.

Table 1. Mix constituents of CARDIFRC (kg/m^3)

Constituent	Mix I	Mix II
Cement	855	744
Microsilica	214	178
Quartz sand:		
9-300 μm	470	166
250-600 μm	470	-
212-1000 μm	-	335
1-2 mm	-	672
Water	188	149
Superplasticiser	28	55
Fibers: - 6mm	390	351
- 13mm	78	117
Water/cement	0.22	0.20
Water/binder	0.18	0.16
Superplasticiser/water	0.15	0.37

2 Influence of the Amount and Size of Steel Fibers

The prohibitive cost of CARDIFRC (80% of the cost is due to the use of 6% by volume brass-coated steel fibers) makes it unattractive for large volume application. For this reason, studies were conducted on CARDIFRC mix II in which the 13 mm long brass-coated fibers (volume fraction 1.5%) were replaced by the much cheaper Dramix fibers (30 mm long and aspect ratio 55 with crimped ends, 2.5% by volume) and the amount of short brass-coated fibers was reduced; it ranged from 0.5 to 2.0% against 4.5% in the original mix II (Table 1).

The influence of the amount and size of fibers on the mechanical and fracture properties of the resulting mixes can be judged from Table 2. It is important to point out that the specific energy G_F reported in Table 2 has been determined by the procedure that ensures that it is independent of the specimen size [15, 16]. Acoustic emission studies have confirmed the validity of this procedure [17]. The size-independent value is generally larger than the traditional size-dependent value, obtained by the RILEM procedure. This size-independent specific fracture energy is practically equal to the area under the tension softening diagram given by equation (4) above. It is for this reason that we have made no distinction between their designations.

Table 2. Influence of the amount and size of fibers

Mechanical Properties	f_c (MPa)	f_t (MPa)	E (GPa)	G_F (N/m)
Mix II (0.5% 6mm, 2.5% 30 mm fibers)	167.1	21.1	46.2	32531
Mix II (1.0% 6mm, 2.5% 30 mm fibers)	171.7	23.0	46.4	31514
Mix II (1.5% 6mm, 2.5% 30 mm fibers)	176.9	23.7	47.4	32059
Mix II (2.0% 6mm, 2.5% 30 mm fibers)	182.4	26.5	47.4	31950

As expected from the micro-mechanical constitutive relations (2)-(4), the short fibers enhance primarily the tensile strength, whereas the long fibers have a much greater influence on the toughness G_F. This is borne out by the entries in Table 2. However, the amount of long fibers cannot be increased indefinitely because they reduce the workability of the mix.

3 Influence of Thermal Cycling

In order to be able to use CARDIFRC in hot arid climatic conditions where the temperature on exposed concrete surfaces can reach 90°C on a typical hot summer day and drop to around 25°C during the evening and night, studies were conducted to investigate the influence of thermal cycling on the mechanical properties of CARDIFRC mix I [18, 19, 20]. The specimens were heated slowly to a maximum temperature of 90°C from the room temperature (about 25°C) in about 20 minutes to avoid any risk of thermal shock during heating, but without any mechanical loading. The maximum temperature was maintained for 8 hours before the specimens were cooled down to the room temperature in 15 hours and 40 minutes. This 24-hour heating, hold and cooling cycle reflects the temperature fluctuations during a day and night in hot arid climate. The mechanical properties of CARDIFRC mix I were measured at room temperature after subjecting it to between 30 and 180 thermal cycles. The results are summarised in Table 3. It is evident from Table 3 that thermal cycling, in the absence of mechanical loading, has little or no effect on the mechanical properties of CARDIFRC. The room temperature properties can be safely used for design purposes.

Table 3. Influence of thermal cycling

Number of Thermal Cycles	Compressive Strength (MPa)	Split Cylinder Strength (MPa)	Modulus of Rupture (MPa)	Elastic Modulus (GPa)
0	200.0	32.2	37.0	48.0
30	224.3	30.7	33.5	49.0
90	225.0	31.0	33.7	49.0
120	206.5	28.4	40.8	48.3
180	205.0	28.3	38.4	48.2

4 Commercial version of CARDIFRC

The eventual commercial version of CARDIFRC contained only 2.5% by volume of 30 mm long Dramix fibers with crimped ends. It had no short steel fibers, but contained 0.2% by volume of sacrificial polymeric fibers to give it the required 120 min fire rating. This version was self-compacting with a horizontal spread of 750 mm in the slump cone test; the spread of 500 mm was reached in the required 3 sec. The mix had not only the required workability, but also the required passing ability as measured in a J-ring test. The self-compacting version was obtained by replacing 36.4% of cement in the original CARDIFRC Mix I with ground granulated blast furnace slag (GGBS) to increase the powder volume, and by using the more effective Glenium Ace 333 superplasticiser instead of the naphthalene-based superplasticiser in the original CARDIFRC mix I. The mix proportions of this commercial version are given in Table 4.

Table 4. Mix constituents of self-compacting commercial version of CARDIFRC mix I (kg/m^3)

Constituent		Amount
Cement		543.5 (1.00)
Micro-silica		214.0 (0.39)
GGBS		311.5 (0.57)
Quartz sand:	9-300μm	470.0 (0.86)
	250-600μm	470.0 (0.86)
Water		188.0
Superplasticiser (SP)		20.68
Fibers:	30 mm Dramix	195.0
	Polymeric	15.6
Water/cement		0.22
Water/binder		0.18
SP/water		0.11
Slump flow (mm)		750
T_{500} (sec)		3

The figures in parenthesis against the various dry mix constituents in Table 4 show the proportions of these constituents relative to the mass of cement. The mechanical properties of the resulting self-compacting commercial version of CARDIFRC mix I are given in Table 5.

Table 5. Mechanical properties of SC commercial version of CARDIFRC Mix I

Mechanical Properties	f_c (MPa)	f_t (MPa)[1]	E (GPa)	G_F (N/m)
Mix I (2.5% 30 mm fibers, 0.2% polymeric)	141	19.5/23.2	44.2	33031

[1] First value is split cylinder strength and the second modulus of rupture

It should be mentioned that in common with the original CARDIFRC the self-compacting commercial version was cured for 7 days at 90°C. Since it contains a large amount of GGBS, the early age strengths reported in Table 5 are low, but they are expected to increase with age.

This industrially competitive self-compacting variant of CARDIFRC is ideal for manufacturing pre-cast pre-stressed concrete elements without shear reinforcement for use in structural applications where steel has often been used. In fact, it will be used for all main and secondary beams of a 14-storey steel frame building in the City of London which is due for total refurbishment. In the absence of any guidelines for design with this material, we have conducted extensive testing on full size pre-cast pre-stressed girders made from this material. Their load bearing capacities at both the serviceability and the ultimate limit states far exceeded the expected values. Details of these tests will be given during the presentation.

References

[1] Bache, H.H.: Densified cement ultra-fine particle-based materials. CBL Report No. 40, Aalborg Portland, Denmark (1981)

[2] Bonneau, O., Poulin, C., Dugat, J., Richard, P., Aitcin, P.: Reactive powder concretes: from theory to practice. Concr. Int. 18, 47–49 (1996)

[3] Karihaloo, B.L., Benson, S.D.P., Alaee, F.J.: Cementitious mixtures and a method of production thereof. UK Patent GB2391010 (2005)

[4] Benson, S.D.P., Karihaloo, B.L.: CARDIFRC – Development and mechanical properties. Part I: Development and workability. Mag. Concr. Res. 57, 347–352 (2005a)

[5] Benson, S.D.P., Karihaloo, B.L.: CARDIFRC — Development and mechanical properties. Part III: Uniaxial tensile response and other mechanical properties. Mag. Concr. Res. 57, 433–443 (2005b)

[6] Benson, S.D.P., Nicolaides, D., Karihaloo, B.L.: CARDIFRC — Development and mechanical properties. Part II: Fibre distribution. Mag. Concr. Res. 57, 421–432 (2005)

[7] Farhat, F.A., Nicolaides, D., Kanellopoulos, A.D., Karihaloo, B.L.: CARDIFRC – Performance and application to retrofitting. Eng. Fract. Mech. 74, 151–167 (2007)

[8] Nicolaides, D., Kanellopoulos, A.D., Karihaloo, B.L.: Fatigue life and self-induced volumetric changes of CARDIFRC. Mag. Concr. Res. 62, 679–683 (2010)
[9] Alaee, F.H., Karihaloo, B.L.: Fracture model for flexural failure of RC beams retrofitted with CARDIFRC. J. Eng. Mech., Am. Soc. Civil Eng. 129, 1028–1038 (2003a)
[10] Alaee, F.H., Karihaloo, B.L.: Retrofitting RC beams with CARDIFRC. J. Compos. Const., Am. Soc. Civil Eng. 7, 174–186 (2003b)
[11] Karihaloo, B.L.: Fracture Mechanics and Structural Concrete. Addison Wesley Longman, Harlow (1995)
[12] Lange-Kornbak, D., Karihaloo, B.L.: Design of fibre-reinforced cementitious composites for minimum brittleness. J. Adv. Cement-Based Mater. 7, 89–101 (1998)
[13] Wang, J., Karihaloo, B.L.: Constitutive modelling of high-performance fibre-reinforced cementitious composites. J. Adv. Eng. Mater. 2, 726–732 (2000)
[14] Mobasher, B., Stang, H., Shah, S.P.: Microcracking in fiber reinforced concrete. Cement Concr. Res. 20, 665–676 (1990)
[15] Abdalla, H.M., Karihaloo, B.L.: Determination of size-independent specific fracture energy from three point bend and wedge splitting tests. Mag. Concr. Res. 55, 133–141 (2003)
[16] Karihaloo, B.L., Abdalla, H.M., Imjai, T.: A simple method for determining the true specific fracture energy of concrete. Mag. Concr. Res. 55, 471–481 (2003)
[17] Muralidhara, S., Raghu Prasad, B.K., Eskandari, H., Karihaloo, B.L.: Fracture process zone size and true fracture energy of plain concrete from acoustic emission catalogue. J. Const. Build. Mater. 24, 479–486 (2010)
[18] Farhat, F.A.: Performance of concrete structures retrofitted with CARDIFRC after thermal cycling, PhD Thesis, Cardiff University, UK (2004)
[19] Farhat, F.A., Nicolaides, D., Kanellopoulos, A.D., Karihaloo, B.L.: Behavior of RC beams retrofitted with CARDIFRC after thermal cycling. J. Mater. Civil Eng. Am. Soc. Civil Eng. 22, 21–28 (2010)
[20] Ewaze, M.A.: Retrofitting of RC beams with CARDIFRC in hot climate. PhD Thesis, Cardiff University, UK (2010)

Static and Dynamic Behavior of Hybrid Precast Bridge Parapet Made of Ultra-High Performance Fiber Reinforced Concrete

J.-P. Charron, F. Duchesneau, and B. Massicotte

École Polytechnique de Montréal, Civil Engineering Department, Montréal, Canada

Abstract. An industrial research project was launched in 2007 at École Polytechnique de Montréal to develop a new generation of precast bridge parapets using nonlinear finite element calculations. The paper presents the design of a hybrid precast parapet made with ultra-high performance fiber reinforced concretes (UHPFRC) and its experimental validation. The hybrid parapet consisted of a thin shell of UHPFRC paired with a normal strength concrete core. The outstanding properties of the UHPFRC permitted the removal of all conventional reinforcement while keeping the cast-on-site parapet thickness. A new anchorage system was optimized for parapet installation on a new bridge deck. Quasi-static tests, and dynamic tests replicating 4 levels of vehicle impacts, were conducted on full-scale parapets. Test results showed that the load-carry capacity of the hybrid UHPFRC parapet exceed the PL-2 or TL-4 parapet specifications of CSA and AASHTO. Moreover it offers a strength level equivalent to standard cast-on-site constructions.

1 Introduction

The Canadian Highway Bridge Design Code [1] presently limits its design recommendations to continuous cast-on-site parapets anchored to bridge decks with conventional reinforcement (Fig. 1a). This method of parapet construction offers good performance in term of impact resistance, but it also presents many drawbacks. The construction sequence is highly time-consuming with the installation of the formwork, the concrete casting, and a minimum curing time of the concrete. This becomes especially problematic in Canada due to its winter climate, which limits the period of casting, and importance of social costs associated with traffic mitigation. Secondly, the shrinkage of built on site parapets is restrained by the bridge deck and leads frequently to early-age cracking [2]. The combination of cracks and severe environmental expositions of the parapet can provoke early deterioration that requires periodical repairs.

G.J. Parra-Montesinos, H.W. Reinhardt, and A.E. Naaman (Eds.): HPFRCC 6, pp. 405–412.

An effective solution to the previous problems is the utilization of precast bridge parapet. A precast parapet is quickly installed on the bridge deck with mechanical anchors and does not present shrinkage induced cracking as its deformations are free to occur before its installation. In this context, an extensive research project led by École Polytechnique of Montreal and involving industrial partners was launched in 2007. The project aimed at developing a new generation of precast bridge parapets that minimizes the construction time while enhancing parapet durability. This paper presents the results of the experimental program carried out on a new design of precast parapet made in ultra-high performance fiber reinforced concrete (UHPFRC). A complete overview of the industrial research program conducted on 6 different parapet designs can be found in thesis of Niamba [3] and Duchesneau [4].

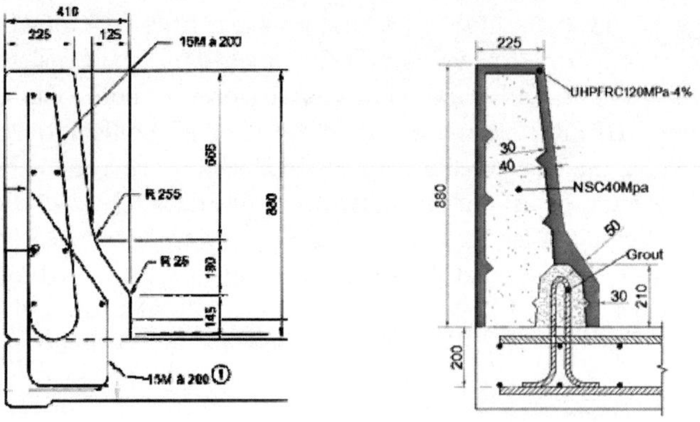

a) b)

Fig. 1. Parapet design, a) Cast-in-place parapet, b) Hybrid UHPFRC precast parapet

2 Conception of Precast Parapet

The precast parapets were designed with the New Jersey profile according to the AASHTO Bridge Design code [5] and the Canadian Highway Bridge Design Code [1]. The intermediate performance level PL-2, which is equivalent to the test level TL-4, was selected for the parapets, since it is the most widely used for Canadian urban bridges.

The CSA-S6 [1] and AASHTO [5] specifications stipulate that PL-2/TL-4 parapets must be designed to resist a factored static transverse force of 170 kN. The minimum ultimate strength for static loading is obtained by dividing the design force by the performance factor, taken equal to 0.75 for concrete material. Consequently, the target ultimate design strength value is equal to 227 kN. Moreover, practical considerations suggested the limitation of the precast element length between 2 and 4 m. The standard PL-2 parapet design that was adopted by the Quebec Ministry of Transportation [6] provides an ultimate bending resistance equal to 350 kN for a 2 m long parapet. In this project, the precast parapet design was made with the most severe criterion corresponding to a maximum transverse force of 350 kN applied at the mid-length of a 2 m long parapet such that the new designs offer a strength level equivalent to cast-on-site constructions.

The precast parapet designs were optimized by conducting nonlinear finite element analyses with the software Atena 3D [7]. The numerical model took into account the parapet and the overhang slab from the external bridge girder, the mechanical connection parapet/slab, the interface parapet/slab, the interface between materials in the parapet, and the non-linear material laws of concrete and steel. Details on the numerical model and the parametric study that was carried out are presented elsewhere [8].

One of the designs retained in this project was a hybrid parapet (Fig. 1b). It comprised a core of unreinforced normal strength concrete (NSC) and a thin shell of UHPFRC. The later material allowed the complete removal of all the conventional reinforcement while keeping the thickness of the reference cast-in-place parapet. UHPFRC thickness was varied in the parapet according to the required strength. Where the load requirements are at a minimum, a thickness of 30 mm was chosen to facilitate the casting of the material into the mold, although a thinner thickness would be sufficient for durability consideration. At the locations where the load effects were at a maximum, above and next to the recess, the numerical optimization indicated that a thickness of 50 mm was required to resist the design load. The hybrid parapet is considered as an optimum solution due to the high energy absorption capability and the exceptional durability of the UHPFRC [3, 9].

The mechanical connection between the slab and the parapet was assured by U-shape N°15 rebar (Fig. 1b) spaced at 200 mm, as in the reference cast-in-place parapet (Fig. 1a). The grout volume required to fill the recess of the parapet on-site was minimized by installing U-shape rebar at 45 degrees with the longitudinal parapet axis. Two shear keys were added to the recess to improve the force transfer from the parapet to the connection grout.

Three materials were used in the project; a UHPFRC and a NSC for the parapet itself, and a grout for the recess filling. The UHPFRC developed at Polytechnique [10] contained 4% of straight steel fiber (l_f=10 mm, ϕ_f=0.2 mm). It was self-levelling and presented a strain hardening and softening behavior in direct tension. The NSC and the grout were concrete mixes commonly produced in the precast industry. The measured mechanical properties of the materials are summarized in Table 1.

Table 1. Mechanical properties of concretes

Property	Grout	NSC	UHPFRC
f'_c (MPa)	65.2	41.5	148.0
f'_t (MPa)	4.1	-	9.0
E_c (MPa)	28 700	27 500	41 400
ν (-)	0.207	0.207	0.273

3 Experimental Program

Two full-scale structural tests were carried out on 2-m length hybrid parapets described in Fig. 1b with the test setup depicted in Fig. 2. The retaining slab was installed to impede the lateral displacement of the anchorage stab during the test. The anchorage and retaining slabs were post-tensioned to the laboratory strong floor by 4 and 2 post-tensioning bars, respectively. A 10 mm layer of grout was spread on the anchorage slab before the installation of the parapet to obtain a good adherence. The same grout with a fiber addition was pumped into the parapet recess afterwards. Many sensors were installed on the specimen and test setup in order to measure displacements of the parapet and the slabs as well as strains in the connection rebar.

The transverse load was applied with a 1000 kN dynamic actuator, equipped with two 1500 l/min 3-stage servo-valves, reacting on a rigid steel frame (Fig. 2). The system was supplied with a 1350 l/min hydraulic unit. The actuator force was applied on a 700 x 350 mm surface by means of a thick steel plate and a mortar spacer attached to the actuator head.

The first hybrid parapet was subjected to a quasi-static loading up to ultimate, whereas the second hybrid parapet was submitted to a dynamic loading equivalent to a vehicle impact. The quasi-static loading was controlled by the LVDT on the actuator and the loading rate was set at 0.01 mm/s to avoid dynamic effect. For the dynamic loading, a high capacity spring system was added between the actuator and the steel plate (Fig. 2). This system intended to eliminate significant inertia effects that impair control signals in dynamic tests. The spring system was calibrated to control the tests with the average displacement of 4 LVDT measuring springs flattening during loading.

The dynamic loading applied to the parapet had the objective to reproduce a realistic crash test. It was decided to apply a normalized impact of a Toyota Echo, hitting the parapet at 110 km/h with an angle of 20° from the road axis [11]. Considering the experimental test setup, only the normal component of the parapet acceleration could be reproduced in the laboratory. The maximal normal force reached 215 kN and the cash test lasts 0.1 s. To study the evolution of the damage in the parapets, 4 levels of impact were applied on the parapet. The maximal force applied was successively 20, 50, 100 and 215 kN in 0.1 s period, like the original crash test signal. After the application of the series of impacts, a quasi-static test was performed to evaluate the residual strength of the hybrid parapet.

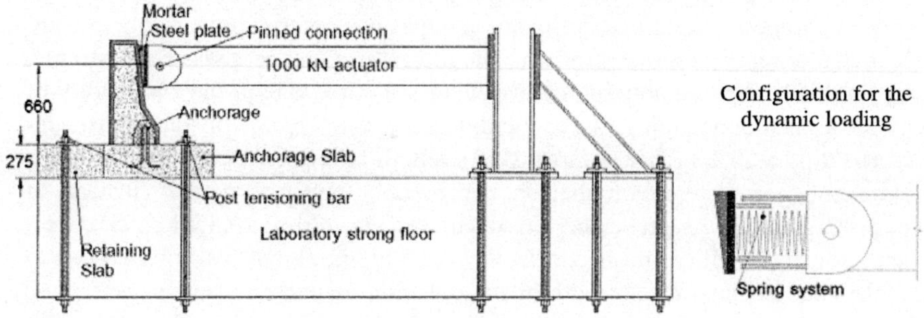

Fig. 2. Test setup

4 Results and Discussion

The quasi-static behavior of the hybrid parapet is shown in Fig. 3, the applied force variation is presented with respect to the parapet horizontal displacement (measured in the actuator axis). The tests were interrupted periodically to draw the crack development on the specimens. These pauses caused small drops in the applied force at each 50 kN intervals and at the peak load. First, the parapet not submitted to dynamic loads is described. The parapet presented an elastic behavior up to 100 kN after which the stiffness decreased with the progression of cracking. Once the maximal strength was reached (344 kN), 3 macrocracks formed and progressed from the recess grout into the NSC core and caused a significant and sudden strength reduction attributed to the absence of reinforcement in the core. The crack pattern at test ending is depicted at Fig. 4a. No cracks were noted on the UHPFRC shell and no sign of core-shell interface failure was observed. A slight debonding occurred along the parapet-recess interface, but only at the very end of the test. The quasi-static behavior obtained for the parapet submitted previously to dynamic loads will be discussed later on.

Fig. 5a presents the series of dynamic loadings applied to the second hybrid parapet. Four successive loadings were applied in 0.1 s period with increasing the peak load at 20, 50, 100 and 215 kN. The last loading corresponded to the real-time normal crash load of a Toyota Echo on a parapet. After each dynamic loading, parapet was examined to drawn cracks. In fact, no crack was observed in all cases, the presence of damage was not visible. Fig. 5b shows the evolution of the force against the horizontal displacement of the parapet for the 3 higher dynamic loadings. One can note that the parapet stiffness remained identical, except for the unloading curve of the 215 kN loading. This observation means that damage was induced in the parapet only by the last loading; this is confirmed by the residual displacement at the end of that test.

A quasi-static test was also conducted on the hybrid parapet submitted previously to the series of dynamic loads to measure its residual load-carry capacity. Fig. 3 illustrates clearly that the mechanical behavior was different between the two tested parapets. The dynamic loaded parapet presented a higher stiffness under load and a lower maximal strength. The increase of stiffness was not expected, it may be due to the combination of two factors. Firstly, due to the availability of equipments, the connection grout of the dynamic loaded parapet had a higher maturity and thus a 20% higher compressive strength at testing. Secondly, the crack development was different in that specimen; only one macrocrack (instead of three) propagated in the grout and led to failure (Fig. 4b). A limited crack development reduces ductility and is consistent with the higher stiffness observed. Besides, the smaller maximal strength of the dynamic loaded parapet can be related to the internal damage produced by the 215 kN impact shown on Fig. 5b.

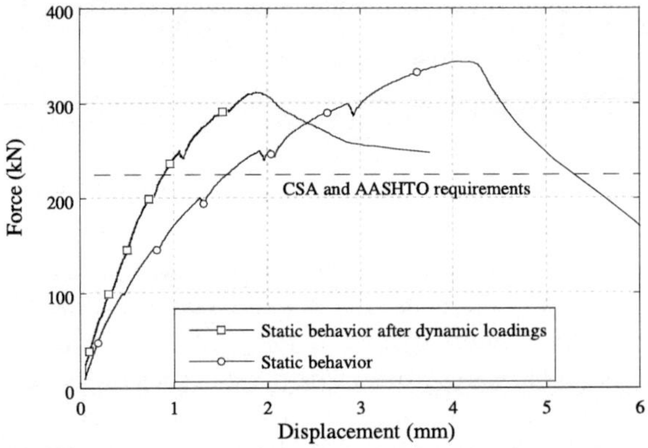

Fig. 3. Force versus displacement behavior of hybrid parapets under quasi-static loading

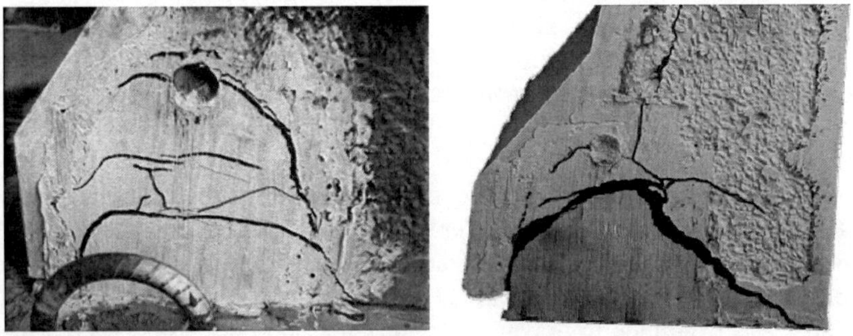

a) b)

Fig. 4. Crack pattern of hybrid parapet at the end of the quasi-static test, a) No dynamic load applied before the quasi-static test, b) With dynamic loads applied before the quasi-static test

Experimental results demonstrated that the hybrid UHPFRC parapet showed an ultimate strength of 344 kN, and a residual ultimate strength of 310 kN after a series of dynamic loads equivalent to a Toyota Echo crash impact. Those results exceed the ultimate design strength of 227 kN specified in CSA and ASSHTO bridge design codes by 52% and 37%, respectively. Moreover, the hybrid parapet offers a strength level equivalent to Quebec Ministry of Transportation cast-on-site constructions (350 kN for 2-m length parapet).

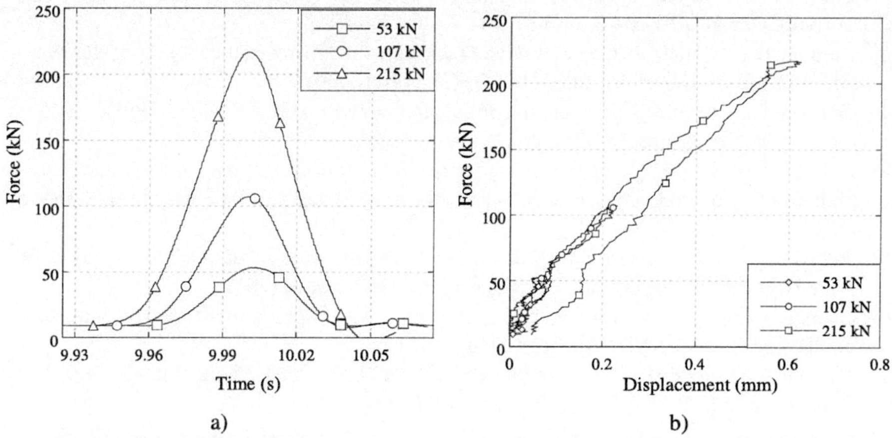

Fig. 5. Dynamic loadings applied to a hybrid parapet, a) Force versus time, b) Force versus parapet displacement

5 Conclusion

The research project aimed to design new generations of precast bridge parapets. This paper described the hybrid UHPFRC parapet and an anchorage system to accelerate bridge construction. The analysis of the experimental results led to the following conclusions:

- The precast hybrid UHPFRC parapet showed a reduction of 100% of conventional reinforcement while keeping the thickness of the reference cast-in-place parapet.
- The load-carry capacity of the hybrid parapet exceeds by 52% the CSA and AASHTO specifications, its ultimate residual strength after a series of realistic dynamic loadings still exceeds the specifications by 37%.
- The proposed anchorage system (U-shape rebar connection) allows parapet installation on a new bridge and performs efficiently under quasi-static and dynamic loadings.

Acknowledgement. The research project was financially supported by the Natural Science and Engineering Research Council of Canada, Béton Brunet and the City of Montréal. Moreover, some materials were graciously provided by Bekaert, Euclid, and Ciment St-Laurent. The authors would like to acknowledge the participation of the industrial partners and the technical staff of École Polytechnique de Montréal.

References

[1] CSA-S6-06 Canadian Highway Bridge Design Code (CHBDC). CSA International, Toronto, Canada (2006)
[2] Cusson, D., Repette, W.L.: Early-Age cracking in reconstructed concrete bridge Barrier Walls. ACI Materials Journal 97(4), 438–446 (2000)
[3] Niamba, E.: Conception of precast bridge parapets with HPFRC and UHPFRC. M.Sc. thesis of Polytechnique of Montréal, Canada (2009)
[4] Duchesneau, F.: Design of hybrid and monolithic precast parapet in high and ultra-high fibre reinforced concrete. M.Sc. thesis of Polytechnique of Montréal, Canada (2010)
[5] AASHTO Bridge Design Specification SI Units 4th edn. American Association of State Highway Transportation Officials, Washington, USA (2007)
[6] Quebec Ministry of Transport (QMT) Normes - Ouvrages d'art tome III. Ministère des Transport du Québec, Québec (2010)
[7] Červenka, V., Jendele, L., Červenka, J.: ATENA Program Documentation. Cervenka Consulting Ltd., Prague, Czech Republic (2009)
[8] Duchesneau, F., Charron, J.-P., Massicotte, B.: Hybrid and monolithic precast bridge parapets in high and ultra high performance fibre reinforced concretes. Accepted for publication in Canadian Journal of Civil Engineering (2010)
[9] Habel, K., Charron, J.-P., Braike, S., Hooton, R.D., Gauvreau, P., Massicotte, B.: Ultra-high performance fibre reinforced concrete mix design in central Canada. Canadian Journal of Civil Engineering 35(2), 217–224 (2008)
[10] Braike, S.: Conception of precast structural elements with HPFRC and UHPFRC. M.Sc. thesis of Polytechnique of Montréal, Canada (2007)
[11] Jiang, T., Grzebieta, R.H., Zhao, L.: Predicting impact loads of a car crashing into a concrete roadside safety barrier. International Journal of Crashworthiness 9, 45–63 (2004)

Shear Strength of Ultra High Performance Fiber Reinforced Concrete (UHPFRC) Precast Bridge Joint

C.H. Lee, Y.J. Kim, W.J. Chin, and E.S. Choi

Structural Engineering and Bridge Research Division,
Infrastructure Research Department, Korea Institute of Construction Technology (KICT),
Goyang, Republic of Korea

Abstract. This paper presents a study of the behavior of ultra high performance fiber reinforced concrete (UHPFRC) precast segmental bridges, focusing on the response of joint types under shear, in service and ultimate limit states. Tests have been performed on UHPFRC segment with different levels of lateral average stress, evaluating the behavior of epoxy joints under direct shear. The possible benefit of using joint method is also evaluated in view of joint efficiency. The results obtained in these tests, as well as those found in the literature, have been compared with several design formulae for assessing the joint shear strength of precast segment, identifying the formula that gives the best predictions.

1 Introduction

The development of ultra high performance fiber reinforced concrete (UHPFRC) has been accompanied by a tremendous growth in the criteria evaluating the performances of structures like high-strength, high-constructability and high-durability [1]. This trend resulted in the development of a material satisfying fundamental mechanical requirements and has turned today toward research focusing on the structural safety and constructability of mega structures [2, 3].

UHPFRC is a material offering outstanding mechanical properties with a compressive strength of 200 MPa, tensile strength of 10 MPa and elastic modulus of 50 GPa, which makes it superior to normal strength concrete and previous high strength concrete [1]. However, since the current design codes for concrete structures in Korea and overseas specify unilaterally limiting values for high strength concrete with compressive strength larger than 70 MPa, excessive safety factors are applied during the design of actual bridge structures. Accordingly, the current formulae appear difficult to apply as they are and the properties of UHPFRC imply the necessity of numerous modifications and supplementations of the design formulae applied in conventional concrete structures [4, 5]. The fundamental studies on the mechanical characteristics of UHPFRC material as

G.J. Parra-Montesinos, H.W. Reinhardt, and A.E. Naaman (Eds.): HPFRCC 6, pp. 413–420.

well as the analysis of the behavioral properties of UHPFRC members have been performed progressively during the last decade by a few universities and research institutes [6, 7]. Recently, focus is done on the applicability and constructability of UHPFRC to actual large scale structures for the purpose of its application to ultra long-span bridges. Especially, research was also extended to the field of the behavior of bridge joints using UHPFRC precast segment. As mentioned above, diversified researches were dedicated to provide theories and empirical formulae related to the shear behavior of joints using previous normal strength concrete and high strength concretes. However, the application of these formulae as they are in the case of UHPFRC bears definitely limitations [8, 9, 10]. Accordingly, this study intends to investigate the relationship between the lateral average stress and the shear strength of UHPFRC joints and to analyze their performances with respect to the joint type by performing tests focusing on the behavioral analysis according to the joint type and lateral average stress of UHPFRC precast bridge joint.

2 Test Set-Up and Method

2.1 Test Planning

In order to simulate the shear behavior of precast joint using UHPFRC, this study implemented tests by distinguishing the shear key joint method during a cast-in-place joint shown in Fig. 1 and the shear key joint method accompanying the epoxy joint of Fig. 2. Where, both shear keys are made of UHPFRC.

The lateral average stress acting on the specimen was set as test variable with values of 2 MPa, 4 MPa and 8 MPa for both cast-in place joint type and epoxy joint type. Moreover, additional values of 0 MPa, 0.1 MPa and 12 MPa were set for the cast-in-place joint considering field constructability.

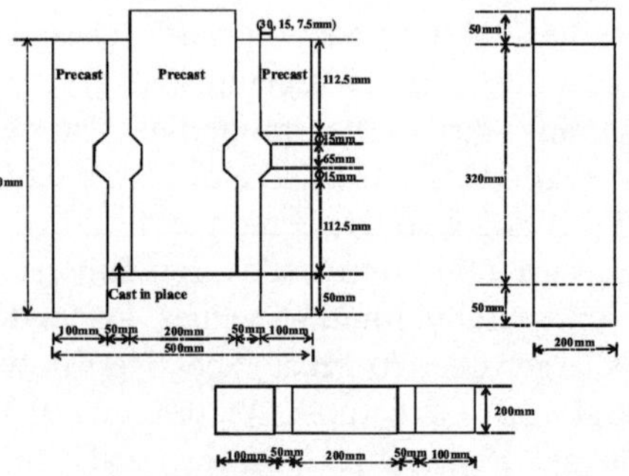

Fig. 1. Details of cast-in-place joint type

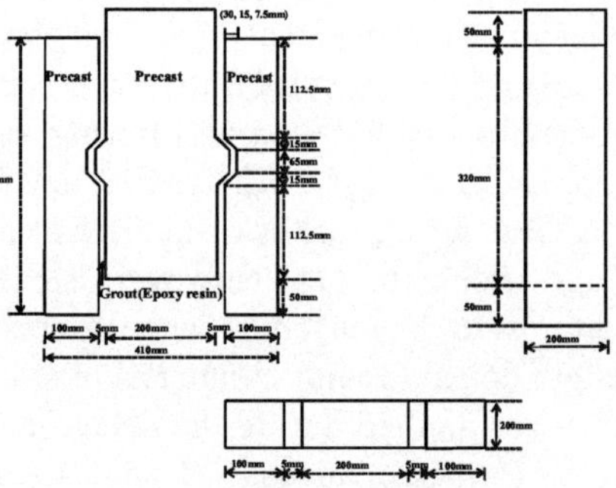

Fig. 2. Details of epoxy joint type

2.2 Composition of the Testing System and Measurement

Fig. 3 illustrates the system set-up of the UHPFRC precast segment joint specimen. As shown in the figure, the lateral average stress was applied using a 4500-kN actuator with the values of 2, 4 and 8 MPa as set initially for the test variable in order to analyze the shear behavior of the precast joint. The variation of the lateral force during the test reached merely ± 0.5 kN showing that the variation of the permanent load itself was insignificant. On the other hand, the vertical compressive load was applied by means of a 3500-kN actuator at speed of 0.4 mm/min. The displacement and strain were measured using concrete gauges and LVDTs installed at the top, center and bottom around the shear key as shown in Fig. 3.

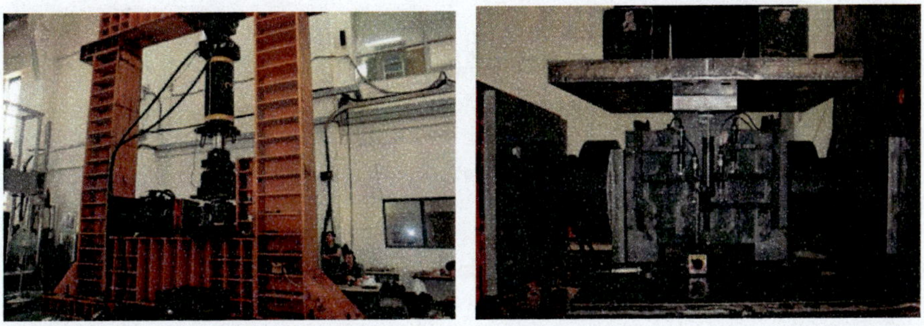

Fig. 3. System set-up for joint shear key behavior

3 Test Results and Analysis

3.1 Joint Type

Fig. 4 plots the applied load with respect to the vertical displacement for the cast-in-place joints and epoxy joints. The failure load reached 2,168 kN for the cast-in place joint while a value of 2,378 kN was measured for the epoxy joint. The epoxy joint exhibits thus larger failure load than the cast-in-place joint. Furthermore, the macro behavior produced by the application of the vertical compressive force shows similar behavior for both joint types. However, initial slop of cast-in-place joint is higher than epoxy joint. Thus, this means that epoxy joint is more resistible than cast-in-place joint in view of ductility.

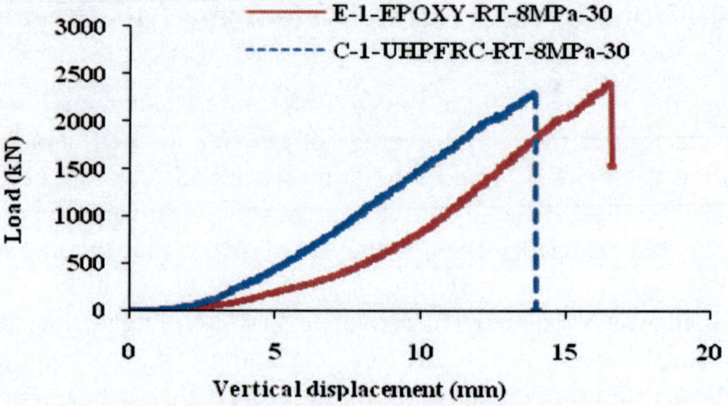

Fig. 4. Load vs. vertical displacement results of UHPFRC precast joint

Fig. 5 plots the shear stress with respect to the vertical displacement for the cast-in place joint and epoxy joints. It should be pointed out that the value of the absolute vertical strain for the epoxy joint is bounded to ± 60ε corresponding to a stress of 16 MPa assumed to be the point at which crack initiated. On the other hand, the strain of the cast-in-place joint increases linearly until 9.1 MPa to reach an absolute value of 370ε after which the strain experiences fluctuations offering resistance until the maximum shear stress of 17.9 MPa. Namely, even if the cast-in-place joint is made of a unique material, the bond stress at the joint interface would on the contrary be smaller than that of the epoxy joint. Accordingly, whether the improvement of the materials used in UHPFRC for the precast joint is achieved, the cast-in place joint will remain disadvantageous compared to the epoxy joint until the development of a material satisfying the mechanics related to compression, tension, shear, bending and warping.

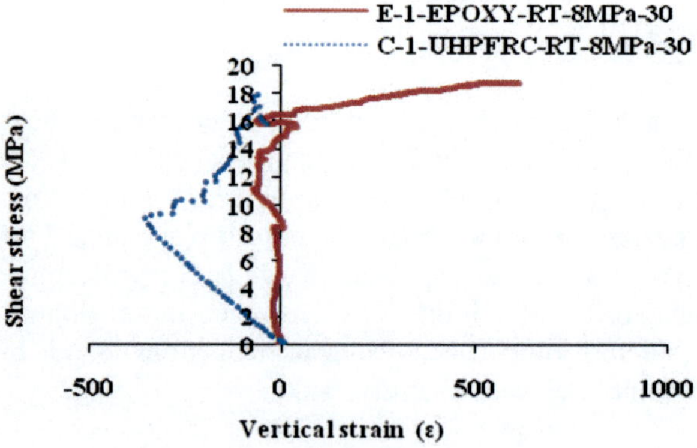

Fig. 5. Shear stress vs. vertical strain results of UHPFRC precast joint

3.2 *Lateral Average Stress*

In general, the increase of the lateral average stress in precast PC bridge joint is known to increase the shear stress of the joint. Fig. 6(a) plots the maximum failure load according to the lateral average stress of the cast-in-place joint increasing as 0, 0.1, 2, 4, 8 and 12 MPa. In the case of UHPFRC with a mean compressive strength of 200 MPa and mean tensile strength of 10 MPa, the increase of the lateral average stress reaches a maximum of 8 MPa at 2,332 kN. This implies the existence of a critical lateral force providing the optimal behavior even if the lateral average stress inside the UHPFRC member is increased. Such feature can also be observed in the shear strength curve according to the lateral average stress in Fig. 6(b). The results showed that the previously suggested equations have difference with experimental value in view of critical lateral average stress.

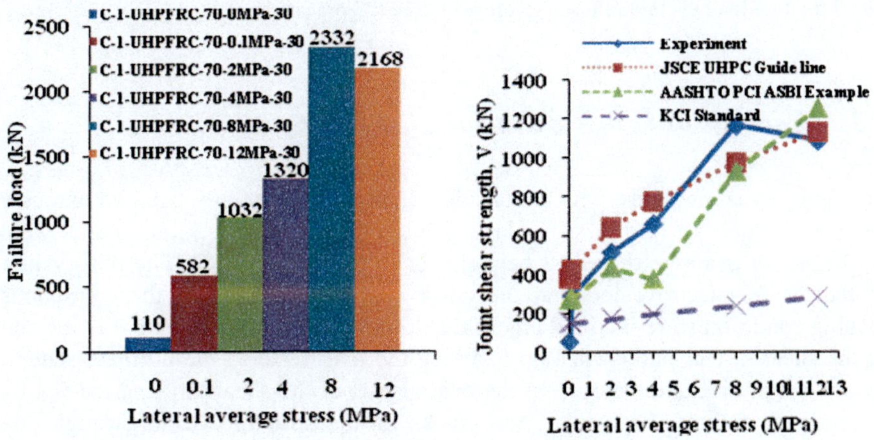

(a) Failure load vs. lateral average stress (b) Shear strength vs. lateral average stress

Fig. 6. Comparison of standard design for shear strength

Fig. 7 presents the regression analysiss of the failure load with respect to the lateral force for each of the cast-in-place joint and epoxy joints. As shown in the figure, the epoxy joint resists a relatively larger failure load than the cast-in-place joint. The failure load increases linearly with larger lateral force whereas that of the cast-in-place joint exhibits linear increase of until 8 MPa to tend to decrease after that value. In particular, in this study, both joint types presented similar trend under a lateral average stress of 8 MPa, which indicates the existence of common optimal conditions within a range of 8 MPa providing the optimal lateral average stress for the precast PC joint using UHPFRC.

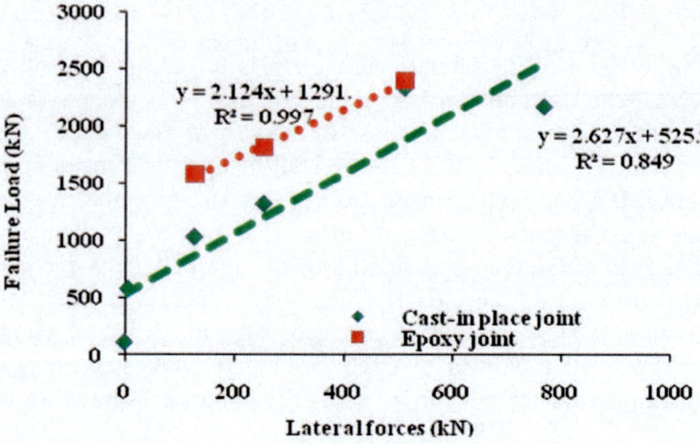

Fig. 7. Failure load vs. lateral force results of UHPFRC precast joint

3.3 Analysis of Crack Behavior

The joint type methods which considered in this study are the continuously monitored each of the cast-in-place joint and the epoxy joint. Tests were performed to examine the shear behavior of these two methods. The observation of the test results revealed that the crack characteristics under the continuous loading condition were inclined cracks due to the concentration of stress at the end of the shear key as shown in Fig. 8. When the load reached the ultimate state, a second set of cracks developed at the central part of the shear key and led finally to shear-off failure. Especially, the cracks of the structure made through one continuous placing were seen to be uniformly distributed inclined cracks due to the uniform application of the lateral force in most of both shear keys despite of the variability of the lateral force. It was also observed that, when shear-off failure occurs in the shear key, the propagation of the inclined cracks developed at first stops and that the crack width narrows. As shown in Fig. 9, the increase of the load in the epoxy joint led to the failure of the epoxy bonded surface below the shear key which, differently to the cast-in-place joint, let the cracks to propagate until the shear key. In addition, differently from the cast-in place joint specimen in which failure occurred through inclined cracks evenly distributed in both shear keys, the epoxy joint showed a pattern in which one of the shear keys became the weak portion. This weak portion propagated from the support along the cracks and led to vertical cracks leading to a behavior where the shear key at the opposite side supported the load. In view of the crack behavior, the propagation of cracks differs according to the joint type but is insensitive to the lateral average stress conditions. The different propagation of cracks in the shear keys in the epoxy joint can be firstly explained by the uneven bond of the epoxy in the joint as well as the action of the unbalanced loading provided by the vertical load actuator. Apart from these reasons, the specimen was dedicated to receive simultaneously lateral force and vertical compression, which made it a member subjected at instant compression, tension, shear, bending and warping. Especially, the presence of epoxy also involved numerous additional parameters like the local concentration of the bond stress.

Fig. 8. Failure of UHPFRC precast joint (Cast-in-place joint)

Fig. 9. Failure of UHPFRC precast joint (Epoxy joint)

4 Conculsions

This study compared experimentally the shear behavior of precast joint using UHPFRC with respect to the joint type and lateral average stress. The following conclusions could be drawn from the test results.

1) The maximum shear stress for the epoxy joint reached 18.7 MPa. Compared to the cast-in-place joint, the initial cracks occurred at about 16 MPa, and the epoxy joint behaved with a sharp increase of the displacement until failure at the limit state. This shows that the epoxy joint is superior to the cast-in-place joint in terms of shear strength and ductility.
2) The experimental results showed that the possibility to suggest a critical lateral average stress providing optimal behavior even if the lateral average stress inside the UHPFRC member is increased.
3) The epoxy joint exhibits thus larger failure load than the cast-in-place joint. Furthermore, the macro behavior produced by the application of the vertical compressive force shows similar behavior for both joint types. However, initial slop of cast-in-place joint is higher than epoxy joint.
4) The joint made through a cast-in-place joint and the epoxy joint exhibited similar trend under lateral force of 8 MPa. This implies the existence of a critical lateral force providing the optimal behavior even if the lateral average stress inside the UHPFRC member is increased.
5) For the crack behavior, the propagation of cracks showed difference due only to the type of joint.

Acknowledgements. The authors would like to acknowledge the research project "development of design and construction system for hybrid cable stayed bridge", financed by the korea institute of construction technology. The institute's support for this research is gratefully acknowledged.

References

[1] FHWA, Material Property Characterization of Ultra High Performance Concrete. Federation Highway Administration Report, VA. 22101-2296. 1(1):1-231 (2006)

[2] Hmosen, A.I., Hiba, A.A.: Structural Behavior of Single Key Joints in Precast Concrete Segmental Bridge. Bridge Engineering 12(3), 315–324 (2007)

[3] Stuart, A., Paul, F., Simon, A., Peter, R.: A Test Method and Deterioration Model for Joints and Cracks in Concrete Slabs. Cement and Concrete Research 35(1), 2371–2383 (2005)

[4] KCI Standard, Commentary of Concrete Structure Design Specifications. Kimoondang Publishing Cooperation, 1(1), 67–68 (2007)

[5] AASHTO, Guide Specifications for The Design and Construction of Segmental Concrete Bridges. AASHTO 2nd Edn., With Interim Revisions 19(1),1–3 (2003)

[6] Kim, T.H., Lee, H.M., Kim, Y.J., Shin, H.M.: Performance Assessment of Precast Concrete Segmental Bridge Columns with a Shear Resistant Connecting Structure. Engineering Structures 32(1), 1292–1303 (2010)

[7] Kim, J.H., LaFave, J.M.: Key Influence Parameters for the Joint Shear Behaviour of Reinforced Concrete (RC) Beam-column Connections. Engineering Structures 29(10), 2523–2539 (2007)

[8] Collepardi, M., Troli, R., Bressan, M., Liberatore, F., Sforza, G.: Crack-Free Concrete for Outside Industrial Floors in The Absence of Wet Curing and Contraction Joints. Cement and Concrete Composites 30(2), 887–891 (2008)

[9] Dias, J.L.M.: Cracking Due to Shear in Masonry Mortar Joints and Around The Interface Between Masonry Walls and Reinforced Concrete Beams. Constructin and Building Materials 21(1), 446–457 (2007)

[10] JSCE UHPC Guide line, Recommendations for Design and Construction of Ultra High Strength Fiber Reinforced Concrete Structures. JSCE Draft Version - Appendix 5, 1(1), 1–5 (2010)

UHPFRC Bolted Joints: Failure Modes of a New Simple Connection System

E. Camacho, P. Serna, and J.A. López

Civil Engineering, Universitat Politècnica de València, Spain

Abstract. To be used for the joint between precast elements made of UHPFRC, a connection system inspired on those for steel structures, based on either high strength screws or bolts, is proposed and analyzed. This process may speed the work units and increase the control. As different failure modes may be expected depending on the connection parameters (screw diameter, cover...), those failure modes are characterized and specific tests set ups are designed and carried out to evaluate the variables effect on the UHPFRC joints behavior. The study variables were the elements dimensions, the diameter of the screws, the thickness of the elements to be connected and the process to prepare the hole. The connection behavior has been analyzed by means of the shear-out, cleavage, net-tension or bearing failure, studying the parameters that affect the load-displacement diagram.

1 Introduction

UHPFRC precast solutions are increasing their market share and have still a big potential for growing [1]. Unidirectional flat roofs, Double T girders, shells, retaining walls, truss structures are some of the applications [2, 3, 4, 5]. One feature for project with this material is the possibility of use a simpler connection method between UHPFRC elements. This option will be analyzed in this paper.

Using the UHPFRC high tensile and compressive strength, ductility, durability and multi-micro cracking, the design of direct bolted or screwed connections, similar to the common practice in steel structures, is allowed. This fast connection could simplify the use of UHPFRC in industrial structures and trusses, reducing the assembly works and reaching a higher industrialization.

Bolted joints have been proposed and tested with materials like composites [8], wood or steel. The failure observed modes for similar sizes depended on the ratio between compressive, tensile and shear strength of every material. The experience with UHPFRC is reduced. Applications with pieces used to guarantee the joint [6] and for shear connectors in composite structures [7] are satisfactorily reported.

G.J. Parra-Montesinos, H.W. Reinhardt, and A.E. Naaman (Eds.): HPFRCC 6, pp. 421–428.

2 Proposed Connection System

The connection proposed and analyzed is done with a High Strength Screw connecting UHPFRC element through holes, which are directly made in the element when casting. The screw prestressing option is opened if required.

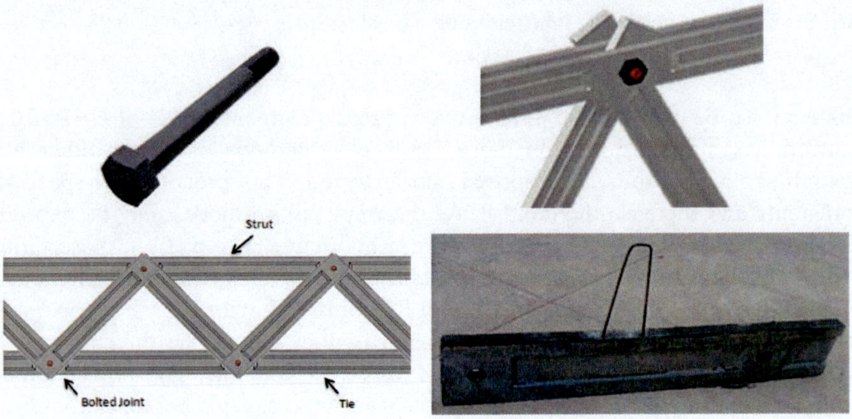

Fig. 1. Example of a screwed joint in UHPFRC truss structure

The cconnection system must be stronger than the weakest element joined. Its possible failure modes are (Fig 2): a) Bearing, b) Net tension, c) Shear out, d) Cleavage. It is also possible the screw failure by shear or flexion.

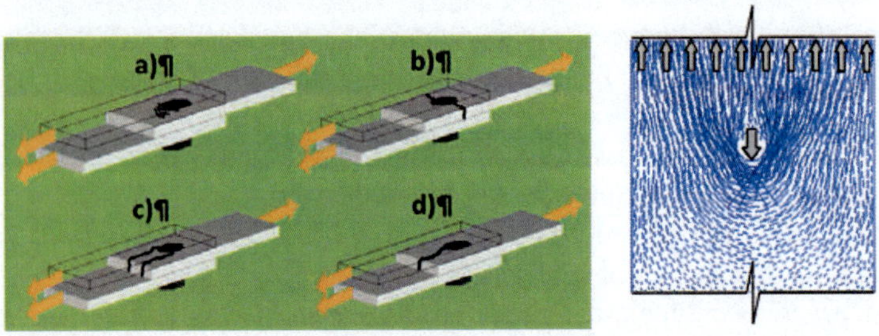

Fig. 2. Possible failures of the screwed joint (left) and generated tensile stress distribution (right)

Initially the desirable failure is the local compressive bearing, being the most ductile [8, 9]. Despite of that shear test showed high ductility in previous research [9]. The weakest failures expected are b) and d), which depends of the UHPFRC tensile strength. It is interesting to point out that the screw induces biaxial compressive stresses condition, with an increase of the compressive strength.

A forced fiber orientation parallel to the hole perimeter results adequate to sew correctly the tensile and shear efforts generated (Fig. 2, right).

3 Test Set Up and Experimental Program

To know the appropriate geometry and material characteristics to guarantee a safety connection is the aim of the developed experimental program. Simulating a tie, an UHPFRC specimen (Fig. 3 left) with two holes is submitted to a direct tension by means of an anchoring system as shown in Fig. 3 right. The test set up is designed to localize a failure area in the top connection (1) of the specimen (2). High strength steel plates connect the specimen to the loading system. To avoid the failure of the other hole a reinforcement when casting the specimen and the 25 mm diameter screw was prestressed. No preload moment or prestressing to the studied screw (3) was introduced. To control the specimen to support system relative displacement four transducers were placed (4). Fiber orientation around the hole was checked my means of X-Ray images obtained of some of the specimens. The pouring system, wall effect and the concrete fluidity lead to an as desired fiber orientation (Fig. 4).

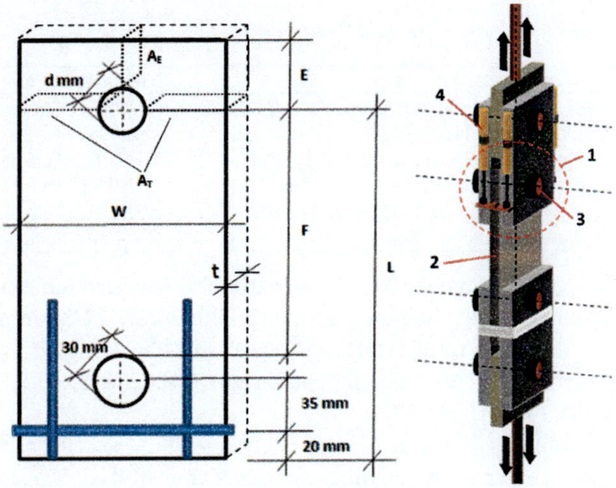

Fig. 3. Test set up scheme

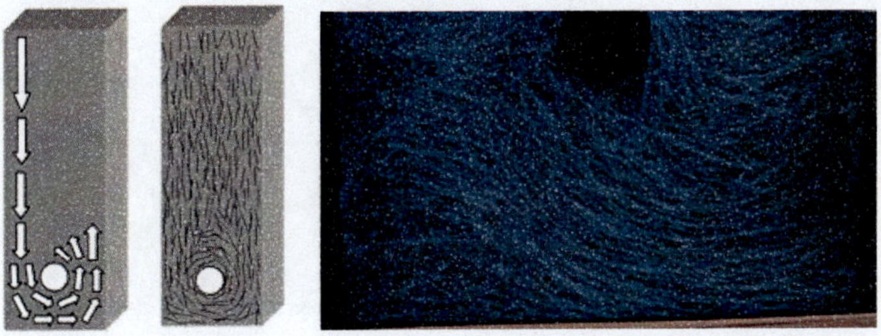

Fig. 4. Casting process, desired/expected orientation and real fiber orientation

The UHPFRC mix design used to cast the specimens and characterize the mechanical/rheological properties was developed in UPV [10].

Four 30 litres batches were casted in a EIRICH R08W mixer and a 28 days 20º air curing with 100% of relative humidity was applied. To characterize the UHPFRC six 100 x 100 x 100 mm cubes and two 100 x 100 x 500 mm prism were cast from each batch. The compressive strength evolution was obtained from the cubes and the flexural strength behavior from three point flexural tests on unnotched prismatic specimens. The tensile stress-strain diagram was deduced by inverse analysis applied to the flexural data. The concrete rheology was measured with the Slump flow test (EN 12350-8), and aggregate or fiber segregation was discarded with visual checking of fresh material and cross sections after flexural test. The Slump flow provided an average value of 820 mm and a T500 of 10 s.

For the connection tests 27 specimens (9 by batch) were cast, with the two holes prepared with the framework.

In Table 1 the considered influencing variables for each failure mode are shown.

Table 1. Failure modes and affecting variables

Type of Failure	Geometrical Parameters	Mechanical Parameters
a, bearing	Screw and hole diameters "D_{screw}; D_{hole}" ; Specimen Thickess "t"	UHPFRC Compressive stress-strain law
b, net tension	Specimen Width and Thickness "W"; "t"	UHPFRC Tensile strength
c, shear out	Cover behind bolt "E", thickness "t", "D_{hole}"	UHPFRC Shear strength
d, cleveage	Cover behind bolt "E", thickness "t"	UHPFRC Tensile strength
e, screw shear	Screw diameter "D_{screw}"	Steel bolt Shear strength

The variables for connection test specimens were: The specimen thickness "t" (20, 30, 40 mm), the diameters of the hole / screw (20/20, 24/22, 27/22 mm), the width specimen vs hole diameter ratio (W/D_{hole}) (4, 8), and, the distance from the hole centre to the specimen end vs hole diameter ratio (E/D_{hole}) (2, 4). The analyzed combinations are shown in Table 2.

An additional compressive bearing test was done (Fig. 5) with a 200 mm distance from the hole center to the loading section. The analyzed diameter equal for hole and screw were 20, 24, 27 and 30mm. No screw prestressing was induced.

Fig. 5. From left to right: Bearing strength test set up, bearing failure mode and connection test

4 Results

The UHPFRC compressive strength at 1, 2, 7 and 28 days was 59, 103.5, 133.1 and 156.3 MPa with a 1.8% coefficient of variation for the last. The flexural behavior is shown in Fig. 6.

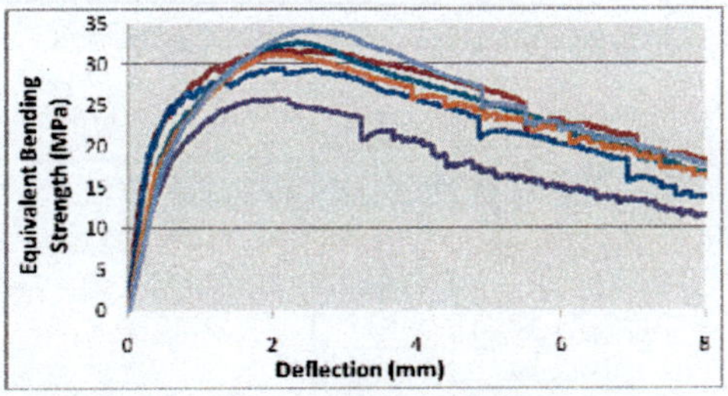

Prism	Min	Max	Average	CV(%)
Bending Strength	25,2	34	30,8	8,60%
f_{tj} (MPa) [11]	10,1	11,8	11,07	4,77%
σ_{bt} (MPa) [11]	10,5	11,8	11,58	4,18%

Fig. 6. Flexural behavior of the tested beams

The connection tests generated mainly two different failure modes (Fig. 7), with a typical stress-displacement behavior shown in Fig. 8, where the stress is evaluated as the Load / A_E for the cleavage failure an Load / A_T for the net tension failure, where A_E and A_T areas are shown in Fig. 4. In four of the tests the failure suffered was in the secondary bolt, being refused. Table 2 shows the obtained results.

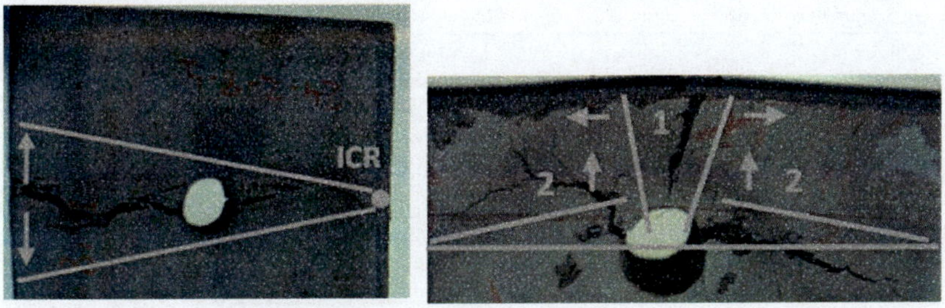

Fig. 7. Typical failure modes (test codes 20-24-8-4 net tension and 20-27-8-2 cleavage)

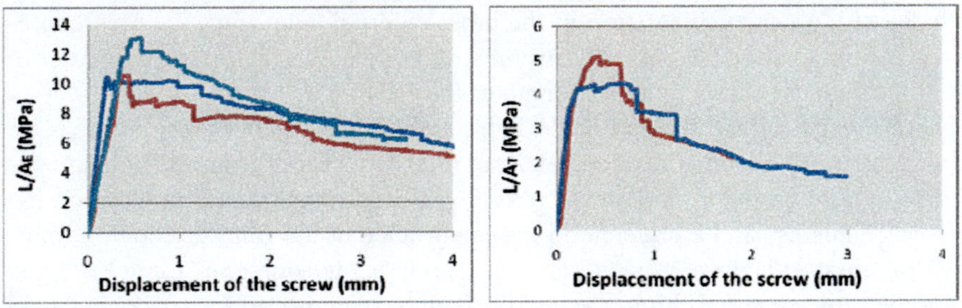

Fig. 8. Characteristic strength-deflection behavior for cleavage and net tension failure

Cleavage failure mode was associated to all large specimens (W/D_{hole}=8) with low cover (E/D_{hole}=2). In specimens with wider cover (E/D_{hole}=4) only net tension were observed even with cleavage strengths higher than in less cover elements cracking by cleavage. For intermediate cases (W/D_{hole}=4 and E/D_{hole}=2) cleavage and net tension failures may occurs. The thickness of the specimen or screw diameter didn't seem to be relevant in the failure mode.

Table 2. Specimens geometry, failure modes and connection strength

Nº test	Code	t	Dscrew	W/D_{screw}	E/D_{screw}	Failure	L/A_E	L/A_T
1	20-20-4-2	20	20	4	2	Cleveage	13,04	>6,47
2	20-20-8-2	20	20	8	2	Cleveage	15,72	>3,50
3	20-20-8-4	20	20	8	4	Net tension	>12,94	6,35
4	20-24-4-2	20	24	4	2	Cleveage	11,18	>5,24
5	20-24-8-2	20	24	8	2	Cleveage	10,30	>2,12
6	20-24-8-4	20	24	8	4	Net tension	>8,87	4,39
7	20-27-4-2	20	27	4	2	Net tension	>7,76	4,33
8	20-27-8-2	20	27	8	2	Cleveage	16,72	>3,64
9	30-20-4-4	30	20	4	4	Net tension	>6,64	9,30
10	30-20-8-2	30	20	8	2	Cleveage	12,99	>2,78
11	30-20-8-4	30	20	8	4	Net tension	>10,30	5,09
12	30-24-4-2	30	24	4	2	Cleveage	10,38	>4,50
13	30-24-8-2 (1)	30	24	8	2	Cleveage	10,57	>2,28
14	30-24-8-2 (2)	30	24	8	2	Cleveage	13,05	>3,41
15	30-27-8-2	30	27	8	2	Cleveage	18,15	>3,83
16	30-27-8-4	30	27	8	4	2º hole	>13,08	>6,41
17	40-20-4-2	40	20	4	2	Cleveage	14,68	>6,61
18	40-20-8-2	40	20	8	2	Cleveage	21,66	>4,77
19	40-24-4-2 (1)	40	24	4	2	Cleveage	11,42	>5,56
20	40-24-4-2 (2)	40	24	4	2	Net tension	>8,62	4,28
21	40-24-8-2	40	24	8	2	Cleveage	19,90	>4,43
22	40-24-8-4 (1)	40	24	8	4	2º hole	>14,02	>7,18
23	40-24-8-4 (2)	40	24	8	4	Net tension	>11,64	5,95
24	40-27-4-2	40	27	4	2	Net tension	>9,94	5,06
25	40-27-8-2	40	27	8	2	Cleveage	11,60	>2,53
26	40-27-8-4 (1)	40	27	8	4	2º hole	>12,18	>6,09
27	40-27-8-4 (2)	40	27	8	4	2º hole	>13,98	>6,97

When a net tension failure mode is produced this failure showed no uniform crack opening which develops from a local damage in one of the external sides to the hole, with a center of rotation in the other external side, (Fig. 7). As a consequence the obtained values for net tension strength were in the magnitude order of half of the deduced direct tensile strength of the UHPFRC.

The cleavage failure seemed to be more stable than the net tension one. One crack starts from the top specimen side, and after two cracks from the hole are developed in the horizontal direction, providing an additional ductility. Considering those mechanisms can be understood the importance of the fiber orientation, most significant in the cleavage mode (3 cracks) than in net tension one (1 crack).

For the same cover (E) higher value of Load/A_E is produced for specimens with W/D_{hole}=8 than for W/D=4 due to the higher length for develop those horizontal cracks (Fig 7). It is also noticed a size effect in specimens failing by cleavage with W/D_{hole}=4 and E/D_{hole}=2, where the cleavage strength is higher as much as lower is the diameter (Fig. 9 left). This means that the failure evolution depends on a local crack and not all the section is working at the same level. With higher ratios of W/D_{hole} and E/D_{hole} the redistribution capacity is higher, and the size effect is not more evident. No clear size effect is found in net tension failures (Fig. 9 right).

As was expected from previous tests [3], no one shearing failure was observed due to the high UHPFRC shear strength.

The compressive bearing failure strength was determined for 12 holes, finding a ductile behavior. The UHPFRC compressive strength (Load/($D_{hole} \cdot$ t)) and the safety factor with regard to the screw failure are shown in the Fig. 10. It can be observed that, the lower the screw, the higher the strength, due to the intensity of biaxial compressive state.

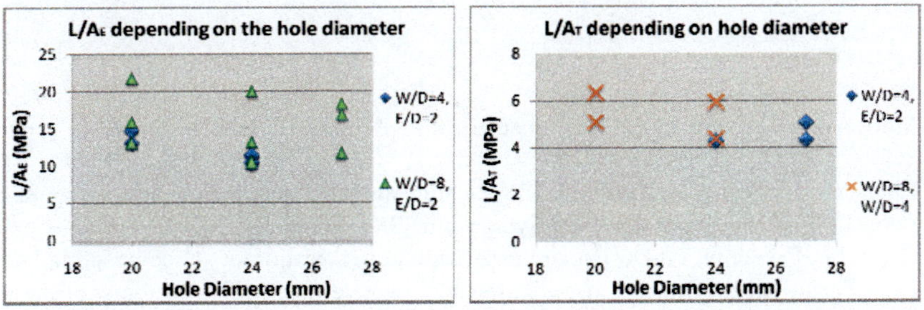

Fig. 9. Strength Load/A_E and Load/A_T vs hole diameter for both failure modes

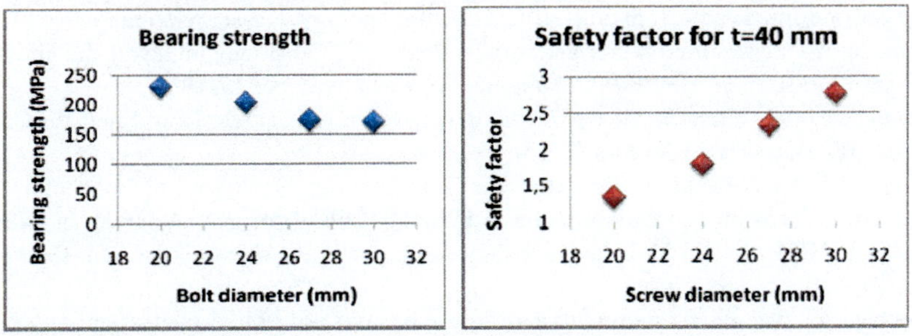

Fig. 10. Compressive Bearing strength and safety factor for the screw joints

5 Conclusions

– The proposed test method is adequate to analyze the projected connection system, but it needs to be verified for most strong specimens. Reinforced or presstresed specimens must be evaluated.

- Failure modes are characterized and evaluated. The weakest failures modes detected are cleavage and net tension.
- Cleavage failure must be controlled by increasing the cover or with local reinforcement. In this failure mode size effect can be appreciated for ratios $W/D_{hole}=4$ and $E/D_{hole}=2$, being negligible for bigger sizes.
- The net tension failure can't be estimated with the maximum tensile strength, due to the propagation of a crack through the section.
- The bearing failure was ductile and the equivalent compressive strength higher than the expected from the compressive tests.

Acknowledgements. The presented results were developed in the context of the research project "FIBAC" with reference BIA2009-12722 financed by Ministerio de Ciencia e Innovación. The support of this project is grateful acknowledged. We also would like to thank Prevalesa S.L. and IMPIVA for their support.

References

[1] McCraven, S.: Ultra High Performance Fiber Reinforced Precast Concrete proves market potential. PCI Journal (2007)

[2] Camacho, E., Serna, P.: Hybrid UHPFRC thin walled beams for flat roofing applications. In: Proceedings of the IASS Symposium 2009, Valencia (2009)

[3] Camacho, E., Serna, P.: Design and experimental verification of selfcompacting Ultra High Performance Hybrid Fiber Reinforced Concrete ties. In: Proceedings of the 6th international RILEM Symposium on SCC 2010, Montreal, Canada, September 26-29, pp. 1283–1291 (2010)

[4] Bendt Aarup, CRC – Applications of Fibre Reinforced High Performance Concrete. In: Second Int. Symposium on UHPC, CRC Technology, Kassel, Germany (2008)

[5] Tanaka, Y., et al.: Innovation and application of UFC bridges in Japan. In: International Workshop on UHPFRC, Marseille, France, November 17-18

[6] Enzo, V., et al.: First use of UHPFRC in Thin Precast Concrete Roof Shell for Canadian LRT station. PCI Journal (2005)

[7] Hegger, J., Gallwoszus, J., Rauscher, S.: Load-Carrying Behaviour of Connectors under Shear, Tension and Compression in Ultra High Performance Concrete. In: NSCC 2009, RWTH Aachen University, Institute of Structural Concrete, Aachen, Germany (2009)

[8] Pekbey, Y.: The bearing strength and failure behavior of bolted E-glass/epoxy joints. Mechanics of Composite Materials 44 (2008)

[9] Camacho, E., Serna, P.: Design and analysis of Ultra High Performance Hybrid FRC ties for truss structures. In: 8th fib Int. PhD Symposium in C. Eng., Denmark (2010)

[10] Camacho, E., Serna, P., López, J.A.: Caracterización de HMAR con Fibras: Estudio de las propiedades reológicas y mecánicas para dosificaciones de distintos niveles de rendimiento. Master Thesis, Universidad Politécnica de Valencia, Spain (January 2011)

[11] SETRA – AFGC Ultra High Performance Fiber-Reinforced Concretes. Interim Recommendations, AFGC Groupe de travail BFUP (January 2002)

Rapid Jacketing Technique by Using UHP-SHCC for Damaged RC Column under Seismic Loading

M. Kunieda, Y. Umeda, N. Ueda, and H. Nakamura

Dept. of Civil Engineering, Nagoya University, Japan

Abstract. This paper introduces a rapid jacketing technique for RC column, which involves spraying UHP-SHCC. Any formwork and/or additional reinforcement are not needed. The technique may be helpful in reducing construction time and cost. Cyclic loading tests were conducted for the column specimens repaired by UHP-SHCC and ordinary polymer cement mortar. It was confirmed experimentally that the developed technique using UHP-SHCC improve not only ultimate load but also ductility of recovered specimen. It seems that the prevention of buckling of longitudinal reinforcement, which was constrained by UHP-SHCC, imparts the mechanical improvement to the column specimens.

1 Introduction

For concrete structures such as super structures in highway, development of rapid repair/retrofitting methods [1, 2] is a very important issue and required in terms of business continuity plan (BCP). There are a wide variety of methods such as jacketing with steel plate, continuous fibers and RC. However, these methods involve many construction processes such as manufacture of parts in steel jacketing, formwork and arrangement of steel reinforcement in RC jacketing. The construction process gives longer downtime of the infrastructure.

Ultra High Performance-Strain Hardening Cementitious Composites (UHP-SHCC) [3, 4], one of the fiber reinforced composites, has high strength in both compression and tension, high strain capacity in tension with pseudo strain hardening behavior. And spraying technique with UHP-SHCC [5] has been also developed for repair applications.

This paper introduces the developed rapid jacketing technique using UHP-SHCC for damaged RC subjected to seismic loading. The RC column specimen repaired by UHP-SHCC was tested, and load carrying capacity and ductility were evaluated comparing to the specimen repaired by ordinary Polymer Cement Mortar (PCM).

G.J. Parra-Montesinos, H.W. Reinhardt, and A.E. Naaman (Eds.): HPFRCC 6, pp. 429–436.

2 Conceptual Idea on Rapid Jacketing Technique and Its Advantages

As shown in Fig. 1, UHP-SHCC is sprayed to a damaged part without any additional reinforcement. Both the novel material properties (i.e. high strength and high strain capacity) and spraying technique enable to develop the rapid jacketing technique, and followings are remarkable features of the developed technique including its advantages.

- A damaged structure has buckling of longitudinal reinforcement with spalling of cover concrete.
- Load carrying capacity including ductility after the jacketing is as well as that before the repair.
- Cross sectional shape after the recovery is the same as that before damage. It means that the damaged concrete corresponding to cover concrete is replaced by the repair material (UHP-SHCC). It is possible to apply the other strengthening (e.g. steel plate jacketing) later.
- No additional reinforcement is utilized in the jacketing.
- No formwork is required because of spraying technique.

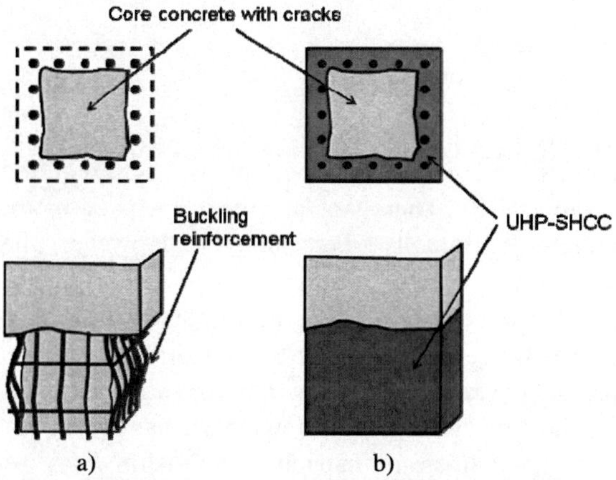

Fig. 1. Construction procedure of developed technique: (a) damaged column, and (b) repaired column

3 Outline of Experimental Program

In this study, a cyclic loading was carried out in order to induce initial damage for column specimen (namely initial loading). After that, repair of the damaged column was conducted using two kinds of repair materials; one was ordinary

PCM, and the other was UHP-SHCC. After the curing of the repaired specimens for 7 days, a cyclic loading was conducted again (second loading).

Shape of column specimens and reinforcement arrangement in the specimens are shown in Fig. 2. Cross sectional size of the specimen was 400 x 400 mm^2, and height of it was 1760 mm. Sixteen longitudinal reinforcements (D13, SD345, f_y=368 MPa) were used, and hoop reinforcement (D13, SD295A, f_y=360 MPa) were also arranged at intervals of 80 mm. Reinforcement ratio of longitudinal reinforcement was 1.27%. In order to measure the strain of reinforcement and concrete, strain gages having the gauge length of 2 mm and 30 mm were attached, as shown in Fig. 2. Two specimens were prepared in this study, because different repair materials were adopted.

Material properties of concrete at the age of 28 days and reinforcements are tabulated in Tables 1 and 2, respectively.

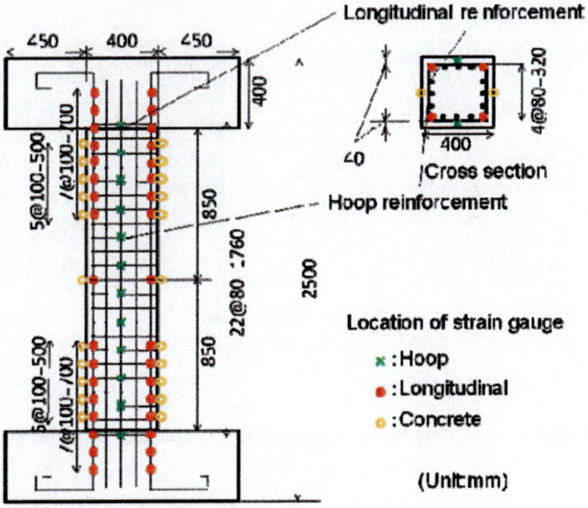

Fig. 2. Specimen and reinforcement arrangement

Table 1. Material properties of concrete (28 days)

Compressive strength (MPa)	Young's modulus (GPa)
29.3	32

Table 2. Material properties of reinforcement

Longitudinal		Hoop	
Yield strength (MPa)	Young's modulus (GPa)	Yield strength (MPa)	Young's modulus (GPa)
368	186	360	190

A cyclic loading was adopted to the column specimens in both initial loading and second one. Axial load of 160 kN (nominal axial stress: 1 MPa) was adopted. The loading was terminated, when the load was decreased up to yielding load after the peak.

4 Initial Loading to Induce Damage

Fig. 3 shows the relationship between load and displacement of the specimen in initial loading. Yielding of longitudinal reinforcement was occurred at load of 160 kN and displacement of 6.9 mm (δ_y). Horizontal load was significantly increased until 2 δ_y. However, the curve represents plateau after 2 δ_y. The peak load of the first specimen was 205 kN at the displacement of 41.4 mm (6 δ_y), and the peak load of the second specimen was 200 kN at displacement of 48.3 mm (7 δ_y). Note that the first and second specimens were used for the repair using PCM and UHP-SHCC, respectively. After that, softening behavior was observed in both specimens, and loading was terminated when the load was same as yielding load (164 kN). Eventually, both specimens exhibited similar load-displacement response, as shown in Fig. 3.

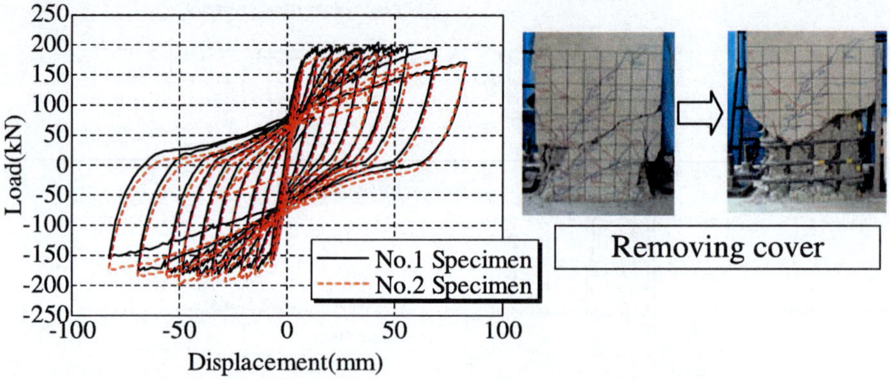

Fig. 3. Load-displacement curves of initial loading and photo of damaged part

Flexural cracks occurred and propagated at hinge part of the column specimens mainly, and the number of cracks increased until the displacement corresponding to 1 δ_y. After that, the number of crack was not increased, however, each crack opened wider.

On the side face of the column specimens, shear cracks occurred within 1D height (D: effective depth) from footing. And spalling of cover concrete was observed partially. Finally, buckling of longitudinal reinforcement was observed after spalling of cover concrete. In addition, fracture of core concrete was also obtained.

5 Repair Materials and Adopted Repair Methods

Two kinds of cement based materials were used in this study; one was ordinary PCM (premixed type) for plastering, and the other was UHP-SHCC for spraying. Where used PCM is not for this usage generally (i.e. no reinforcement is used).

For UHP-SHCC, water to binder ratio (W/B) was 0.22, and volume fraction of PE fiber (high strength Polyethylene fiber, tensile strength: 2700MPa, Young's modulus: 88 GPa, length: 12 mm, diameter: 0.012 mm) was 2.5% [5].

Table 3 tabulates the material properties of each repair material at the age of 7 days, which is corresponding to the loading age of second loading. Compressive strengths of PCM and UHP-SHCC were 23.2 MPa and 79.7 MPa, respectively. The compressive strength was obtained from the cylindrical specimens having the size of 50 mm in diameter and 100 mm in length. The compressive strength after 28 days ranged between 131 and 150 MPa.

Table 3. Material properties of repair materials (7 days)

PCM		UHP-SHCC			
Compressive strength (MPa)	Young's modulus (GPa)	Compressive strength (MPa)	Young's modulus (GPa)	Tensile strength (MPa)	Strain at tensile strength (%)
23.2	19.2	79.7	23.7	7.7	1.22

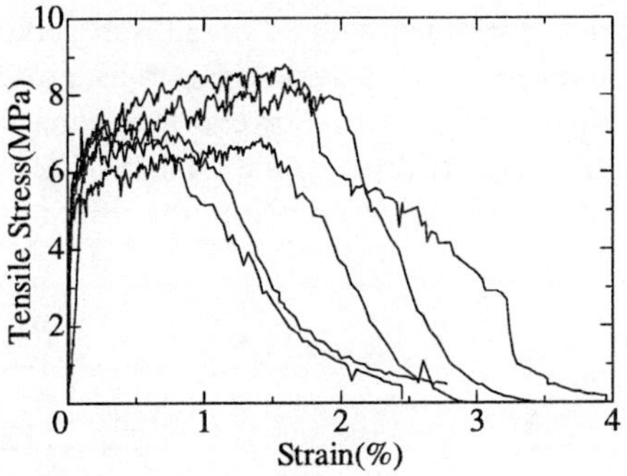

Fig. 4. Tensile stress-strain curves of UHP-SHCC (7 days)

Fig. 4 illustrates the tensile test results of UHP-SHCC dumbbell shaped specimens (5 specimens), which were also made by spraying. Note that tested cross section was 30 x 13 mm^2. Averaged tensile strength of UHP-SHCC was 7.7 MPa, and pseudo strain hardening and multiple fine cracking were observed, in addition to higher strain capacity over 1%.

The target of the repair was the damaged column specimens that were brought to the original position (residual displacement: 0 mm). The damaged cover concrete was removed by using a hammer. And water spraying was also applied to prevent the water absorption of repair material itself. PCM was applied by using plaster finishing. Note that the finished cross section was almost the same as the un-damaged one (i.e. 400 mm x 400 mm), and the approximate thickness of the repaired layer was equal to about 60 mm. Regarding UHP-SHCC, it was only sprayed to the damaged part. After the repair, curing was conducted for 7 days, and cyclic loading was carried out again for repaired specimens.

6 Second Loading for Repaired Specimens

6.1 Load-Displacement Curves and Failure Behavior

Fig. 5 shows the load-displacement curves of the specimens repaired by PCM and UHP-SHCC. Regarding the specimen repaired by PCM, maximum load was about 120 kN and it was dramatically decreased comparing to that of initial loading. Ductility of the specimen was also decreased significantly. In the initial loading stage (up to 2 δ_y), splitting cracks occurred adjacent to longitudinal reinforcement of tensile side, and the cracks induced the buckling of reinforcement under compressive stress.

For the specimen repaired by UHP-SHCC, maximum load was about 210 kN, and it was higher than that of initial loading. Load was significantly decreased at 12 δ_y, and ductility of the specimen was increased comparing to that of initial loading, eventually. However, initial stiffness was not recovered because of the crack located at the joint between column and footing.

Photos of the damaged part of the specimen with PCM and UHP-SHCC are shown in Fig. 6. In the case of specimen with PCM, splitting cracks were observed along longitudinal reinforcement, and repaired part was crushed and spalling. Regarding the specimen with UHP-SHCC, however, diagonal multiple cracking was observed in the front side. No crushing of the UHP-SHCC and no buckling of the longitudinal reinforcement were observed before the peak load.

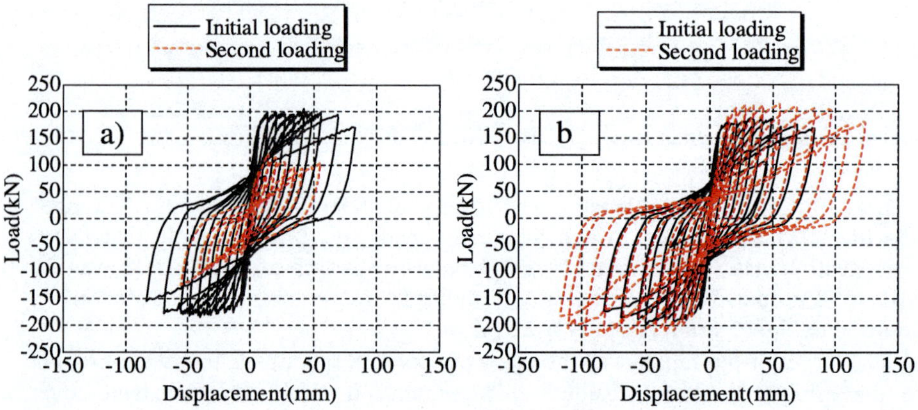

Fig. 5. Load-displacement curves of each repaired specimens: a) PCM, b) UHP-SHCC

a)  b) 

Fig. 6. Photos of repaired specimens after second loading: a) PCM, b) UHP-SHCC

After the peak load, concrete just above the interface between UHP-SHCC and concrete was damaged, and delamination of the repaired layer was observed finally. Regarding the side surface that is perpendicular to loading direction, although multiple cracking was expected, no crack was observed. Since the weakest cross section was the joint between UHP-SHCC and footing, only interfacial crack opening was observed during the loading.

6.2 Relationship between Vertical and Horizontal Displacement

Fig. 7 shows relationship between vertical and horizontal displacement of the columns repaired by PCM and UHP-SHCC. In the case of PCM, it was indicated that vertical displacement of second loading was decreased with increasing of loading cycles, because buckling of longitudinal reinforcement progressed. It seems that PCM could not resist against progress of buckling of the longitudinal reinforcement. Regarding the specimen with UHP-SHCC, vertical displacement was increased with increasing of loading cycles. It means that UHP-SHCC imparts higher resistance against buckling of the longitudinal reinforcement to the column specimen.

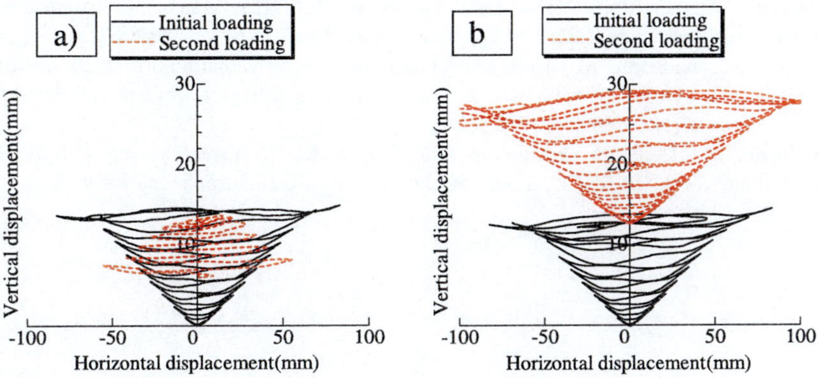

Fig. 7. Vertical displacement: a) PCM, and b) UHP-SHCC

7 Conclusions

The rapid jacketing technique using UHP-SHCC for damaged RC column was developed, and cyclic loading tests were conducted. Following conclusions were obtained;

- The rapid jacketing technique involves only spraying of UHP-SHCC. Any formwork and additional reinforcement are not required. It might be helpful to reduce a construction process that is corresponding to reduction of cost.
- It was confirmed experimentally that the developed technique using UHP-SHCC improve not only ultimate load but also ductility of recovered specimen. It seems that the prevention of buckling of longitudinal reinforcement, which was constrained by UHP-SHCC, imparts the mechanical improvement to the column specimens.

Acknowledgements. A part of this research was supported by Committee on Advanced Road Technology in established by Road Bureau of Ministry of Land, Infrastructure, Transport and Tourism. The authors would like to express their sincere thanks to their financial support.

References

[1] Inaguma, H., Seki, M.: Study on Restoration Work Method of Steel Jacketing Retrofit Column of Railway Viaduct. Proc. of the Japan Concrete Institute 27(2), 1057–1062 (2005) (in Japanese)

[2] Ishibashi, T., Tsuyoshi, T., Kobayashi, K., Kobayashi, M.: An experimental study on damage levels and repairing effects of reinforced concrete columns subjected to reversal cyclic loading with large deformations. JSCE Journal of Materials, Concrete Structures and Pavements 47(648), 55–69 (2000)

[3] Kunieda, M., Denarié, E., Brühwiler, E., Nakamura, H.: Challenges for Strain Hardening Cementitious Composites – Deformability Versus Matrix Density. In: Proc. of the Fifth International RILEM Workshop on HPFRCC, pp. 31–38 (2007)

[4] Kamal, A., Kunieda, M., Ueda, N., Nakamura, H.: Evaluation of crack opening performance of a repair material with strain hardening behavior. Cement and Concrete Composites 30(10), 863–871 (2008)

[5] Kunieda, M., Ueda, N., Nakamura, H., Tamakoshi, T.: Development of Spraying Technique for UHP-SHCC. Proc. of the Concrete Structure Scenarios 9, 349–354 (2009)

Structural Design and Previous Tests for a Retaining Wall Made with Precast Elements of UHPFRC

J.A. López, P. Serna, and E. Camacho

Civil Engineering, ICITECH, Universitat Politècnica de València, Spain

Abstract. An alternative to the initial design made with reinforced concrete is proposed using UHPFRC in order to build a retaining wall located in an aggressive environment. A structural design that aims to take advantages of the specific qualities of the UHPFRC and enables minimize the concrete section is analyzed and justified. The proposed solution consists of vertical prestressed beams designed with a lightened section and fastened to the soil with three levels of anchors. The wall is completed by placing slender curved shells elements between the beams and supported on them. Mechanical test of beam and shell are included as well as an economic valuation and comparison that takes into account both the material production costs and the size reduction advantages in the transport and placement.

1 Introduction

UHPFRC has already been used to build some structural applications due to high strength and ductility. Moreover, the excellent durability of UHPFRC, even in an aggressive environment, leads to a reduction of the maintenance costs. Since durability and strength properties are improved, the only limiting factor is the high cost of this material. In order that UHPFRC can be considered as a competitive alternative in structural applications it is necessary to optimize structural systems. That is only possible by understanding its mechanical behavior and by adapting its design to more efficient cross-sectional shapes and details.

It has been developed lots of applications which use UHPFRC especially on bridge girders and decks [1]. However, there are few elements for geotechnical applications in UHPFRC: prestressed sheet piles has been developed successfully in The Netherlands [2]; and prestressed piles for deep foundation has been developed according to de AASHTO Strategic Plan in Iowa [3]. This paper shows an alternative solution for a retaining wall designed with precast UHPFRC elements. Design criteria, structural evaluation, mechanical tests and an economic comparison with a previous traditional solution are shown in this paper.

G.J. Parra-Montesinos, H.W. Reinhardt, and A.E. Naaman (Eds.): HPFRCC 6, pp. 437–444.

2 Retaining Wall Design

The initial retaining wall design with conventional concrete consists of reinforced precast plates, 0.25 m thick, 2.5 m wide and variable 5.5-10 m length fastened to the soil. In Fig. 1 can be seen the wall geometry scheme and a picture of a previous construction where the traditional solution was applied. The rebars amount is 220 kg/m^3 and the weight of one element is 125 kN. The retaining wall is located in an aggressive environment (close to the sea). As an alternative of that element a new solution using UHPFRC has been proposed which consists of (Fig. 2) vertical precast prestressed beams designed with a lightened section and fastened to the soil with the same levels of anchors. The wall is completed by slender curved shells elements between the beams and supported on them.

The prestressed beam is 300 mm depth and includes a total of six 15.2 mm diameter 1860 N/mm^2 prestressed strands. The beam is lightened with a 200 mm diameter hole except at the anchored area. The curve shell is 30 mm thick with a ratio rise/span of 1/7.

The solution using UHPFRC lets to reduce four times the total concrete volume, approximately, with the same structural capacity and improving the concrete durability. In Fig. 2 UHPFRC section is compared with the one in conventional reinforced concrete.

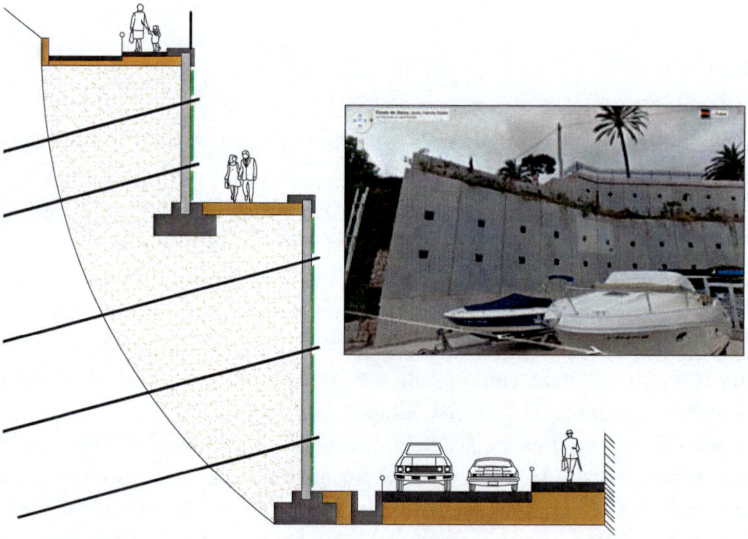

Fig. 1. Retaining wall scheme and reinforced concrete solution view

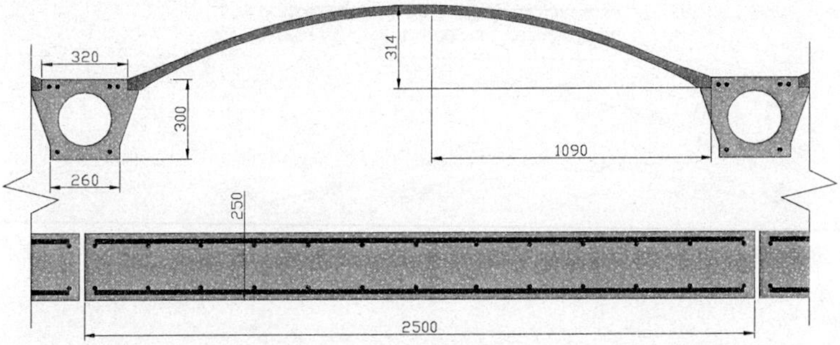

Fig. 2. Comparison between both UHPFRC design and conventional reinforced concrete one

3 UHPFRC Mixture and Properties

The UHPFRC used in this work was chosen as a result of a previous study [4] that aimed to optimize the concrete behavior at both fresh (flow test) and hardened state (compressive and flexural strength). The optimal dosage obtained is shown in Table 1. In order to characterize the mechanical properties of this dosage, six mixes were made. From each one four 100 x 100 x 100 mm cubic specimens and one 100 x 100 x 500 mm prismatic specimen were cast and a flow test was made (no heat treatment was applied). The average slump flow was 760 mm and the average compressive strength from cubic specimens was 150 N/mm^2. Flexural tests were made on unnotched prismatic specimens. Linear equivalent tensile stress-deflection curves from the six tests are shown in Fig. 3.

Based on flexural tests and following the formulation developed by Spasojevic, tensile stress-strain law can be obtained using an inverse analysis [5]. While Spasojevic assumes a horizontal plastic branch after the yielding tensile stress, in this work a hardening branch has been accepted (Fig. 4 right) and evaluated following the steps outlined by Spasojevic using an Excel worksheet.

Table 1. Optimal dosage use for cast UHPC elements

	kg/m^3		kg/m^3
Cement	1000	Fibers OL13/0.16 (1)	78.1
FC3R (3)	150	Fibers RC80/40 BP (1)	78.1
Silica Sand 0.5 mm	702	Superplasticizer (2)	9.06
Silica Sand 1.8 mm	380	w/c	0.213
Total water	213	w/b	0.185

(1) 1% in volumen of each kind of fiber from Bekaert

(2) Solid fraction of the superplasticizer

(3) Fluid catalytic cracking catalyst residue from petrol refinery which has a great pozzolanic activity in Portland cement pastes [6]

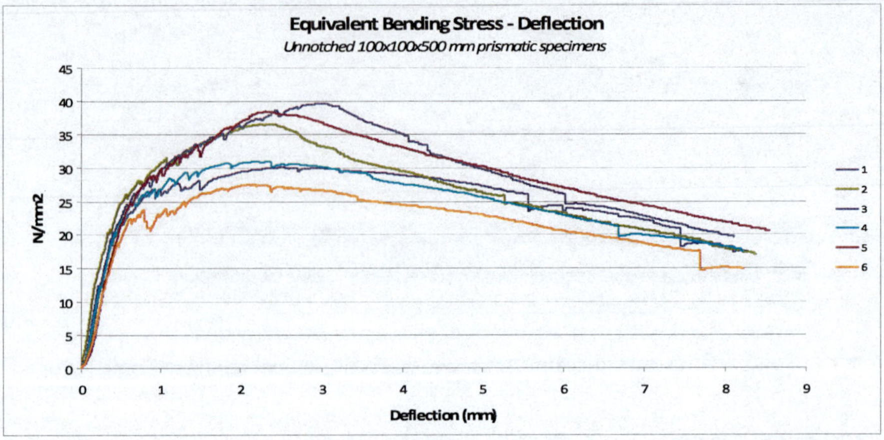

Fig. 3. Flexural test on unnotched prismatic specimens

For the inverse analysis the elastic modulus, strain at point B and ultimate strain were fixed. A 44 000 N/mm^2 elastic modulus was obtained in previous test [4] close to which is proposed by Graybeal [7] for a 150 N/mm^2 UHPFRC. The stress at point B was estimated in 0.3% and ultimate strain has been obtained according to French recommendations [8]. Stress at points A and B was modified in order to get the best approximation to the experimental test. Using this method, the average stress obtained at points A and B were 11.5 N/mm^2 and 13 N/mm^2 respectively. The respective variation coefficients were 5% and 11%. These results are similar to those obtained by Markovic with the same percentage and kind of fibers [9]. That shows good production stability.

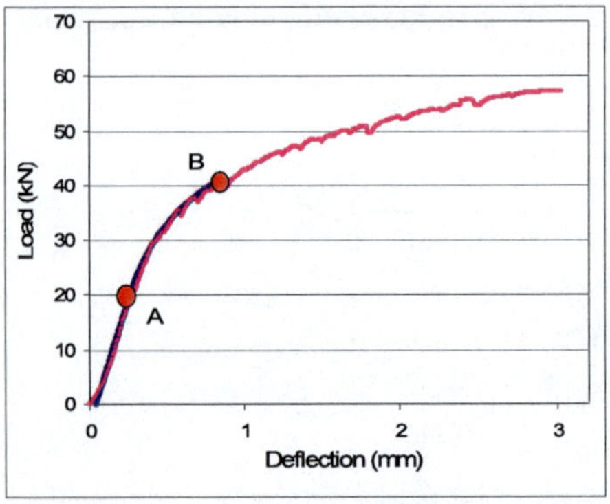

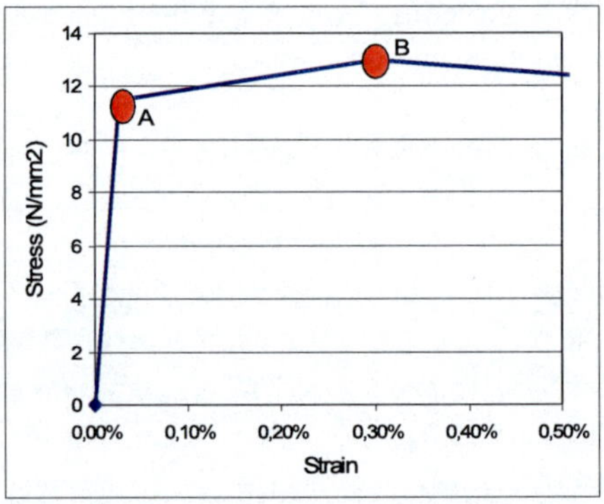

Fig. 4. Comparison between theoretical and experimental load-deflection graphics (left); stress-strain tensile law estimated (right)

4 Structural Evaluation

For the structural evaluation anchors were considered as joint loads applied directly on the wall and considering the wall supported on the earth modelled by springs with a coefficient of subgrade reaction of 100 000 kN/m^3. A surface load of 10 kN/m^2 was also considered. The beam was considered as a linear element and the shell as a bi-dimensional one. The model was checked using SAP2000. The beam prestressing strand release was considered two days after casting. Ultimate Limit State (ULS) and Service Limit State (SLS) were checked using French recommendations.

In SLS, a linear analysis was made limiting the tensile stress to the characteristic yielding tensile stress and the compression stress to the 60% of the characteristic compression strength. On this way, the beam prestressing level was designed.

In ULS a non-linear plastic analysis was considered. In the case of the beam the interaction diagram axial-bending was made and it was checked that all efforts were inside. In the case of the shell, on each point was checked the tensional state and due to the redundant conditions of the shell, the plastic redistributed was considered on areas where the tensile stress reached the yielding tensile stress.

5 Experimental Beam and Shell Test

With the purpose to check the design criteria, a reinforced UHPFRC three metres length beam with the designed cross-section was cast. The traditional reinforcement was evaluated to guaranty a tensile failure with a compressive stress level close to the maximum concrete capacity.

A four-point-flexural test of the beam was performed. It was simply supported with a 2.36 m span applying the load in two centred points separated 300 mm.

The structural beam capacity was evaluated according with the French recommendations and the average tensile stress-strain law obtained. The maximum moment resisted by the mid span section was estimated in 291 kNm with a central load of 565 kN.

The experimental load / deflection curve at mid-span is shown in Fig. 5. On it, points A and B indicate the load in which, theoretically, the tensile strain at the bottom beam line corresponds respectively to the tensile stress at A and B point in Fig. 4. It is considered that cracking process starts at point A. However, between points A and B cracks were only visible if the surface was wetted. A large number of microcracks were then visible with an opening lower than 0.1 mm. In this range of loads a shear crack appeared but it did not develop.

When load reached point B (at 90% of the bearing capacity) the macrocrack at mid-span section became visible. The experimental ultimate load was 545 kN, nearby the theoretical one.

A half meter length shell with the designed cross-sectional shape was also cast and a four-point flexural test was performed (Fig. 6) allowing a maximum horizontal displacement of 6 mm.

The shell framework was made using acrylic plates and this kind of material results in an improvement on the final aesthetics of the concrete surface. Fig. 6 shows a surface similar to a polished material. The visible pores can be avoided by improving the pouring process and the framework in order to avoid that the air remains inside it.

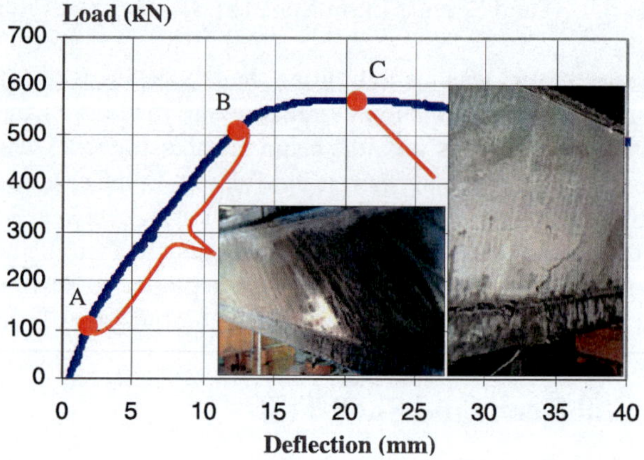

Fig. 5. Beam test result; beam appearance until point B (left); beam at point C (right)

Fig. 6. Shell four-point flexural test (left); surface aspect (right)

The theoretical collapse load was estimated using the plasticity theory assuming the formation of two plastic hinges, first at mid-span and then between mid-span and the support. The theoretical collapse load was estimated in 55 kN and the experimental bearing load was 47 kN. The difference between these two values could be explained since fiber segregation in the upper shell surface was noticed.

6 Economic Comparison between Both Alternatives

As the precast UHPFRC elements designed can bear the same loads than the reinforced concrete ones, an economic comparison has been valued between both solutions. Table 2 shows the price by meter of both alternatives as shown in Fig. 2. In order that a good valuation could be made the parameters considered on the study were those that differed in both alternatives, on this case, the materials used in the mix of UHPFRC, the steels bars and strands, the transport and placement.

- Although the fibers price is in the range of 3 times that of steel bars, the reduction of concrete volume and the high rebar demand in the traditional solution make the manufacturing process more economic in the UHPFRC solution.
- The framework used in UHPFRC elements casting is more expensive. Although there are no rebar jobs in this case, the casting process cost are not compensated.
- The weight of the UHPFRC elements does not exceed 20 kN, while the elements of the initial alternative weight 125 kN. As a consequence, transportation needs decreases and the cranes use to handle the elements are smaller. In addition, the time spent placing elements is lower.

As a consequence the final result shows that the UHPFRC alternative is 25% cheaper than the conventional one. With a reduction of four times the shape of the original solution UHPC becomes in an economic solution.

Table 2. Comparative between both alternatives by linear meter of wall

	Reinforced Concrete €/m	UHPFRC €/m	Ratio UHPFRC to Reinforced Concrete
Materials in concrete mixture	72.16	131.25	1.8:1
Steel bars/Prestressed strands	127.39	21.52	1:5.9
Framework	1.36	8.72	6.4:1
Transport and placement	24.18	7.86	1:3.1
Total	225.09	169.35	1:1.3

7 Conclusions

- The UHPFRC produced has shown a reliable mechanical behavior.
- The two designed elements can be manufactured in a conventional precast company, and show a very ductile behavior.
- The method used to obtain the UHPFRC behavior and the structural evaluation has been adequate to predict the beam and shell tests response.
- The design proposed made with precast elements of UHPFRC is suitable for its use on a retaining wall and lets to reduce the final costs, mainly in the casting, transporting and placing jobs cost.

- The possibilities of UHPFRC in civil applications are large and its development depends on the ability of engineers in order to find newer shapes and structural systems that optimize this material features.
- Aesthetics properties of UHPC can be improved by the use of plastics materials covering the framework

Acknowledgements. The presented results were developed in the context of the research project "FIBAC" with reference BIA2009-12722 financed by Ministerio de Ciencia e Innovación. The support of this project is gratefully acknowledged. We also would like to thank Prevalesa S.L. and IMPIVA for their support.

References

[1] Walraven, J.C.: Designing with ultra high strength concrete: basics, potential and perspectives. In: Proceedings of the International Symposium on Ultra-High Performance Concrete, Kassel, Germany, September 13-15, pp. 853–864 (2004)
[2] Grünewald, S.: Performance-based design of self-compacting fibre reinforced concrete. Delft University Press, Delft, The Netherlands (2004)
[3] Van de Voort, T., Suleiman, M., Sritharan, S.: Design and Performance Verification of Ultra-High Performance Concrete Piles for Deep Foundations (2008)
[4] López, J.A., Serna, P., Camacho, E.: Diseño estructural de un muro de contención de tierras para la defensa de taludes en el puerto de Jávea (Alicante) con la utilización de hormigón de muy alto rendimiento. Final Project Degree, Universidad Politécnica de Valencia, Spain (2010)
[5] Spasojevic, A.: Possibilities of Innovation in Bridge Design. École Polytechnique Fédérale de Lausanne, IS-Beton (2006)
[6] Payá, J., Monzó, J., Borrachero, M.V.: Fluid catalytic cracking catalyst residue (FC3R). An excellent mineral by-product for improving early strength development of cement mixtures. Cem. Concr. Res. 29, 1773–1779 (1999)
[7] AFGC (Association Francaise de Génie Civil). Recommendations provisoires. Bétons fibrés à ultra-hautes performances (2002)
[8] Graybeal, B.: Compressive Behavior of Ultra-High-Performance Fiber-Reinforced Concrete. ACI Materials Journal, 146–152 (March-April 2007)
[9] Markovic, I.: High performance hybrid reinforced fibre concrete. Development and utilization. PhD Thesis, Delft university of Technology (2006)

Design for Serviceability of Ultra High Performance Concrete Structures

T. Leutbecher and E. Fehling

University of Kassel, Institute of Structural Engineering

Abstract. The combination of steel bar reinforcement with fibers causes differences in the structural behavior compared to reinforced and prestressed concrete. This fact is of special interest for ultra-high-performance concretes (UHPC), because fibers are added to these concretes generally to improve ductility. Based on the well-known relations concerning calculation of crack width for reinforced concrete, the influence of the fibers on the cracking process is introduced considering equilibrium and compatibility. For the calculation of deformations a modification of the stress-strain relationship is derived considering the tension stiffening due to the contribution of fibers within the cracked sections and in between them. Because of their mechanical basis, the equations are not limited to UHPC but generally applicable to all types of concrete.

1 Introduction

With regard to durability, the positive influence of the fibers on the crack widths is significant. Especially for enhanced durability requirements, an essential improvement compared to reinforced concrete can be achieved by combined reinforcement. Experimental investigations [1] confirm that economical fiber contents below 1 vol.-% enable equally distributed cracking with very small crack spacings and crack widths. Thereby, the purely fiber-reinforced UHPC does not need to show hardening behavior itself. The only necessity is that the combined reinforcement consisting of fibers plus rebars enables hardening behavior.

To take advantage of these findings in structural design, the current design rules for the serviceability limit state have to be modified considering the influence of the fibers on the cracking process as well as on the increase in stiffness. Doing so, the mechanical relationships presented in the following can be derived. They are now part of the current draft of the *fib* Bulletin on Ultra High Performance Fiber Reinforced Concrete [2].

Validation in extensive test series on UHPC-members with combined reinforcement subject to tension [1, 3] and bending [4] confirmed suitability of the theoretically obtained relations.

G.J. Parra-Montesinos, H.W. Reinhardt, and A.E. Naaman (Eds.): HPFRCC 6, pp. 445–452.

2 Contribution of Fibers to the Load Transfer in the Crack

In the cracked cross-section of concrete elements reinforced with rebars and fibers, the fibers contribute to the load transfer and thus relieve the bar reinforcement. For members subject to tension or bending, Fig. 1a and b give a schematic view of the equilibrium of forces in the crack. Based on the stress-crack opening relationship of the fiber-reinforced concrete, the contribution of fibers in tension can be determined depending on the actual crack width (Fig 1c).

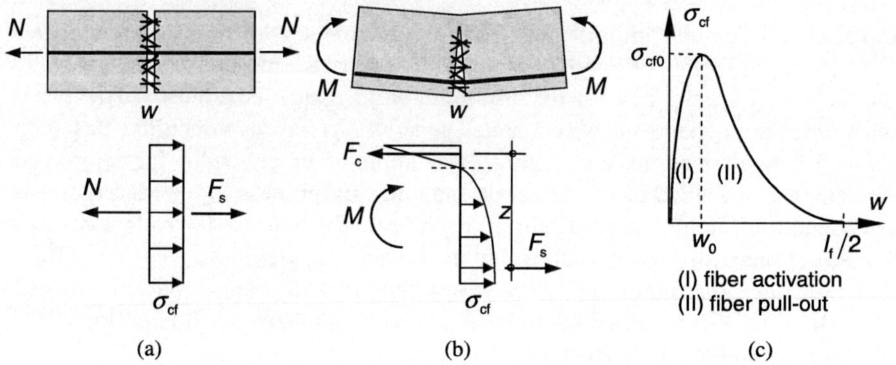

Fig. 1. Contribution of fibers to the load transfer in a cracked cross-section (a), (b) Schematic view of the equilibrium of forces for members subject to tension or bending (c) Typical stress-crack opening relationship of fiber-reinforced concrete

Presuming compatibility of deformations, the relative displacement between rebars and matrix on the one hand as well as between fibers and matrix on the other hand must lead to the same crack width. Thus, distribution of internal forces can be determined definitely.

For pure tension, the tensile force in the bar reinforcement amounts to

$$F_s = N - \sigma_{cf} \cdot A_c \tag{1}$$

where N is the axial force and A_c is the cross-sectional area.

For cross-sections subject to bending or bending with axial force, the tensile force in the bar reinforcement can be determine according to Eq. (2) generally.

$$F_s = \frac{M_s}{z} + N - \alpha_f \cdot \sigma_{cf} \cdot A_c \tag{2}$$

where M_s is the bending moment referred to the center line of the tensile reinforcement; z is the lever arm, that can be assumed to approximately $5/6$ times the effective depth of the cross-section; and α_f is a coefficient defining the contribution of fibers in the cracked cross-section, that can be determined by analysis of equilibrium of internal forces (Fig. 1b).

Thereby, the contribution of the matrix in tension may be neglected, following the common design philosophy of reinforced concrete.

Evaluation of some limit cases shows that, for rectangular cross-sections predominantly subject to bending (load eccentricity $M/N > 3.5\,h$), α_f results in values close to approximately 0.4 that may be used simplifying without detailed consideration.

The tensile stress σ_{cf} in Eq. (1) and (2), that is transferred by the fibers in the crack, is related to the crack width w in the center line of the bar reinforcement and may be determined as follows [1, 5]:

$$\sigma_{cf} = \sigma_{cf0,k} \cdot \left(2 \cdot \sqrt{\frac{w}{w_0}} - \frac{w}{w_0} \right) \quad \text{(fiber activation phase; } w \leq w_0\text{; see Fig. 1c)} \qquad (3a)$$

$$\sigma_{cf} = \sigma_{cf0,k} \quad \text{(fiber pull-out phase; } w > w_0\text{; simplification for SLS)} \qquad (3b)$$

In Eq. (3a) and (3b) w is the actual crack width in the center line of the bar reinforcement. $\sigma_{cf0,k}$ represents the characteristic value of the fiber efficiency (maximum tensile stress transferred by the fibers in the cracked state, see Fig. 1c). It is reached when the fiber with an embedment length of $l_f/2$ is fully activated. $\sigma_{cf0,k}$ as well as the corresponding crack width w_0 (Fig. 1c) can be obtained experimentally. Following theoretical considerations [1, 5] w_0 amounts to

$$w_0 = \frac{\tau_{fm} \cdot l_f^2}{E_f \cdot d_f} \qquad (4)$$

where τ_{fm} is the average bond stress between fiber and matrix, assuming a rigid-plastic bond law; l_f is the length of the fiber; d_f is the diameter of the fiber; and E_f is the modulus of elasticity of the fiber material.

Regarding the serviceability limit state, the crack width w should be limited to a small value which in most practical cases will be comparable to the order of magnitude of w_0. The stress-crack opening relationship can therefore be treated to be approximately constant at the beginning of the pull-out phase (see Eq. (3b)).

3 Crack Width Control

The characteristic value of the crack width w_k of concrete members with combined reinforcement subject to tension or bending can be determined using the basic crack width formula known from reinforced concrete (e. g. Eq. (7.8) in [6]).

$$w_k = s_{r,max} \cdot (\varepsilon_{sm} - \varepsilon_{cm}) \qquad (5)$$

In Eq. (5) $s_{r,max}$ represents the maximum crack spacing in the state of stabilized cracking (maximum length over which slip between steel and concrete occurs). For single cracks, $s_{r,max}$ is equal to the double of the load transmission length l_{es}

of the bar reinforcement. ε_{sm} and ε_{cm} are the average steel strains of bar reinforcement and concrete within $s_{r,max}$.

Fig. 2 shows qualitatively for a single crack and for the state of stabilized cracking the distribution of strains for the bar reinforcement and for the matrix, considering the influence of shrinkage, which is comparatively high for UHPC.

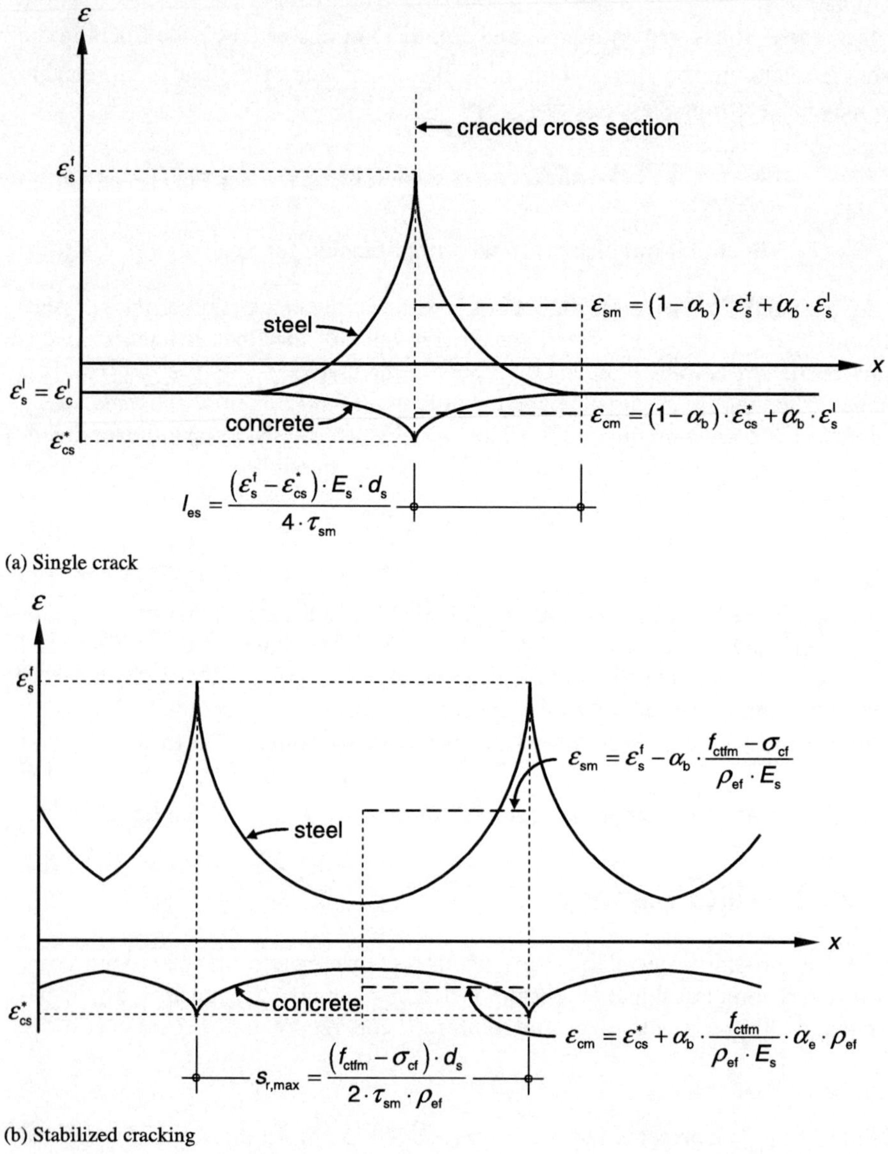

Fig. 2. Qualitative distribution of strains for the bar reinforcement and for the matrix, considering the influence of shrinkage

Thus, the difference of average steel and concrete strains results in

$$\varepsilon_{sm} - \varepsilon_{cm} = \varepsilon_s^f - \varepsilon_{cs}^* - \alpha_b \cdot \frac{f_{ctfm} \cdot (1 + \alpha_e \cdot \rho_{ef}) - \sigma_{cf}}{\rho_{ef} \cdot E_s} \geq (1 - \alpha_b) \cdot (\varepsilon_s^f - \varepsilon_{cs}^*) \tag{6}$$

where ε_s^f is the steel strain in the crack according to analysis of cross-section under consideration of fibers; α_b is the shape coefficient of strain courses ($\alpha_b = 0.6$ for short term loading, $\alpha_b = 0.4$ for long term or repeated loading); f_{ctfm} is the mean value of the elastic tensile strength of concrete (un-cracked state); ρ_{ef} is the effective reinforcement ratio of the bar reinforcement $\rho_{ef} = A_s / A_{c,ef}$; A_s is the cross-sectional area of the bar reinforcement; $A_{c,ef}$ is the effective tension area, accounting for the non-uniform normal stress distribution by bond forces into the concrete cross-section at the end of the transmission length, e. g. according to Fig. 7.1 in [6]; σ_{cf} is the tensile stress transferred by the fibers in the crack according to Eq. (3a) and (3b); ε_{cs}^* is the strain of concrete due to shrinkage at the crack surface after cracking (simplifying, it may be taken equal to the free shrinkage strain of the concrete ε_{cs}); and α_e is the ratio E_s / E_c (E_s and E_c are the moduli of elasticity of reinforcing steel and concrete).

The steel strain in the crack amounts to $\varepsilon_s^f = F_s / (A_s \cdot E_s)$. For pure tension, F_s can be determined according to Eq. (1). For bending with/without axial force, F_s is obtained using Eq. (2).

The length $s_{r,max}$ can be calculated as follows:

$$s_{r,max} = \frac{(f_{ctfm} - \sigma_{cf}) \cdot d_s}{2 \cdot \tau_{sm} \cdot \rho_{ef}} \leq \frac{(\varepsilon_s^f - \varepsilon_{cs}^*) \cdot E_s \cdot d_s}{2 \cdot \tau_{sm}} \tag{7}$$

where d_s is the diameter of the bar reinforcement and τ_{sm} is the average bond stress of the rebars within $s_{r,max}$.

For practical use, the bond stress often is assumed to be constant (rigid-plastic bond law), e.g. $\tau_{sm} = 1.8 \cdot f_{ct}$ for NSC. For UHPC, the reference values according to Table 1 may be used depending on the crack width w_k and the relative rib area of the rebars f_R. If concrete splitting is avoided before reaching w_k, the proposed values do not depend significantly on the fiber content.

Table 1. Reference values for the average bond stress τ_{sm}

w_k [mm]	Average bond stress τ_{sm} for rebars with $f_R = 0.024$	$f_R = 0.072$
0.05	$1.2 f_{ctm}$ *)	$2.0 f_{ctm}$ *)
0.10	$1.7 f_{ctm}$ *)	$3.3 f_{ctm}$ *)

Intermediate values may be found by linear interpolation.

*) f_{ctm} is the tensile strength of the matrix

Using Eq. (5) to (7), the crack width w_k can only be found by iteration because ε_{sm} and ε_{cm} as well as $s_{r,max}$ depend on the crack width itself. However, the bar reinforcement A_s, that is necessary to limit on a certain crack width w_k can be determined directly, transforming Eq. (5) to (7) as follows:

For single cracks:

$$A_s = \frac{F_s}{\sqrt{\frac{2 \cdot E_s \cdot \tau_{sm} \cdot w_k}{(1-\alpha_b) \cdot d_s}} + \varepsilon_{cs}^* \cdot E_s} \tag{8}$$

In the state of stabilized cracking:

$$A_s = \Omega^f \cdot \left(-\varepsilon_{cs}^* + \sqrt{\varepsilon_{cs}^{*\,2} + 2 \cdot \frac{F_s - \alpha_b \cdot (F_{f,cr,ef} - F_{f,ef})}{\Omega^f \cdot E_s}} \right) \tag{9}$$

where $\Omega^f = (F_{f,cr,ef} - F_{f,ef}) \cdot d_s / 4 \cdot w_k \cdot \tau_{sm}$; $F_{f,cr,ef}$ is the cracking force of the effective tension area of the fiber-reinforced concrete $F_{f,cr,ef} = A_{c,ef} \cdot f_{ctfm,el}$; and $F_{f,ef}$ is the tensile force carried by the fibers in the crack in the effective tension area $F_{f,ef} = A_{c,ef} \cdot \sigma_{cf}$.

Simplifying, in Eq. (8) and (9) the concrete strain caused by the action of loading is neglected.

The equations presented within this paper can be extended to consider also the use of two types of continuous reinforcement with different bond behavior (e.g. non-prestressed together with bonded prestressed reinforcement) in combination with fibers [2].

4 Deformations due to Tension

Based on the approach of the CEB-FIP Model Code 1990 [7], Fig. 3 exemplifies the principle stress-strain relationship of a reinforced concrete tensile member in the serviceability range. Compared to the fully-cracked state, the contribution of concrete between the cracks leads to a reduction of the average tensile strain, the so-called tension stiffening.

If shrinkage is to be considered, the stress-strain curve – including the corresponding crack spacings and crack widths – is shifted towards negative stresses and strains. Due to shrinkage the tensile member receives a pre-strain. However, the concrete cannot deform freely because of the restraint due to the reinforcement. This leads to compressive stresses in the steel bars and tensile stresses in the concrete. As a consequence, the steel stress in the crack, when the first crack has formed, has decreased from σ_{sr1} to $\sigma_{sr,cs}$ (point B in Fig. 3). In contrast, shrinkage influences only marginally the tension stiffening in the state of stabilized cracking.

If fibers are added to a reinforced concrete tie, tension stiffening increases due to the contribution of fibers in and between the cracks. If distribution and orientation of the fibers are considered with their mean values along the specimen's length, the average tensile strain in the state of stabilized cracking (phase D in Fig. 3) can be determined according to Eq. (10).

$$\varepsilon_{\rm sm} = \varepsilon_{\rm s} - \frac{\beta_{\rm t} \cdot (f_{\rm ctfm} - \sigma_{\rm cfm}) + \sigma_{\rm cfm} \cdot \rho_{\rm ef} / \rho_{\rm s}}{\rho_{\rm ef} \cdot E_{\rm s}} \tag{10}$$

where $\varepsilon_{\rm s}$ is the strain of bare steel in the crack (fully-cracked state) neglecting the participation of fibers; $\beta_{\rm t}$ describes the contribution of the (fiber-reinforced) concrete in tension in and between the cracks, depending on the duration of load and repeated loading, respectively ($\beta_{\rm t} = 0.4$ for short term monotonic loading, $\beta_{\rm t} = 0.25$ for long term or repeated loading); $\rho_{\rm s}$ is the geometrical reinforcement ratio $\rho_{\rm s} = A_{\rm s} / A_{\rm c}$; and $\sigma_{\rm cfm}$ is the tensile stress transferred over a macro crack, depending on the mean crack width for a specific load level, considering the influence of long term or repeated loading if necessary.

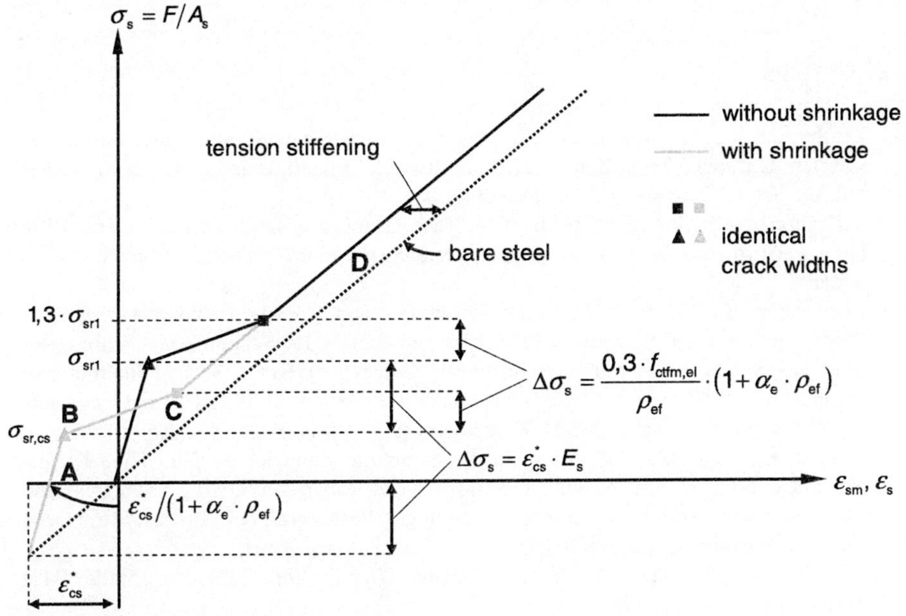

Influence of shrinkage:
A: Negative steel strain (restraint due to shrinkage)
B: Reduced cracking load due to internal restraint
C: Reaching the state of stabilized cracking at lower tensile stress due to internal restraint
D: Only small influence of shrinkage on tension stiffening in the state of stabilized cracking

Fig. 3. Average stress-strain relationship of (fiber-)reinforced concrete

σ_{cfm} can be determined according to Eq. (3a) and (3b) replacing the characteristic value of the fiber efficiency $\sigma_{cf0,k}$ by its mean value $\sigma_{cf0,m}$. As a simplification for smooth straight steel fibers with geometries typically used for UHPC, the tensile stress σ_{cfm} may be taken as $\sigma_{cfm} = 0.8 \cdot \sigma_{cf0,m} = \text{const}$. If in Eq. (10) σ_{cfm} is set to zero, the result known from reinforced concrete (without fibers) is obtained [7].

5 Conclusions

In structural members with combined reinforcement, fibers contribute to the load transfer in the crack and thus reduce the load transmission length of the bar reinforcement. Following the design philosophy of reinforced concrete, the equations presented in this paper quantify the positive influence of the fibers on the crack spacings and crack widths as well as on the deformation behavior. With Eq. (8) to (10) easily manageable relations for practical use have been derived that provide satisfying behavior in serviceability limit state by sufficiently designed bar reinforcement and without uneconomically high fiber contents. Experimental validation of the proposed design procedure as well as design examples can be found in [1].

References

[1] Leutbecher, T.: Rissbildung und Zugtragverhalten von mit Stabstahl und Fasern bewehrtem Ultrahochfesten Beton (UHPC). Structural Materials and Engineering Series, vol. (9). Kassel University Press, Kassel (2008)

[2] fédération internationale du béton (fib), Task Group 4.6, Draft of Bulletin on Ultra High Performance Fiber Reinforced Concrete (November 2010) (previously unreleased)

[3] Leutbecher, T., Fehling, E.: Crack Formation and Tensile Behaviour of UHPC Reinforced with a Combination of Rebars and Fibres. In: Schmidt, M., Fehling, E., Stürwald, S. (eds.) Ultra High Performance Concrete (UHPC), Second International Symposium on Ultra High Performance Concrete. Structural Materials and Engineering Series, vol. (10), pp. 497–504. Kassel University Press, Kassel (2008)

[4] Fehling, E., Leutbecher, T., Stürwald, S.: Structural Behavior of Ultra High Performance Concrete Reinforced with Steel Fibers and Rebars. In: 2010 fib Congress and PCI Convention Bridge Conference Proceedings, Washington D.C. Precast/Prestressed Concrete Institute, Chicago (2010)

[5] Pfyl, T.: Tragverhalten von Stahlfaserbeton. Dissertation ETH No. 15005, ETH Zürich, Switzerland (2003)

[6] Eurocode 2, Design of concrete structures - Part 1-1: General rules and rules for buildings, European Committee for Standardization (December 2004)

[7] CEB-FIP Model Code 1990, Design Code, Comité Euro-International du Béton. Thomas Telford Services Ltd., London (1993)

Part 7

Textile Reinforced Concrete (TRC) and Hybrid Composites

Influence of Textile Alignment, Moisture and Shape of Specimens on First Crack Load and Load Bearing Behavior of Textile Reinforced Concrete Containing Short Fibers

M. Hinzen, A. Hatting, and W. Brameshuber

Institute of Building Materials Research (ibac), RWTH Aachen, Germany

Abstract. For more than ten years researchers have concentrated on the development as well as on the mechanical properties of textile reinforced concrete. The tensile load bearing behavior of this new composite material is generally determined with tensile specimens in most different ways. The variety of specimens and test configurations lead however to the fact that results are not always comparable. Moreover, own results with uncoated AR-glass textiles showed relatively high scattering in the past. Some influences on the load bearing behavior could already be clarified. However, many questions are still unanswered. Therefore, it is the scope of this paper to investigate a set of influencing factors with regard to the first crack strength and the load bearing behavior of textile reinforced concrete. The investigated parameters include the alignment of the textile reinforcement, the moisture content of the specimens during testing and in some cases the specimen shape. Special attention was given to the repeatability of the results.

1 Introduction

The positive influences of short fibers on the load bearing behavior of textile reinforced concrete have already often been phenomenologically described [1, 2]. An examination of the effectiveness of the short fibers for example regarding an increase in the first crack load, the crack formation or even the load bearing capacity requires a reproducible load bearing behavior of the concrete with only textile reinforcement. Within the framework of the Collaborative Research Center 532 "Textile Reinforced Concrete", up to now uncoated textiles made of AR glass with a yarn fineness of 1200 tex and tricot binding have been used since they yield higher bond strengths than for instance textiles with 2400 tex and chain stitch binding. Furthermore, these textiles are especially suited for the applied laminating technique. Dumbbell-shaped specimens (Fig. 1) have so far been used which were either stored in water for 7 days and afterwards in a standard climate

G.J. Parra-Montesinos, H.W. Reinhardt, and A.E. Naaman (Eds.): HPFRCC 6, pp. 455–462.

(20 °C/65% RH) for 21 days or which were only stored in water. This procedure led to relatively high scattering at the first crack load as well as at the load bearing behavior. Therefore, the effects resulting from the reinforcement alignment and the moisture and shape of the specimens with and without short fibers were examined in small test series. The results are partly presented in this paper.

2 Materials and Specimens

A mix especially developed for the use of short fibers was applied as fine-grained concrete. It is shown in Table 1 and specified in detail in [3]. The short fibers used play only a minor role in the tests. They serve to determine if the detected effects on first crack load and load bearing capacity are changed by adding short fibers. The type and dosage of short fibers are therefore kept constant within the scope of this paper. 6 mm long, integral, AR glass fibers yielding good results regarding workability and load bearing effect are used. On the one hand, yarns with a fineness of 1200 tex and 2400 tex and on the other, textiles with a yarn fineness of 1200 tex and tricot binding were chosen as textile reinforcement. The dumbbell specimens hitherto applied and, as an alternative, rectangular specimens with the same cross sectional area were used as tensile specimens for textile reinforced concrete. The shape and dimensions of the dumbbell specimens are illustrated in Fig. 1. Small-sized cylinders with a diameter of 50 mm and a height of 100 mm were used as reference specimens for the concrete.

Table 1. Mix design of mix Fil-05-10

raw material	amount kg/m³
cement CEM I 52.5 N	500.9
fly ash	179.0
silica fume	63.1
superplasticizer	8.7
lime powder	278.1
sand 0.2 - 0.6	369.4
sand 0.06 - 0.18	452.3
water	303.8

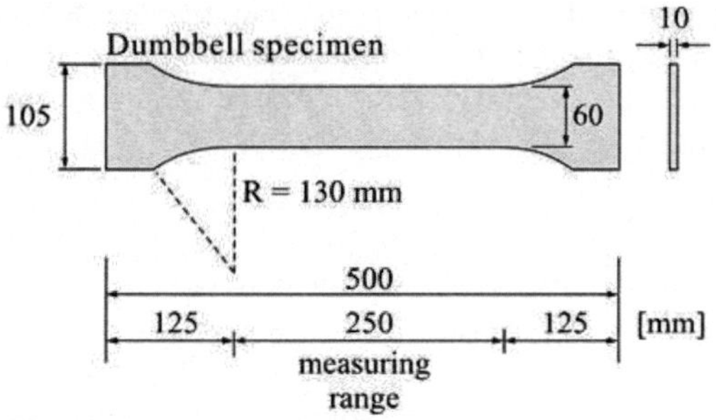

Fig. 1. Dumbbell specimen

3 Manufacturing, Storage and Test Setup

All mixes were mixed in a mortar mixer for five minutes. The short fibers of the fiber reinforced fine-grained concretes were added in a mixing break. The cylinders were manufactured in accordance with DIN 1048-5 [4]. Both types of strip specimens were produced horizontally with the laminating technique. A thin layer of concrete and the reinforcement with and without alignment were alternatingly placed. The specimens with the aligned textile or yarn were manufactured with a newly developed technique in which the concrete layers are first accurately trowelled and then the reinforcement is placed and manually slightly prestressed. In all cases, two reinforcement layers were inserted. After 24 hours of wet storage, all specimens were demoulded. Specimens that were tested in a dry state were first stored in water at 20 °C for another 20 days and afterwards in a standard climate at 20 °C and 65% RH up to a test age of 28 days. Specimens tested in a wet state were stored in water until testing. All tests were performed on a universal testing machine. The cylinders were loaded at a test rate of 0.1 kN/s in a force controlled way and the strip specimens were examined at a test rate of 1 mm/min in a displacement controlled way. At the dumbbell specimens, the load was introduced with steel flanges exactly adjusted to the radius of the load introduction zone, at the rectangular specimens, the load was introduced into the top and bottom area with hydraulic clamping devices.

4 Influences on the First Crack Load

4.1 *Specimen shape*

The investigations on the specimen shape were conducted in order to analyze which test variation was better suited to determine the tensile strength. Moreover, the tensile strengths of the strip specimens should be compared to the tensile strength of a cylinder as reference specimen to quantify possible influences resulting from load introduction and size effect. The tests were performed with and without short fibers as well as strictly on water-stored specimens. The results illustrated in Fig. 2 show that, at the fine-grained concrete without short fibers, both types of strip specimens yield the same tensile strengths and similar scattering as the cylinder.

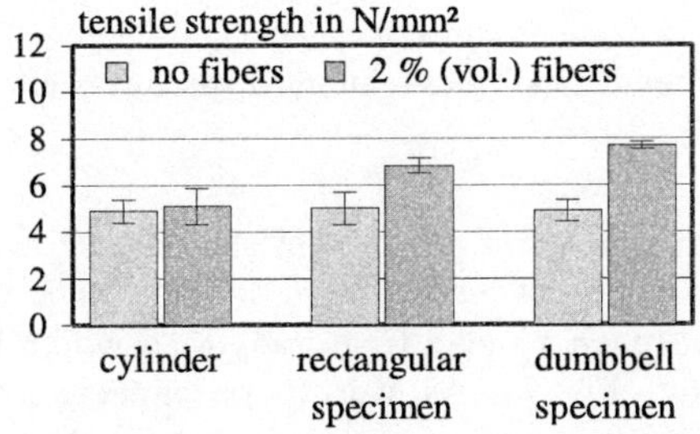

Fig. 2. Tensile strength depending on short fibers and specimen shape

When the fiber reinforced concrete is used, the tensile strength of the strip specimens is significantly higher. Due to the laminating technique, the short fibers of the strip specimens are almost orientated two-dimensionally so that many fibers can participate in the load transfer. The cast cylinder, however, has fibers with a three-dimensional orientation. The tensile strengths obtained with the dumbbell specimens were almost 1 N/mm^2 higher and the scattering slightly lower than those yielded with the rectangular specimens. All in all it can be stated that the influence of the geometry of both strip specimens is only marginal and that the measured strengths are comparable to the reference strengths of the cylinder, at least when measured on wet specimens. As it is well known, this is not the case in the dry state because of the warping of the specimens [5].

4.2 *Moisture of Specimens*

The influence of the specimen moisture was only determined on dumbbell specimens because the specimen shape had only a minor influence on the first crack load. Three different reinforcement variations were examined. Specimens with 2 layers of textile or with short fibers or with a combination of both were tested. Within the scope of this examination, it should be found out whether the first crack load is mainly determined by the short fibers or also by the textile. The results of this test series are illustrated in Fig. 3.

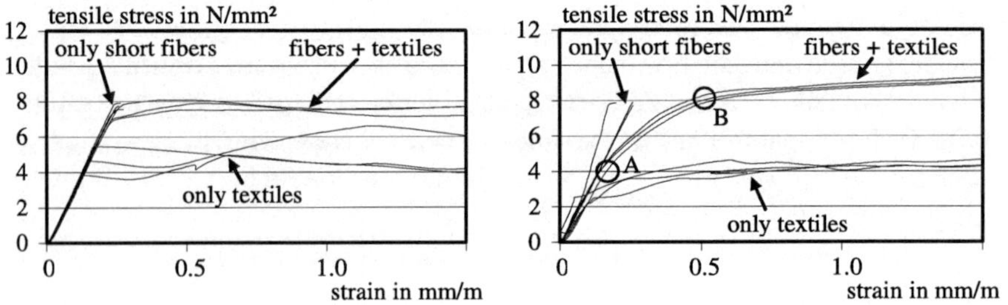

Fig. 3. First crack load of textile and fiber reinforced specimens after water storage (left) and dry storage (right)

On the left, the specimens tested in the wet state are shown and on the right, there are the respective specimens tested in the dry state. The wet specimens with the textile reinforcement reach a mean first crack load which lies on the level of unreinforced concrete (cf. Fig. 2). As was to be expected, this crack level is considerably increased by the use of short fibers. Furthermore, it turns out that the short fibers determine the level of the first crack load. The specimens with additional textile reinforcement did not lead to any further load increase within the framework of this examination. In the respective literature, however, different experiences have been made regarding the influence of textile on the first crack load [5, 7]. Depending on the bond quality between concrete and textile, increases in the crack load could be observed. If this behavior is compared to the dry but otherwise similar specimens in Fig. 3, right, another behavior appears. The crack level of the concrete reinforced by textile only is significantly lower which is ascribed to the imperfections due to drying [5]. Warping could however not be discerned on the specimens examined in this context. Moreover, the dry fiber reinforced concrete reaches the same first crack strength as the wet one. It is especially obvious that there is a clear deviation from the linear course in state I at the dry textile reinforced concrete containing short fibers. When compared to the other results, this can only be caused by the combination of textile and drying. In this connection, the question arises how the first crack load of textile reinforced concrete with short fibers is defined. First, the non-linear course hints at a pronounced micro cracking.

To confirm this assumption, two different methods were applied. On the one hand, it was tried to visualize the beginning crack formation with a step-by-step loading. On the other hand, acoustic events were compared to the course of the stress-strain curve by means of an acoustic emission analysis. Both methods confirm that the crack formation starts already at the section of the stress-strain curve marked with "A" (see Fig. 3, right). However, in addition to the acoustic events, the crack visualization showed only cracks on the cast side in the zone between "A" and "B", i. e. no full cracks, yet. Full cracks start to form in the area marked "B". It could not be clarified definitely why this observation was only made on dry textile reinforced specimens containing short fibers. In this respect, further investigations with the acoustic emission analysis must be performed.

5 Influences on the Load Bearing Behavior

The examinations of the load bearing behavior were performed on rectangular and dumbbell specimens. In the following, only the dumbbell specimens are dealt with. Within the scope of this investigation, there were all in all no significant differences between both strip specimens. Presently, an international round robin test on the load bearing behavior of both variations is performed within the framework of the RILEM TC 232-TDT "Test methods and design of textile reinforced concrete (TRC)".

5.1 Reinforcement Alignment

The investigations on the alignment of the reinforcement are conducted to determine if a possible waviness of the reinforcement layers which is inherent to the laminating technique has a decisive effect on the load bearing capacity and the test scattering.

Since filaments are not loaded simultaneously in wavy reinforcement layer, a tendency towards larger deformations must be anticipated. In different test series, dumbbell specimens with and without aligned reinforcement were compared with each other. Fig. 4, left, exemplarily shows such a comparison with textile reinforced specimens (1200 tex). It can be discerned that this kind of reinforcement does not present any basic advantages due to the alignment. The expected higher stiffness at an aligned reinforcement or a changed crack behavior did not occur. It was completely different with the specimens reinforced with 2400 tex yarns (Fig. 4, right). There are, however, only minor differences in stiffness but, against the expectations, the load bearing capacity of the aligned yarns is significantly lower.

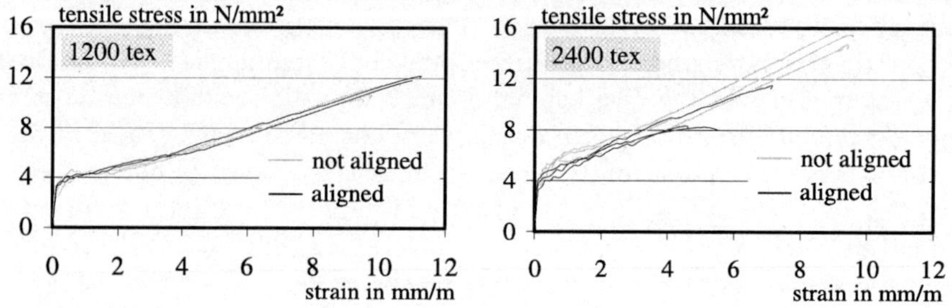

Fig. 4. Influence of the reinforcement alignment at 1200 tex textiles (left) and 2400 tex yarns (right)

It can be assumed that the bond area between concrete and reinforcement is changed by the small tensile load. Moreover, it is known [6] that the type of textile binding exerts a decisive influence on the bond area. In order to assess the bond area, the specimens were sawn after the test and the cross-sectional areas were examined under an optical microscope and photographed. Afterwards the yarn cross sections were marked. Fig. 5 contains a comparison of the bond areas depending on the yarn fineness and the alignment.

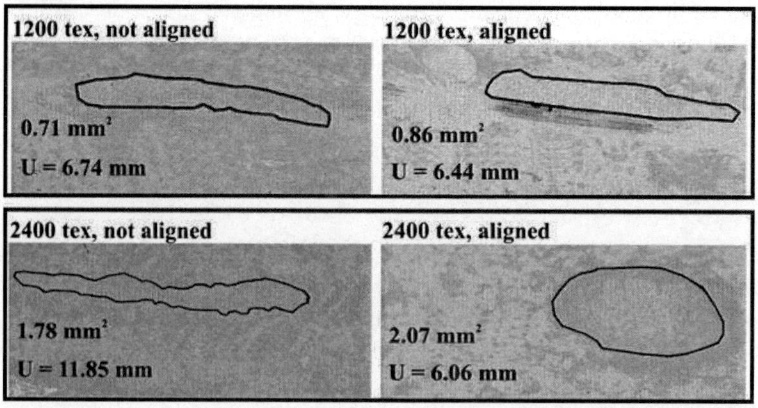

Fig. 5. Change of the bond areas of 1200 and 2400 tex yarns due to alignment

The images show that the 1200 tex yarn hardly changes its shape after the alignment. Above all, this is due to the fact that the yarn was processed in a textile with tricot binding and is therefore dimensionally stable. At the higher fineness of 2400 tex, single yarns were used. In the laminating process, the yarn is flatly rolled into the concrete without alignment and thus reaches a high bond area. If the yarn is aligned, the bond area is reduced by the slight tensile stress and thus also the load bearing capacity decreases. Therefore, it can be noted that the influence of the bond area of the yarn is stronger than the positive influence of the alignment. All in all, the alignment of the reinforcement led to a smaller scattering of the results. However, it makes only sense when the bond area is not reduced.

5.2 *Moisture of Specimens*

The pre-storage and the moisture content of specimens with textile reinforcement were both varied in order to examine the influences resulting from warping and micro cracking due to drying. In previous tests, fiber reinforced concretes that were stored in water and tested in the wet state reached higher first crack loads and smaller scattering. In this test series, some of the dumbbell specimens were stored dry for 7 days before the test and the others were stored in water until the test date. In Fig. 6, the results are compared. As was to be expected, the specimens stored dry show a lower first crack load, nevertheless, they feature a typical post-cracking behavior. Conversely, the first crack load of the water-stored specimens is much higher because these specimens are straight and not pre-damaged. However, in the water saturated state, the textile is unable to carry the occurring loads and it fails at about 70 % of the possible maximum load. Moreover, previous investigations showed that a drying of the specimens for only a few hours before the test was already sufficient to avoid any negative effect on the load bearing capacity. Hence, this is a phenomenon that only appears at total water saturation. Therefore, the exposure of textile reinforced concrete components containing uncoated glass fibers must be taken into account at the design.

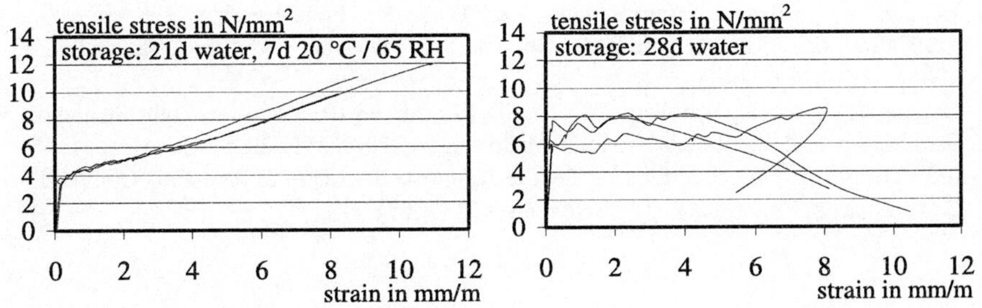

Fig. 6. Stress-strain curves of dry (left) and water-saturated (right) dumbbell specimens with textile reinforcement

6 Summary

The test series showed that the tensile strength of fine grained concrete can also be determined on strip specimens without any negative effects resulting from size effects or the load introduction as long as the specimens are stored and tested in the wet state. Dry concretes with a combined reinforcement of short fibers and textiles show cracks at a significantly earlier stage than dry concretes with only short fibers or wet concretes in general. The alignment of the reinforcement can reduce scattering. But it turned out to be only an advantage if the bond areas of the yarns are not reduced by the slight tension in the alignment process. This is especially a problem when using yarns with a high fineness. Water-saturated specimens with uncoated glass reinforcement failed at about 70 % of the maximum load as compared to dry specimens.

References

[1] Hinzen, M., Brameshuber, W.: Influence of Short Fibers on Strength, Ductility and Crack Development of Textile Reinforced Concrete. In: Reinhardt, H.W., Naaman, A.E. (eds.) Proceedings of the Fifth International RILEM Workshop on High Performance Fiber Reinforced Cement Composites (HPFRCC5), RILEM Proceedings PRO 53, Mainz, July 10-13, pp. 105–112 (2007), ISBN 978-2-35158-046-2

[2] Barhum, R., Mechtcherine, V.: In: Oh, B.H., et al. (eds.) Fracture Mechanics of Concrete and Concrete Structures, Korea Concrete Institute, pp. 1498–1503 (2010)

[3] Hinzen, M., Brameshuber, W.: Influence of Matrix Composition and Short Fibres on the Workability of Fine Grained Fibre Concrete. Bagneux: RILEM, 2010. In: Brameshuber, W. (ed.) Proceedings of the International RILEM Conference on Materials Science (MatSci). 2nd ICTRC Textile Reinforced Concrete, Aachen, September 6-8, vol. I, pp. 978–972 (2010), ISBN 978-2-35158-106-3

[4] DIN 1048-5:1991-06 Prüfverfahren für Beton; Festbeton, gesondert hergestellte Probekörper

[5] Jesse, F.: Tragverhalten von Filamentgarnen in zementgebundener Matrix. PhD Thesis, Dresden: Faculty of Civil Engineering, Technische Universität Dresden, urn:nbn:de:swb: 14-1122970324369-39398 (2005)

[6] Hegger, J., Will, N., Bruckermann, O., Voss, S.: Load-Bearing Behaviour and Simulation of Textile Reinforced Concrete. In: Materials and Structures (RILEM) 39(8), 765–776 (2006)

[7] Hegger, J., Voss, S., Scholzen, A.: Textile-Reinforced Concrete for Light Structures. Farmington Hill: American Concrete Institute, ACI SP-251, 2008. In: Aldea, C.-M. (ed.) Design and Applications of Textile Reinforced Concrete, ACI Fall Convention 2007, Fajardo, Puerto Rico, 14, October 14-18 (2007)

Tensile Behavior of Textile: Influence of Multilayer Reinforcement

I. Colombo, M. Colombo, A. Magri, G. Zani, and M. di Prisco

Politecnico di Milano, Milano, Italy

Abstract. The interest in the use of textile fabrics as reinforcement for cement based composite materials has been increasing in the last few years. Textile cementitious composites show improved tensile performances; as a matter of fact in addiction to a considerable peak strength, these composites show a strain hardening behavior even when the reinforcing yarns have a low modulus of elasticity. This was explained by the enhancement in bonding due to the mechanical anchoring provided by the non-linear geometry of the individual yarn within the fabric.

The paper investigates the tensile behavior of textile reinforced mortar as influenced by different geometrical reinforcement ratio in prismatic specimens. In all the experimental tests the load direction will be parallel to the fabric warp. Three different geometrical reinforcement ratio were investigated in order to evaluate their effectiveness and their influence both on tensile strength and on crack pattern.

1 Introduction

Modern architecture shows a definite trend towards thinner structures made by high-performance materials which demand a continuous development of their mechanical properties. Nowadays thin, light-weight and cost saving elements with improved durability by using non-metallic reinforcement for concrete are an open issue [1].

Textile Reinforced Mortar (TRM) is recently seen as a performing solution in this framework [2, 3, 4]. The use of fabrics, mainly made by alkali resistant (AR) glass, aligned to principal stress direction, allows the designer to optimize structural performance.

The application here considered refers to a layered prefabricated structure where the TRM is located on the tensile side. The production procedure of the TRM has to guarantee the shear stress transfer through the layer interface and a relative simplicity in the assembling of the whole structure.

The property, the content and the layout of the fabric have a great influence on the characteristics of the composite material. The fabric texture structure and the special geometry of its yarns have a significant effect on bonding [5]. The

G.J. Parra-Montesinos, H.W. Reinhardt, and A.E. Naaman (Eds.): HPFRCC 6, pp. 463–470.

geometrical characteristics to be considered are the nature of the yarns and the various geometries by which these yarns are combined together in the fabric. The structure of fabrics can be modified by changing the warp and weft spacing. This paper analyzes the influence of different reinforcement ratio on the tensile strength and ductility of the composite.

2 Materials

2.1 *Matrix*

The matrix of the TRM investigated is a high strength mortar characterized by a water to binder ratio equal to 0.19. The maximum aggregate size was taken equal to 600 μm in order to allow the matrix to flow through the mesh of the fabric. The mix design is summarized in Table 1.

Table 1. Mix design

Component	Content
Cement I 52.5	600 kg/m³
Sand 0 - 600μm	957 kg/m³
Water	209 l/m³
Superplasticizer	55.8 kg/m³
Slag	500 kg/m³

Table 2. Bending and compressive strength

Specimen	f_{ctf} [MPa]	f_{cc} [MPa]	
N1	13.82	98.69	114.29
N2	17.45	103.23	107.47
N3	14.46	89.74	102.61
Average	15.24	102.67	
STD	1.94	8.26	
STD %	12.73	8.05	

The mechanical properties of the matrix were characterized according to EN 196 Standard for mortar. Table 2 shows the bending tensile strength (f_{ctf}) together with the cubic compressive strength (f_{cc}).

2.2 *Fabric*

The fabric used was chosen after performing several investigations in order to optimize performances in terms of ductility of the composite material, bond between matrix and fabric and internal filaments slip. The variables considered were the fabric geometry (warp and weft spacing and their cross section), fabric weaving and fabric coating. The final purpose was to achieve in the TRM the largest strength and ductility. The geometrical characteristics are summarized in Table 3.

Table 3. Geometrical characteristic of fabric

Warp wire spacing [mm]	5.0
Weft wire spacing [mm]	7.1
Warp [Tex]	640
Weft [Tex]	1200
Fabrication tecnique	Leno weave

Ten uniaxial tensile tests were performed to characterize the fabric. The tensile tests were carried out using an electromechanical press INSTRON 5867 with a maximum load capacity of 30 kN. The specimens were 400 mm x 70 mm and were clamped to the machine. At the upper and lower ends of the sample ten layers of adhesive paper tape were applied to prevent the slip between the clamps and the fabric. The pressure applied to the pneumatic clamps was 3.4 bar. The tests were displacement-controlled by imposing a constant stroke rate of 1.67 mm/sec. The set-up is shown in Fig. 1a, while the results in terms of load-displacement curves are shown in Fig. 2.

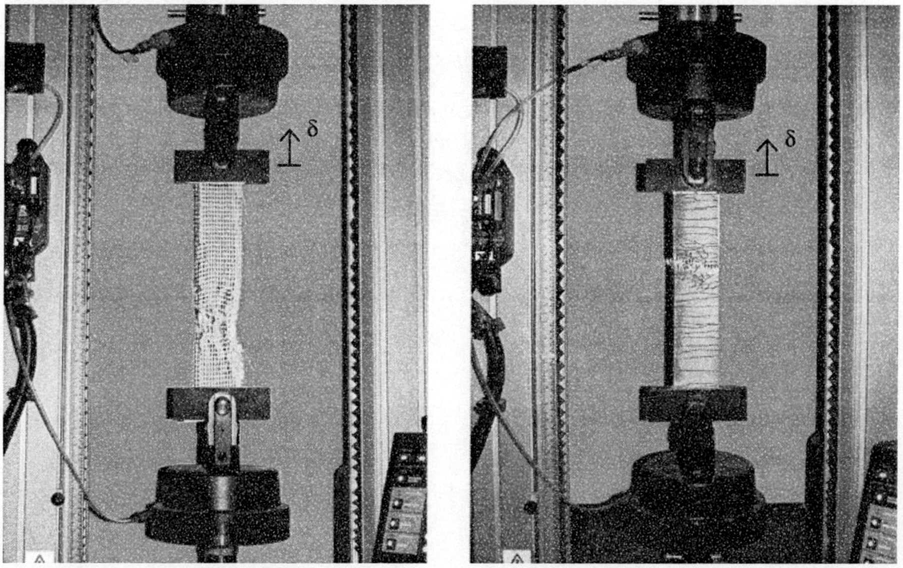

(a) (b)

Fig. 1. Set-up tensile test of fabric (a) and TRM (b)

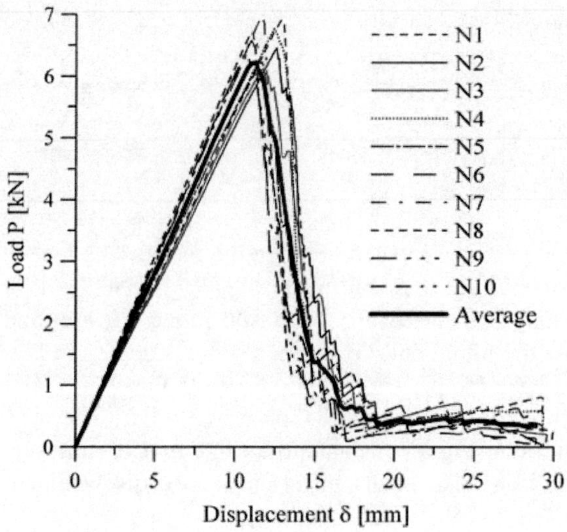

Specimen	$Load_{peak}$ [kN]
N1	6.91
N2	6.45
N3	6.12
N4	6.77
N5	6.31
N6	6.58
N7	6.58
N8	6.71
N9	6.39
N10	6.97
Average	6.58
STD	0.27
STD %	4.10

Fig. 2. Load vs displacement of fabric under uniaxial tension

2.3 Specimens Preparation

The specimens used in the experimental investigation were 400 mm long, 70 mm wide and with different thickness varying between 6 and 12 mm. The different reinforcement ratios investigated, summarized in Table 4, were achieved by introducing different layers of the same fabric. Due to the production technology, different thicknesses were obtained (Table 4).

Fig. 3. Specimen preparation

Table 4. Geometrical characteristics (average of 3 samples)

ρ_f [%]	Layer [-]	Thickness [mm]
1.63	1	6.3
2.54	3	12.1
3.30	2	6.2

The specimen preparation was made by using a proper formwork with the transparent bottom plate in order to check the penetration of the matrix into the fabric mesh. Some overlapping steel rails (3 mm thick) were designed in order to interpose different textile layers. The mortar was spread and smoothed with a roller. After that, the reinforcement was tight and fixed at the edge. Repeating the procedure it was possible to create a multilayer specimen. The layers of textile were divided by about 2 mm thickness of mortar. In all the cases the warp was parallel to the long side of the specimen.

3 TRM Tensile Behavior

3.1 Test Set Up

The same set-up used to test fabric specimens was adopted (Fig. 1b). 70 mm x 55 mm steel plates were glued on the surface of the specimen at the grips to minimize damage. To prevent torsional and bending moments owing to a misalignment of the constraints, spherical joints were placed at the ends. The pressure applied to clamp the edge zone was 4 bar and the tests stroke rate imposed was equal to 0.02 mm/sec. For each reinforcement ratio, 3 nominally identical tests were carried out.

3.2 Experimental Results

The experimental results are reported in the paragraph for all the specimens investigated by means of the nominal stress (σ) vs strain (ε) curves (Figs. 4a, 5a, 6a). The stress was obtained by dividing the load by the nominal area of the specimen cross section and the strain was evaluated by the ratio between the elongation (δ) of the specimen and its initial length. A common trend can be observed: three sections characterize the non linear response. The first cracking was reached when the tensile strength of the matrix was exceeded and then multicracking strain hardening behavior occurred. A final branch in which no further cracks develop takes place: its extension depends on the reinforcement ratio. It is worth noting that the scattering of the experimental results is really small (less than 7%).

The crack distance decreases with higher reinforcement ratios (Figs. 4b, 5b, 6b).

In order to compare the results, the average stress-strain curves are plotted in Fig. 7 showing for all the cases a good repeatability of the initial un-cracked stiffness and a cracked regime that becomes stiffer with increasing reinforcement ratios.

Table 5 shows the average working ratios of the different cases investigated. This parameter is computed as the ratio between the strength of the TRM composite and the reinforcement only. The solution with ρ_f =3.30% has a working ratio equal to 81% that shows that the fabric mechanical properties are quite well used by the TRM composite.

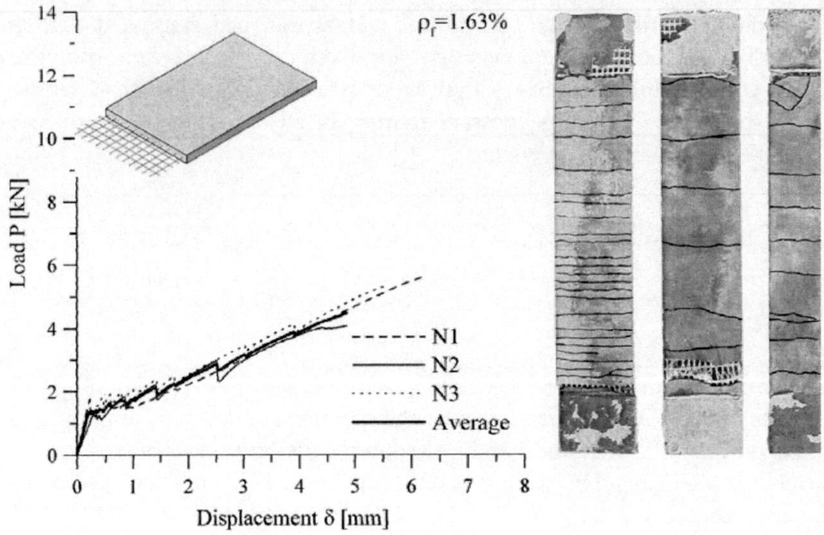

(a) (b)

Fig. 4. Load vs displacement curve in uniaxial tension (ρ_f = 1,63%) (a) and specimen crack pattern (b)

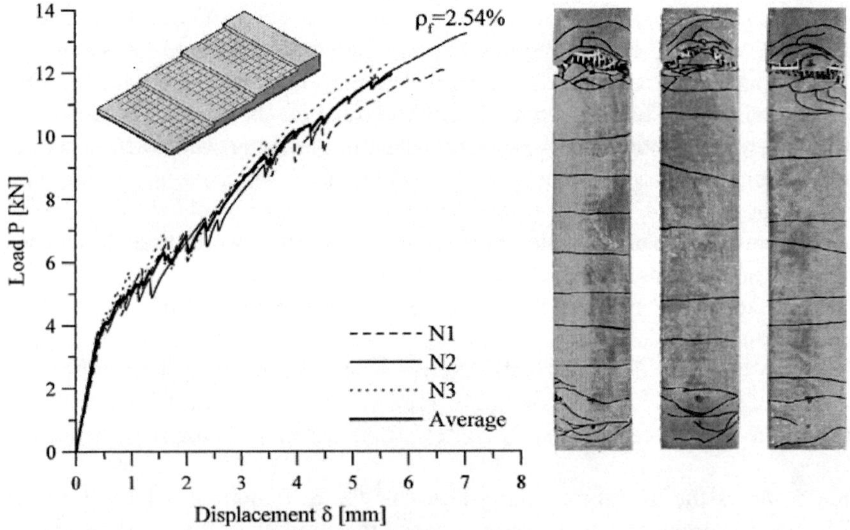

(a) (b)

Fig. 5. Load vs displacement curve in uniaxial tension (ρ_f = 2.54%) (a) and specimen crack pattern (b)

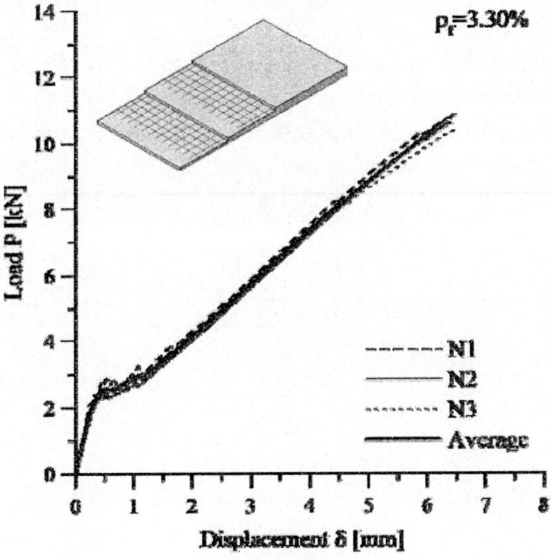

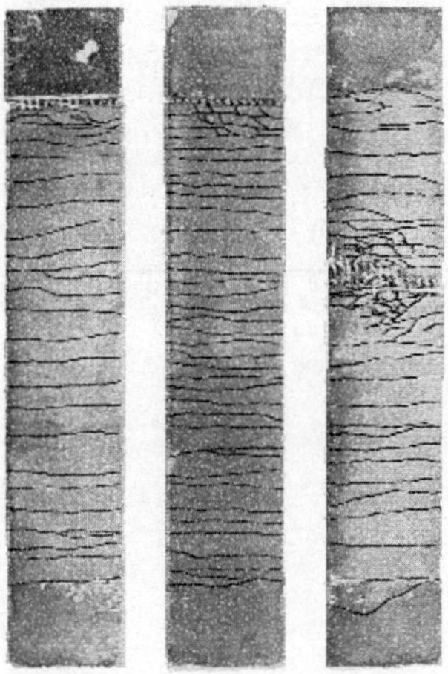

Fig. 6. Load vs displacement curve in uniaxial tension (ρ_f = 3,30%) (a) and specimen crack pattern (b)

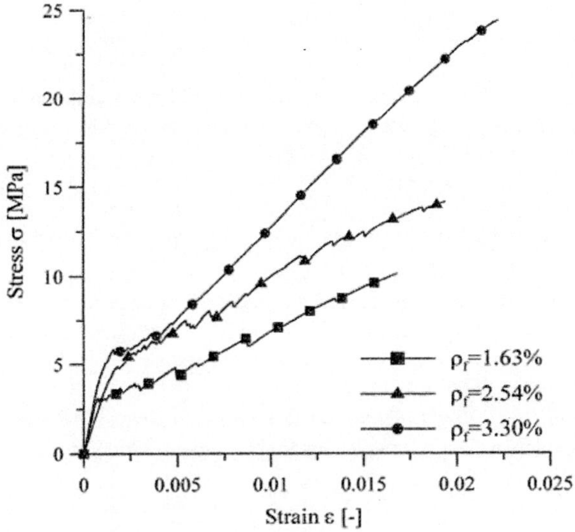

Fig. 7. Comparison between average curves

4 Conclusions

The experimental investigation indicates that the strength grows with geometrical reinforcement ratio increasing. In particular, in the case of ρ_f equal to 3.30% the strength is around 25 MPa. A large ductility was guaranteed by the small crack spacing which decreases for growing reinforcement ratios. The cracking occurs at each weft wire (approximately 7 mm). The ductility behavior was influenced by a slip of the inner filaments getting a telescopic break.

Table 5. Test result

ρ_f [%}	σ_{peak} [MPa]	Ductility $\varepsilon_u/\varepsilon_1$ [-]	Working ratio [%]
1.63	11.36	24.71	76
2.54	14.82	13.43	64
3.30	24.42	24.68	81

The solution with ρ_f equal to 3.30% shows the best performance also in terms of working ratio.

Acknowledgements. The authors would like to thank Gavazzi company for its cooperation and supplying the AR glass fabrics. The research was financially supported by European INTERREG IT/CH 2006_2013 project ACCIDENT ID7629770, Measure 2.2.

References

[1] Roye, A., Gries, T., Peled, A.: Spacer fabrics for thin walled concrete elements. In: de Plizzari, P.F. (ed.) 6th RILEM Symposium on Fiber-reinforced Concretes (FRC), BEFIB, Varenna, Italy, pp. 1505–1514 (2004)

[2] Brameshuber, W. (ed.) Textile Reinforced Concrete state of art report of RILEM technical committee 201-TRC, Aachen (2006)

[3] Hegger, J., Brameshuber, W., Will, N. (eds.): 2nd ICTRC Textile Reinforced Concrete Proceedings International RILEM Conference on Material Science, MatSci (2010)

[4] Jesse, F., Curbach, M.: Strength of continuous AR-glass fiber reinforcement of cementitious composites. In: Jesse, F., Curbach, M. (eds.) Fourth International Workshop on High Performance Fiber Reinforced Cement Composites (HPFRCC 4), pp. 337–348. RILEM Publications (2003)

[5] Häußler-Combe, U., Hartig, J.: Bond and failure mechanisms of textile reinforced concrete (TRC) under uniaxial tensile loading. Cement Concrete & Composites 29, 279–289 (2007)

Optimization of Quasi-isotropic Formulation of Fiber-Cement Laminates: Polar Method and Experimental Validation

P. Hamelin, A. Gabor, T. Q. Bach, and A. Si Larbi

Civil Engineering and Environmental Laboratory (LGCIE),
Université Claude Bernard Lyon 1, Université de Lyon, France

Abstract. The development of Textile Reinforced Concrete (TRC) requires the development of design methods. As cement based composites present anisotropic behavior at different levels, we suggest to consider the initial elastic stage of the elementary layer behavior law in tension, compression and shear, limited by a crack opening threshold. Then, we apply the polar method, allowing to optimize a multilayer stacking sequence in order to obtain an overall quasi-isotropic behavior of a composite plate.

1 Introduction

The serviceability limit state design of structures made of textile reinforced cement composites requires the knowledge of the non-cracked and cracked behavior and more particularly the admissible cracking level, in order to minimize the effect on the structure of the environmental impact. Moreover, it is essential to get behavioral information for the ultimate limit state design by the analysis of the different failure criteria.

It is particularly complex to establish feasible mechanical data set for textile reinforced cement composites given the anisotropy of the composite induced by the heterogeneity of the behavior of the two components and the orientation of the textile fabrics.

The scope of the present research work is to optimise the number and orientation of the textile/cement matrix stacking sequences in order to obtain an overall quasi-isotropic membrane and bending behavior.

G.J. Parra-Montesinos, H.W. Reinhardt, and A.E. Naaman (Eds.): HPFRCC 6, pp. 471–478.

2 Formulation and Tensile Behavior of the Basic Unidirectional Layer

The matrix of the considered composites is Inorganic Phosphate Cement (IPC) and the textile reinforcement is made of E-glass. The IPC matrix is a non-alkaline polymer which is obtained by mixing a powder (calcium inosilicate) and a liquid component (about 60% of phosphoric acid) [1]. The textile reinforcement is a unidirectional (UD) textile, which is woven of continuous unidirectional 9 μm diameter fibers. The surface weight of the fabrics is 220 g/m² and 97% of the fibers are oriented in the main direction. Table 1 presents the main characteristics of the matrix and of the reinforcement.

Table 1. Mechanical properties of the IPC matrix and E-glass

	IPC matrix	E-glass
Density(kg/m^3)	1900	2540
Elastic modulus (MPa)	10000	85000
Shear modulus (MPa)	3845	34000
Poisson's coefficient	0.30	0.22
Tensile strength (MPa)	8-10	1500
Shear strength (MPa)	4-5	-

For the manufacturing of the composites the hand lay-up procedure is used, in similar conditions as for fiber reinforced polymers.

The tensile behavior law is established using a testing procedure developed by Si Larbi et al. [2]. The obtained constitutive equation shows three stages of the behavior. The first stage (non-cracked) can be considered as elastic, with insignificant damaging. The second one, where cracks start to develop, is the transition between the first, non-cracked stage and the completely cracked, third stage. In this latter stage only the fibers support the applied load.

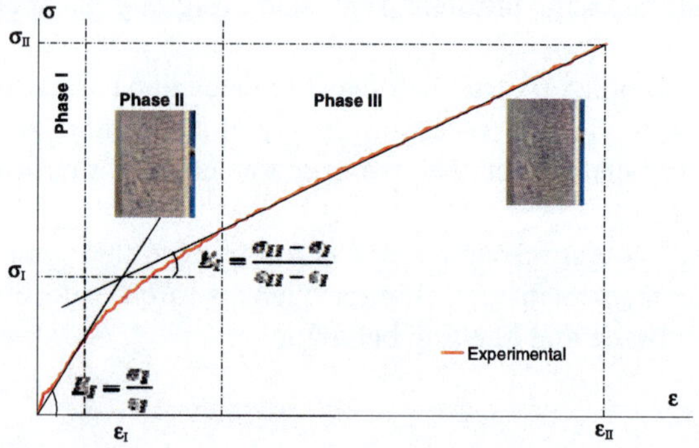

Fig. 1. Tensile behavior (stress vs. strain curve) for a UD layer

In some recent works of Hamelin [3] it was shown that the use of a specific surface treatment of the textiles contributes to the improvement of the mechanical impregnation, leading to the reduction of the second stage of the behavior. Therefore, a bilinear behavior can be taken into account for the overall tensile behavior. For further developments, we will consider the first elastic domain.

3 Preliminary Design of the Elastic Constants for an Elementary Layer Sequence

In the case of an elementary layer constituted from one or several impregnated layers of textile fabrics (for a volume fraction ratio ranging from 10% to 30%) it was demonstrated that it is possible to use with good accuracy the classical formula of composite theory. For a fiber volume fraction V of 20% we have:

$$E_L = V \times E_F + (1-V)\times E_M \quad \Rightarrow E_L = 25\,000 \text{ MPa}$$

$$\frac{1}{E_T} = \frac{V}{E_F} + \frac{(1-V)}{E_M} \quad \Rightarrow E_T = 12\,143 \text{ MPa}$$

$$\nu_{LT} = \nu_{12} = V \times \nu_F + (1-V)\times \nu_M \Rightarrow \nu_{LT} = 0{,}282$$

$$\frac{1}{G_{LT}} = \frac{V}{G_F} + \frac{(1-V)}{G_M} \Rightarrow G_{LT} = 4\,503 \text{ MPa}$$

The stiffness matrix of an elementary layer is given by:

$$[Q]_{LT} = \begin{bmatrix} \frac{E_L}{1-\nu_{LT}\nu_{TL}} & \frac{\nu_{TL}E_L}{1-\nu_{LT}\nu_{TL}} & 0 \\ \frac{\nu_{LT}E_T}{1-\nu_{LT}\nu_{TL}} & \frac{E_T}{1-\nu_{LT}\nu_{TL}} & 0 \\ 0 & 0 & G_{LT} \end{bmatrix} = \begin{bmatrix} 26004 & 3562 & 0 \\ 3562 & 12631 & 0 \\ 0 & 0 & 4503 \end{bmatrix} (MPa)$$

The comparison between the calculated elastic constants using the above relation and those obtained experimentally was done in a previous paper by Promis et al. [4]. The validation was done for several types of textile reinforcement and also in compression and plane shear.

4 Preliminary Design of the Elastic Constants of a Multilayer Composite

In this section we suggest to calculate the terms of the stiffness matrix of a multilayer composite using invariants as proposed by Tsaï and Wu [5] and after we use a polar method in order to identify the isotropy of the multilayer composite.

4.1 The Method of Invariants

The overall behavior of a multilayer composite plate is described by the well known relation:

$$\begin{bmatrix} N_i \\ M_i \end{bmatrix} = \begin{bmatrix} A & B \\ B & D \end{bmatrix} \cdot \begin{bmatrix} \varepsilon^0_i \\ \kappa_i \end{bmatrix}$$ A_{ij}- membrane stiffness, D_{ij} – flexural stiffness

In the present paper we focus only on the membrane behavior considering a symmetrical stacking sequence (B=0), but the presented methodologies are valid also in bending behavior. In the case of the membrane behavior the stiffness terms A_{ij} are given by the relations:

$$A_{11} = U_1.V_{0A} + U_2.V_{1A} + U_3.V_{3A} \qquad V_{0A} = \sum_1^n (z_k - z_{k-1})$$

$$A_{22} = U_1.V_{0A} - U_2.V_{1A} + U_3.V_{3A} \qquad V_{1A} = \sum_1^n (z_k - z_{k-1})\cos(2\alpha)_k$$

$$A_{12} = U_4.V_{0A} - U_3.V_{3A}$$

$$A_{66} = U_5.V_{0A} - U_3.V_{3A} \qquad V_{2A} = \sum_1^n (z_k - z_{k-1})\sin(2\alpha)_k$$

$$A_{16} = \frac{1}{2}U_2.V_{2A} + U_3.V_{4A} \qquad V_{3A} = \sum_1^n (z_k - z_{k-1})\cos(4\alpha)_k$$

$$A_{26} = \frac{1}{2}U_2.V_{2A} - U_3.V_{4A} \qquad V_{4A} = \sum_1^n (z_k - z_{k-1})\sin(4\alpha)_k$$

Where U_i are the invariants depending only on the mechanical characteristics of an elementary layer and V_{iA} are geometrical constants calculated in function of the orientation of the elementary layer and its position to the reference plane. The terms U_i are calculated considering the terms Q_{ij} of the stiffness matrix of the elementary layer (Table 2).

Table 2. Computation of the invariant terms U_i

Relation for terms U_i	Values of U_i for 20% of volume fraction of UD textile fabric (MPa)
$U_1 = \frac{1}{8}(3Q_{11}+3Q_{22}+2Q_{12}+4Q_{66})$	17630
$U_2 = \frac{1}{2}(Q_{11}-Q_{22})$	6687
$U_3 = \frac{1}{8}(Q_{11}+Q_{22}-2Q_{12}-4Q_{66})$	1588
$U_4 = \frac{1}{8}(Q_{11}+Q_{22}+6Q_{12}-4Q_{66})$	5250
$U_5 = \frac{1}{8}(Q_{11}+Q_{22}-2Q_{12}+4Q_{66})$	6190

Using the above terms it is possible to establish the stiffness terms A_{ij} for the considered multilayer composites. Three types of stacking sequences where realized: [0 45 90 -45], $[0\ 90]_s$, $[-60\ 0\ 60]_s$. Table 3 presents the obtained values for the membrane stiffness:

Table 3. Values of the membrane stiffness terms A_{ij} calculated using the invariants U_i

	[0 45 90 -45]	$[0\ 90]_s$	$[-60\ 0\ 60]_s$
A11	70520	77270	105779
A22	70520	77270	105779
A12	20998	14247	31497
A66	24761	18010	37141
A16	0	0	0
A26	0	0	0

4.2 The Polar Method

The polar method is based on the representation of the six independent terms terms of a fourth order tensor T_{ijkl} by six polar invariants: two scalars T_0 and T_1, two moduli R_1 and R_2, and two angular coefficients θ_0 and θ_1:

$$T_{1111} = T_0 + 2T_1 + R_0 \cos 4\theta_0 + 4R_1 \cos 2\theta_1$$
$$T_{1122} = T_0 + 2T1 - R_0 \cos 4\theta_0$$
$$T_{1112} = R_0 \sin 4\theta_0 + 2R_1 \sin 2\theta_1$$
$$T_{2222} = T_0 + 2T_1 + R_0 \cos 4\theta_0 - 4R_1 \cos 2\theta_1$$
$$T_{1222} = -R_0 \sin 4\theta_0 + 2R_1 \sin 2\theta_1$$
$$T_{1212} = T_0 - R_0 \cos 4\theta_0$$

For the polar method, it is possible to introduce: two scalars T_0 and T_1 and two complex variables $R_0.e^{4i\theta_0}$ and $R_1.e^{2i\theta_1}$. Considering inverse relation the polar components can be written as:

$$8\ T_0 = T_{1111} + T_{2222} - 2T_{1122} + 4\ T_{1212} \qquad 8\ T_1 = T_{1111} + T_{2222} + 2T_{1122}$$
$$8\ R_0.e^{4i\theta_0} = (T_{1111} + T_{2222} - 2T_{1122} - 4\ T_{1212}) + 4i\ (T_{1112} - T_{1222})$$
$$8\ R_1.e^{2i\theta_1} = (T_{1111} - T_{2222}) + 2i\ (T_{1112} + T_{1222})$$
$$4\ \theta_0 = \arctan\frac{4.(T_{1112} - T_{1222})}{T_{1111} + T_{2222} - 2T_{1122} - 4T_{1212}} \qquad 2\ \theta_1 = \arctan\frac{2.(T_{1112} + T_{1222})}{T_{1111} - T_{2222}}$$
$$8\ R_0 = \frac{4.(T_{1112} - T_{1222})}{\sin\theta_0} = \frac{T_{1111} + T_{2222} - 2T_{1122} - 4T_{1212}}{\cos\theta_0}$$
$$= \sqrt{(T_{1111} + T_{2222} - 2T_{1122} - 4T_{1212})^2 + 16.(T_{1112} - T_{1222})^2}$$
$$8\ R_1 = \frac{2.(T_{1112} + T_{1222})}{\sin\theta_1} = \frac{T_{1111} - T_{2222}}{\cos\theta_1} = \sqrt{(T_{1111} - T_{2222})^2 + 4.(T_{1112} + T_{1222})^2}$$

The diagram of a tensor in function of the six polar components is given in Fig. 2.

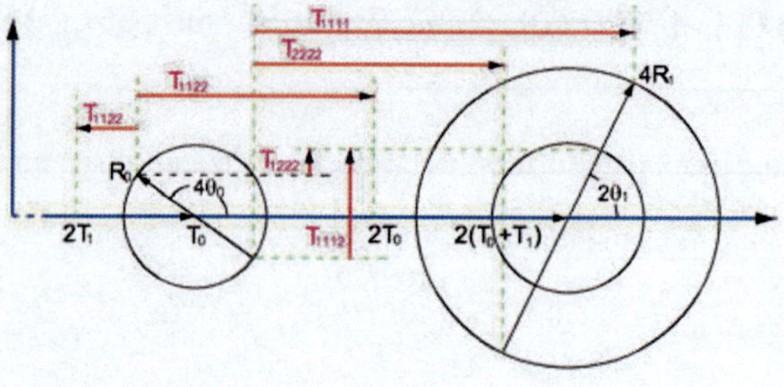

Fig. 2. Polar representation of a fourth order tensor

The application of this method in the case of a UD composite proposed by G. Verchery et al. [6], yields the values presented in Table 4. The graphical representation of the stiffness matrix is given in Fig. 3.

Table 4. Terms of the polar components in the case of an elementary layer

Stiffness terms Q_{ij}		Polar components
Q_{11}	23713	T_0=5545
Q_{12}	3230	T_1=5117
Q_{22}	10764	θ_0=0
Q_{16}=Q_{26}=Q_{61}=Q_{62}	0	θ_1=0
Q_{66}	4085	R_0=1460
		R_1=1619

Applying the same method to multilayer composites having variable stacking sequences it is possible to obtain the terms A_{ij} of the membrane stiffness:

$$T_0 = 1/8\,(A_{11} + A_{22} - 2\,A_{12} + 4\,A_{66}) \qquad T_1 = 1/8\,(A_{11} + A_{22} + 2\,A_{12})$$

$$\theta_0 = 1/4 \cdot \arctan \frac{4.(A_{16} - A_{26})}{A_{11} + A_{22} - 2A_{12} - 4A_{66}} \qquad \theta_1 = 1/2 \cdot \arctan \frac{2.(A_{16} + A_{26})}{A_{11} - A_{22}}$$

$$R_0 = \frac{A_{11} + A_{22} - 2A_{12} - 4A_{66}}{8\cos\theta_0} \qquad R_1 = \frac{A_{11} - A_{22}}{8\cos\theta_1}$$

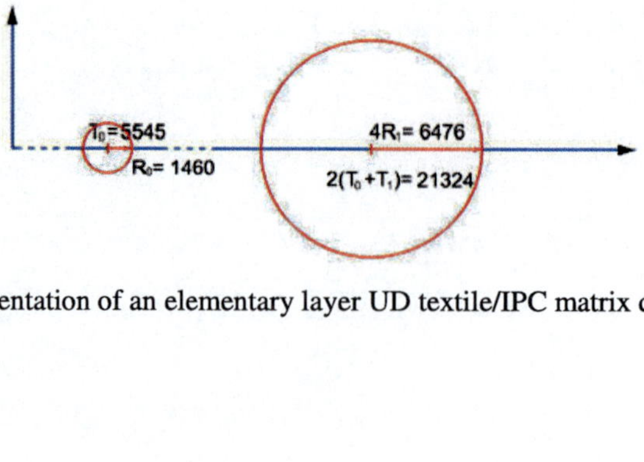

Fig. 3 Polar representation of an elementary layer UD textile/IPC matrix composite

As it can be seen in Table 5, for different stacking sequences it is possible to calculate the different polar invariants in function of the stiffness terms of the multilayer composite.

Moreover, we compared the predicted values by calculation with experimental measurements of the global tensile elastic modulus (see Table 6) [3]. The analysis of the results of Tables 5 and 6 confirms that the membrane behavior is quasi-isotropic.

Table 5. Stiffness and polar invariant terms and geometrical diagram for the studied multilayer composites

	[0 45 90 -45]	[0 90]$_s$	[-60 0 60]$_s$
A11	70520	77270	105779
A22	70520	77270	105779
A12	20998	14247	31497
A66	24761	18010	37141
A16	0	0	0
A26	0	0	0
To	24761	24761	37141
T1	22879	22879	34319
R0	0	6751	0
R1	0	0	0
a0	-	0	-
a1	-	-	-

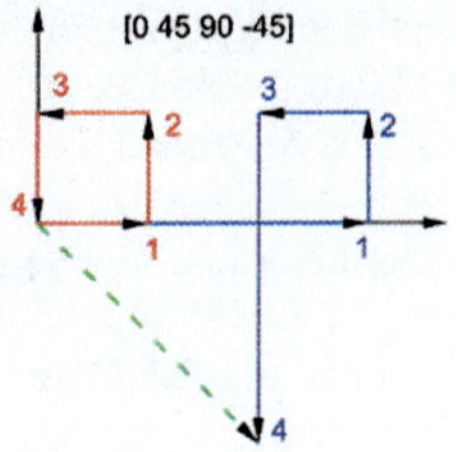

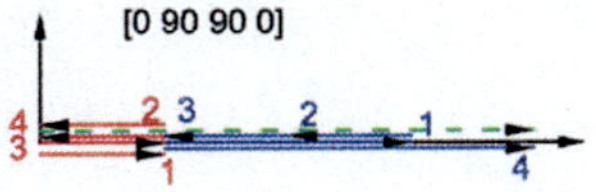

Table 6. Comparison of experimental and theoretical results

Stacking sequence	Experimental results (MPa)	Preliminary design (MPa)	Difference (%)
[0 90]$_s$	11370	13196	16%
[-60 0 60]$_s$	10799	12084	12%
[-60 0 60 60 0 -60]$_s$	12150	13168	14%

5 Synthesis-Conclusions

The polar method is particularly useful for the designers beacuse, by simple geometrical representation, it is possible to visualize if the stacking sequence of a multilayer composite is quasi-isotropic or not. This method is available for membrane and flexural behavior. For example the combination [-60 0 60 60 0 -60]$_s$ corresponds to this purpose.

References

[1] Remy, O., Wastiels, J.: High performance textile reinforced cements: tensile hardening behavior and modeling. In: Marques, A., Juvandes, L. (eds.) Proc. of the Inetrnational Conference Challenge for Civil Construction (CCC), Portugal, pp. 116–117 (2008)

[2] Contamine, R., Si Larbi, A., Hamelin, P.: Tensile identification of textile reinforced concrete behavior. In: Contamine, R., Si Larbi, A., Hamelin, P. (eds.) International Conference on Material Science, Aachen, Germany, RILEM no. 64 (2010)

[3] Promis, G., Gabor, A., Hamelin, P.: 2D Braiding technology for composite beam confinement. In: International Conference on Textile Composites, TEXCOMP 2010, Lille, France, October 26-28 (2010)

[4] Promis, G., Gabor, A., Hamelin, P.: Assessment and modeling of the behavior of IPC composites in tension, compression and shear. In: 2nd ICTRC, International RILEM Conference on Material Science, pp. 407–417 (2010)

[5] Tsai, S.W., Wu, E.: A general theory of strength of anisotropic materials 5, 58–80 (1971)

[6] Verchery, G.: Les invariants des tenseurs d'ordre quatre du type de l'élasticité. Mechanical behavior of anisotropic solids. In: Verchery, G. (ed.) Proc. of the Euromech Colloqium, pp. 93–94. published by J.P. Boehler-CNRS editions (1982)

Bond Behavior of Textile Reinforcements - Development of a Pull-Out Test and Modeling of the Respective Bond versus Slip Relation

E. Lorenz and R. Ortlepp

Institut für Massivbau, TU Dresden, Germany

Abstract. For strengthening existing concrete structures with textile reinforced concrete (TRC), anchoring of the TRC layer to the old concrete is crucial. Owing to their very high tensile strength, especially the anchorage lengths of textiles consisting of carbon fibers are to be determined from the pull-out of the filament yarns of the fine-grained concrete. For this reason, further research was necessary to clarify the failure criteria caused by yarn pull-out, which necessitated the development of a pull-out test for textile reinforcements. With the help of the force-crack opening relationships determined within these experimental investigations, the respective bond stress versus the slip relationship (BSR) of the textile reinforcement could be modeled analytically.

1 Introduction

The safe introduction and transfer of the forces is required for the functionality of composite constructions. In respect to a bond failure within textile reinforced concrete (TRC) layers, the delamination in the surface of textile reinforcements and the pull-out of the yarns from the fine-grained concrete are to be differentiated according to Ortlepp [1]. For this, the tendency toward the bond failure of coated textile fabrics is essentially determined by the bond behavior between the filament yarn and the fine-grained concrete. Several test set-ups for the examination of the bond characteristics of endless filament yarns embedded in a concrete matrix are already known from the literature. Generally, one distinguishes between one-sided and double-sided pull-out tests. In the one-sided pull-out tests as described e.g. by Banholzer [2], the to be tested yarns are embedded one-sidedly in the pull-out area of the concrete matrix. At the opposite end, the filament yarn is poured into an epoxy resin block. The tests were carried out only on unmanufactured single yarns. Based on the great influence of textile processing on the bond characteristics, the results obtained from single yarns are not necessarily to be transferred to textile fabrics. In double-sided pullout tests as e.g. mentioned by Butler [3] and Krüger [4] a direct force introduction into the filament yarn takes place via the

G.J. Parra-Montesinos, H.W. Reinhardt, and A.E. Naaman (Eds.): HPFRCC 6, pp. 479–486.

bond with the matrix. In the test set-up described by Butler [3], the filament yarn is anchored symmetrically into the fine-grained concrete double-sidedly with anchoring lengths of 100 mm, respectively. Whereas, according to Ortlepp [1], for uncoated textiles as a rule very long end anchoring lengths of more than 100 mm are necessary, double-sided symmetrical anchoring with lengths of 100 mm are problematic, especially for coated filament yarns necessitating anchoring lengths of less than 100 mm. In this case, there is no pull-out failure but a yarn crack when the yarn tensile strength is reached. In addition, the one-sided yarn pull-out can only be described very insufficiently by assuming half the crack width w as a pull-out displacement of the yarn in the crack. In contrast, the double-sided test specimen described by Krüger [4] shows unsymmetrical anchoring lengths. Therefore, on one side of the specimen, a yarn pull-out is secured by choosing a short embedding length of 20 mm. On the opposite side, the anchoring of the yarn into the fine-grained concrete matrix can be guaranteed by an existing embedding length of 120 mm. Accordingly, the test set-up developed by Krüger [4] proves to be best suited to a component fair and as exact as possible determination of the bonding characteristics of textile reinforcement structures in the fine-grained matrix. Based on this, it was modified and developed further for carrying out the textile pull-out tests with short bond lengths, especially for the examination of coated carbon fabrics.

2 Test Set-Up

Based on the test specimen designed by Krüger [4], one-layer reinforced, disk-shaped test specimens measuring 278 mm × 60 mm × 8 mm having unsymmetrical anchoring lengths are used for the textile pull-out tests (Fig. 1a).

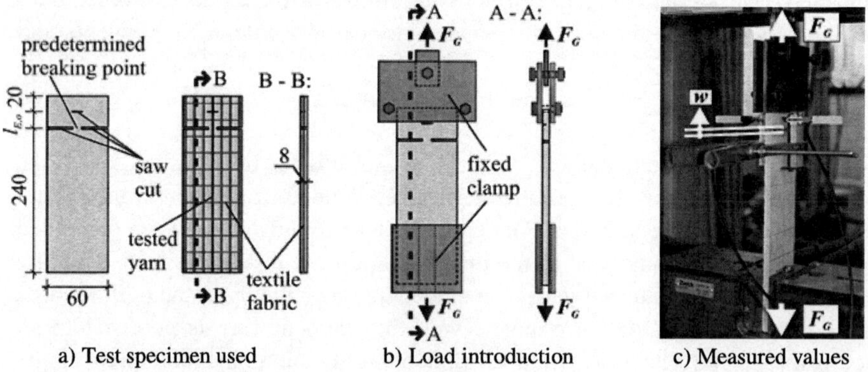

Fig. 1. Test set-up

Per specimen, exactly one yarn will be tested. The upper anchoring length $l_{E,o}$ necessary for the examination of the bond characteristics is chosen with the single and more numerous distance of the cross threads. The determination of the anchoring length takes place in the upper part of the test specimen by way of a saw-cut below the clamp with a separation of the yarn to be examined. The upper anchoring area is restricted by a specific positioning of a predetermined breaking point with the help of a double-sided saw-cut. Since the examined undisturbed section of the filament yarn has all the influences of textiles such as yarn waviness and cross-thread influence in a smeared way, a representative examination of the bond behavior of the textile is possible. The anchoring length of 240 mm chosen in the lower part of the specimen is, as a rule, sufficient for the securing of the anchoring. In the upper area, the specimen is clamped to a length of 20 mm above the to be examined anchoring length $l_{E,o}$ into a clamp mechanism (Fig. 1b). A constraint-free and hinged connection of the clamp mechanism to the test machine is secured via a long hole in the steel sheet. In the lower area of the test set-up, directly clamping into the test machine takes place with a clamping length of 70 mm. Thus, a sufficient anchoring of the specimen into the test machine is secured. By using double-sided cutouts in the lower load transfer plates, a direct lateral pressure onto the examined filament yarn is avoided.

3 Experimental Research

For the matrix, the high strength concrete mixture with a maximum grain size of 1 mm as described by Curbach & Jesse [5] was used. The, here, exemplarily examined textile reinforcement used was a stitch-bonded biaxial fabric (Fig. 2).

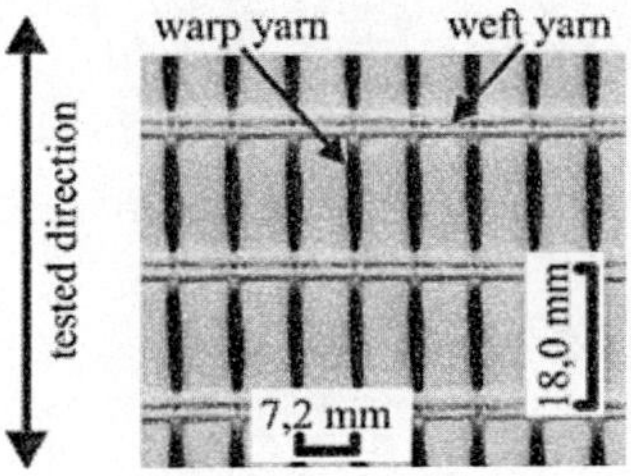

Fig. 2. Textile fabric used

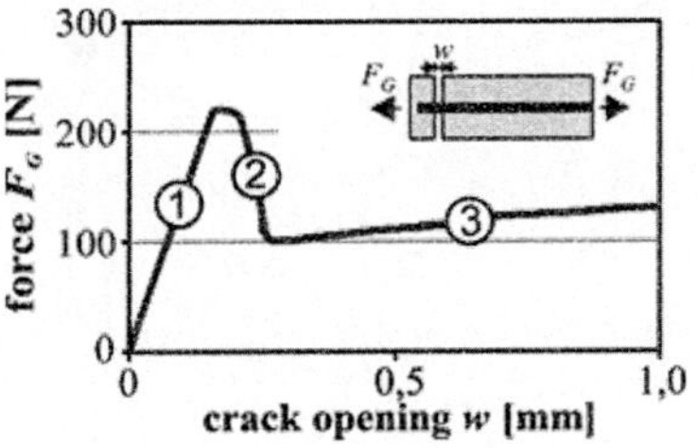

Fig. 3. Force-crack opening-relationship

The textile consisted of carbon filament yarns with a fineness of 800 tex in warp (= research) direction. In weft direction, alkali-resistant glass fibers with a fineness of 640 tex were processed. The textiles were modified with a coating on a polymer basis. The deflection-controlled pull-out tests were carried out with a loading rate of 0.15 mm/sec (1 mm/min). Aside of the machine force F_G ,the crack opening w in the area of the predetermined breaking point was measured by way of external by clip gauges DD1 (Fig. 1c). In the textile pull-out tests, the force-crack opening-relationship shown in Fig. 3 could be determined. It was defined as the average value from eight pull-out tests. The three different bonding areas were clearly visible. Whereas in the increasing area 1 the adhesive bond was activated, the decreasing section 2 experienced a destruction of the adhesive bond with a debonding of the yarn from the fine-grained concrete matrix. In the consecutive 3rd area, the remaining pull-out force was only determined by friction. Processing related influences at growing crack opening led to additional yarn pull-out resistance. In contrast, a steady, uniform decrease in the friction area can be assumed for unmanufactured textile yarns.

4 Simplified Analytical Determination of the BSR

4.1 Description of the Solution Approach

The determination of the bond stress versus the slip relationship (BSR) is based on the analytical modeling of the force-crack opening-relationships determined in the pull-out tests based upon the multilinear, segmentally closed solutions of the bond differential equation described by Richter [6]. With the help of a simple parameter variation of the supportive points of the BSR, the experimentally determined force-crack opening-relationship is experimentally brought closer.

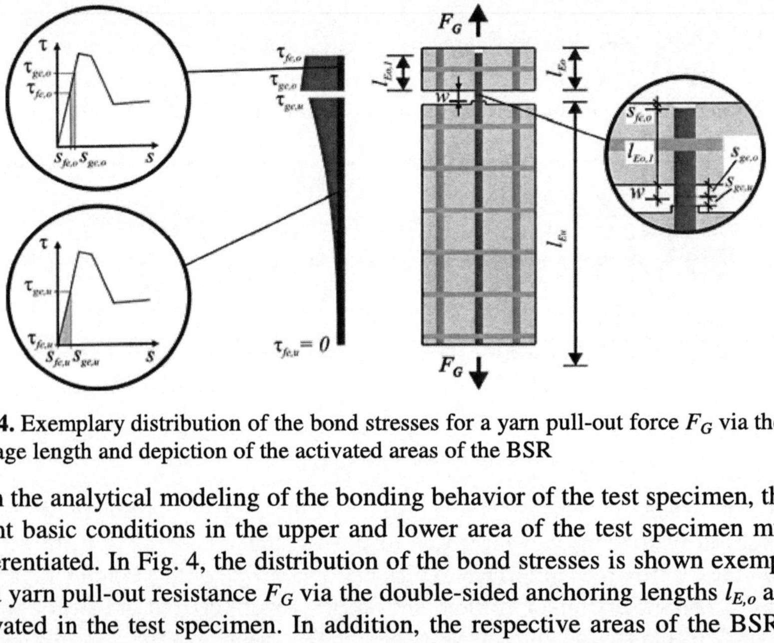

Fig. 4. Exemplary distribution of the bond stresses for a yarn pull-out force F_G via the anchorage length and depiction of the activated areas of the BSR

In the analytical modeling of the bonding behavior of the test specimen, the different basic conditions in the upper and lower area of the test specimen must be differentiated. In Fig. 4, the distribution of the bond stresses is shown exemplarily for a yarn pull-out resistance F_G via the double-sided anchoring lengths $l_{E,o}$ and $l_{E,u}$ activated in the test specimen. In addition, the respective areas of the BSR used are depicted by means of the relative slip belonging to it in the predetermined breaking point ($s_{ge,o}$ and $s_{ge,u}$) as well as the unloaded yarn ends that go with it ($s_{fe,o}$ and $s_{fe,u}$). In the upper area of the specimen, due to the short anchoring length $l_{E,o}$ a pull-out of the filament yarn from the fine-grained concrete occurs. A calculation of the pull-out resistance F_G in dependence on the slip of the free yarn end $s_{fe,o}$ is possible. In this, the distribution of the bond stresses via the anchoring length $l_{Eo,1}$

results in the respective relative slips $s_{fe,o}$ and $s_{ge,o}$ during yarn pull-out. By means of the integral of the bond stress via the anchorage length, the yarn pull-out force F_G can then be calculated for every phase of the yarn pull-out. Due to the two functions $F_G(s_{fe,o})$ and $s_{ge,o}(s_{fe,o})$ a respective area-wise closed solution with a formulation of the function $s_{ge,o}(F_G)$ becomes possible. Due to the short anchoring lengths chosen as well as to the high strain-rigidities of coated carbon-yarns, a simultaneous activation of several sections of the BSR can, as a rule, be avoided. Only in the transition areas between two linear sections does it come to an activation of two sections of the BSR in small areas. The calculation shows here a local discontinuity. However, due to their insignificance the resultant discontinuity can usually be smeared in simplified fashion. In contrast to the yarn pull-out with a partial activation of the BSR in the upper section of the test specimen, a yarn pull-out can be avoided in the lower section of the specimen due to the clearly greater anchoring length $l_{E,u}$ chosen. The displacement of the free yarn end $s_{fe,u}$ results in zero. For anchoring the yarn pull-out forces into the fine-grained concrete determined by the pull-out resistance of the short upper specimen part, only the activation of the adhesive stress is necessary. Therefore, elastic behavior of the yarn can be assumed in the lower part of the specimen used. The corresponding displacement in the crack $s_{ge,u}$ can be calculated in dependence on the yarn pull-out force F_G. For the consideration of the yarn strain in the area of the crack, this load-dependent additional strain $\Delta l(F_G)$ is simplifyingly determined in the crack by means of the strain rigidity of the yarn and the sum of the relative slips $s_{ge,o}$ and $s_{ge,u}$. After the determination of the single values of the relative slips of the pulled ends of the upper and lower part of the specimen as well as the additional strain of the yarn in the crack, the corresponding crack opening $w(F_G)$ can be calculated.

4.2 Assumptions and Calculative Basis

The determination of the statically effective reinforcement area A_G of the yarn is carried out on the basis of the density of the yarn material and the yarn fineness

$$A_G = \frac{fineness}{\rho} = \frac{803tex}{1.79\frac{g}{cm^3}} = 0.45mm^2\ . \qquad (1)$$

Due to the almost straight course of the stress-strain curves of coated filament yarns determined in yarn tensile tests, a linear-elastic material behavior of the yarns with an elastic modulus of 204 400 N/mm² is assumed.

In general, the fine-grained concrete matrix (E_M, A_M) of the TRC components is clearly more rigid than the yarn (E_G, A_G). Therefore, by assuming linearly elastic material behavior and neglecting the matrix, the constant substitute strain rigidity $(E \times A)^*$ becomes simplifyingly assumed according to equation (2) [6].

$$(E\times A)^* = E_G \times A_G \qquad (2)$$

The terms used for the individual sections of the BSR consisting of the four linear sections with a linearly increasing and decreasing bond cited in the equations that follow for the modeling of the pull-out behavior are explained in Fig. 5a.

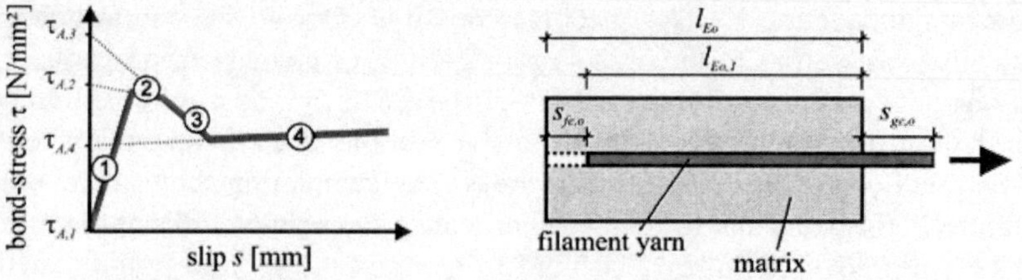

a) Terms of BSR b) Relative displacement of the yarn

Fig. 5. Explanation of the terminology

The increase of the BSR within the respective tested segment (here 1 to 4) is here called Z_x. For this, the increasing and decreasing areas must be differentiated. According to equation (3) [6] for the determination of the constant k_x within the respective linear area then applies:

$$k_x = \sqrt{\frac{u_G}{(E \times A)^*} \times |Z_x|}\,. \tag{3}$$

4.3 Analytical Modeling of the Test Specimen

In the following, the solutions for the areas of the kind of BSR chosen in the framework of this study are described (Fig. 5a).

For the analytical consideration of the yarn pull-out in the upper specimen part, the reduction of the existing anchoring length l_{Eo} via displacement of the free yarn end s_{fe} must be considered during the yarn pull-out according to equation (4) (Fig. 5b). The reduced anchoring length is then called $l_{Eo,1}$.

$$l_{Eo,1} = l_{Eo} - s_{fe,o} \tag{4}$$

According to equation (5), in the increasing parts 1 and 4 of the BSR, the displacement $s_{ge,o}$ at the pulled yarn end in the upper specimen part with a known anchorage length $l_{Eo,1}$ for an assumed displacement $s_{fe,o}$ results in

$$s_{ge,o}(s_{fe,o}) = \frac{\tau_{A,x} + Z_x \times s_{fe,o}}{Z_x} \times cosh\left(k_x \times l_{Eo,1}\right) - \frac{\tau_{A,x}}{Z_x}\,. \tag{5}$$

The corresponding force F_G in the yarn can then be calculated for the assumed displacement $s_{fe,o}$ at the free yarn end according to equation (6).

$$F_G(s_{fe,o}) = \frac{u_G}{k_x} \times \left(\tau_{A,x} + Z_x \times s_{fe,o}\right) \times sinh\left(k_x \times l_{Eo,1}\right) \tag{6}$$

In the decreasing parts 2 and 3 of the BSR in the upper specimen part, the displacement $s_{ge,o}$ at the pulled yarn end for an assumed displacement $s_{fe,o}$ can be calculated according to equation (7).

$$s_{ge,o}(s_{fe,o}) = \frac{\tau_{A,x} + Z_x \times s_{fe,o}}{Z_x} \times cos\left(k_x \times l_{Eo,1}\right) - \frac{\tau_{A,x}}{Z_x} \quad (7)$$

The respective force F_G in the yarn can be calculated according to equation (8).

$$F_G(s_{fe,o}) = \frac{u_G}{k_x} \times \left(\tau_{A,x} + Z_x \times s_{fe,o}\right) \times sin\left(k_x \times l_{Eo,1}\right) \quad (8)$$

By using the pull-out load F_G determined by applying the analytical considerations in the upper pull-out area, the corresponding final displacement $s_{ge,u}$ can be determined for the lower specimen part using equation (9) [6].

$$s_{ge,u}(F_G) = \frac{1}{k_1} \times \sqrt{\frac{k_1^2}{Z_1^2} \times \tau_{A,1}^2 + \left(\frac{F_G}{(EA)^*}\right)^2} - \frac{\tau_{A,1}}{Z_1} \quad (9)$$

According to equation (10), the strain portion □$l(F_G)$ of the yarn in the crack is simplified calculated by means of the sum of the relative displacements $s_{ge,o}$ and $s_{ge,u}$ in the crack.

$$\Delta l\left(F_G\right) = \frac{F_G}{A_G} \times \frac{1}{E_G} \times \left(s_{ge,o}(F_G) + s_{ge,u}(F_G)\right) \quad (10)$$

Hence the crack opening w in the area of the predetermined breaking point as the sum of the relative displacements of the strained yarn ends ($s_{ge,o}$ and $s_{ge,u}$) and the yarn strain in the crack results from the pull-out load F_G (equation (11).

$$w(F_G) = s_{ge,1}(F_G) + s_{ge,2}(F_G) + \Delta l\left(F_G\right) \quad (11)$$

4.4 Results and Discussion

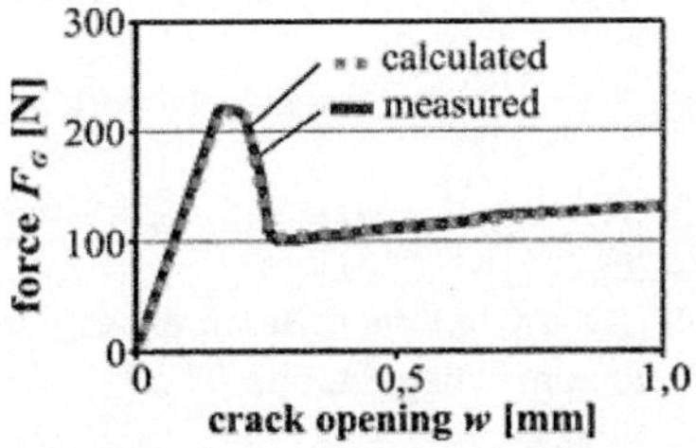

Fig. 6. Measured and analytically approached force-crack opening-relationship

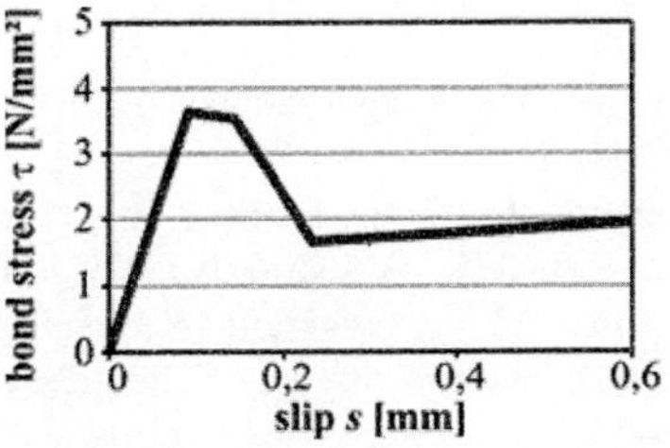

Fig. 7. Determined BSR

In Fig. 6, the measured and analytically approached force-crack opening-relationship is shown for the textile fabric examined. The analytically determined BSR is depicted in Fig. 7. With the help of the introduced calculation model, a good approach to the test results could be reached. The examination of the calculated BSR for the calculation of the end anchoring lengths of textile fabrics in reinforcement layers made of TRC as e.g. described in Lorenz & Ortlepp [7] confirms the suitability of the presented calculation approach.

5 Summary and Outlook

In connection with the calculation method shown, the presented textile pull-out tests make it possible to reliably examine the bonding behavior of coated textile carbon fabrics and single yarns. Besides the simple specimen production and test execution, advantages are gained for the determination of the BSR on account of the chosen test specimen geometry. Hence, a specific one-sided pull-out failure can be safe-guarded by means of the kind of test specimen chosen. The presented calculation model makes a simple determination of the BSR possible by means of the experimentally measured force-crack opening relationship. By this, an additional calculative use for the experimentally determined bond properties, for instance, for the calculation of the end anchoring and lap lengths, can be obtained.

Acknowledgements. The authors would like to thank the Deutsche Forschungsgemeinschaft (DFG) for founding the Collaborative Research Centre 528 at TU Dresden.

References

[1] Ortlepp, R.: Untersuchungen zur Verbundverankerung textilbewehrter Feinbetonverstärkungsschichten für Betonbauteile. TU Dresden, Dissertation (2007)
[2] Banholzer, B.: Bond of a strand in a cementitious matrix. Materials and Structures 39(10), 1015–1028 (2006)
[3] Butler, M.: Zur Dauerhaftigkeit von Verbundwerkstoffen aus zementgebundenen Matrices und alkaliresistenten Glasfaser-Multifilamentgarnen. TU Dresden, Dissertation (2009)
[4] Krüger, M.: Vorgespannter textilbewehrter Beton Universität Stuttgart, Dissertation (2004)
[5] Curbach, M., Jesse, F.: Beton im Textilbeton - Entwicklung. In: Lohaus, L. (Ed.) Ingenieurbaustoffe - konstruktive Wege in die Zukunft, pp. 29–44 (2001)
[6] Richter, M.: Entwicklung mechanischer Modelle zur analytischen Beschreibung der Materialeigenschaften von textilbewehrtem Feinbeton. Dissertation, TU Dresden (2004)
[7] Lorenz, E., Ortlepp, R.: Anchoring Failure Mechanisms of Textile Reinforced Concrete Strengthening of RC Structures. In: Proceedings of the ACI 2010 Fall Convention, Pittsburgh, October 24-28 (2010)

Effect of Short Fibers on the Behavior of Textile Reinforced Concrete under Tensile Loading

R. Barhum and V. Mechtcherine

Institute for Construction Materials, TU Dresden, Germany

Abstract. This paper addresses the influence of the addition of short fibers made of alkali-resistant glass on the fracture behavior of textile reinforced concrete (TRC). A series of uniaxial, deformation-controlled tension tests was performed to observe the strength, deformation and fracture behavior of thin, narrow plates made of TRC both with and without the addition of short fibers. Furthermore, multifilament-yarn pullout and single-fiber pullout tests were carried out to gain a better understanding of crack bridging behavior which suppresses crack growth and widening. Various effects of the addition of short fiber on the stress-strain relationship and cracking behavior of TRC were observed and discussed with reference to microscopic investigation of fractured surfaces.

1 Introduction

Textile reinforced concrete (TRC) is a composite material consisting of a finely grained cement-based matrix and high-performance, continuous multifilament yarns made of alkali-resistant (AR) glass, carbon, or polymer. The major advantages of TRC are its high tensile strength and pseudo-ductile behavior, characterised by large deformations due to its tolerance of multiple cracking [1]. In recent years it could be shown that the addition of short fibers can have a positive effect on various properties of textile reinforced concrete [2, 3]. However, the mechanisms of the joint action between short fiber and textile reinforcement are still not fully understood. In this treatise experimental investigations are presented which give a deeper insight in the specific behavior of individual components of the complex composite material TRC with short fiber. In addition to various mechanical experiments on TRC and its components, visual inspection of the specimens' surfaces, and microscopic investigation of the fractured surfaces obtained from various experiments were performed and evaluated.

G.J. Parra-Montesinos, H.W. Reinhardt, and A.E. Naaman (Eds.): HPFRCC 6, pp. 487–494.

2 Materials

In previous investigations it was found that matrices with slag furnace cement (CEM III) and the addition of pozzolans show favourable properties regarding the durability of glass fiber as well as of the bond between fiber and cementitious matrix [4]. Because of the small diameter of both the continuous filaments and the short fibers, the maximum aggregate diameter had to be small as well (< 1 mm). One such fine-grained, cement-based concrete was chosen for this investigation. The binder was composed of cement, fly ash and micro-silica. Two designated mixtures M030 and M045, having water-to-binder ratios of 0.30 and 0.45, respectively, were used in this investigation. Finally, a super-plasticizer with a basis of naphthalene-sulfonate was added in order to achieve sufficient flowability. Table 1 lists the matrices' compositions. The average slump flow value obtained with a small cone was 200 mm.

Table 1. Matrix composition [kg/m³]

Matrix	w/b ratio	Cement CEM III B 32.5	Fly ash	Micro-silica suspension*	Fine sand	Water	Super-plasticizer
M030	0.30	632	265	101	947	234	11
M045	0.45	554	233	89	832	330	2

* solid:water = 1:1

One type of coated, biaxial fabric made of alkali-resistant glass was used as textile reinforcement for TRC specimens as well as for the yarn pullout tests. The weft and warp threads had a fineness of 2 x 640 tex[1]; the spacing between yarns was 7.2 mm. 2 layers of textile were used as reinforcement.

Dispersed and integral short AR glass fibers (SGF) both with a length of 6 mm were chosen and combined with textile layers for this investigation. Dispersed short fibers act as such in water, and they are distributed and spread within the mixture as thousands of single mono-filaments. The integral short fibers remain stuck together and act as a short piece of "roving" in the mixture. Short fibers of AR-glass (dispersed and integral) have high strength and relatively high stiffness; however, they are brittle. Fiber content of 1.0% by volume was chosen for this investigation. Table 2 lists the tensile strength and Young's modulus for the reinforcement used.

Table 2. Selected properties of multifilament-yarn of the textile fabric and the short fibers used

	Yarn of the textile fabric	Dispersed short fibers	Integral short fibers
Tensile strength [MPa]	1,008.91	1,700	1,700
Young's modulus [MPa]	64,760	72,000	72,000

[1] Mass is g of 1 km yarn; (tex = g/km).

3 Preparation of Specimens and Test Setup

Rectangular plates (500 mm x 100 mm x 12 mm) reinforced by 2 layers of textile were produced for uniaxial tension tests using the lamination technique explained in [3]. Additionally, specimens of the same dimensions made of plain mortar (matrix) as well of mortar reinforced with short fiber only were tested. The plates were demoulded at a concrete age of 2 days and then stored in water until reaching the age of 7 days. Subsequently, the plates were stored in a climate-controlled room at 20°C/65% RH until the age of 28 days. The tests were performed with a controlled deformation rate of 0.5 mm/min. The force was introduced to the specimens via non-rotatable steel plates glued to the TRC plates, cf. Fig 1a. Deformation was measured by 2 LVDTs.

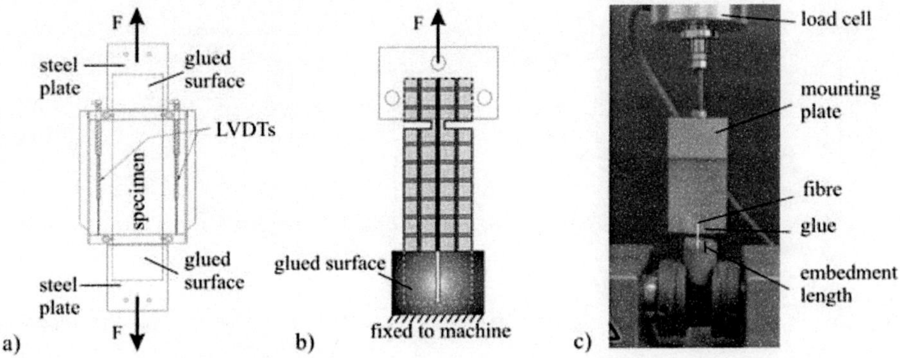

Fig. 1. Schematic view of tests setup, a) uniaxial tensile, b) yarn pullout, c) single-fiber pullout

For multifilament yarn pullout tests, rectangular specimens (200 mm x 70 mm x 12 mm) were cut from larger plates produced and cured in the same manner as those for uniaxial tensile tests. One layer of textile was used as reinforcement. The anchorage length was determined by specific arrangement of a "should" crack-position using a saw cut on both sides, cf. Fig 1b. In the notched cross-section only one multifilament yarn will be active and connect the two parts of the plates to each other. More details may be found in [5].

A series of single-fiber pullout tests was performed on specimens prepared according to [6, 7]. In this setup, the short fiber was inserted into a hollow medical cannula. The position of the fiber was fixed in wax when the embedment length of 3 mm, half the length of the short glass fiber, was approximately attained. A frame with a number of prepared cannulas was fixed to the mould, and matrix was added. The specimens were demoulded at the age of 2 days and stored in a climate-controlled room at 20°C and 65% RH until the age of 28 days. Finally, the specimen was fixed to the clamps, and the fiber was glued to the upper mounting plate of a testing machine with 10 N cell using commercial fast glue as it is shown in Fig. 1c.

4 Experimental Results

The stress-strain curves obtained from the uniaxial tensile tests on TRC are presented in Fig. 2. An obvious increase in the first-crack stress could be clearly observed in all experiments with the addition of short dispersed and integral glass fibers in comparison to the results obtained for TRC plates without short fibers. In the case of dispersed short glass fibers the increase was by factor of 2. The effect of this kind of fiber reinforcement was, however, less pronounced with increasing strain and stress levels. Integral short fibers improved the load-bearing capacity of TRC over the entire strain range, but they were less effective in comparison to dispersed fibers in respect of increasing the first-crack stress.

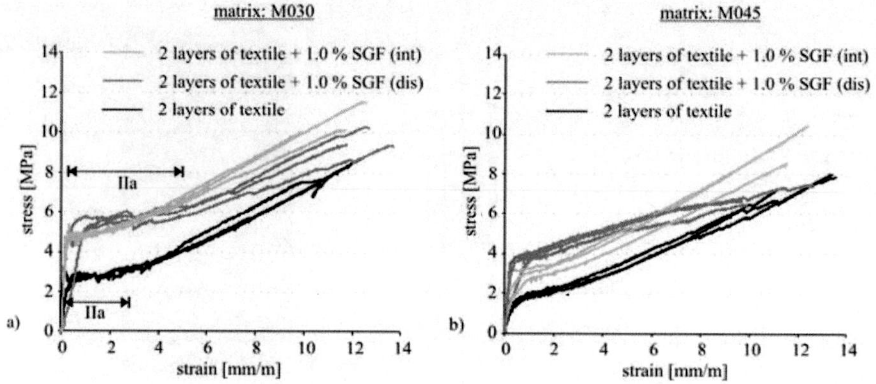

Fig. 2. Results of tension tests on TRC plates without and with short (dispersed and integral) glass fibers with the matrix a) M030 and b) M045

The formation of multiple cracks began directly after the first crack appeared. At this stage the effect of short fiber addition consists in the expanding of the strain region IIa, where multiple cracks form. This expansion is particularly pronounced in the case of the TRC with short dispersed glass fibers: the width of the region IIa is more than doubled, cf. Fig. 2a. Moreover, since the stress-strain curves for TRC with short fiber are always above the corresponding curves for TRC without short fiber, it can be concluded that the energy absorption (area under the stress-strain curve) increases significantly due to the addition of short fibers.

Fig. 3 represented the results of uniaxial tension tests performed on rectangular plates without textile reinforcement. The curves were obtained for the matrices M030 and M045 without and with the addition of 1.0% by volume of short glass fibers. The low w/b ratio in the matrix M030 leads to a better bond of matrix to fibers and, hence, to an increase in the tensile strength of the composite. However, due to the failure of many fibers there is a rather steep softening branch in the curve, which indicates relatively brittle failure of the composite (Fig. 3a). In contrast, for matrix M045 pronounced fiber pullout behavior was observed with only moderate fiber failure. This is obviously due to a weaker matrix-fiber bond resulting from a higher water-to-binder ratio. Here, no improvement of first-crack stress was noticed as a result of the addition of short fibers (Fig. 3b). These results are in agreement with the findings of pullout tests on single glass fiber shown in Fig. 4.

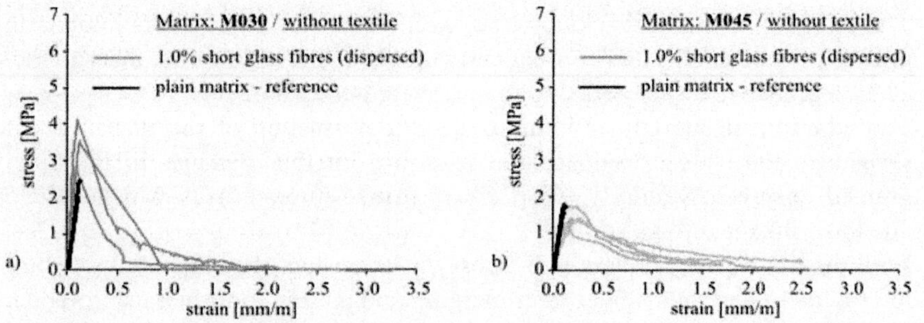

Fig. 3. Results of tension tests on plates made of the matrix a) M030 and b) M045, reinforced with 1.0% by volume short glass fiber

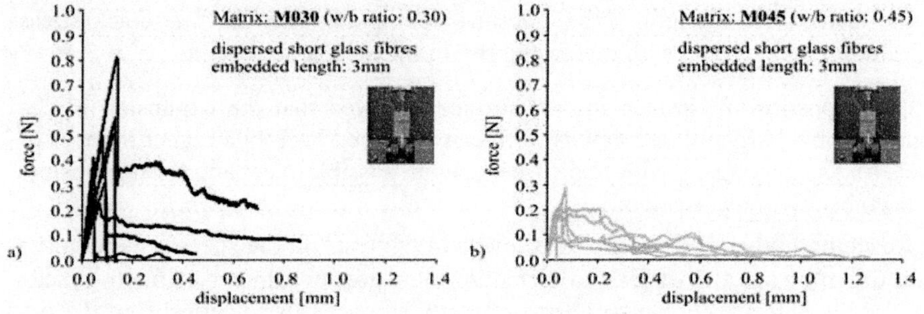

Fig. 4. Force-displacement curves obtained from single glass fiber pullout tests with the matrix a) M030 and b) M045

The influence of the addition of short fibers and water-to-binder ratio on the pullout behavior of multifilament yarns is shown in Fig. 5. Obviously, the addition of short fiber leads to a higher ultimate force, hence to a better bond between the yarn and matrix. This holds true for both matrices M030 and M045.

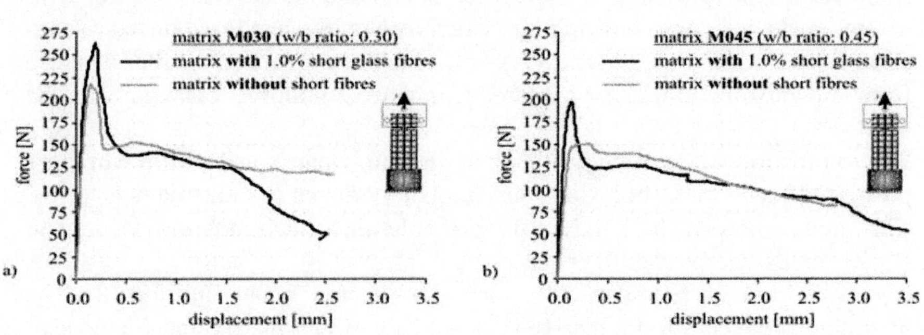

Fig. 5. Results of multifilament-yarn pullout tests with the matrix a) M030 and b) M045

5 Micro-structural Investigations and Discussion

The general enhancement in the magnitude of first-crack stress by the addition of short fibers can be likely ascribed to four mechanisms:

1) The bridging of micro-cracks by fine, well distributed short fibers inhibits the growth of such cracks and consequently the formation of the first macro-crack. Thus, a higher stress is needed to induce macro-cracking.
2) The addition of short fiber minimizes the deformation of the matrix due to shrinkage and, hence, reduces the resulting internal damage in the finely grained concrete. Similarly, short fibers bridge micro-cracks which develop due to shrinkage (cf. also point 1).
3) Well distributed short fibers will arrest the extension of micro-cracks to other micro-cracks and thus hinder their merging and uniting to form a macro-crack.
4) As a consequence of the addition of short fiber, the overall degree of reinforcement increases. Since the strength and stiffness of these fiber reinforcements are considerably higher than the corresponding material parameters of the matrix, the strength of the crack-free composite (i.e. the first-crack stress value) must increase with increasing fiber reinforcement.

Visual inspection of the specimens´ surfaces showed that the expansion of the strain region where multiple cracks form can be traced back to a higher number of cracks in the specimens with short fibers. Four possible mechanisms responsible for such behavior are assumed:

1) Reaching higher first-crack stress levels in the tests on the specimens with the addition of short fiber lead to formation of a greater number of micro-cracks over the entire specimen volume or length, respectively, even prior to the development of the first macro-crack. Beginning with the first-crack stress, the macro-cracks develop from these micro-cracks. A greater number of finely distributed micro-cracks offer more nuclei for macro-crack formation, leading to more pronounced multiple cracking.
2) The formation of a macro-crack results in a decrease of matrix stress in the vicinity of the crack. The next crack may not form at a distance below a threshold value. The presence of short fiber causes additional stress transfer over macro-cracks by means of bridging, which results in a less pronounced relaxation of the matrix in their vicinity. A new crack can form at a smaller distance from an existing crack, thus more pronounced multiple cracking can be observed.
3) Due to the production technique applied, in which matrix and textile fabric are pressed to each other, short fibers are frequently linked to multifilament yarns. This could improve the bond between textile and matrix, thereby leading to smaller crack widths and higher cracking density.
4) The bond between filament and cementitious matrix is non-uniform [8]. The hydration products form adhesive cross-links between matrix and filament or filament and filament, respectively. Microscopic investigation showed that

short fibers can improve the multifilament yarn-matrix bond by means of new "special" cross-links. By their random positioning on the yarn surface, the short fibers provide extra connecting points to the surrounding matrix, as can be seen in micro-graphs in Fig. 6.

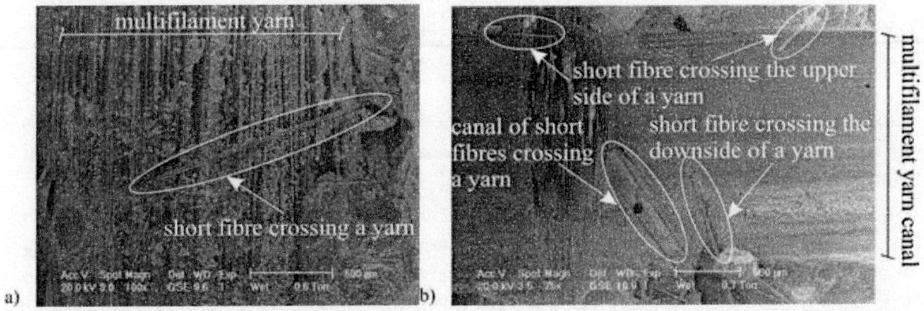

Fig. 6. ESEM images of multifilament yarn in matrix a) M030 and b) M045 with addition of short glass fibers

Micro-structural investigations of the multifilament yarns' surfaces show a very good bond between the yarn and the matrix M030. A very nearly complete layer of hydration products (C-S-H phases) covers the yarn's surface, cf. Fig. 7a. This is not the case when matrix M045 is used. Only in widely scattered areas are such hydration products to be observed, cf. Fig. 7b.

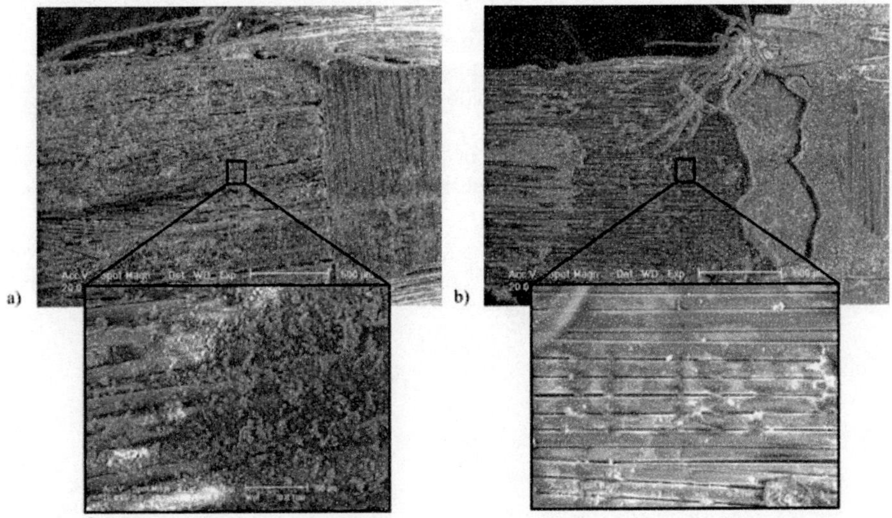

Fig. 7. ESEM images on multifilament yarn' surface in matrix a) M206 and b) M207

Acknowledgement. The results were obtained in a project initiated in the Collaborative Research Centre SFB 528 ''Textile Reinforcement for Structural Strengthening and Retrofitting" financed by the German Research Foundation "DFG". The authors acknowledge with gratitude the foundation's financial support.

References

[1] Curbach, M., Jesse, F. (eds.): Textile reinforced structures: Proc. of the 4th colloquium on textile reinforced structures (CTRS4), Dresden, SFB 528, Technische Universität Dresden (2009)

[2] Butler, M., Hempel, R., Schiekel, M.: The influence of short glass fibres on the working capacity of textile reinforced concrete. Bangneux: RILEM. In: Hegger, J., Brameshuber, W., Will, N. (eds.) Proc. of the 1st International RILEM Symposium on Textile Reinforced Concrete, Aachen, pp. 45–54 (2006)

[3] Barhum, R., Mechtcherine, V.: Effect of short fibres on fracture behaviour of textile reinforced concrete. In: The 7th International Conference on Fracture Mechanics of Concrete and Concrete Structures (FramCos 7), Jeju, Korea, pp. 77–88 (2010)

[4] Butler, M., Mechtcherine, V., Hempel, S.: Experimental investigations on the durability of fibre-matrix interfaces in textile reinforced concrete. Cement and Concrete Composites 31, 221–231 (2009)

[5] Lorenz, E., Ortlepp, R.: Berechnungsalgorithmus zur Bestimmung der Verankerungslänge der textilen Bewehrung in der Feinbetonmatrix. In: Curbach, M., Jesse, F. (eds.): Textile Reinforced Structures: Proc. of the 4th Colloquium on Textile Reinforced Structures (CTRS4), Dresden, SFB 528, Technische Universität Dresden, pp. 491–502 (2009)

[6] Kabele, P., Novak, L., Nemecek, J., Kopecky, L.: Effects of chemical exposure on bond between synthetic fibre and cementitious matrix. In: Textile reinforced concrete – Proc. of the 1st international RILEM Conference, pp. 91–99. RILEM publications, Cachan (2006)

[7] Jun, P., Mechtcherine, V.: Behaviour of strain-hardening cement-based composites (SHCC) under monotonic and cyclic tensile loading. Cement and Concrete Composites 32, 801–809 (2010)

[8] Butler, M., Hempel, S., Mechtcherine, V.: Modelling of ageing effects on crack-bridging behaviour of AR glass multifilament yarns embedded in cement-based matrix. Cement and Concrete Research (in print)

High Performance Light-Weight Cement Composite Plates Using Wastepaper Fibers and Wire Mesh

C. Sujivorakul[1], T. Muhummud[1], and N. Dokkhan[2]

[1] Department of Civil Technology Education,
King Mongkut's University of Technology Thonburi, Bangkok, Thailand
[2] Department of Highways, Ministry of Transport, Thailand

Abstract. The main objective of this study is to investigate parameters that affect the flexural behavior and shrinkage properties of light-weight cement composite plates using wastepaper fibers and wire mesh. Wastepaper fibers were prepared from disposed office papers which were shredded and immersed into water for a period of time and then they were grinded into small fibers. The properties of the composite plate were studied, i.e. flexural behavior, first-cracking and post-cracking strengths, energy absorption, and shrinkage. Parameters used in this study were Portland cement to paper ratio of the mixture by volume as well as shape, size, and layers of wire mesh used for reinforcement. The experimental results showed that layers of wire mesh and shape of wire mesh significantly influenced the flexural behavior and shrinkage of the composite plates. To achieve the highest performance in bending and the least shrinkage of the composite plates, the two layers of square wire mesh with the smallest opening size should be used for reinforcements.

1 Introduction

The development of composite plates made of cement and wastepaper fiber has been done for many years. Different kinds of fibers, including natural fibers from plants [1] and cellulose fibers obtained from wastepaper, have been used in cement mortar or concrete for reinforcement as they can effectively improve flexural behavior, ductility, toughness, and shrinkage characteristics of the cement composite product. Esmeraldaet al. [2], Ellis et al. [3], Massijaya and Okuma [4], and Rowell and Harrison [5] studied the production of low density boards from shredded wastepaper bonded with various types of resins. Mechanical properties of the boards were influenced by the percentage of bonding resins. Thomas et al. [6], Coutts [7], and Soroushian et al. [8, 9] proposed the use of cellulose fiber reclaimed from wastepaper as reinforcement materials in thin-sheet cement products. The finding of these researches was revealed a similar trend that is the

G.J. Parra-Montesinos, H.W. Reinhardt, and A.E. Naaman (Eds.): HPFRCC 6, pp. 495–502.

physical and mechanical properties of the composite products depending on the mixing proportions and mixing techniques. The long-term durability of wastepaper fiber-cement composites was also studied by Soroushian et al. [10, 11] and they were found that carbonation was an important disadvantage in the natural aging of cellulose fiber cement composites. Sujivorakul et al. [12] studied the parameters that affected the properties of wastepaper-cement composites and they found that both physical and mechanical properties of the composite plates also depended on the preparation of the paper fibers, namely, immersing time of paper into water and their flexural behavior was improved drastically when discrete natural fibers or continuous natural fiber meshes were used for reinforcement.

This research continued the study on enhancing flexural behavior of wastepaper-cement composite materials which can be used as interior cost-effective products for interior wall or indoor furniture. In the composite plate, steel wire meshes replaced continuous natural fiber meshes used in previous study and parameters that could affect wastepaper-cement composite plates were studied and reported herein.

2 Experimental Program

The experimental program comprised of a set of tests to determine physical and mechanical properties of the wastepaper-cement composites plate reinforced with steel wire mesh. These properties were flexural behavior, first-cracking and post-cracking strength, energy absorption, and shrinkage. Four main parameters influencing on the properties of the wastepaper-cement composites were investigated. They were (a) cement to paper ratio in the mixture (1:0.5, 1:1.0 and 1:1.5 by volume), (b) shape of the wire mesh opening (square and hexagon), (c) sizes of the opening (#1, #2, and #3) and (d) number of wire mesh layers used for reinforcement (1 and 2 layers) (see wire mesh properties in Table 1)

3 Preparation of Plate Specimens

Wastepaper fibers were prepared from disposed office papers which were shredded and immersed into water for a period of time and then they were grinded to obtain paper fibers or paper pulps. The plate specimens were prepared by mixing wastepaper fibers and Portland cement together and then poured the mixtures into steel molds. The mold had a size of 150 mm in width and 2.0 m in length, while the final thickness was controlled at 15 mm after pressing with the compressive stress of 0.5 MPa for about 2 minutes. The composite plates were removed from the mold immediately and placed under the sunlight until the specimens dry, and then kept in laboratory for 28 days before testing. The density of the composite plates varied from 1.2 to 1.4, 0.82 to 0.89, and 0.64 to 0.71, for the mixture with cement to paper ratio of 1:0.5, 1:1.0, and, 1:1.5, respectively.

Table 1. Size and weight per square meter of wire mesh

Shape of Mesh	No.	Diameter (mm)	Mesh size or Spacing (mm.)	Weight (g/m^2)
Square Wire Mesh (SM)	#1	0.60	12.70	452
	#2	0.75	19.05	478
	#3	1.00	25.40	491
Hexagonal Wire Mesh (HM)	#1	0.30	12.70	199
	#2	0.35	19.05	145
	#3	0.45	25.40	159

4 Experiments

4.1 Bending Tests

Each composite plate with a dimension of 150 mm in width, 2.0 m in length, and 15 mm in thickness was cut into smaller piece of specimens with a dimension of 100 mm in width, 330 mm in length, and 15 mm in thickness and these specimens were used for the bending tests. The specimens were tested under 4-points bending tests with a span length of 300 mm (Fig. 1). All specimens were loaded continuously until they failed or the deflection was so large. The applied load was monitored by a load cell, and the corresponding deflection was measured using a LVDT placed under the mid-span section of the specimen. The equivalent bending stress σ_b was calculated using the following equations:

$$\sigma_b = \frac{6M}{bh^2} \tag{1}$$

where $M = \dfrac{PL_S}{6}$ (2)

where M is the bending moment, b is the width of the plate, h is the thickness of the plate, P is applied load, and L_S is the span length. Load-deflection measurements obtained from the tests are presented as equivalent elastic bending stress versus deflection curves. Energy absorption of the cement composite plates can also obtained from the area under the curve of the equivalent bending stress versus deflection relationship.

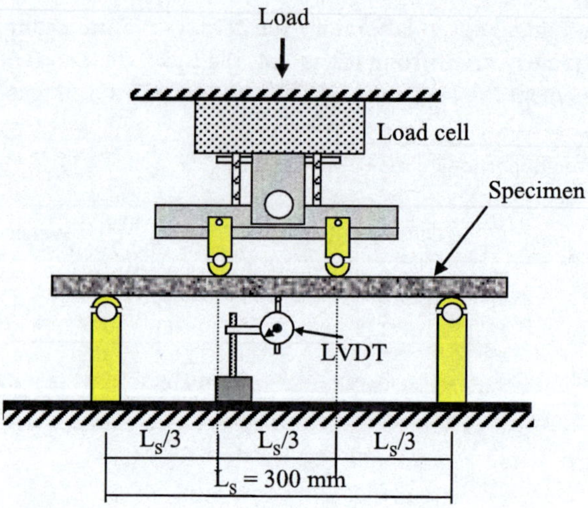

Fig. 1. Bending test set-up

4.2 Shrinkage Test

After being removed from the mold, each composite plate with a dimension of 150 mm in width, 2.0 m in length, and 15 mm in thickness was cured in the air and measured its dimension weekly to monitor its shrinkage comparing to its original length.

5 Results and Discussion

5.1 Flexural Behavior of Specimens

Fig.2 shows the flexural behavior of different composite specimens. It is observed that the addition of steel wire mesh significantly enhances the flexural behavior of the wastepaper-fiber cement plates. It is noted that the symbol "None" refers to no reinforcement; "SM" and "HM" refer to square and hexagonal wire mesh, respectively; and 1 and 2 refer to one and two layers of wire mesh, respectively. It is observed from specimens with wire mesh that the equivalent bending stress increases linearly up to the point where first-cracking strength of composites takes place, after that it drops down to certain level and increase again until it reaches the second peak, called post-cracking strength. The comparison of the first and the post-cracking strength as well as the energy absorption and shrinkage of all specimens will be discussed in the next section.

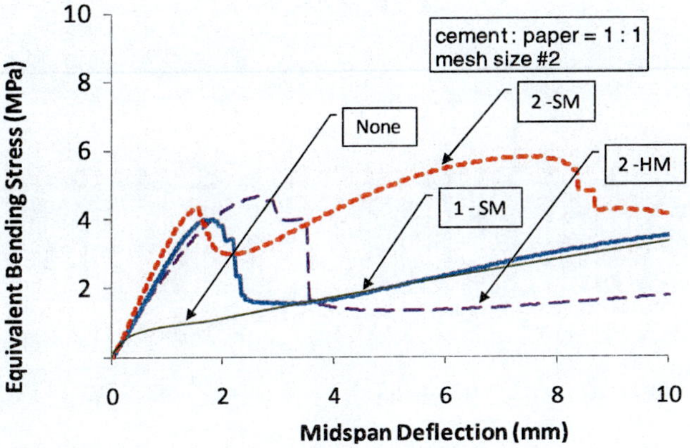

Fig. 2. Comparison of relationship between equivalent bending stress versus midspan deflection

5.2 *First-Cracking Strength of Specimens*

Fig.3a presents the effect of all parameters on the first-cracking strength of wastepaper-fiber cement plates reinforced with steel wire mesh. It is seen that there is some variation in the effect of cement to paper ratio in the mixture on the first-cracking strength. However, for specimens with square and hexagonal mesh whose opening sizes were smallest (i.e. #1), the higher strength was obtained when less amount of paper was used in the mixture while the strength of those with other opening size is inconclusive. When comparing the effect of mesh shape, it was found that the specimens with hexagonal mesh outperform those with square mesh. In addition, it was observed that for the specimens with hexagonal mesh, the use of smaller mesh opening and two layers of reinforcement resulted in a higher first-cracking strength for the composite plates. However, the effect of the opening size and number of layers are almost the same for specimens with square mesh.

5.3 *Post-Cracking Strength of Specimens*

Fig.3b presents the effect of all parameters on the post-cracking strength of composite plates reinforced with steel wire mesh. It is obviously seen that the opening shape and the number of layers of wire mesh affect the post-cracking strength of the composite plates. The use of square mesh exhibits higher post-cracking strength than that of hexagonal mesh due to higher in steel content.

Similarly, the use of two layers of wire mesh, leading to higher steel content, is significantly better than that of one layers of mesh. However, they still shows the variation in the test result when the effect on cement to paper ratio of the mixture and the size of mesh opening are considered.

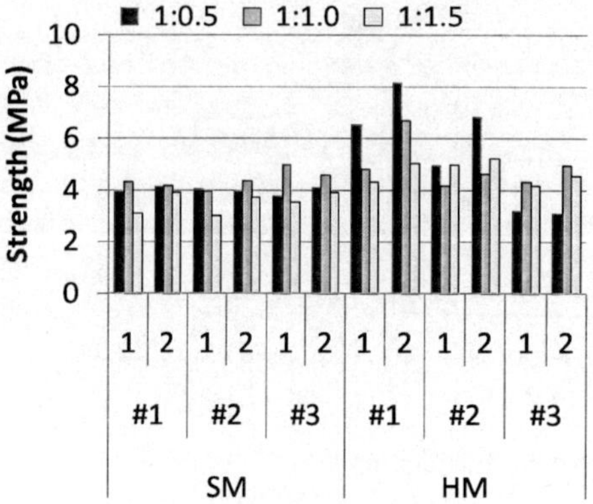

(a) First-cracking strength

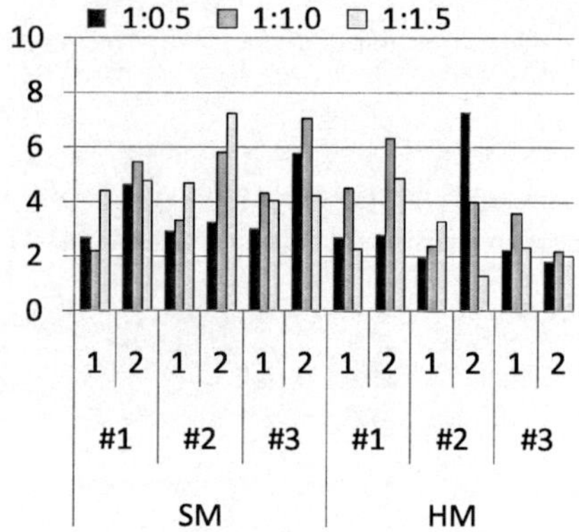

(b) Post-cracking strength

Fig. 3. Comparison of first-cracking and post-cracking strengths of all specimens

5.4 Energy Absorption of Specimens

Fig.4 shows the effect of all parameters on the energy absorption of wastepaper-fiber cement plates reinforced with steel wire mesh. The energy absorption was calculated from area under stress-deflection curve up to the mid-span deflection of 4 mm. It is seen that the number of layers of wire mesh significantly affects the energy absorption of composites, i.e. the use of two layers presents higher energy absorption. When comparing the shape of wire mesh, it found that for one layer reinforcement, the energy absorptions of specimens with both shapes are similar, but for two layer reinforcement, those of specimens with square mesh are higher. However, it is inconclusive in the effects of cement to paper ratio of the mixture and the size of mesh since it is observed some variation in the results.

5.5 Shrinkage of Specimens

The effect of parameters on the shrinkage of cement composites reinforced with steel wire mesh is shown in Fig.5. It is observed that the use of steel wire meshes leads to the reduction of shrinkage in composites. The smaller shrinkage was found when the less amount of paper was used in the mixture and when the smaller size of the mesh opening was used for reinforcement. The shape of the opening also affects the shrinkage of specimens. The specimens with square mesh have less shrinkage than those with hexagonal mesh. However, the number of layers seems to have little effect on the shrinkage of composite.

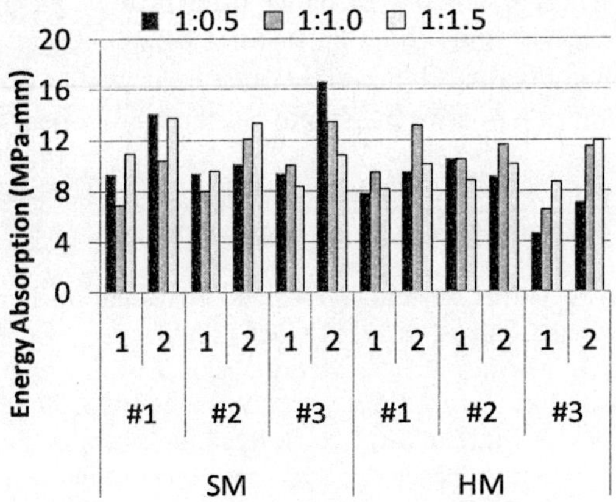

Fig. 4. Comparison of energy absorption of all specimens at the deflection of 4 mm

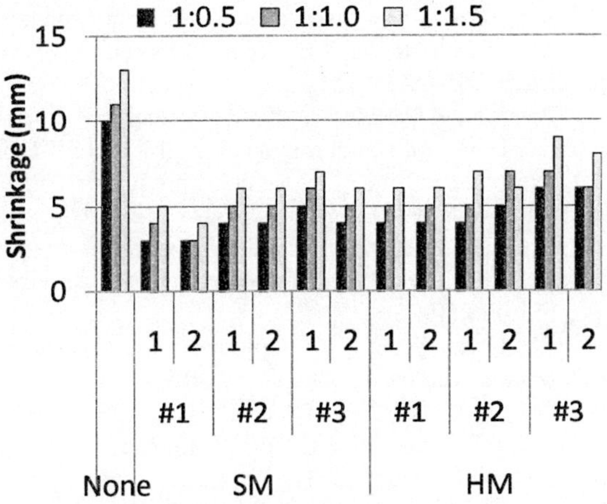

Fig. 5. Comparison of the shrinkage of all specimens

6 Conclusions

This paper presents parameters affecting flexural behavior and shrinkage properties of light-weight cement composite plates made of cement, wastepaper fibers, and wire mesh. Parameters investigated in this study were mix proportion of wastepaper-cement composites and shape, size, and number of layers of wire mesh. The experimental results showed that shape, size, and number of layers of wire mesh significantly influenced the flexural behavior and shrinkage properties of the composite plates. The use of two layers of square wire mesh with smallest opening size for reinforcement exhibited the best flexural behavior and the least shrinkage. While the larger amount of paper used in the mix proportion led to larger shrinkage, but it had little effect on flexural behavior, and it could not lead to explanation. In summary, the high performance light-weight cement composites can be obtained by addition of wire mesh into wastepaper-cement composites.

References

[1] Coutts, R.S.P.: A review of Australian research into natural fiber cement composites. Cement and Concrete Composites 27, 518–526 (2005)

[2] Esmeralda, Y.A.O., Marcos, A.E.S., MarioRabelo, S.: Utilization of wastepaper to manufacture low density boards. Bioresource Technology 73, 77–79 (2000)

[3] Ellis, S.C., Ruddick, J.N.R., Steiner, P.R.: A feasibility study of composites produced from telephone directory paper, plastic, and other adhesives. Forest Product Journal 43(7/8), 23–26 (1993)

[4] Massijaya, M.Y., Okuma, M.: Development of boards made from waste newspapers I. Production and fundamental properties of waste-newspaper boards. MokuzaiGakkaishi 42(12), 1243–1249 (1996)

[5] Rowell, R.M., Harrison, S.: Fiber based composites from recycled mixed paper and magazine stock. In: Rowell, R.M., et al. (eds.) Proceeding of Materials Research Society Symposium, pp. 65–72 (1992)

[6] Thomas, C.O., Thomas, R.C., Hover, K.C.: Wastepaper fibers in cementitious composites. Journal of Environmental Engineering 113(1), 16–31 (1987)

[7] Coutts, R.S.P.: Wastepaper fibers in cement products. Cement and Concrete Composites 11(3), 143–147 (1989)

[8] Soroushian, P., Arola, R., Shah, Z.: Recycling of wood and paper in cementitious materials. In: Rowell, R.M., et al. (eds.) Proceeding of Materials Research Society Symposium, pp. 165–175 (1992)

[9] Soroushian, P., Shah, Z., Won, J.P.: Optimization of wastepaper fiber-cement composites. ACI Materials Journal 92(1), 82–92 (1995)

[10] Soroushian, P., Won, J.P., Hsu, J.W.: Control of aging process in wastepaper fiber reinforced cement. In: Basham, K. (ed.) Proceeding of 3rd Materials Engineering Conference, pp. 539–544 (1994)

[11] Soroushian, P., Shah, Z., Won, J.P., Hsu, J.W.: Durability and moisture sensitivity of recycled wastepaper-fiber-cement composites. Cement and Concrete Composites 16(2), 115–128 (1994)

[12] Sujivorakul, C., Muhummud, T., Narmluk, M., Yodnumkhum, W.: Development of Cement Composite Plates Reinforced with Wastepaper Fibers. In: Nimityongskul, P., et al. (eds.) Proceeding of Ferro – 8 Symposium and Workshop, pp. 241–252 (2006)

Textile Reinforced Cementitious Composites for Retrofit and Strengthening of Concrete Structures under Impact Loading

A. Katz[1], M. Tsesarsky[2], A. Peled[2], and I. Anteby[3]

[1] Civil and Environmental Engineering, Technion - Israel Institute of Technology
[2] Structural Engineering Department, Ben Gurion University of The Negev
[3] Mechanical Engineering Department, Ben Gurion University of The Negev

Abstract. Textile Reinforced Concrete (TRC) became a common method lately for the production of thin elements having excellent properties. These elements can be used for strengthening concrete elements by applying them on the surface of these elements. This study examined concrete beams strengthened by carbon, glass or polyethylene TRC for their behavior under static and dynamic loads. Different static and dynamic behavior was identified for the different TRC materials. Significant differences between the carbon, glass and PE were observed in the static tests but not at the dynamic ones.

1 Introduction

Structural elements made of concrete can be subjected to a wide range of loading conditions during their service life, from very low strain rate up to severe dynamic events such as blast explosions, missiles, earthquakes, projectiles, and more. Due to the inherent brittleness and low tensile strength of most cement-based elements they do not always possess enough strength, toughness and ductility to maintain integrity without collapse under impact and other dynamic loads. As a result, extreme events can cause severe damage to the building [1].

The mechanical performance of concrete can be improved substantially by fiber reinforcement [2]. A rather new development of an innovative composite material is Textile Reinforced Concrete (TRC), where multi-axial fabrics are used in combination with fine grained cementitious matrix. Research findings show that these TRC composites significantly improve tensile strength, ductility, and energy absorption, the properties which are critical for the dynamic and impact resistance of the concrete element. The use of technical textiles which are placed in the main stress directions of the composite, leads to high effectiveness in comparison to the use of randomly distributed short fibers. These properties suggest a potential use

G.J. Parra-Montesinos, H.W. Reinhardt, and A.E. Naaman (Eds.): HPFRCC 6, pp. 503–510.

of TRC as strengthening materials for concrete members under static loading [3-7].

The goal of the current research was to develop and examine in-situ technology where textile fabrics impregnated in cement-based material (TRC layers) are used to strengthen concrete structural elements in order to improve their resistance to impacts during extreme events. The wide experimental program included concrete beams strengthened by various thin TRC layers made of alkali resistance glass, polyethylene and carbon fabric systems. The surface of the original concrete was either rough or smooth. The thin TRC layer was applied in two ways (Fig. 1): (i) around the original concrete, providing full confinement around the concrete, and (ii) as separated TRC layers at the top and at the bottom of the concrete element, providing strengthening layer only at the two surfaces of the concrete beam. The load vs. time and deflection was recorded and the impulse (force·time) and energy (area under force-deflection curves) were calculated. This paper presents partial results comparing the behavior of rough-surface beams under impact or static loads.

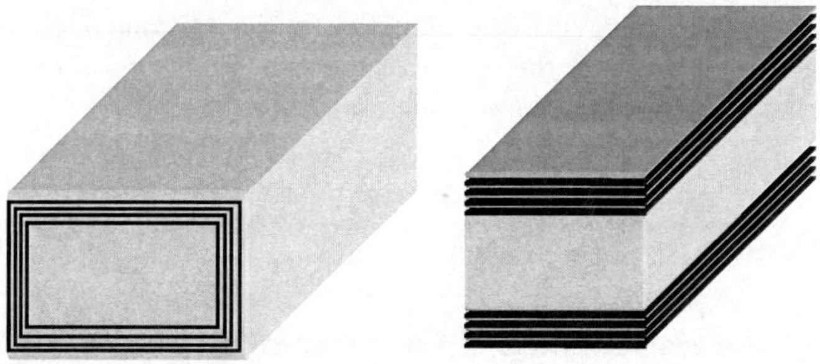

Fig. 1. Two types of fabric application (a) full wrap, (b) top and bottom

2 Experimental Setup

Concrete elements having a characteristic strength of 30 MPa and 400 x 400 x 50 mm in size were prepared. These elements were sectioned to produce two types of beams: 400 x 100 x 50 mm which were used for impact tests, and 200 x 50 x 50 mm which were used for static flexural tests. Two surface treatments were used: smooth, as-is surface finish and rough finish (denoted R in the followings). The fabrics tested were made of Glass-fiber (G), Polyethylene (PE) and Carbon (CA).

At age 28 days the beams were strengthened by TRC. The fabrics were immersed in a cement paste (water to cement ratio of 0.45) and hand laid over the surface of the beams. Two types of setup were used: i. full wrap of the beam by four layers of the fabric; and ii. applying four layers of fabrics at the top and bottom of beam (not reported in this paper). The strengthened beams were cured in water for additional 28 days and then in air until testing.

The strengthened beam was tested for impact by a falling pendulum having a mass of 150 kg dropped from a distance of 60 mm. The beam simply support over a span of 350 mm (Fig. 2a). At the static tests the beam was supported over a span of 180 mm and centrally loaded at a rate of 1 mm/min (Fig. 2b). The load and deflection of the beam was monitored continuously throughout the experiment enabling the extraction of flexural stress, energy (area under load-deflection curve) and impulse (area under load-time curve in the dynamic tests).

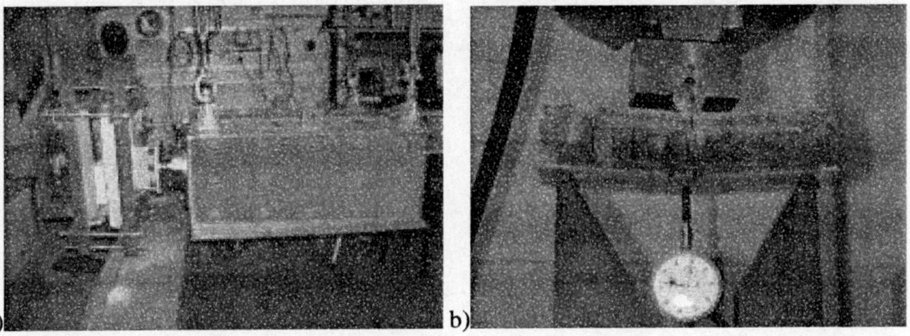

a) b)

Fig. 2. Experimental setup for (a) dynamic and (b) static loading

3 Results and Discussion

3.1 Dynamic Behavior of Reinforced and Non-reinforced Beam

The results from dynamic tests of concrete elements reinforced with glass fiber fabric are presented in Fig. 3 for energy, impulse, load and stress development. It is seen that the behavior of the reinforced beam follows quite well that of the non-reinforced beam for the 2-3 initial m-seconds. At that point of time the concrete beam fails under flexure and the properties are affected by the fibers. After some reduction, the load increases significantly to a higher level, thus the energy, impulse and stress increase similarly. Despite the similarity in the load-time history the stress does not follow exactly the same pattern as the load since the calculation is done based on the actual dimensions of the beam including the reinforcement when applied.

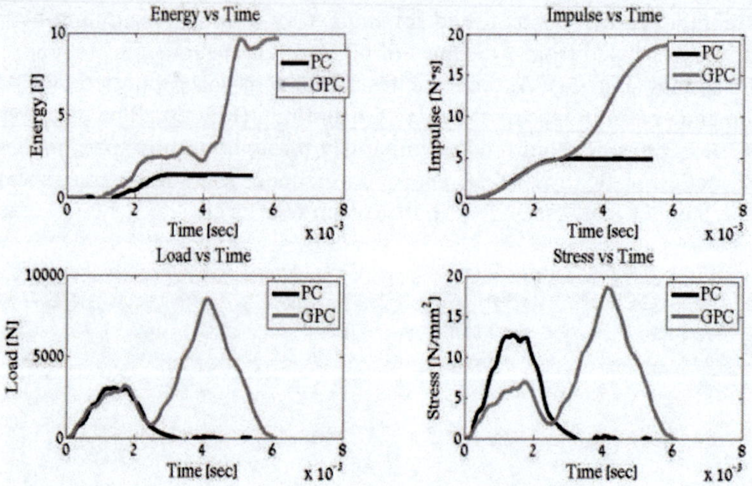

Fig. 3. Test results comparing dynamic behavior of concrete beam, non-reinforced (dark line) and reinforced with glass fiber fabric (gray line)

3.2 *Load Deflection Behavior in Static and Dynamic Loading*

Significantly different load-deflection behavior was seen when the specimens were loaded by static or dynamic loads. Figs. 4 and 5 present the results for static and dynamic loading, respectively, for beams having rough surface.

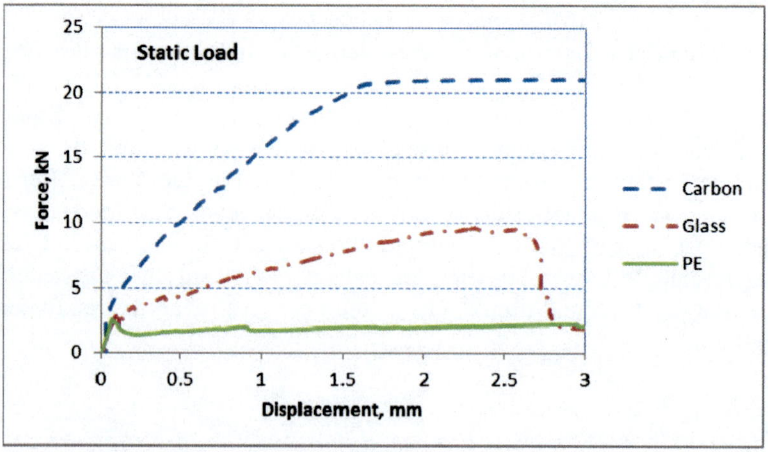

Fig. 4. Load deflection behavior of the tested fabrics under static loading

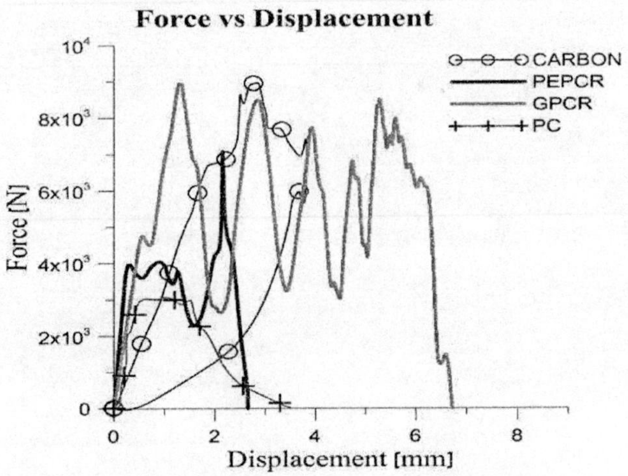

Fig. 5. Load deflection behavior of the tested fabrics under dynamic loading

An initial increase in the load is seen in the static loading setup, until cracking of the concrete beam (Fig. 4). The deflection continued to increase while supporting additional load when the reinforcement was carbon fiber and glass fibers, but smaller loads were observed for PE reinforcement. The load bearing capacity of the carbon fabric was approximately twice compared with the glass fabric whereas that of the PE fabric was about 25% only. Significant post-peak deflection was identified for all types of fibers, whether involved with post-peak hardening or softening. It should be noted that the unreinforced concrete beam fully fractured when reaching the initial peak load.

The load-deflection behavior under dynamic loading was significantly different compared to the behavior observed in the static loading setup. Some displacement was seen for the non-reinforced concrete with a certain load capacity before fracture (Fig. 5). Somewhat higher load bearing capacity was seen for the PE reinforced beams with slightly more deflection. The load capacity of the glass reinforced beam was about twice then the formers while exhibiting much larger deflection. The carbon reinforced beam exhibited the same load capacity as the glass but also full recovery of the deflection. Visual observation showed full fracture of the non-reinforced concrete and glass reinforce concrete, significant residual deflection of the PE reinforced beam and full recovery of the carbon reinforced beam (Fig. 6). It should be noted that a second loading peak was observed for the PE reinforced beam. This peak was at much lower loads but extended at much longer period of time (~250 m-sec. compared with 3-6 m-sec. for the other types of reinforcements). This effect is not shown in Fig. 5 that presents the initial peak only. Thus, when considering the total amount of energy absorbed by the specimen (the area under the load-deflection curve), it is the largest for the PE, despite the lower load capacity.

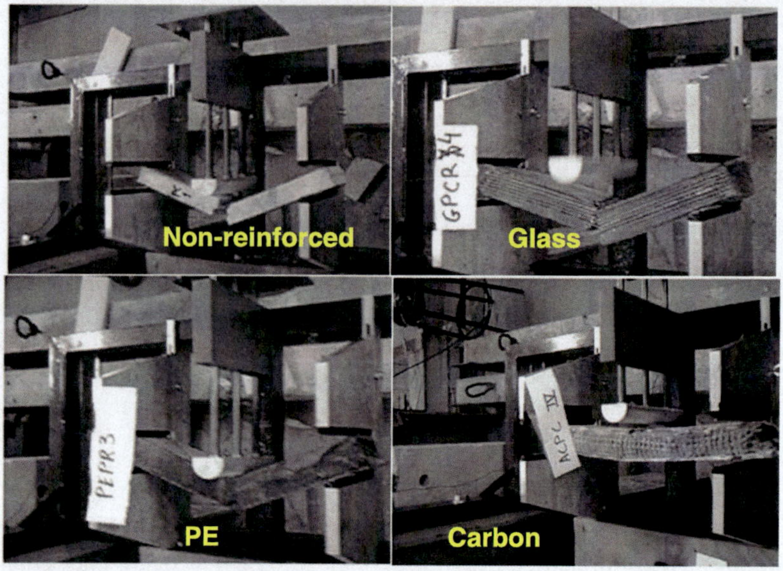

Fig. 6. Specimens after dynamic tests

3.3 Effect of Surface Treatment

Two surface treatments were tested: smooth and rough. Fig. 7 presents the effect of surface treatment on the energy absorption in the static and dynamic tests. As discussed before, concrete elements reinforced with carbon fabrics absorbed much more energy than the others, in the static tests but not in the dynamic tests. Negligible change in the energy absorption was identified in PE and ~50% increase in glass reinforced elements, when strengthening was done over a rough surface, in the static tests (Fig. 7a). In the dynamic tests, significantly higher energy absorption was identified in rough-surface beams reinforced with glass fabrics (+115%) while no effect of surface treatment was seen in PE systems. It is possible that better bond of the TRC layer to the surface of the concrete prevented local crack development. In the case of brittle fibers, as glass, the dispersion of cracks leads to better energy absorption. In ductile fibers, like PE, this phenomenon is less important. The effect of surface treatment was not studied on carbon fabric systems.

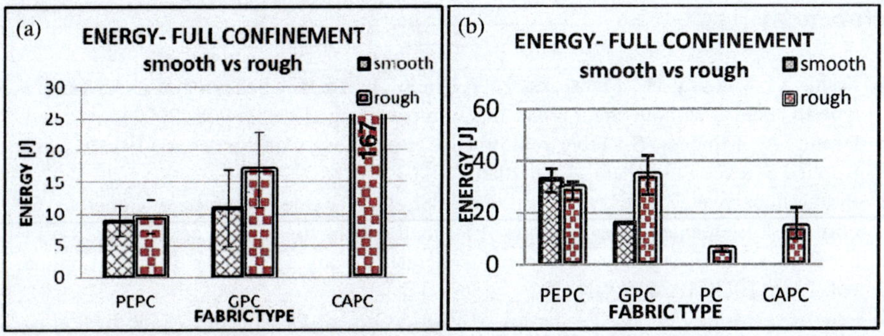

Fig. 7. Effect of surface treatment in (a) static and (b) dynamic tests

4 Summary and Conclusions

Three types of fabrics were used to strengthen concrete beams: carbon, glass and polyethylene. The fabrics have been applied in 4 layers, wrapping the beams, thus providing some confinement and full anchorage to the fabrics. The strengthened elements were tested for flexure by static and dynamic loading.

Significant increase in stress resistance was observed when the beams were strengthened by either glass or carbon fabrics, compared with plain concrete. This increase is expected due to the high mechanical properties of the fibers. However, the load capacity in static loading for carbon fabric was higher than the glass but was quite similar under dynamic loading. The polyethylene exhibited lower load capacity of ~25% compared with glass under static loading, but performed better (50% only) under dynamic loading. It seems that polyethylene fabrics exhibit the highest energy absorption under dynamic loading. Surface treatment to improve bond between the TRC layer and the concrete seems to improve the energy absorption capacity, mostly in brittle fiber systems subjected to dynamic loads. These properties emphasize the potential of textile reinforced concrete (TRC) in strengthening concrete elements for improved resistance to static and dynamic loads and the possibility to "tailor" desired properties by different types of fibers.

Acknowledgement. Partial support by the Israeli Ministry for Construction and Housing is gratefully acknowledged.

References

[1] Kishi, N., Mikami, H., Matsuoka, K.G., Ando, T.: Impact behavior of shear-failure-type RC beams without shear rebar. Int. J. Impact. Eng. 27, 955–968 (2002)
[2] Bentur, A., Mindess, S.: Fiber reinforced Cementitious Composites, p. 191. Elsevier Applied Science Publishers, Amsterdam (1990)
[3] Brameshuber, W., Brockmann, J., Roessler, G.: Textile Reinforced Concrete for Formwork Elements– Investigations of Structural Behaviour. In: Proc., FRPRCS-5 Fiber Reinforced Plastics for Reinforced Concrete Structures, Thomas Telford, London, vol. 2, pp. 1019–1026 (2001)
[4] Peled, A., Mobasher, B.: Properties of Fabric-Cement Composites Made by Pultrusion. Materials and Structure, RILEM, 1359–5997 (Print) 1871–6873 (Online) (2006)
[5] Peled, A., Mobasher, B.: Tensile Behavior of Fabric Cement-Based Composites: Pultruded and Cast. ASCE, J. of Materials in Civil Engineering 19(4), 340–348 (2007)
[6] Peled, A., Mobasher, B.: Pultruded Fabric-Cement Composites. ACI Materials J. 102(1), 15–23 (2005)
[7] Peled, A.: Confinement of Damaged and Non-Damaged Structural Concrete with FRP and TRC Sleeves. ASCE J. of Composites for Construction 11(5), 514–523 (2007)

Industrial Processing Technique for Textile Reinforced Cement Composites with Structural Use

J. Wastiels and O. Remy

Department of Mechanics of Materials and Constructions (MEMC),
Vrije Universiteit Brussel, Belgium

Abstract. The post-cracking hardening behavior under tensile loading of high performance fiber reinforced cement composites is determined as well by the fiber volume fraction as by the reinforcement structure and its impregnation by the matrix. High volume fractions can be obtained by using a continuous fiber structure in textile form, as is usual for polymer matrix composites. The corresponding processing techniques should thus be similar to polymer matrix composite manufacturing. However, due to the presence of an important powder fraction in the matrix raw material mix, the impregnation of the fiber structure as a whole or on the level of fiber bundles, is difficult.

The impregnation process has to be considered as an independent processing step. A new device, called Self Compacting Impregnator (SCI) has been developed. It consists of two parallel grooved cylinders rotating in opposite sense, which squeeze the cement matrix in and through the textile fiber structure by applying a controllable pressure, thus enabling the continuous industrial production of well impregnated textiles, which can be processed and shaped further. Fiber volume fractions of more than 20% can be obtained.

Test plates were manufactured by careful hand lay-up as reference, and by SCI impregnation as comparison. Mechanical testing (static tensile strength and fatigue loading) indicate that the proposed processing technique yields superior results to hand lay-up.

1 Introduction

All cementitious materials are inherently brittle and present a modest and unreliable tensile strength. Structural applications thus require reinforcement in the zones where tensile stresses may occur. In conventional concrete this is achieved with steel rebars, leading generally to massive sections, taking into account the required cover. Lightweight structural design with slender elements, as is possible with composite materials, is hardly or not achievable. A merging of composite and concrete technologies, where a cement based matrix is reinforced with strong

G.J. Parra-Montesinos, H.W. Reinhardt, and A.E. Naaman (Eds.): HPFRCC 6, pp. 511–518.

fibrous materials, could in some cases provide a solution: the diameter of the reinforcement reduces thousand fold. This means that the resulting HPFRCC material can be considered as homogenous on a much smaller scale, and that the crack pattern can be dramatically finer. Moreover, corrosion problems are defined in a quite different way.

Common manufacturing processes for cement composites, like premix casting or spray-up, present the advantages of being economical and near to traditional concrete technology. The use of (bundles of) short fibers limit the control of their orientation after placing, while workability issues limit the maximum fiber fraction V_f to several percent. As a consequence, while it is possible to obtain remarkable strain capacity (e.g. [1]), a strain hardening behavior under tensile loading is far less obvious. This actually limits their suitability for structural applications, since a composite for structural use must meet the highest performance requirements, as defined in [2]:

- multiple cracking with crack width control;
- post-cracking stiffness, or strain hardening;
- high tensile strength.

In order to obtain this, it is necessary to have a good bond between fibers and matrix, while the fibers should be strong, stiff and present a large length to diameter ratio. Moreover a high amount of fibers is required, as well as their controlled alignment. This implies the use of a preformed continuous fibrous structure (in which the fibers themselves can be continuous or discontinuous), usually termed as textile reinforcement. Independently of the way of assembling the fibers (weaving, knitting, bonding, braiding, …), the reinforcement structure is essentially two dimensional although not necessarily plane. The above mentioned common manufacturing processes are not adapted, and production techniques of polymer matrix composites, capable of handling continuous fibrous structures and high fiber fractions, have to be adopted.

2 Processing Issues of Textile Reinforced Cement Composites

Production techniques of polymer matrix composites with continuous fiber structure involve a wet out step before or together with the shaping and curing step. In pultrusion and filament or tape winding, this is achieved by pulling the fibers through a resin bath. The resin is picked up in a rather uncontrolled amount, which is subsequently adjusted by a roller or calendaring system that also improves the impregnation of the fiber bundles which consist of hundreds or more individual filaments. A fresh cement or mortar mixture presents, compared to a resin, a highly viscous non-Newtonian behavior with a non-zero yield stress. This results in an insufficient matrix pick-up and bundle impregnation: the liquid matrix phase is easily wetting out the fibers, but the tight fiber structure linked to the wanted high fiber content actually filters out the solid particles which do hardly penetrate inside the bundle.

In other production techniques such as Resin Transfer Molding (RTM) or Vacuum-Assisted RTH (VARTM), the matrix is forced under pressure to travel through the fiber structure, and wet it out in this way. The filtering effect is even more dramatic: a complete segregation occurs.

The filtering of the cement particles at the outer edge of the fiber bundles lead to a weak fiber-matrix bond, and the composite presents a low strength due to premature debonding and pull-out. A partial solution can be found in using an open fiber grid structure, where the fiber bundles are pre-impregnated with a polymer: the cement can flow without segregation through the open cells. Unfortunately, the consequence is that the fiber fraction is reduced to amounts which hardly lead to strain hardening, while the cracks become more localised.

3 Self Compacting Impregnator (SCI)

The only valid processing technique (besides the Hatschek process) to obtain a higher amount of fibers seems to be the low volume manual (and thus high cost) hand lay-up. The experience of the authors has indicated that the standard hand lay-up technique, where an amount of matrix is spread in which the fiber structure is placed and subsequently impregnated by using relatively sharp de-airing rollers, can be used for manufacturing cement matrix composites. Contrary to polymer matrix composites, one needs to work layer by layer and exert an important pressure on the rollers in order to wet out sufficiently the fiber bundles. As a result, it is possible to obtain laminates with a fiber volume content of up to 25%, using woven fibers or chopped strand mats of a surface density of up to 450 g/m^2 [3].

Vrije Universiteit Brussel has developed a production technique for cement matrix composites, which is simulating this hand lay-up technique in an automated way: the textile reinforcement is impregnated by pressing the matrix in a controlled, uniform and continuous way through the fiber structure and compacting in the same movement the impregnated layer(s). The device is called Self Compacting Impregnator (SCI), which overcomes, and paradoxically even takes advantage of the above mentioned filtering process [4]. Its working principle is shown in Fig. 1 (left). Two cylinders are rotating in opposite sense around parallel horizontal axes, leaving no or only a limited opening between their surfaces. In this way, a receptacle can be created by placing side walls between the upper half of the cylinders at the desired width. This receptacle is filled with fresh matrix mixture, which is continuously consumed during the rotation of the cylinders, thus avoiding setting issues with rapid hardening cements.

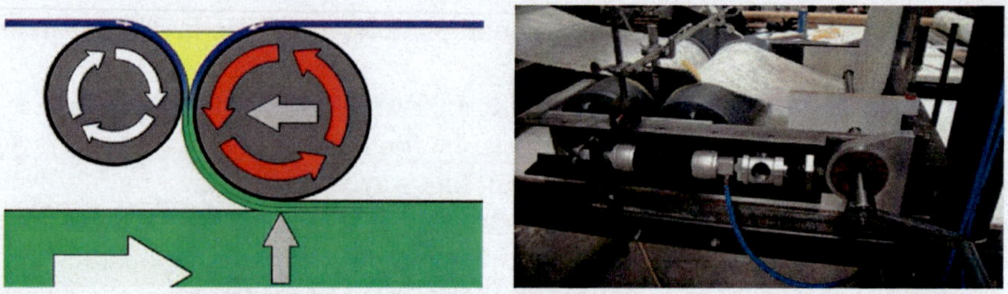

Fig. 1. Working principle and prototype of the Self Compacting Impregnator (SCI)

In analogy with the de-airing rollers used with hand lay-up, it is essential that the surface of the cylinders is adequately grooved, in order to impregnate the fiber fabric at the line of smallest distance between the rolling cylinders, where the highest pressure is generated. The grooves allow for the full impregnation, by creating a space for the matrix mixture which passes completely through the fiber structure. The right cylinder is driving the system, and is pressed against the left freewheeling cylinder with a controllable and adjustable pressure. As such, two or more textile structures, which are fed from rolls located above the cylinders, are continuously impregnated. The right cylinder can pivot around the axis of the left cylinder, thus resting on a movable support on which the impregnated textiles are deposited, and which speed is controlled by the right cylinder. Its weight, eventually increased, assures the compaction of the impregnated textiles. After this, the semi-finished product, or "wet pre-preg", can be further processed using composite production techniques like pultrusion, tape winding, (warm) compression moulding and others, without encountering the above mentioned problems.

The support with the wet pre-preg can be re-entered in the SCI process (e.g. in a subsequent pair of cylinders) in order to obtain the desired thickness. In this way, plates can be produced in a continuous way.

A suitable choice of depth, distance and shape of the grooves, as well as the applied pressure, allows for the desired uptake of matrix component, depending on the cement or mortar composition used, the type of fiber structure, and the required fiber volume content.

4 Fabrication of Test Specimens

Fiber reinforced laminates of 500 x 400 mm^2, composed of four layers, were produced using two manufacturing methods: hand lay-up (HLU) and SCI. The fiber structure of a layer is a standard powder bonded chopped E-glass strand mat (Vetrotex M5) with a surface density of 300 g/m^2. The cement matrix is a two component (wollastonite powder, and liquid based on phosphoric acid) chemically bonded phosphate ceramic (CBPC), available under the brand name vubonite. It presents the rather unique feature of being fully compatible with standard glass fibers due to its pH neutral environment after hardening, and is a good testing object since excess of liquid component must be absolutely avoided, thus not tolerating any filtering during impregnation.

The HLU laminates were build up layer per layer, by forcing each time the required amount of matrix through each fiber layer with a ribbed Teflon de-airing roller. The hand operated SCI prototype has a working width of 600 mm, and a continuous production capacity of around 5 m per minute. The pressure is generated by two pneumatic muscles (Type: DMSP-40 from Festo) with a controlled force of maximum 6000 N each, hence a pressure of 20 N/mm width when working on the full width of the machine (Fig. 1 right). This pressure, which is roughly 20 times higher than the pressure generated by hand lay-up, was used for the SCI manufacturing. The cylinders have a diameter of 200 mm and are confectioned in PVC, which was chosen for the ease of implementing the grooves. The grooves in this prototype, rounded V-shaped, have a depth of 1 mm and a step of 1.5 mm. Two fiber fabrics were impregnated in one pass, following which the movable supporting plate was reintroduced below the SCI after cutting the fibers at the edge of the plate.

After production, the laminates were covered by hardboard plates, on which a mass of approx. 100 kg was placed to obtain a minimum of pressure during setting. All specimens were produced at ambient conditions. The curing conditions were equivalent for all laminates: after 1 day in ambient conditions, the laminates were cured at 60°C in a Heraeus oven for 24 hours while evaporation of matrix water was avoided. After curing, the test specimens with nominal length of 250 mm and width of 40 mm were obtained using a water fed circular saw with diamond blade. All HLU and SCI specimens presented a measured volume fiber fraction of around 18%.

5 Experimental Results and Discussion

As well for the HLU as for the SCI production method, three identical specimens were tested statically in uniaxial tension, to determine their constitutive behavior. The tests were performed on an INSTRON 5885H test bench with a capacity of 250 kN, and the strains were measured by a clip-on double-side averaging INSTRON extensometer with 50 mm gage length. The obtained stress-strain behavior is represented as dotted lines on Fig. 2 left (HLU) and right (SCI). The typical nonlinear behavior presents three stages: uncracked, gradual development of matrix multiple cracking, and finally fully cracked [5]. The final crack spacing (failure occurs at a strain of around 1%) amounts 0.8 mm for HLU and 0.6 mm for SCI, which means that the mean crack opening is even at full load less than 10 μm. The measured mechanical properties are given in Table 1, as mean and standard deviation values out of three specimens: tensile strength σ_{max}, initial E-modulus E_1 (between 0.015 and 0.025% strain) and final E-modulus E_3 (between 0.4 and 0.6% strain). It appears that the initial (uncracked) stiffness is far superior for SCI compared to HLU manufacturing. From Fig. 2, it is clear also that the deviation of the initial linear behavior (due to the onset of multiple cracking) occurs at lower stresses for HLU than for SCI. This probably indicates a better impregnation of the fiber bundles with SCI, which applies a significantly higher pressure at manufacturing. Post cracking stiffness E_3 and strength σ_{max} are 10% higher for SCI, indicating again a better bundle impregnation.

Table 1. Static tensile mechanical properties

Property	HLU	SCI
σ_{max} (MPa)	45.6 ± 2.6	50.5 ± 1.9
E_1 (GPa)	14.5 ± 0.2	19.0 ± 1.8
E_3 (GPa)	3.53 ± 0.3	3.75 ± 0.2

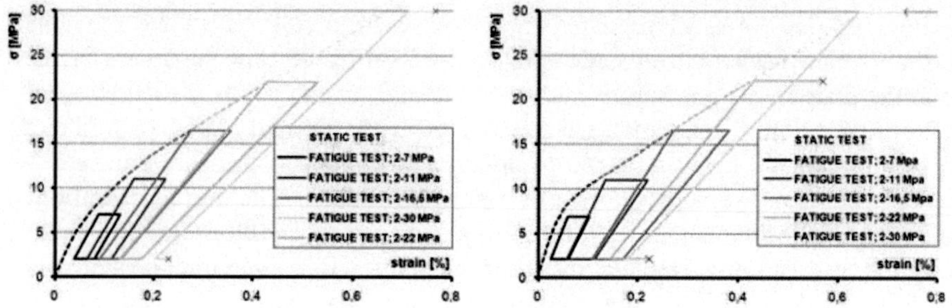

Fig. 2. Tensile stress-strain diagram with HLU (left) and SCI (right)

Tensile fatigue testing was performed on a hydraulic MTS test bench with a capacity of 100 kN. A sinusoidal cyclic load was imposed with a frequency of 10 Hz, up to a total number of one million cycles. The lower load level corresponds to a stress of 2 MPa, the highest to 7 MPa, 11 MPa, 16.5 MPa, 22 MPa and 30 MPa. Three identical specimens were tested at the same maximum stress level $\sigma_{c,max}$. The strain was again measured with an extensometer. The test results are indicated on Fig. 2. The quadrangles represent for each maximum stress level the evolution of the deformations: the two left corner points represent the deformation at the beginning and the end of the unloading part of the first cycle, while the two right corner points give the same deformations after one million cycles. A quadrangle which is open to the right represents the deformation at failure (cross point) when this occurred before one million cycles.

Table 2 summarises the stiffness data obtained after the first ($E_{cycle,1}$) and after the millionth ($E_{cycle,10E6}$) cycle, thus the slope of the line between the two left resp. right corners of the quadrangles of Fig. 2 (mean values of three specimens). Numbers between brackets are values at the last cycle before failure. It is clear that the higher the maximum stress, the lower the cycle stiffness. Moreover, the cycle stiffness decreases continuously during the cyclic loading. However the decrease is lower with increasing maximum stress, while the cycle stiffness always remains significantly higher than the tangential stiffness E_3 from Table 1. This indicates that the fully cracked matrix contributes even after one million cycles to the composite stiffness, mainly through friction with the fibers. The higher stiffness in all cases for SCI is again an indication of the better impregnation which can be obtained.

Table 2. Stiffness values after one and one million cycles

$\sigma_{c,max}$ (MPa)		7	11	16,5	22	30
$E_{cycle,1}$ (GPa)	HLU	10.6	9.7	7.9	5.8	5.5
$E_{cycle,10E6}$ (GPa)	HLU	8.3	7.0	5.9	(4.7)	(4.7)
$E_{cycle,1}$ (GPa)	SCI	14.6	12.1	9.2	7.2	6.5
$E_{cycle,10E6}$ (GPa)	SCI	10.9	7.8	6.5	(5.5)	(5.2)

Fig. 3 gives an overview of the number of cycles after which failure did occur. An arrow pointing to the right indicates that no failure occurred after one million cycles. While the cycling at a maximum stress of 30 MPa leads to failure after only 10,000 cycles, the maximum cyclic stress of 22 MPa (roughly 50% of the static strength) nearly reaches one million cycles.

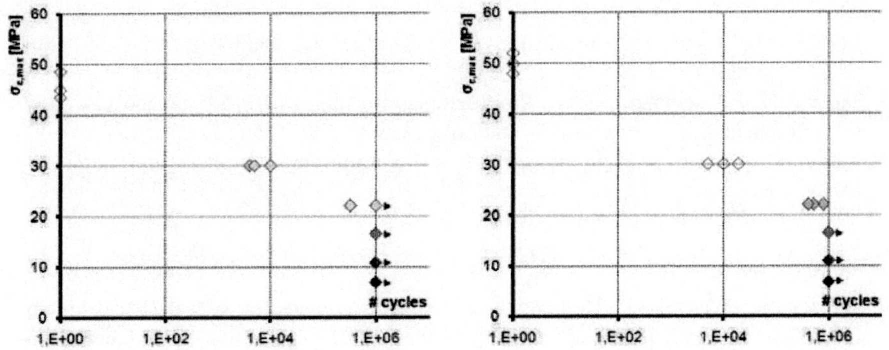

Fig. 3. Number of cycles as function of maximum stress for HLU (left) and SCI (right)

The specimens who did not fail within one million cycles were statically loaded up to failure. Strength values seem hardly affected (mean value around 45 MPa) except for the not failed specimen at a maximum cycling stress of 22 MPa, which failed at 32 MPa, confirming that one million cycles is the limit for this cycling stress.

6 Conclusions

Two processing techniques were used for the impregnation of chopped glass fiber mats to produce textile reinforced cement composites: hand lay-up and a mechanised device called Self Compacting Impregnator. Both techniques allow for the inclusion of high volume fractions of fibers, 18% in this study. Tensile specimens show a remarkable strain hardening behavior with a strength of 45 to 50 MPa, after a gradual development of a very fine crack pattern with less than 1 mm spacing, leading to a mean crack width of less than 10 μm at ultimate load. Strength and stiffness properties of SCI specimens are superior to HLU specimens, indicating a better impregnation of the fiber bundles.

Repeated cyclic loading results show that specimens subjected to a maximum cycling stress of up to 22 MPa can withstand one million cycles. This stress is about 50% of the static strength, and well beyond the cracking load of the matrix. Further tests under environmental conditions (humidity and freezing) are on the way to assess the durability and serviceability limit state of this kind of composites.

Acknowledgements. The help of FP6 EU contract 26574 "Textile architecture – Textile structures and buildings of the future" (Contex-T) is greatly acknowledged through a grant for the second author, and the financial means to build prototype V of the SCI. Thanks to René Heremans for the manufacturing of the specimens and his never ending enthusiasm for cement composite research.

References

[1] Li, V.C.: On engineered cementitious composites (ECC). Journal of Advanced Concrete Technology 1(3), 215–230 (2003)

[2] Naaman, A.E., Reinhardt, H.W.: Proposed classification of HPFRC composites based on their tensile response. Materials and Structures 39, 547–555 (2006)

[3] Remy, O., Wastiels, J.: High performance textile reinforced cements: tensile hardening behaviour and modeling. In: Marques, T., et al. (eds.) Challenges for Civil Construction, p. 10 (2008) ISBN 978-972-752-100-5

[4] Vanherck, J., Wastiels, J., Remy, O., Dannau, M.: Impregnation method. WO 2009/030710

[5] Cuypers, H., Wastiels, J.: Thin and strong concrete composites with glass textile reinforcement: about modelling the tensile response. In: Dubey, A. (ed.) ACI Special Publication 250, pp. 131–148 (2008)

A Mixed Pultrusion and Braiding Process Adapted to the Production of High Performance Cement Composite Beams

A. Gabor, P. Hamelin, and G. Promis

Civil Engineering and Environmental Laboratory (LGCIE),
Université Claude Bernard Lyon 1, Université de Lyon, France

Abstract. The objective of the present paper is the optimization of the behavior of structural elements for civil engineering applications. A specific manufacturing process has been developed (Lamtextress patent) allowing to produce hollow beams made of mineral based pultruded composite plates. The beams are hollow with square cross section; they are realized by bonding together composite plates having the dimensions 200 x 10 x 1 cm^3. In order to improve the overall performances of the beams, they are reinforced by braiding technology and hand lay-up wrapping. The shear and flexural performance of the final composite beams can be calculated taking into account different process parameters (e.g. nature of fibers, number of filaments, braiding angle).

1 Introduction

Newly developed composites used in civil engineering have to combine mechanical performances and architectural freedom, satisfying environmental and sustainable development criteria. Mineral matrix composites could be a solution for these requirements [1]. The applications using fiber reinforced mineral matrix composites or textile reinforced concrete are increasing and focus also on loadbearing structural elements. Consequently detailed information is needed for the safe design of these loadbearing structures [2].

Mineral matrices are characterised by a fragile failure. Even if they have a good mechanical behavior in compression, they have to be strengthened in order to ameliorate their characteristics in tension or shear. The observed damage mechanisms occur at different levels: impregnation [3], influence of the geometry of the textile reinforcement [4], cracking of the composite material [5].

G.J. Parra-Montesinos, H.W. Reinhardt, and A.E. Naaman (Eds.): HPFRCC 6, pp. 519–526.

2 Intrinsic Characteristics of the Composite

The matrix of the composite is inorganic phosphate cement (IPC), a new material combining the flexibility of polyester resins with the properties of the ceramics [5]. The obtained ceramic paste can be generally reinforced by glass fibers or textiles in order to obtain a composite with increased toughness, tensile strength and stiffness. The cure can be done at ambient or high temperature (approximately 60-80°C), but confinement is necessary in order to avoid evaporation.

2.1 Manufacturing Procedure

For the realization of the beams, plates have been manufactured using pultrusion technique. The pultrusion process is similar to that used in the case of organic polymer matrix composites (e.g. epoxy, polyurethane) with some modifications of the impregnation bath and the shaping die. The scheme of the pultrusion process is presented in Fig. 1.

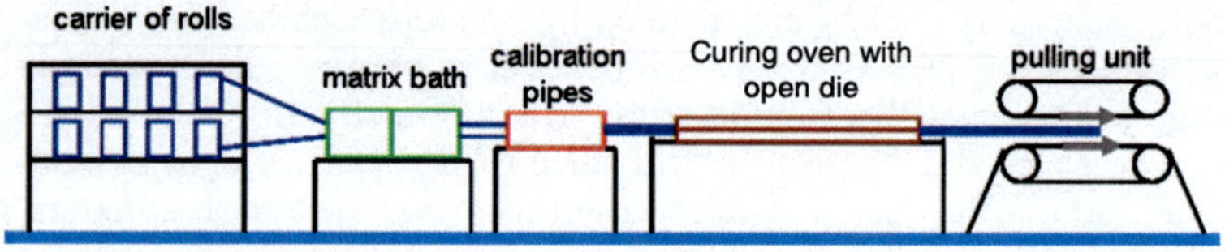

Fig. 1. Outline of the pultrusion process

The reinforcement is made of classical unidirectional E-glass rovings having a linear weight of 735 tex, which pass trough an impregnation bath. After the impregnation of the filaments in the bath, they transit through calibration pipes. The next step is the final shaping and curing in an oven: the composite passes first through a classical die at 40°C and through an open die at a temperature of 70°C. The pulling unit ensures the advancement of the process at a speed of 20 cm/min. The fiber volume fraction ratio obtained here is estimated to 22%. The characteristics of the matrix and E-glass rovings are summarized in Table 1.

Table 1. Physical properties of the composite's components

	IPC matrix	E-glass roving
Density(kg/m^3)	1 700	2 540
Linear weight (tex)	-	735
Elastic modulus (MPa)	10 000	85 000
Shear modulus (MPa)	3 845	30 000
Poisson's coefficient	0.30	0.22
Tensile strength (MPa)	8-10	2 500
Shear strength (MPa)	4-5	-

2.2 *Mechanical Properties of the Composite in Tension, Compression and Shear*

Fig. 2 presents the tensile/compression constitutive equation of the composite. Shear behavior is considered as linear, with a shear strength defined by the

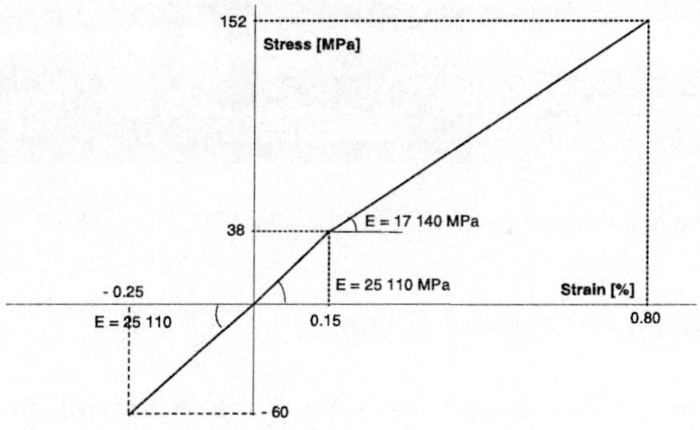

Fig. 2. Constitutive equation of the composite in tension/compression

strength of the matrix (Table 1). The detailed testing and results are given in a previous article [6].

3 Characteristics of Beams

3.1 *Unstrengthened Beam*

A first beam was realized assembling together the plates obtained by the pultrusion process in a manner to have a hollow square cross section. The four plates forming the beam were bonded together using a bi-component epoxy resin, hardening at room temperature. This first beam is referenced B1 (Fig. 3).

Knowing that the shear characteristics of the composite material are not satisfactory, three other beams, having the same geometrical configuration as the initial beam B1, were realized with different types of reinforcements.

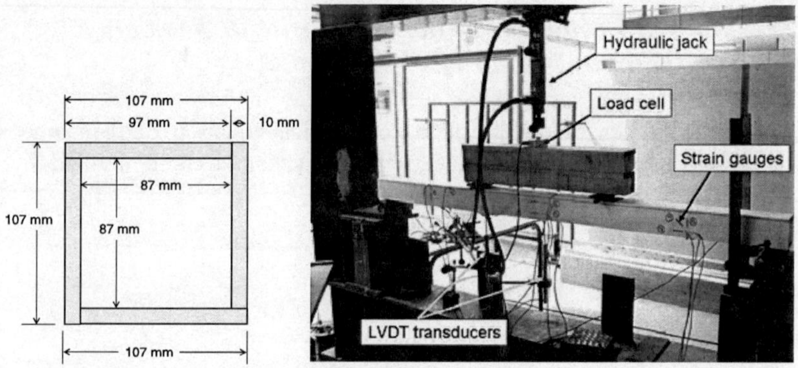

Fig. 3. Cross section and experimental setup of the beam B1

3.2 Internally Strengthened Beam

Is obtained by the bonding of vertical struts at the interior of the beam, cut-out from the pultruted plates, having a spacing of 200 mm (Fig. 4). The struts were placed in a way that the fiber reinforcement is oriented vertically in order to have a maximum efficiency. The reference of this beam is BS2.

3.3 Beam Strengthened by 2D Braiding

The braiding process is a classical one, used in the industry. The different operations are described on Fig. 5. The transverse filaments are winded on spindles (a), which are moving on the circular plate of the braiding machine. Additional longitudinal textile fibers can be introduced directly through the circular plate of the braiding machinery. The transverse reinforcements are passing through a braiding ring (b) and are winded on a mandrel (c) which is pulled in the longitudinal direction. The rate of displacement of the mandrel and the rotational speed of the spindle permits to control the spacing distribution of the shear reinforcement. The main parameters which can be adapted for each braiding operation are the following: fiber nature (glass, carbon, aramid, …); filament size, sizing, number of filaments (Tex), braiding angle, braiding spacing, tension force in filament. In the present case 36 carriers were used for supporting the roving rollers. The beam was fixed on a support and travelled across the braiding device (Fig. 6).

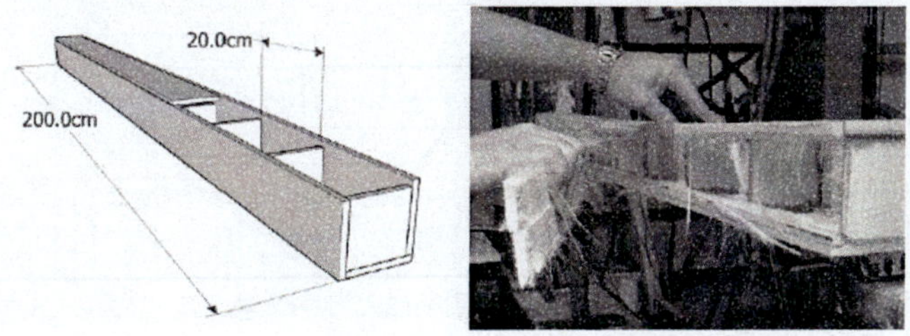

Fig. 4. Overview of the beam BS2, strengthened with struts (geometry and failure after test)

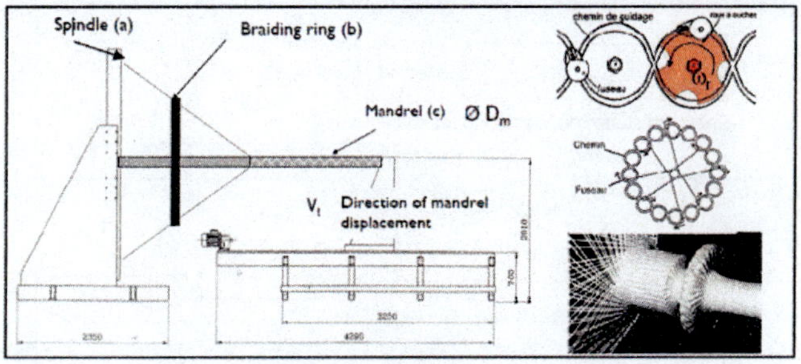

Fig. 5. Different operations associated to 2D braiding process

Fig. 6. Overview of the braiding process used for the beam BC3

In the present case, the external 2D braiding was realized with carbon strands (1000 dtex, 4.1 mm width) spaced at 20 mm and inclined at 45°. The average thickness of the braiding was about 1 mm. This beam is referenced BC3.

For the estimation of the shear reinforcement capacity of the braiding, the steps are the following (Fig. 7):

1. The maximum shear force which can be supported by the section of the beam is calculated, considering a shear stress limit of the composite equal to 4 MPa ($V_{Rd,c}$)

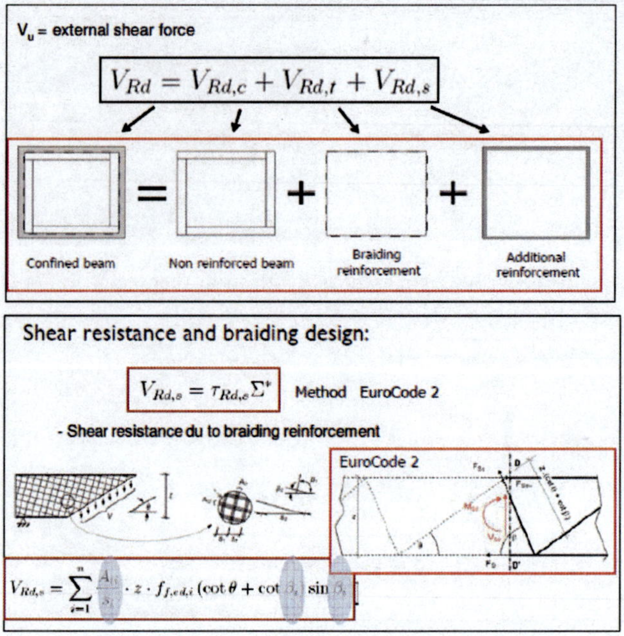

Fig. 7. Setting up of braiding parameters for shear design

2. The shear force ($V_{Rd,t}$) , which must be supported by the braided glass reinforcement is determined as the difference between the external shear force applied to the structure (V_u) and the shear force step 1: $V_{Rd,t}=V_u-V_{Rd,c}$
3. Consequently, the number of yarns, their orientation, and their distance will be calculated from the expression defined in Fig. 7.
4. If necessary, it is possible to complete the reinforcement by additional layers.

3.4 Beam Strengthened by External Wrapping

An un-strengthened beam was wounded around with a continuous layer of IPC composite reinforced with a 300 g/m² E-glass chopped strand mat (CSM). A hand lay-up technique similar to that used for classic FRP composites is employed for the realization of the winding directly on the beam. The additional thickness of this reinforcement is about 4.1 mm. This latter beam is referenced BW4.

4 Experimental Study

All the four beams have been tested in four point bending with a usual instrumentation in order to realize the analysis of the mechanical behavior. The beams having a length of 2000 mm, the distance between the bearings was established at 1890 mm; loads are applied at the third of the span (630 mm), as shown in Fig. 3.

The instrumentation of the beams was made in order to ensure the acquisition of the following parameters: applied load, mid span displacement, strain fields in the different zones of the beams. The test was force controlled.

Fig. 8 illustrates the behavior of the beams in terms of applied load and mid-span deflection: it seems to be a linear relation between the two parameters until the maximum load supported by the beams, which in the case of beams B1 and BC3 corresponds to the failure load. Concerning the evolution of stiffnesses, for the beams B1, BS2 and BC3, they are quasi identical (1160 N/mm) with low discrepancy (less than 5%). This is fully understandable since little material was added, with low effect on the geometrical and mechanical configuration of the beam. On the other hand, for the last beam, an increase of 12% of the stiffness is noticed since in this case the total thickness of the external layers is about 8.2 mm.

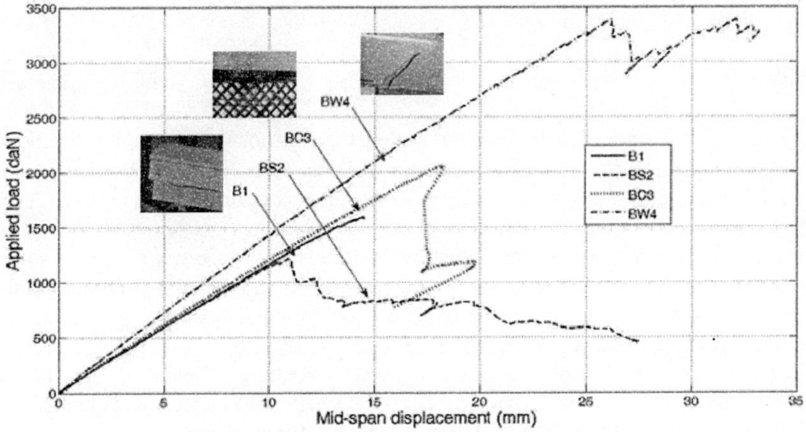

Fig. 8. Global response of beams in terms of shear force – mid-span displacement curves

The failure of the beam B1 was due exclusively to shear, with a quasi instantaneous development of a horizontal crack at the mid-height of the beam, parallel to the orientation of the reinforcing roving strands. The slipping between the upper and lower part of the beam was about 4.7 mm, measured at the level of the bearings. The maximum deflection is equal to 14.5 mm (0.73% of the span). The beam BS2 showed a reduced strength compared to that of the reference beam B1, its failure was due also to shear, with the same longitudinal crack at the mid-height of the beam. In fact, the struts had a damaging effect instead of reinforcing: the struts reduced the buckling length of the fibers in the compression zone, inducing supplementary longitudinal stresses. The carbon fiber braided beam BC3 shows an increase in strength of about 30%, the maximum deflection is 18.2 mm (1% of the span). In this case, the shear failure was accompanied by the local debonding of the braiding. For the fourth beam BW4, wrapped with external layers, the increase in ultimate load is very important, approximately 135%. The beam still failed in shear: the crack started at the zone where the load was applied and developed at 45°.

5 Conclusions

In order to realize structural elements it is important to develop adequate manufacturing process to the composite material. For this, the pultrusion technique has been adapted to the use of IPC composite material. The pultruded plates showed some deficiency in terms of shear capacity, therefore an improvement of its performance was necessary. The external confinement by carbon fiber braiding increased the shear capacity in the range of 25-30%, but the weak adhesion between the beam and the braiding (only mechanical confinement) limited its effect. The external wrapping with CSM reinforced IPC composite was the most effective, with an increase of the shear load in the range of 135%. The maximum flexural loadbearing capacity is not reached, thus an improvement of the shear behavior is still recommendable.

References

[1] Brockmann, T., Brameshuber, W.: Matrix development for the production technology of textile reinforced concrete (TRC) structural elements. In: Third International Conference Composites in Con-struction (CCC 2005), France, Lyon, pp. 1165–1172 (2005)

[2] Magalhaes, A., Marques, A., Oliveira, F., Soukatchoff, P., Castro, P.: Mechanical behaviour of cementitious matrix composites. Cement & Concrete Composites 18 (1996); Peled, A.: Textile cement based composites, effects of fabric geometry, fabric type and processing. In: Third International Conference Composite in Construction, France, Lyon, vol. 9-22, pp. 1129–1138 (2005)

[3] Hegger, J., Voss, S.: Textile reinforced concrete – Bearing behavior, design, applications, Composite in Construction. In: Third International Conference Composites in Construction (CCC 2005), France, Lyon, pp. 1139–1146 (2005)

[4] Peled, A.: Textile cement based composites, effects of fabric geometry, fabric type and processing . In: Third International Conference Composite in Construction , France, Lyon, pp. 1129–1138 (2005)

[5] Cuypers, H., Van Itterbeeck, P., De Bolster, E., Wastiels, J.: Durability of cementitious composites. In: Third International Conference Composites in Construction (CCC 2005), France, Lyon, pp. 1205–1212 (2005)

[6] Promis, G., Gabor, A., Hamelin, P.: Assessment and modeling of the behavior of IPC composites in tension, compression and shear. In: 2nd ICTRC, International RILEM Conference on Material Science, pp. 407–417 (2010)

TRC and Hybrid Solutions for Repairing and/or Strengthening Reinforced Concrete Beams

A. Si Larbi, R. Contamine, and P. Hamelin

LGCIE Site Bohr, Université Claude Bernard Lyon I INSA de Lyon, Villeurbanne, France

Abstract. Over the last twenty years, composite materials, especially carbon epoxy, have developed as part of the repair and strengthening of reinforced concrete structures. This is mainly due to their very good mechanical performance as well as their easy implementation, which is fast and unobtrusive. However, composites also have a number of limitations that make the alternatives advantageous, for example their very high price, their incompatibility with sustainable development, and their very low efficiency ratio.

Also, although they are adapted to various solicitations (and corresponding structural elements) FRP are used essentially for flexural strengthening. The reliability of the strengthening method is generally study for the ultimate performance (ultimate load and ductility) or to the performance at the service limit state (crack opening, deflection). The objective of this study is to assess potential alternative solutions based on textile reinforced concrete (TRC) mainly in the control of cracking in reinforcement (undamaged beams), or hybrid solutions combining TRC and rods (carbon, glass or both) when it is important to satisfy the two limit states (ultimate and service) as part of the repair (previously damaged beams). The resulting performance will be compared to externally bonded FRP solutions.

1 Introduction

Although sometimes incomplete and uncertain, our knowledge of TRC has made significant progress in recent years; the main related studies have focussed on its mechanical characterization and the identification of interaction mechanisms and the transfer of effort between the weave fabric and the matrix [1, 2, 3]. Few studies have examined the suitability of TRC in structural applications [4, 5] and more specifically in its resistance to bending moment [6].

This work is based on five 2 m beams and aims to study the technological feasibility and the mechanical performance of a solution based on TRC. For this purpose, two variants, associated with different modalities of implementation and different objectives will be considered. The first variant concerns the strengthening of two undamaged beams corresponding to a direct implementation of TRC with the main objective being to control cracking. The second variant

G.J. Parra-Montesinos, H.W. Reinhardt, and A.E. Naaman (Eds.): HPFRCC 6, pp. 527–534.

focuses on three beams which had been damaged to the point of longitudinal steel yielding, and were then repaired (one with CFRP) with plates using CRT + rods (carbon and / or glass) and glued onto the substrate using a structural resin on the tensile face with the primary objective of achieving a significant increase in ultimate performance.

2 Experimental Approach

2.1 Properties of Reinforcing Materials, Specimen Geometry and Instrumentation

It should be noted that reinforcements adopted in the case of damaged beams were designed (first approach) with reference to the AFGC recommendations (French Recommendations, 2010), relating to the repair and strengthening of reinforced concrete structures using composite materials, so as to obtain equivalent ultimate capacities.
TRC: Emaco R315 from BASF was used. The textile reinforcement is a knitted textile grid (warp knitting). The meshes are rectangular and "open", at 3x5mm (frame * array). The warp (direction of loading) is AR glass (for strength and durability) and the weft is polyester. Table 1 completes the details. To optimize the performance of the TRC, a coating product designed to facilitate the impregnation of fibers was applied to the fabric before the mortar was added. The strengthening method to apply the TRC is the wet lay up method.

Table 1. Technical characteristics of AR glass

	Number of filament per roving	Roving titer (tex)	Filament diameter	Roving strength
AR glass	1600	25000	700	1102

Two variants were selected: the first, TRC, consisted of three layers of alkali-resistant fabric embedded in mortar and the second, of two layers of alkali-resistant fabric used with carbon rods (TRC + JV) or a combination of carbon and glass (CRT + JVC). The rods were treated (surface scattered with silica) to improve the roughness of their surface. See Table 2 for the mechanical and geometrical properties of the rods.

Table 2. Properties of the rods

	Diameter (mm)	E (MPa)	Tensile strength (MPa)
Glass rod	2	25000	700
Carbon rod	2	140000	2240

The static characterization of the TRC and the hybrid solutions was established with a suitable tensile test (Fig. 1).
CFRP: The main properties of the CFRP are (thickness: 0.4 mm, tensile strength: 700 MPa, Young's modulus: 80000 MPa).
Steel and concrete: The steel is grade E500. The concrete used was a ready-mixed concrete; this limited the disparity between different batches. The compressive strength of concrete laid at 28 days is 30 MPa.

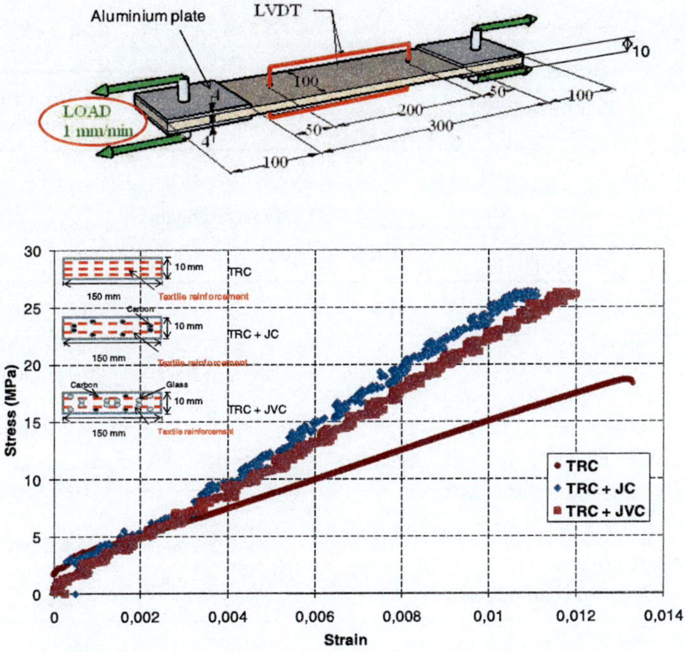

Fig. 1. Tensile specimen (geometry, instrumentation, and loading) behavior law

Specimen definition: The beam had an overall length of 2.3 m with a clear span of 2 metres (Fig. 2). To evaluate the performance of the reinforcement, three reinforced concrete beams were loaded before being strengthened (CFRP, TRC + JC and TRC + JVC).

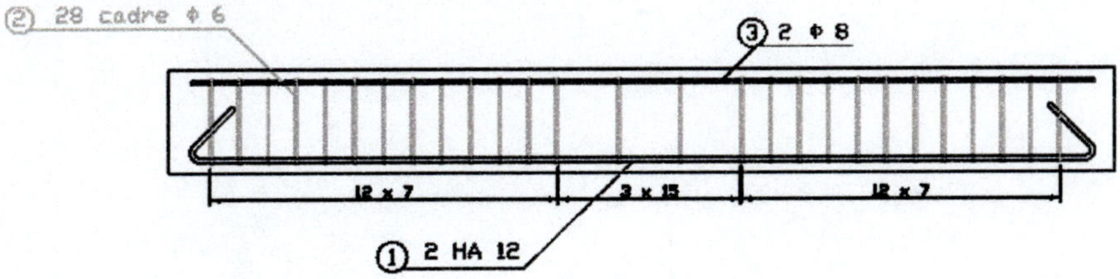

Fig. 2. Geometry of the beams

First maximum loads before strengthening applied to obtain the longitudinal steel yielding. These loads are controlled via a strain gauge glued on steel. Subsequently, this type of beam is said damaged. The beam strengthened by TRC has not been previously damaged (undamaged beam).

Strengthening plates have a length of 1.95 meters and are cut to pass between the support beams during testing. Table 4 summarizes the solutions adopted for this study.The TRC reinforcement was a sanded plate (casted 7 days before) glued to the sanded underside of the reinforced concrete beam (undamaged).

Table 4. Beams - definition

<table>
<tr><th></th><th>Reinforcement</th><th>Axial stiffness E.A. (MN)</th><th>Damage level</th><th>Dimensions of reinforcement</th></tr>
<tr><td>Beam 1 (Reference beam)</td><td>Reference beam</td><td>-</td><td rowspan="2">Undamaged beams</td><td>-</td></tr>
<tr><td>Beam 2 (TRC)</td><td>TRC (3 glass AR fabric)</td><td>1.8</td><td>10 x 150 (mm)</td></tr>
<tr><td>Beam 3 (CFRP)</td><td>CFRP</td><td>4.8</td><td rowspan="3">Damaged beams</td><td>0.4 x 150 (mm)</td></tr>
<tr><td>Beam 4 (TRC + JC)</td><td>TRC (2 glass AR fabric)
+
JC
Carbon rod</td><td>3.5</td><td>10 x 150 (mm)
+
8 φ 2 (carbon)</td></tr>
<tr><td>Beam 5 (TRC + JCV)</td><td>TRC (2 glass AR fabric)
+
(JVC)
Glass and carbon rod</td><td>3.3</td><td>10 x 150 (mm)
+
4 φ 2 (carbon)
+
12 φ 2 (glass)</td></tr>
</table>

Instrumentation: The hybrid reinforcement (TRC + JC; TRC + JVC) and CFRP were casted in place on the underside of a returned beam (damaged). Test specimens were subjected to four-point bending load tests. The clear span of the beams was 2 m. A static monotonous load was applied at 60 cm intervals until specimen failure.

To measure specimen sustained displacement, a LVDT Racing ± 100 mm was placed at the centre of the beam

3 Results

3.1 Analysis of Load-Displacement Curves

The load-deflection curves (Fig. 3) bring out the difference in performance between the initially undamaged beams and the previously damaged and repaired ones.

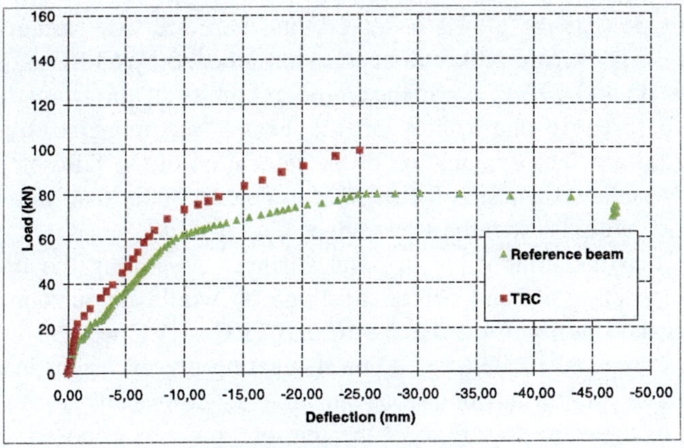

(a) Initially undamaged beams

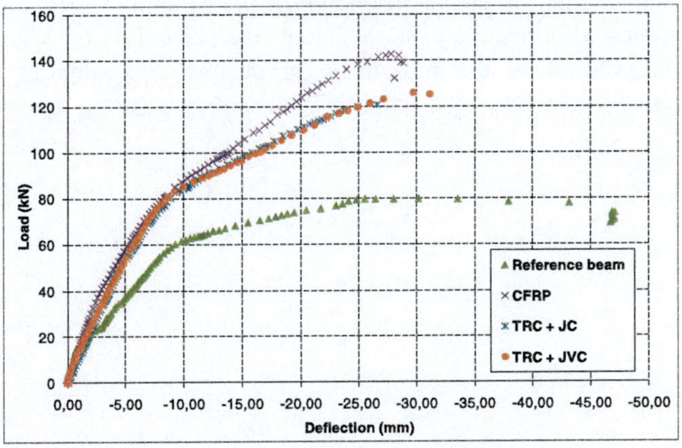

(b) Initially damaged beams

Fig. 3. Load-displacement curve

In the case of the undamaged beams there are three phases. The first reflects the integrity of the beam in that none of the materials of the beam is damaged. The second is the propagation of cracks and their multiplication along the beam; the last phase is that of the steel yielding. The previously damaged beams did not have a phase 1. As for the gains in terms of ultimate and service load obtained through TRC reinforcement compared to the reference beam, we see that hybrid solutions based on TRC associated with carbon and / or glass are well placed in relation to the CFRP solution although the ultimate load level is lower.

3.2 Failure Mode

One failure mode (peeling-off) is observed, but there are two variants. The first (Fig. 4b), is a characteristic failure in beams strengthened by CFRP and concerned beams TRC + JC and CFRP, occurring in the extremity of the beam. The second (Fig. 4 c) occurred in the centre of the beam, accompanied by the plate strengthening failure. The explanation of the relocation of the failure could be the relatively low axial stiffness of the reinforcement plate that allows greater flexibility which would soften the stress edges.

It follows that the limit value of axial stiffness separating the two areas of strength (edge peeling-off and central peeling-off) would be surrounded by the values of the axial rigidity of the TRC + JC and TRC + JVC which would explain their different modes of failure. Central peeling-off is, significantly, very uncommon; it is more advantageous in that it allows us to envisage an improvement in the efficiency ratio of the reinforcement as opposed to the fairly weak rate of the CFRP reinforcement. This type of failure is also independent of the mode of implementation chosen (bonding of plates using a structural resin or used without the use of a resin) and so this opens the possibility of doing without a bonding resin, as long as the mortar is thixotropic and suitable for in-situ implementation.

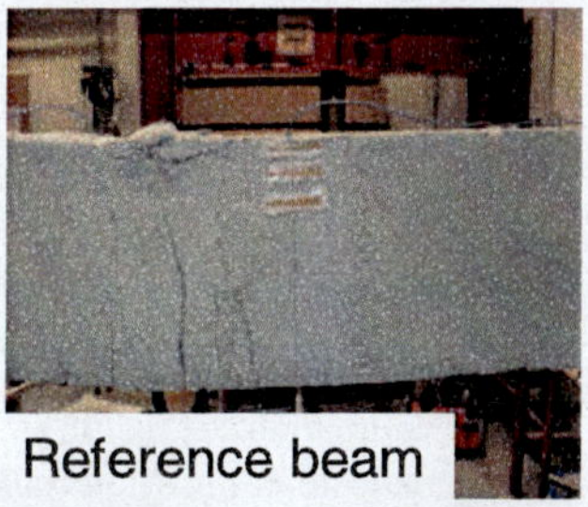

(a) Reference steel yielding

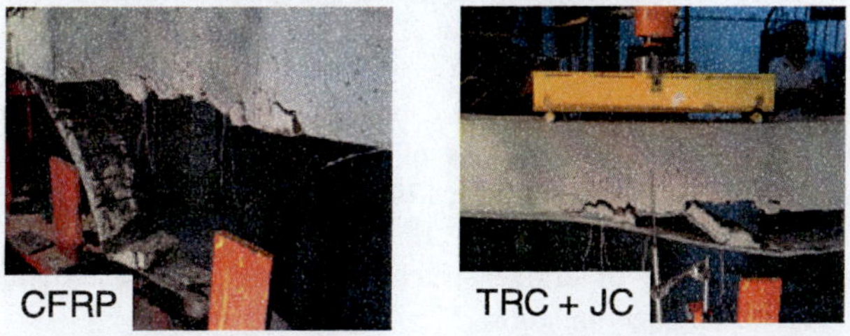

(b) Peeling-off in the end of the plate

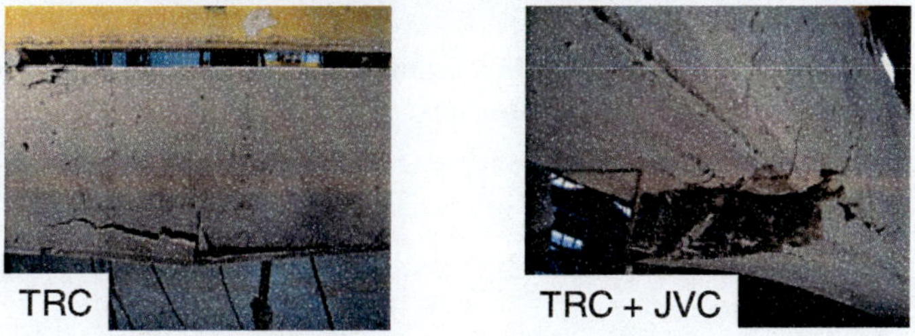

(c) Peeling-off in the center of the beam

Fig. 4. Failure modes

4 Conclusions

The experimental part of this article has brought out the positive factors of TRC hybrid solutions in the repair of reinforced concrete beams, in terms of both ultimate and service behavior, with very similar performances to that of traditional solutions such as CFRP. The use of TRC alone to strengthen a reinforced concrete beam seems less advantageous and while it significantly improves the ultimate performance, it cannot be considered in the context of increasing the bearing capacity. It does, nevertheless, appear interesting for the control of cracking in the service phase.

It is important to outline the change of failure mode in the TRC and TRC + JVC solutions; this allows a better use of the reinforcement and that can be explained by the relatively low axial stiffness of reinforcement which can be provided there is sufficient interaction with the concrete surface.

References

[1] Cuypers, H., Wastiels, J.: Stochastic matrix-cracking model for textile reinforced cementitious composites under tensile loading. Materials and structures 39, 777–786 (2005)

[2] Hegger, J., Voss, S.: Investigations on the bearing behaviour and application potentialof textile reinforced concrete. Engineering structures 30(7), 2050–2056 (2008)

[3] Peled, A., Cohen, Z., Pérez-Garcia, Di capua, D., Pujades, L.: Influences of textile characteristics on the textile properties of warp knitted cement based composites. Cement and Concrete Composites 30(3), 174–183 (2008)

[4] Si Larbi, A., Contamine, R., Ferrier, E., Hamelin, P.: Large scale hybrid FRP compositegirders for use in bridge structures—theory, test and field application. Composites Part B: Engineering 36(8), 573–585 (2005)

[5] Triantafillou, T.C., Papanicolaou, C.: Shear strengthening of reinforced concrete members with textile reinforced mortar (TRM) jackets. Materials and structures 39(1), 93–103 (2006)

[6] Brückner, A., Ortlepp, R., Curbach, M.: Textile reinforced concrete for strengthening in bending and shear. Materials and Structures 39(8), 741–748 (2006)

Hybrid Fiber Reinforcement and Crack Formation in Cementitious Composite Materials

E.B. Pereira[1], G. Fischer[2], and J.A.O. Barros[1]

[1] ISISE, University of Minho, Guimaraes, Portugal
[2] Technical University of Denmark, Lyngby, Denmark

Abstract. The use of different types of fibers simultaneously for reinforcing cementitious matrices is motivated by the concept of a multi-scale nature of the crack propagation process. Fibers with different geometrical and mechanical properties are used to bridge cracks of different sizes from the micro- to the macro-scale. In this study, the performance of different fiber reinforced cementitious composites is assessed in terms of their tensile stress-crack opening behavior. The results obtained from this investigation allow a direct quantitative comparison of the behavior obtained from the different fiber reinforcement systems. The research described in this paper shows that the multi-scale conception of cracking and the use of hybrid fiber reinforcements do not necessarily result in an improved tensile behavior of the composite. Particular material design requirements may nevertheless justify the use of hybrid fiber reinforcements.

1 Introduction

Strain-hardening in tension is a relevant material property for structural applications of cementitious composites. Commonly known as Strain hardening cement composites (SHCC), these materials show benefits in terms of durability and preservation of functional properties of the structural elements. Examples of SHCC exist which have been designed and proven to gather these requisites in different perspectives [1]. Engineered Cementitious Composites (ECC), a class of cement based materials typically reinforced with Polyvinyl Alcohol (PVA) fibers, is one of the examples of SHCC showing relatively high ductility (between 3% and 7% of strain) and average tensile strength of 5 MPa [2, 3].

The recent technological development of a wide variety of fibers has created new opportunities for the development of SHCC. Envisaging an optimal utilization of the materials, either in an economical or environmental perspective, hybridization of the fiber reinforcement has been studied in recent past. The use of fibers of different natures, with distinct geometrical and material properties, has been reported to improve the material features of SHCC [4-7].

G.J. Parra-Montesinos, H.W. Reinhardt, and A.E. Naaman (Eds.): HPFRCC 6, pp. 535–542.

One of the main advantages usually assigned to the hybridization of fiber reinforcement is the possibility to restrain cracking at different scales [5]. Concerning the micromechanics of the composite, the commonly recognized multi-scale nature of concrete should be reflected in the cracking processes. In a simplistic perspective, one can assume that the micro-cracks generated during cracking processes are bridged by smaller fibers and the macro-cracks bridged by the larger fibers [8]. A crack necessarily undergoes different stages, from onset to the fully separated state, and a visible crack may be seen as the result of the coalescence of previous randomly oriented and diffusely distributed micro-cracks. It seems, though, that the unambiguous evidence to this mechanism is not easy to unveil. The relation between this hierachized conception of the cracking process in a composite, the different material scales and the mechanisms of crack restraining by the fiber reinforcement and stiffening mechanisms in tension are not clearly established.

The assessment of the tensile stress – crack opening law may help to quantify the contribution of each fiber reinforcing system to the overall tensile composite behavior. The potential synergistic effect resulting from fiber hybridization may also be directly and explicitly quantified. For this purpose, the single crack tension test (SCTT) setup may be used to allow the direct assessment of the tensile stress – crack opening law. As shown in previous work, this procedure focuses on the assessment of the tensile stress-crack opening law of SHCC [9, 10]. This is done in a direct fashion with the main advantage of allowing the direct characterization of the tensile constitutive behavior of the FRCC without the need of knowing the shape of the tensile stress-crack opening law in advance, as opposed to indirect methods where the constitutive tensile stress-crack opening law is derived by inverse analysis [11-13]. This feature may become especially relevant for composites with hybrid reinforcement, where fibers are activated at different stages of the cracking sequence which may result in more complex shapes of the tensile stress – crack opening law. This procedure may contribute to a deeper and more accurate characterization of the tensile behavior of hybrid FRCC with evident benefits for the structural design with SHCC. Additionally, most of the numerical models for structural design rely on the fracture mode I constitutive law for the material modeling [14, 15]. Furthermore, the material design process also benefits from the direct assessment of the tensile stress-crack opening law. The procedure for the optimization of the different composite components becomes more efficient with the direct quantification of differences between distinct composite systems.

The direct assessment of the tensile stress – crack opening law is expected to allow the direct quantification of the contribution of each fiber reinforcing system to the global tensile behavior of the composite. The main objective is to understand the contribution that the Single Crack Tension Test (SCTT) setup can give towards a better understanding of fracture micro-mechanisms and the tensile performance of hybrid fiber reinforced cement composites.

2 Experimental Program

In the present work, the tensile stress – crack opening law of different hybrid reinforced cement composites was assessed, focusing mainly on the usefulness of the SCTT test-setup to support the material design of hybrid fiber reinforced cement composites.

Fibers of three different natures were used: PVA (polyvinyl alcohol), PAN (polyacrylnitrile) and PP (polypropylene). The main geometrical and mechanical properties of these fibers are presented in Table 1. The proportions of each fiber reinforcement type in terms of volume percentage used in all the six composites (C1 to C6) are also represented in Table 1.

Table 1. Main physical properties of fibers and composition of the six formulated composites

Fiber	Tensile strength	Length	Diam	Fiber reinforcement (%vol)					
	MPa	mm	μm	C1	C2	C3	C4	C5	C6
PVA 15	1600	8	40.0	1	2	0	0	1	1
PP	900	12	40.0	0	0	2.5	0	1.25	0
PAN 1.5	826	6	12.7	0	0	0	2	0	1

For each composite, six specimens were tested in direct tension using the SCTT setup (see [10]), as represented in Fig. 1 and Fig. 2.

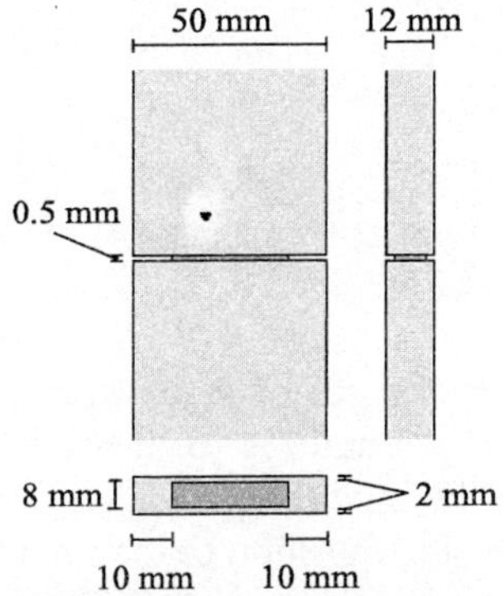

Fig. 1. Geometry used for the assessment of the tensile stress – crack opening law

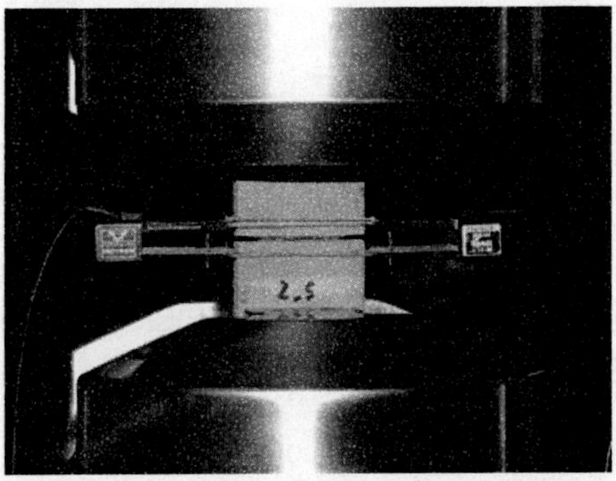

Fig. 2. Tensile test setup, including supports and clip gages

3 Results and Discussion

During the mixing stage, it became evident that the PVA fibers were easier to mix compared to the PAN and PP fibers. As a consequence of the different surface properties of PP fibers, their mixing and dispersion was more difficult. Among all, the PAN fibers were the most difficult to disperse. Likewise, the fibers in the hybrid mixes showed to be easier to mix compared to the single-type fiber reinforced compositions containing PP or PAN fibers.

The test results obtained in terms of tensile stress versus the crack mouth opening displacement (CMOD) for all the composites are presented in Fig. 3 to Fig. 6. The two composites reinforced with PVA fibers (1% and 2%) can be considered as a performance reference given the extensive experience available with PVA ECC [3]. For clarity, only three of the six test results for each composite are presented. The extreme curves were selected for all the composites. The nominal value of the tensile stress was determined by computing the ratio between the registered tensile load during testing and the notched cross section of each specimen.

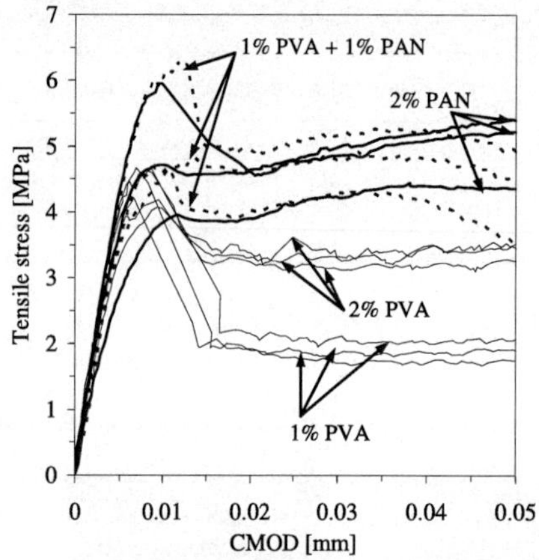

Fig. 3. Preliminary stages of cracking for the composites reinforced with PVA and PAN

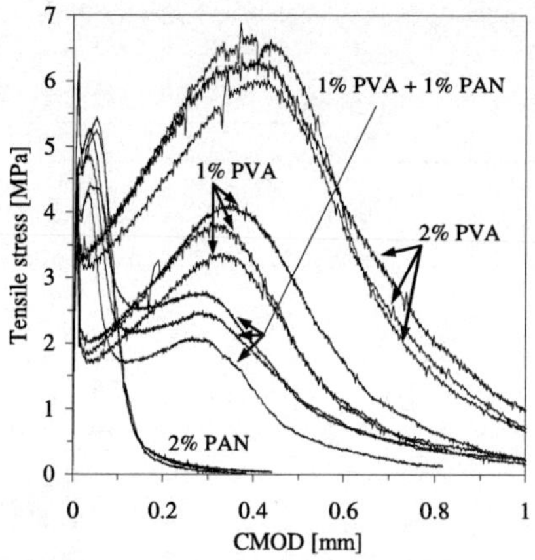

Fig. 4. Overall tensile behavior for the composites reinforced with PVA and PAN

Relevant information can be extracted from several stages of crack initiation and further opening. In a first overall observation (Fig. 4 and Fig. 6) it becomes apparent that the composite behavior resulting from the mixture of two very distinct types of fibers corresponds approximately to the superposition of the original single-type fiber reinforced composites. At an approximately similar reinforcing percentage, the hybrid composite performs figuratively speaking in between the corresponding original single-type fiber reinforced composites. The different types of fibers, either due to distinct geometrical/mechanical properties or due to different interaction with the surrounding matrix (bonding), showed to be activated distinctively and at different stages of the tensile tests. In Fig. 3 and Fig. 5, the results obtained for all the composites for the preliminary stages of cracking are shown. In Fig. 4 and Fig. 6, the overall tensile behavior observed is compared for each group of fibers (PVA/PAN and PVA/PP respectively).

When using PAN fibers, the fiber reinforcement in the composite is activated at relatively small crack openings. In Fig. 3, one can observe that the composite reinforced with 2% PAN fibers showed higher first cracking stresses, reaching peak stresses of up to 6 MPa at this stage. The small fiber diameter and the improved bonding with the matrix resulted in a pronounced crack restraining for CMOD smaller than 10 μm. The subsequent gradual stress decay and the following strain hardening branch developing up to an approximate CMOD of 50 μm are a consequence of the gradual mobilization of the PAN fibers present in the composite. After the peak bridging stress is reached, the subsequent stress decay is rapid, as shown in Fig. 4. When half of the volume fraction of the PAN fiber reinforcement is replaced by PVA fibers, some of the crack restraining effect at the early cracking stage is lost. However, the PVA + PAN composite develops a second strain hardening branch for higher CMOD values, replicating closely the strain hardening branch observed in the PVA fiber reinforced composite. It seems that the hybrid fiber composite is able to combine the features of each of the single-type fiber composites. However it always underperforms the single-type fiber composites at the cracking stages where the latter individually revealed to be more effective. For the PAN + PVA composite, this effect was less significant for smaller CMOD (up to 50 μm). For higher CMOD, the PAN + PVA composite underperformed even the composite containing 1% of PVA. The loss of performance of the fresh mixes due to the addition of PAN fibers may have contributed for these results.

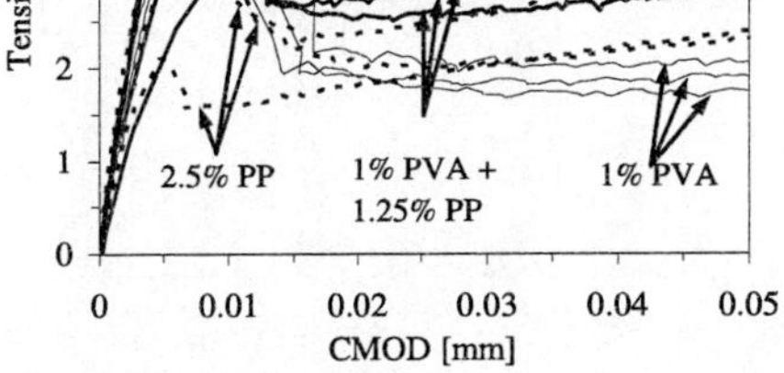

Fig. 5. Preliminary stages of cracking for the composites reinforced with PVA and PP

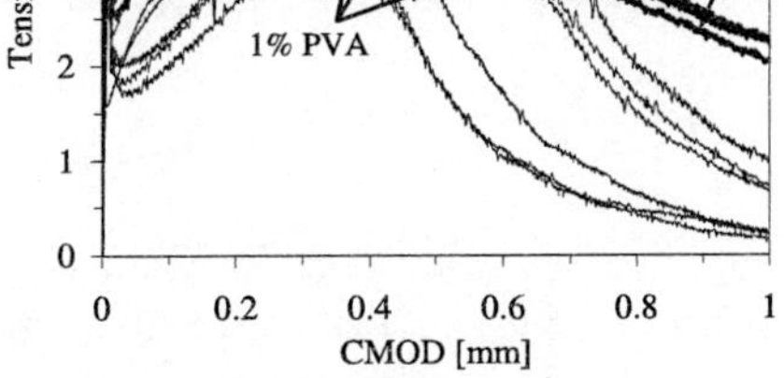

Fig. 6. Overall tensile behavior observed for the composites reinforced with PVA and PP

On the contrary, the use of the PP fibers resulted in a delayed mobilization of the fiber reinforcement. As shown in Fig. 6, the tensile stress – crack opening curve of the composite reinforced with 2.5% PP fibers showed a prolonged tail and a high remaining residual tensile stress for a CMOD of 1 mm. Due to the reduced chemical bonding of the fibers with the surrounding matrix, the prevailing pull-out mechanism of PP fibers manifests itself both in the early stages of cracking (Fig. 5) and in the overall behavior (Fig. 6). During the early stage of cracking, the contribution of the PP fibers to the first cracking strength was slightly smaller than the one observed for the PVA reinforced composites. The differences in the fiber properties become more evident beyond the first cracking stage with a more pronounced post-peak load decay observed for the composites reinforced with PP fibers. This may be an indicator of the sudden debonding of the PP fiber at this stage, which is more pronounced than the one taking place in the PVA fibers. The subsequent stress recovery was also steeper for the PP reinforced composites. The premature debonding of the PP fibers was again revealed by the rapid loss of the initial bridging stiffness in the hardening stage. The hybrid composite behavior was found to be a combination of the tensile features observed in the respective single-type fiber composites. As in the case of PAN + PVA, the hybrid composite of PP and PVA underperformed the single-type fiber reinforced composites at the cracking stages where the latter individually revealed to be more effective.

To allow a clearer interpretation of the results above discussed, in Fig. 7 the average of the experimental results obtained for each composite are shown. This was done by averaging the tensile stress obtained for all the six specimens at each CMOD value. An important indicator of the tensile performance of the tested composites is the fracture energy (G_f), approximately assessed by computing the area limited under the experimental tensile stress – crack opening curves. These values were computed for the average curves of each composite (see Fig. 8).

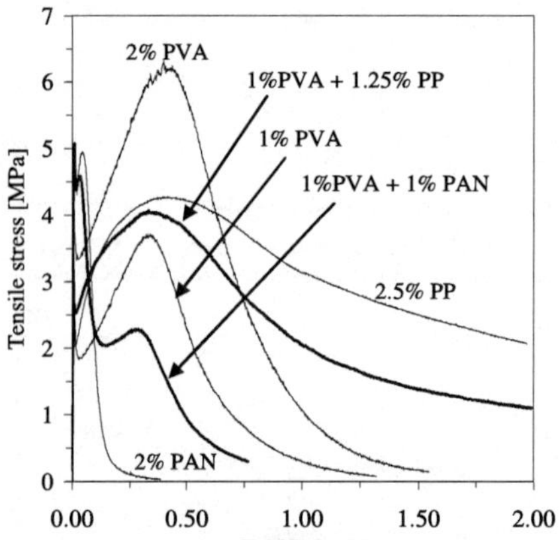

Fig. 7. Overall tensile behavior (average curves)

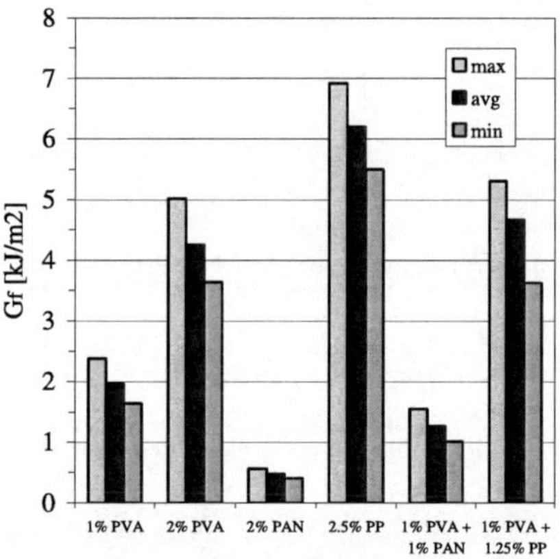

Fig. 8. Fracture energy for the minimum, average and maximum experimental curves

In general, the obtained values show a dispersion that is similar among all the composites, indicating that the parameters adopted for the SCTT setup are adequate and lead to consistent results. The hybrid formulations resulted in tensile performances which are positioned between the ones obtained for the single-type fiber reinforced composites, suggesting that the principle of superposition of the tensile stress – crack opening curves may be used as an underlying mechanism in hybrid FRCC. Considering the potential synergistic effect of having multiple-fiber reinforcements, the hybrid FRCC were expected to outperform the superposition of the respective single-type fiber reinforced composites, which was not observed in this study. In the special case of the hybrid composite reinforced with PVA and PAN fibers, the hybrid even underperformed the 1% PVA reinforced composite for average CMOD between 0.1 and 0.5 mm. One explanation for this may be the unfavorable effect that PAN fibers have on the fresh properties of the composite.

4 Conclusions

In the present work, the single crack tension test (SCTT) setup allowed the unique identification of the mechanical contribution of each fiber reinforcement type on the overall tensile behavior of the hybrid composites. The consistent experimental results support its use for the standard mechanical characterization and design of hybrid FRCC.

The quantitative estimation of the contribution of different fiber types to the overall behavior of hybrid FRCC allowed a better understanding of the principle of superposition of the tensile stress – crack opening curves and the synergistic effect of multi-scale fiber reinforcement for the hybrid tensile micro-mechanisms. The results obtained in this study suggest that the effect of these principles may be influenced by other factors. In particular, the fresh properties showed to have strong influence on the overall tensile behavior of the hybrid composites.

Acknowledgements. The first author thanks the Portuguese National Science Foundation for the financial support, through grant SFRH / BD / 36515 / 2007, funded by POPH - QREN, the Social European Fund and the MCTES, and DTU-Byg for their support of the work as part of this project.

References

[1] Brandt, A.M.: Fibre reinforced cement-based (FRC) composites after over 40 years of de-velopment in building and civil engineering. Comp. Struct. 86, 3–9 (2008)

[2] Fischer, G., Li, V.C.: Effect of fiber reinforcement on the response of structural members. Eng. Frac. Mech. 74, 258–272 (2007)

[3] Li, V.C.: On Engineered Cementitious Composites (ECC) - A Review of the Material and Its Applications. J. Adv. Conc. Tech. 1(3), 215–230 (2003)

[4] Ahmed, S.F.U., Maalej, M., Paramasivam, P.: Flexural responses of hybrid steel–polyethylene fiber reinforced cement composites containing high volume fly ash. Con. Build Mat. 21, 1088–1097 (2007)

[5] Banthia, N., Gupta, R.: Hybrid fiber reinforced concrete (HyFRC): fiber synergy in high strength matrices. Mat. Struct. 37, 707–716 (2004)

[6] Lawler, J.S., Wilhelm, T., Zampini, D., Shah, S.P.: Fracture processes of hybrid fiber-reinforced mortar. Mat. Struct. 36, 197–208 (2003)

[7] Rossi, P.: High performance multimodal fiber reinforced cement composites (HPMFRCC): The LCPC experience. ACI Mat. J. 94, 478–483 (1997)
[8] Betterman, L.R., Ouyang, C., Shah, S.P.: Fiber-matrix interaction in microfiber reinforced matrix. Adv. Cem. Bas. Mat. 2, 52–61 (1995)
[9] Fischer, G., Stang, H., Dick-Nielsen, L.: Initiation and development of cracking in ECC materials: Experimental observations and modeling. In: Carpinteri, G.F.A., Gambarova, P. (eds.) Proc. Int. Symp. on High Performance Concrete, Brick- Masonry and Environmental Aspects, Ia-FraMCos, vol. 3. Taylor & Francis (2007)
[10] Pereira, E.B., Fischer, G., Barros, J.A.O., Lepech, M.: Crack formation and tensile stress-crack opening behavior of fiber reinforced cementitious composites (SHCC). In: Oh, B.H., Choi, O.C., Chung, L. (eds.) Proceedings of FraMCoS-7, Jeju, Korea, May 23-28 (2010)
[11] Chanvillard, G.: Caractérisation des performances d'un béton renforcé de fibres à partir d'un essai de flexion. Partie 2: Identification d'une loi de comportement intrinsèque entraction. Mat. Struct. 32, 601–605 (1999)
[12] Lofgren, I., Stang, H., Olesen, J.F.: The WST method, a fracture mechanics test method for FRC. Mat. Struct., 197–211 (2008), doi:10.1617/s11527-007-9231-3
[13] Zhang, J., Leung, C.K.Y., Xu, S.: Evaluation of fracture parameters of concrete from bending test using inverse analysis approach. Mat. Struct. 43, 857–874 (2010)
[14] Kabele, P.: Multiscale framework for modeling of fracture in high performance fiber reinforced cementitious composites. Eng. Frac. Mech. 74, 194–209 (2007)
[15] Stang, H., Li, V.C., Krenchel, H.: Design and structural applications of stress-crack width relations in fibre reinforced concrete. Mat. Struct. 28, 210–219 (1995)

Impact Behavior of 3D Fabric Reinforced Cementitious Composites

A. Peled[1], D. Zhu[2], and B. Mobasher[2]

[1] Structural Engineering Department, Ben Gurion University of The Negev
[2] Department of Civil and Environmental Engineering, Arizona State University

Abstract. The objective of this research was to study the behavior of 3D AR glass fabric cement-based composites under impact loading. It was found that 3D fabrics significantly improve the toughness and energy absorption of cement-based composites under impact loading, compared to short AR glass fibers reinforcement. The 3D fabric improves the toughness in as high as 200 folds compared to short fiber composites. The energy absorption was highly affected by the thickness of the element and the location of the 3D fabric faces. Greater toughness was obtained when the fabric faces were located in the direction of the hammer drop.

1 Introduction

Cementitious materials may be subjected to dynamic loading in a variety of situations. Due to the inherent brittleness and low tensile strength of most cement-based elements, impact loadings can cause severe damage, resulting in extensive cracking. Since cement based composites are rate-dependent, their mechanical properties are highly dependent on the loading rate. Characterization of the impact response of the concrete is important for planning this activity. Textiles reinforced cement-based elements (TRC) have been intensively investigated in recent years [1]. Superior tensile strength, toughness, ductility and energy absorption were reported with TRC [2-3], properties which are important under dynamic loading. Several researchers recently showed the high potential of cement-based composites reinforced with fabrics under high loading rate [4-6]. Modern textile technology enables wide variety of fabric structures which allows great flexibility in fabric design. It is also possible to produce three-dimensional (3D) fabrics, providing reinforcement in the plane normal to the panel. The research to date focused mainly on the mechanical behavior of cement-based composites reinforced with two dimensional fabrics (2D). 3D fabrics having reinforcement in three orthogonal directions can limit failure by delamination, enhance shear strength and therefore expected to improve the mechanical properties of cement composites under dynamic and impact loads. Recently 3D spacer fabrics were developed for use in cement–based products. Several studies reported the

G.J. Parra-Montesinos, H.W. Reinhardt, and A.E. Naaman (Eds.): HPFRCC 6, pp. 543–550.

efficiency of this fabric technology [7]. Limited studies dealt with the behavior of cement-based composites with 3D fabric as reinforcement [8-9], demonstrating mainly the potential of using these types of reinforcements in the cements field.

The objective of this research was to study and compare the behavior of fabric-cement composites reinforced with 3D fabrics, under impact loading. The influences of fly ash as well as specimen position were studied. The results were compared to cement composites reinforced with short AR glass fibers.

2 Experimental

2.1 Preparation of Specimens

3D warp knitted fabric structures were used in this work, in which two sets of independent 2D knitted fabrics were connected together with a third set of yarns along the thickness, Z direction, of the fabric, referred as spacer yarns. The 3D fabric is presented in Fig. 1a. The yarns in the X and Y directions were made from alkali resistance (AR) glass bundles of 1200 tex with tensile strength of 1325 MPa and 67 GPa modules of elasticity, these yarns were the reinforcing yarns and connected together by stitches (loops) made of fine polyester. The volume content, V_f, of AR glass reinforcement yarns was of ~1% at the composite plain at both directions. The spacer yarns in the Z direction were used mainly for stabilization of the 3D fabric. The 3D fabrics were developed and produced by ITA, RWTH Aachen.

Fig. 1. (a) 3D fabric structure and composite testing arrangements: (b) horizontal, (c) vertical

Two 3D fabric composites were prepared one with cement paste (water and cement only, 3DGPC) of 0.4 water/cement ratio and another with replacement of 20% by volume of cement by fly ash (3DGFA). The 3D fabric specimens were prepared by placing the fabric in a mold and then casting the cement matrix on top of the fabric, until complete covering of the fabric. Vacuum was applied during casting to allow good penetrability of the cement matrix into the opening of the 3D fabric. 24 hours after casting the specimens were demolded and cured in 100% relative humidity for 7 days and then at room environment until testing in impact,

28 days after casting. Before testing, each composite plate was cut to slices providing specimens with a 200 x 50 x 25 mm of length, width and thickness, respectively. For comparison, composites with short AR glass fibers of 12 mm length were prepared (GF12) with similar dimensions and procedure as the 3D fabric composites. The fiber content was 0.3% by weight.

2.2 *Impact Test Procedure*

An impact test set-up based on a free-fall drop of an instrumented hammer on a three point bending specimen was used with a span of 178 mm and load cell with a range of 90 kN. The hammer weight was 134 N with a drop height of 152 mm. A linear variable differential transformer (LVDT) with a range of ± 10 mm was connected to the specimen by means of a lever arm. A high speed digital camera (Phantom v.7) was used to capture pictures of the samples during the impact tests. The damage caused in the samples was then compared by visual examination. The reinforcing direction of the fabric was along the stitches for all systems.

The toughness, maximum stress and rigidity were calculated. The toughness was calculated by the energy dissipated in the specimen using the area under the load-deflection curves. The maximum impact stress was calculated using linear elastic small displacement bending equation and the rigidity was calculated by the slope of load-deflection curve in the linear elastic region. Typical stresses vs. time and stresses vs. deflection curves were chosen to compare between the different tested systems. For more details on the impact testing see [5].

3 Results and Discussion

3.1 *Test Direction*

Two loading arrangements were carried out: parallel to fabric layers, horizontal arrangements (Fig. 1b) and perpendicular to the fabric layers, vertical arrangement (Fig. 1c). In the horizontal arrangement the two faces of the 3D fabric are located at the top and bottom of the specimen (as it appears in Fig. 1a) relative to the drop direction of the hammer, where the spacer yarns are passing through the thickness from top to bottom. In the vertical arrangement the two faces of the 3D fabric are located at the sides of the specimen relative to the drop of the hammer, and the spacer yarns are passing through the width of the composite between the two sides of the composite. The average results are presented in Table 1.

The impact stress vs. deflection curves of the horizontal and vertical arrangement specimens are presented in Figs. 2a and 2b. A uniform impact behavior is observed, as for the horizontal arrangement this uniformity is even clearer. This indicates the uniformity of the composites and the ability of using this test to study the impact behavior of the TRC systems. It is also seen that the deflection of the TRC systems is very high exhibiting values of up to about 16 mm. A rebound behavior of the specimens tested at vertical arrangement is clearly seen, i.e., after reaching the maximum deflection the specimen was shifted back to some extent and no failure has occurred.

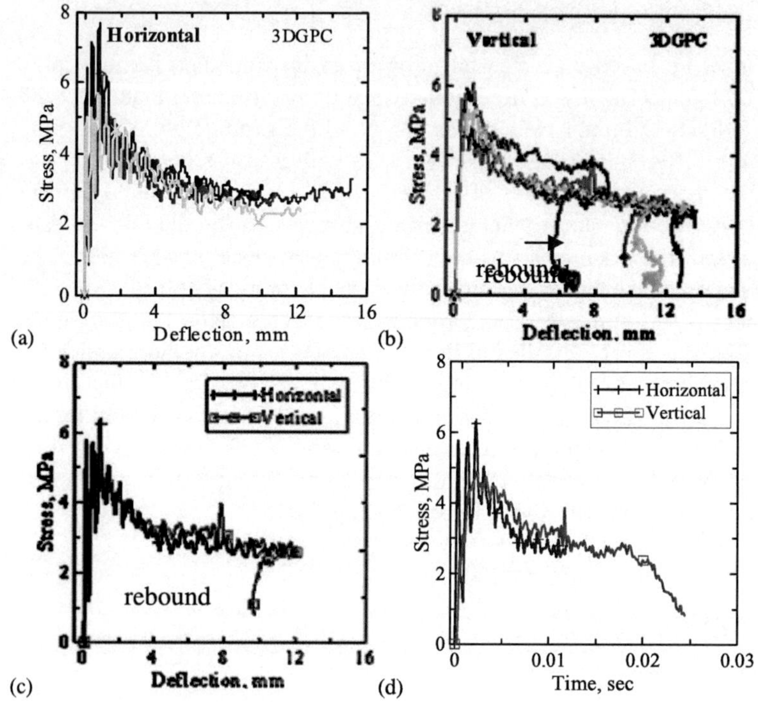

Fig. 2. Impact behavior of 3D fabric composites tested at (a) horizontal and (b) vertical arrangements, comparison of testing arrangements, (c) stress vs. deflection and (d) stress vs. time

Comparison of the typical curves of each system is presented in Figs. 2c and 2d, showing the impact stresses vs. deflection and vs. time. The rebound mechanism is obvious for the vertically tested system. On the other hand not such rebound, i.e. shifting back behavior, is seen for the composites tested at horizontal arrangement. Due to this rebound behavior the entire duration of the test is greater, about twice as much for the composite tested vertically as compared to the composite tested horizontally. The toughness of the vertical system is also much greater than that of the horizontal system (Table 1), providing much better energy absorption of the vertically tested composite. However, the horizontal system exhibits greater maximum stress as compared to the vertical system. But the improved strength by the horizontal arrangement is much smaller, of about 15%, than the improved toughness by the vertical system, of about 50%. So when energy absorption is considered the thickness of the element and location of the fabric faces relative to the impactor, are important factors.

Table 1. Impact properties of the composites

Compo-site	Sample dir.	Rigidity, N/mm	Max. Force, N	Max. Stress, MPa	Max. Deflection, mm	Deflection at Max. Force, mm	Toughness, N.mm
3DGFA	Horiz.	27223±4273	744±73	5.15±0.30	5.98±0.53	0.143±0.179	2255±379
3DGPC	Horiz.	11879±1075	876±49	6.33±0.93	12.05±2.21	0.639±0.338	5460±1110
	Vert.	11014±245	1281±124	5.38±0.49	11.90±2.21	0.824±0.288	8100±961
GF12	Horiz.	29685±5147	730±67	5.68±0.55	0.077±0.005	0.037±0.005	31.3±3.6

The crack pattern and development of these two systems is compared in Fig. 3. When tested horizontally, at duration of 0.75 ms the developed crack is getting almost through the entire thickness of the composite (Fig. 3a). At latter stages of the impact test the crack is widening, and the opening of the crack is relatively large at duration of 13.25 ms (Fig. 3b). The reinforcing yarns of the 3D fabric are pulling out and holding the specimen from a complete failure as seen in Fig. 4a, a major damage of the whole composite is obvious. Broken bundles and filaments at the tensile zone of the composite can easily be seen in Fig. 4b. These last two images were taken after the impact test was ended. However for the vertically tested system the crack development is different, at early test duration of 0.75 ms very fine crack is observed which developed only up to the middle of the composite thickness (Fig. 3c). For this composite at late stages the crack developed through most of the composite thickness but the reinforcing yarns of the fabric are bridging the crack through the specimen thickness and the composite did not fail (Fig. 3d).

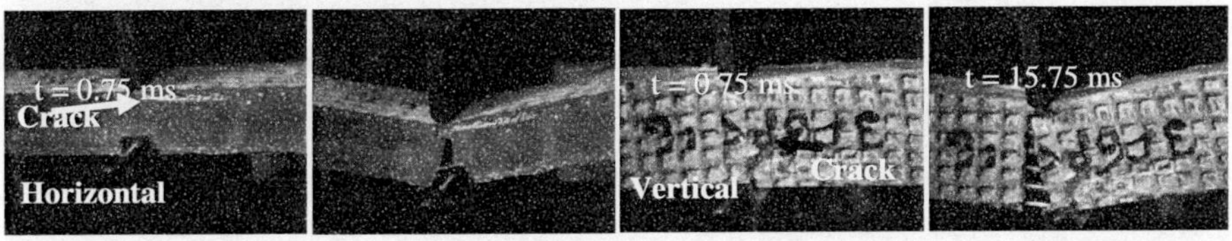

(a) (b) (c) (d)

Fig. 3. Comparison of 3D fabric composites tested at two arrangements at different durations

When observing the vertically tested composite after the impact testing (Fig. 4c), the crack is much finer, as part was closed due to the rebound mechanism, discussed above. No significant damage of the fabric and yarn brakeage was obtained with this composite at the end of testing, with no major failure of the composite. This composite can still hold loads and remain generally in its original shape after ending the test with the specific test parameters discussed above.

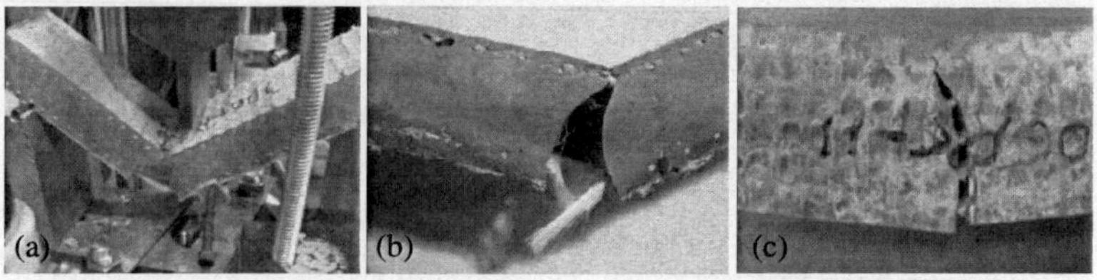

Fig. 4. Comparison of 3D fabric composites at two test arrangements at the end of testing: (a) horizontal, (b) horizontal after impact test (c) vertical after impact test

3.2 *3D Fabrics vs. Short Yarns*

Comparison of the short fiber composites with the 3D fabric composites is shown in Fig. 5. This figure compares typical curves of the 3D fabric composite with the two matrices: with fly ash (3DGFA) and without fly ash (3DGPC) and the short fibers (GF12), presenting stress vs. deflection at log scales and stress vs. time linear scale. The test arrangement in all is horizontal. The significant brittle behavior of short fiber composites is obvious as compared to much more ductile behavior of the 3D fabric composites. Also the duration of the test is much smaller for the fiber composites as compared with the 3D fabric composites. The short fiber composite duration last less than 0.1 ms with a maximum deflection of about 0.08 mm, as for the 3D fabric composites the test duration is about 0.01 ms with deflection of about 10 mm.

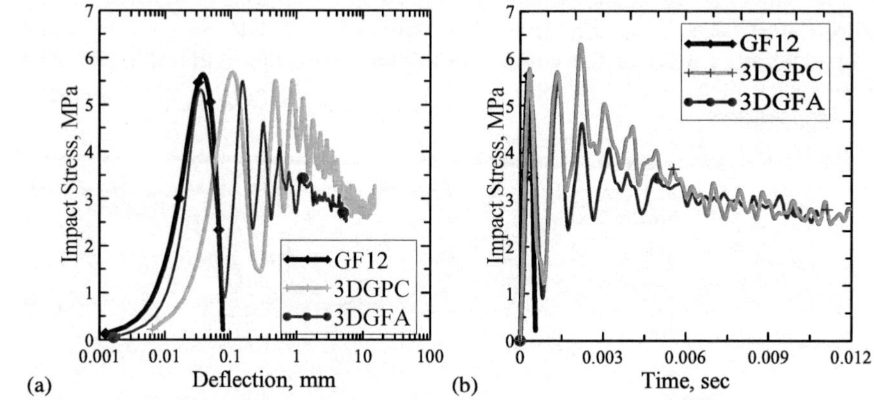

Fig. 5. Comparison of composites impact behavior with short glass fibers and 3D fabrics with plain cement and fly ash matrices: (a) stress vs. deflection logarithm scale, (b) stress vs. time

Such small deflection leads to very low toughness of the short fiber composites of 31 N mm as compared to much greater toughness of 4560 N mm of the 3D fabric composite with the plain matrix (Table 1). Crack pattern is presented in Fig. 6. Wide crack through the entire thickness of the short fiber composite was developed at ~9 ms of test duration (Fig. 6b). No fibers bridging the cracks are seen at this stage of loading. A complete failure of the composite has occurred at the end of the impact test of the short fiber composite (Fig. 6c) correlating with the brittle behavior obtained with these composites. Note that the 3D fabric also slightly improves the maximum impact stress, of about 10%, compared to the short fiber composite, but this improvement is significantly smaller than that of the toughness of almost 200 folds. Based on that, the benefit of the 3D fabric as reinforcement for cement-based composites is improving the energy absorption of the composite.

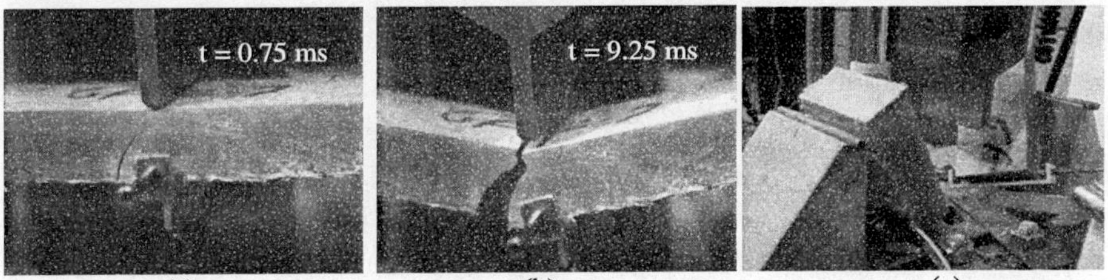

(a) (b) (c)

Fig. 6. Comparison of images of short fiber composite at different duration of testing: (a) 0.75 ms, (b) 9.25 ms, and (c) end of testing.

Addition of fly ash results in reduction in composite properties. The most significant reduction was obtained for the toughness; the fly ash composites exhibit almost half of the toughness as that of the plain paste composites, with values of 2255 N.mm and 5460 N.mm for the fly ash and no fly ash composites, respectively. Similar trend was also observed for the deflection. Further the presence of the fly ash slightly reduces the maximum stress from 6.33 MPa of the plain paste matrix to 5.13 MPa of the fly ash matrix. The fly ash might cause reduction in the bonding between the 3D fabric and the cement matrix, leading at least to some extent to the reduction in composite performance. However further work is required to better understand this behavior.

4 Summary and Conclusions

In this work the impact properties of 3D AR glass fabric cement-based composites were studied and compared to short AR glass fiber composites. It was found that 3D fabrics significantly improve the toughness and energy absorption of cement-based composites to as high as 200 folds compared to short fiber composites.

The arrangement of the fabric within the composite and composite dimension relative to the dropped hammer found to have significant influence on the impact behavior. Improved toughness in about 50% was obtained when the fabric was placed in the direction of the dropped hammer as compared to the horizontal arrangement. When energy absorption is considered the thickness of the element and location of the 3D fabric faces relative to the impactor, are important factors.

Based on this work in can be concluded that 3D fabric can be beneficial as reinforcement for cement-based composites. Further work is required in order to better understand and design these composites. In this work the yarns in the Z direction were spacer yarns with no real reinforcing abilities, for getting better performance stronger yarns are suggested along the Z direction.

Acknowledgements. The authors would like to thank ITA at RWTH Aachen University for their cooperation for providing the 3D fabrics used in this study.

References

[1] Brameshuber, W. (ed.): Textile reinforced concrete. State of the art report. RILEM TC 201-TRC, RILEM (2006)

[2] Peled, A., Cohen, Z., Pasder, Y., Roye, A., Gries, T.: Influence of textile characteristics on the tensile properties of warp knitted cement based composites. Cement and Concrete Composites J. 32, 174–183 (2008)

[3] Peled, A., Mobasher, B.: Tensile Behavior of Fabric Cement-Based Composites: Pultruded and Cast. ASCE, J. of Materials in Civil Engineering 19(4), 340–348 (2007)

[4] Zhu, D., Peled, A., Mobasher, B.: Dynamic tensile testing of fabric-cement composites. Construction and Building Materials J. 25(1), 385–395 (2011)

[5] Zhu, D., Gencoglu, M., Mobasher, B.: Low velocity flexural impact behavior of AR glass fabric reinforced cement composites. Cement and Concrete Composites J. 31, 379–387 (2009)

[6] Haim, E., Peled, A.: Impact behavior of fabric-cement hybrid composites. ACI Materials J. 108(02) (March - April 2011)

[7] Hanisch, V., Kolkmann, A., Roye, A., Gries, T.: Influence of machine settings on mechanical performance of yarns and textile structures. In: Hegger, J., Brameshuber, W., Will, N. (eds.) Proceedings of the 1st International RILEM Symposium (Textile Reinforced Concrete ICTRC) RILEM TC201-TRC, pp. 13–22 (2006)

[8] Roye, A., Gries, T., Peled, A.: Spacer fabric for thin walled concrete elements. In: Di Prisco, M., Felicetti, R., Plizzari, G.A. (eds.) Fiber Reinforced Concrete – BEFIB, PRO 39, RILEM, pp. 1505–1514 (2004)

[9] Naaman, N.: Textile reinforced cement composites: competitive status and research directions. In: Brameshuber, W. (ed.) International RILEM Conference on Materials Science (MatSci) I, pp. 3–22 (2010)

Author Index

Subject Index